Student Solutions

Elementary and Intermediate Algebra
A Combined Course
THIRD EDITION

Charles P. McKeague
Cuesta College

Prepared by

Ross Rueger
College of the Sequoias

THOMSON
BROOKS/COLE

Australia • Brazil • Canada • Mexico • Singapore • Spain • United Kingdom • United States

© 2008 Thomson Brooks/Cole, a part of The Thomson Corporation. Thomson, the Star logo, and Brooks/Cole are trademarks used herein under license.

ALL RIGHTS RESERVED. No part of this work covered by the copyright hereon may be reproduced or used in any form or by any means—graphic, electronic, or mechanical, including photocopying, recording, taping, Web distribution, information storage and retrieval systems, or in any other manner—without the written permission of the publisher.

Printed in the United States of America
1 2 3 4 5 6 7 11 10 09 08 07

Printer: Thomson/West

ISBN-13: 978-0-495-38300-0
ISBN-10: 0-495-38300-7

For more information about our products,
contact us at:
Thomson Learning Academic Resource Center
1-800-423-0563

For permission to use material from this text or product, submit a request online at
http://www.thomsonrights.com.
Any additional questions about permissions can be submitted by email to **thomsonrights@thomson.com**.

Thomson Higher Education
10 Davis Drive
Belmont, CA 94002-3098
USA

Contents

Preface		**vii**
Chapter 1	**The Basics**	**1**
	1.1 Notation and Symbols	1
	1.2 Real Numbers	2
	1.3 Addition of Real Numbers	4
	1.4 Subtraction of Real Numbers	5
	1.5 Properties of Real Numbers	6
	1.6 Multiplication of Real Numbers	7
	1.7 Division of Real Numbers	9
	1.8 Subsets of the Real Numbers	11
	1.9 Addition and Subtraction with Fractions	13
	Chapter 1 Review/Test	15
Chapter 2	**Linear Equations and Inequalities**	**19**
	2.1 Simplifying Expressions	19
	2.2 Addition Property of Equality	21
	2.3 Multiplication Property of Equality	24
	2.4 Solving Linear Equations	28
	2.5 Formulas	32
	2.6 Applications	36
	2.7 More Applications	40
	2.8 Linear Inequalities	45
	Chapter 2 Review/Test	49
Chapter 3	**Linear Equations and Inequalities in Two Variables**	**53**
	3.1 Paired Data and Graphing Ordered Pairs	53
	3.2 Solutions to Linear Equations in Two Variables	57
	3.3 Graphing Linear Equations in Two Variables	60
	3.4 More on Graphing: Intercepts	65
	3.5 Slope and the Equation of a Line	71
	3.6 Solving Linear Systems by Graphing	76
	3.7 The Elimination Method	80
	3.8 The Substitution Method	85
	3.9 Applications	89
	Chapter 3 Review/Test	93

Chapter 4 Exponents and Polynomials 105

4.1	Multiplication with Exponents	105
4.2	Division with Exponents	108
4.3	Operations with Monomials	111
4.4	Addition and Subtraction of Polynomials	114
4.5	Multiplication with Polynomials	116
4.6	Binomial Squares and Other Special Products	119
4.7	Dividing a Polynomial by a Monomial	121
4.8	Dividing a Polynomial by a Polynomial	124
	Chapter 4 Review/Test	127

Chapter 5 Factoring 131

5.1	The Greatest Common Factor and Factoring by Grouping	131
5.2	Factoring Trinomials	133
5.3	More Trinomials to Factor	134
5.4	The Difference of Two Squares	136
5.5	The Sum and Difference of Two Cubes	138
5.6	Factoring: A General Review	139
5.7	Solving Equations by Factoring	141
5.8	Applications	145
	Chapter 5 Review/Test	148

Chapter 6 Rational Expressions 153

6.1	Reducing Rational Expressions to Lowest Terms	153
6.2	Multiplication and Division of Rational Expressions	157
6.3	Addition and Subtraction of Rational Expressions	160
6.4	Equations Involving Rational Expressions	165
6.5	Applications	170
6.6	Complex Fractions	174
6.7	Proportions	177
	Chapter 6 Review/Test	179

Chapter 7 Transitions   185

7.1	Review of Solving Equations	185
7.2	Equations with Absolute Value	189
7.3	Compound Inequalities and Interval Notation	193
7.4	Inequalities Involving Absolute Value	197
7.5	Review of Systems of Linear Equations in Two Variables	200
7.6	Systems of Linear Equations in Three Variables	208
7.7	Linear Inequalities in Two Variables	218
7.8	Systems of Linear Inequalities	221
	Chapter 7 Review/Test	224

Chapter 8 Equations and Functions   231

8.1	The Slope of a Line	231
8.2	The Equation of a Line	235
8.3	Introduction to Functions	240
8.4	Function Notation	245
8.5	Algebra and Composition with Functions	248
8.6	Variation	249
	Chapter 8 Review/Test	255

Chapter 9 Rational Exponents and Roots   259

9.1	Rational Exponents	259
9.2	More Expressions Involving Rational Exponents	261
9.3	Simplified Form for Radicals	264
9.4	Addition and Subtraction of Radical Expressions	267
9.5	Multiplication and Division of Radical Expressions	269
9.6	Equations with Radicals	273
9.7	Complex Numbers	278
	Chapter 9 Review/Test	280

Chapter 10 Quadratic Functions   283

10.1	Completing the Square	283
10.2	The Quadratic Formula	287
10.3	Additional Items Involving Solutions to Equations	294
10.4	Equations Quadratic in Form	297
10.5	Graphing Parabolas	302
10.6	Quadratic Inequalities	306
	Chapter 10 Review/Test	311

Chapter 11   Exponential and Logarithmic Functions   319

11.1	Exponential Functions	319
11.2	The Inverse of a Function	322
11.3	Logarithms Are Exponents	326
11.4	Properties of Logarithms	329
11.5	Common Logarithms and Natural Logarithms	331
11.6	Exponential Equations and Change of Base	333
	Chapter 11 Review/Test	336

Chapter 12   Conic Sections   341

12.1	The Circle	341
12.2	Ellipses and Hyperbolas	344
12.3	Second-Degree Inequalities and Nonlinear Systems	350
	Chapter 12 Review/Test	354

Chapter 13   Sequences and Series   359

13.1	Sequences	359
13.2	Series	360
13.3	Arithmetic Sequences	362
13.4	Geometric Sequences	365
13.5	The Binomial Expansion	368
	Chapter 13 Review/Test	371

Preface

This *Student Solutions Manual* contains complete solutions to all odd-numbered exercises and all chapter review/test exercises of *Elementary and Intermediate Algebra, A Combined Course* by Charles P. McKeague. I have attempted to format solutions for readability and accuracy, and apologize to you for any errors that you may encounter. If you have any comments, suggestions, error corrections, or alternative solutions please feel free to drop me a note or send an email (address below).

Please use this manual with some degree of caution. Be sure that you have attempted a solution, and re-attempted it, before you look it up in this manual. Algebra can only be learned by *doing*, and not by observing! As you use this manual, do not just read the solution but work it along with the manual, using my solution to check your work. If you use this manual in that fashion then it should be helpful to you as you do homework and study for tests.

I would like to thank a number of people for their assistance in preparing this manual. Thanks go to Laura Localio at Thomson Brooks/Cole Publishing for her valuable assistance and support. Special thanks go to Matt Bourez of College of the Sequoias for his meticulous error-checking of my solutions, and prompt return of my manuscript under tight deadlines. Also thanks go to Patrick McKeague for keeping me updated to manuscript corrections and prompt response to my emails.

I wish to express my appreciation to Pat McKeague for asking me to be involved with this textbook. This book provides a complete course in beginning algebra, and you will find the text very easy to read and understand. Good luck!

Ross Rueger
College of the Sequoias
matmanross@aol.com

February, 2007

Chapter 1
The Basics

1.1 Notation and Symbols

1. The equivalent expression is $x+5=14$.
3. The equivalent expression is $5y<30$.
5. The equivalent expression is $3y \le y+6$.
7. The equivalent expression is $\dfrac{x}{3}=x+2$.
9. Expanding the expression: $3^2 = 3 \cdot 3 = 9$
11. Expanding the expression: $7^2 = 7 \cdot 7 = 49$
13. Expanding the expression: $2^3 = 2 \cdot 2 \cdot 2 = 8$
15. Expanding the expression: $4^3 = 4 \cdot 4 \cdot 4 = 64$
17. Expanding the expression: $2^4 = 2 \cdot 2 \cdot 2 \cdot 2 = 16$
19. Expanding the expression: $10^2 = 10 \cdot 10 = 100$
21. Expanding the expression: $11^2 = 11 \cdot 11 = 121$
23.   a. Using the order of operations: $2 \cdot 3 + 5 = 6 + 5 = 11$
      b. Using the order of operations: $2(3+5) = 2(8) = 16$
25.   a. Using the order of operations: $5 + 2 \cdot 6 = 5 + 12 = 17$
      b. Using the order of operations: $(5+2) \cdot 6 = 7 \cdot 6 = 42$
27.   a. Using the order of operations: $5 \cdot 4 + 5 \cdot 2 = 20 + 10 = 30$
      b. Using the order of operations: $5(4+2) = 5(6) = 30$
29.   a. Using the order of operations: $8 + 2(5+3) = 8 + 2(8) = 8 + 16 = 24$
      b. Using the order of operations: $(8+2)(5+3) = (10)(8) = 80$
31. Using the order of operations: $20 + 2(8-5) + 1 = 20 + 2(3) + 1 = 20 + 6 + 1 = 27$
33. Using the order of operations: $5 + 2(3 \cdot 4 - 1) + 8 = 5 + 2(12-1) + 8 = 5 + 2(11) + 8 = 5 + 22 + 8 = 35$
35. Using the order of operations: $4 + 8 \div 4 - 2 = 4 + 2 - 2 = 4$
37. Using the order of operations: $3 \cdot 8 + 10 \div 2 + 4 \cdot 2 = 24 + 5 + 8 = 37$
39.   a. Using the order of operations: $(5+3)(5-3) = (8)(2) = 16$
      b. Using the order of operations: $5^2 - 3^2 = 5 \cdot 5 - 3 \cdot 3 = 25 - 9 = 16$
41.   a. Using the order of operations: $(4+5)^2 = 9^2 = 9 \cdot 9 = 81$
      b. Using the order of operations: $4^2 + 5^2 = 4 \cdot 4 + 5 \cdot 5 = 16 + 25 = 41$
43. Using the order of operations: $2 \cdot 10^3 + 3 \cdot 10^2 + 4 \cdot 10 + 5 = 2000 + 300 + 40 + 5 = 2,345$
45. Using the order of operations: $10 - 2(4 \cdot 5 - 16) = 10 - 2(20 - 16) = 10 - 2(4) = 10 - 8 = 2$
47. Using the order of operations: $4[7 + 3(2 \cdot 9 - 8)] = 4[7 + 3(18-8)] = 4[7 + 3(10)] = 4(7+30) = 4(37) = 148$

2   Chapter 1   The Basics

49. Using the order of operations: $3(4 \cdot 5 - 12) + 6(7 \cdot 6 - 40) = 3(20 - 12) + 6(42 - 40) = 3(8) + 6(2) = 24 + 12 = 36$
51. Using the order of operations: $3^4 + 4^2 \div 2^3 - 5^2 = 81 + 16 \div 8 - 25 = 81 + 2 - 25 = 58$
53. Using the order of operations: $5^2 + 3^4 \div 9^2 + 6^2 = 25 + 81 \div 81 + 36 = 25 + 1 + 36 = 62$
55. Using the order of operations: $20 \div 2 \cdot 10 = 10 \cdot 10 = 100$
57. Using the order of operations: $24 \div 8 \cdot 3 = 3 \cdot 3 = 9$
59. Using the order of operations: $36 \div 6 \cdot 3 = 6 \cdot 3 = 18$
61. Using the order of operations: $16 - 8 + 4 = 8 + 4 = 12$
63. Using the order of operations: $24 - 14 + 8 = 10 + 8 = 18$
65. Using the order of operations: $36 - 6 + 12 = 30 + 12 = 42$
67. Computing the value: $0.08 + 0.09 = 0.17$
69. Computing the value: $0.10 + 0.12 = 0.22$
71. Computing the value: $4.8 - 2.5 = 2.3$
73. Computing the value: $2.07 + 3.48 = 5.55$
75. Computing the value: $0.12(2,000) = 240$
77. Computing the value: $0.25(40) = 10$
79. Computing the value: $510 \div 0.17 = 3,000$
81. Computing the value: $240 \div 0.12 = 2,000$
83. Using a calculator: $37.80 \div 1.07 \approx 35.33$
85. Using a calculator: $555 \div 740 = 0.75$
87. Using a calculator: $70 \div 210 \approx 0.33$
89. Using a calculator: $6,000 \div 22 \approx 272.73$
91. There are $5 \cdot 2 = 10$ cookies in the package.
93. The total number of calories is $210 \cdot 2 = 420$ calories.
95. a. The amount of caffeine is: $6(100) = 600$ mg
    b. The amount of caffeine is: $2(45) + 3(47) = 90 + 141 = 231$ mg
97. Completing the table:

Activity	Calories burned in 1 hour
Bicycling	374
Bowling	265
Handball	680
Jogging	680
Skiing	544

99. The next number is 5.
101. The next number is 10.
103. The next number is $5^2 = 25$.
105. Since $2 + 2 = 4$ and $2 + 4 = 6$, the next number is $4 + 6 = 10$.

1.2 Real Numbers

1. Labeling the point:

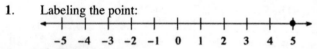

3. Labeling the point:

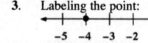

5. Labeling the point:

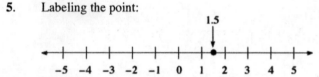

7. Labeling the point:

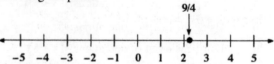

9. Building the fraction: $\frac{3}{4} = \frac{3}{4} \cdot \frac{6}{6} = \frac{18}{24}$
11. Building the fraction: $\frac{1}{2} = \frac{1}{2} \cdot \frac{12}{12} = \frac{12}{24}$
13. Building the fraction: $\frac{5}{8} = \frac{5}{8} \cdot \frac{3}{3} = \frac{15}{24}$
15. Building the fraction: $\frac{3}{5} = \frac{3}{5} \cdot \frac{12}{12} = \frac{36}{60}$
17. Building the fraction: $\frac{11}{30} = \frac{11}{30} \cdot \frac{2}{2} = \frac{22}{60}$
19. Building the fraction: $\frac{1}{2} \cdot \frac{2}{2} = \frac{2}{4}$, so the missing numerator is 2.
21. Building the fraction: $\frac{5}{9} \cdot \frac{5}{5} = \frac{25}{45}$, so the missing numerator is 25.
23. Building the fraction: $\frac{3}{4} \cdot \frac{2}{2} = \frac{6}{8}$, so the missing numerator is 6.
25. The opposite of 10 is -10, the reciprocal is $\frac{1}{10}$, and the absolute value is $|10| = 10$.
27. The opposite of -3 is 3, the reciprocal is $-\frac{1}{3}$, and the absolute value is $|-3| = 3$.

29. The opposite of x is $-x$, the reciprocal is $\frac{1}{x}$, and the absolute value is $|x|$.

31. The correct symbol is $<$: $-5 < -3$
33. The correct symbol is $>$: $-3 > -7$

35. Since $|-4| = 4$ and $-|-4| = -4$, the correct symbol is $>$: $|-4| > -|-4|$

37. Since $-|-7| = -7$, the correct symbol is $>$: $7 > -|-7|$

39. The correct symbol is $<$: $-\frac{3}{4} < -\frac{1}{4}$ 41. The correct symbol is $<$: $-\frac{3}{2} < -\frac{3}{4}$

43. Simplifying the expression: $|8-2| = |6| = 6$

45. Simplifying the expression: $|5 \cdot 2^3 - 2 \cdot 3^2| = |5 \cdot 8 - 2 \cdot 9| = |40 - 18| = |22| = 22$

47. Simplifying the expression: $|7-2| - |4-2| = |5| - |2| = 5 - 2 = 3$

49. Simplifying the expression: $10 - |7 - 2(5-3)| = 10 - |7 - 2(2)| = 10 - |7 - 4| = 10 - |3| = 10 - 3 = 7$

51. Simplifying the expression:
$$\begin{aligned} 15 - |8 - 2(3 \cdot 4 - 9)| - 10 &= 15 - |8 - 2(12 - 9)| - 10 \\ &= 15 - |8 - 2(3)| - 10 \\ &= 15 - |8 - 6| - 10 \\ &= 15 - |2| - 10 \\ &= 15 - 2 - 10 \\ &= 3 \end{aligned}$$

53. Multiplying the fractions: $\frac{2}{3} \cdot \frac{4}{5} = \frac{8}{15}$ 55. Multiplying the fractions: $\frac{1}{2}(3) = \frac{1}{2} \cdot \frac{3}{1} = \frac{3}{2}$

57. Multiplying the fractions: $\frac{4}{3} \cdot \frac{3}{4} = \frac{12}{12} = 1$ 59. Multiplying the fractions: $3 \cdot \frac{1}{3} = \frac{3}{1} \cdot \frac{1}{3} = \frac{3}{3} = 1$

61. a. Multiplying the numbers: $\frac{1}{2}(4) = \frac{1}{2} \cdot \frac{4}{1} = \frac{4}{2} = 2$

    b. Multiplying the numbers: $\frac{1}{2}(8) = \frac{1}{2} \cdot \frac{8}{1} = \frac{8}{2} = 4$

    c. Multiplying the numbers: $\frac{1}{2}(16) = \frac{1}{2} \cdot \frac{16}{1} = \frac{16}{2} = 8$

    d. Multiplying the numbers: $\frac{1}{2}(0.06) = 0.03$

63. a. Multiplying the numbers: $\frac{3}{2}(4) = \frac{3}{2} \cdot \frac{4}{1} = \frac{12}{2} = 6$

    b. Multiplying the numbers: $\frac{3}{2}(8) = \frac{3}{2} \cdot \frac{8}{1} = \frac{24}{2} = 12$

    c. Multiplying the numbers: $\frac{3}{2}(16) = \frac{3}{2} \cdot \frac{16}{1} = \frac{48}{2} = 24$

    d. Multiplying the numbers: $\frac{3}{2}(0.06) = 0.09$

65. Expanding the exponent: $\left(\frac{3}{4}\right)^2 = \frac{3}{4} \cdot \frac{3}{4} = \frac{9}{16}$ 67. Expanding the exponent: $\left(\frac{2}{3}\right)^3 = \frac{2}{3} \cdot \frac{2}{3} \cdot \frac{2}{3} = \frac{8}{27}$

69. a. Substituting $x = 5$: $2x - 6 = 2(5) - 6 = 10 - 6 = 4$

    b. Substituting $x = 10$: $2x - 6 = 2(10) - 6 = 20 - 6 = 14$

    c. Substituting $x = 15$: $2x - 6 = 2(15) - 6 = 30 - 6 = 24$

    d. Substituting $x = 20$: $2x - 6 = 2(20) - 6 = 40 - 6 = 34$

71. a. Substituting $x = 10$: $x + 2 = 10 + 2 = 12$ b. Substituting $x = 10$: $2x = 2(10) = 20$

    c. Substituting $x = 10$: $x^2 = (10)^2 = 100$ d. Substituting $x = 10$: $2^x = 2^{10} = 1{,}024$

73. a. Substituting $x = 4$: $x^2 + 1 = (4)^2 + 1 = 16 + 1 = 17$

    b. Substituting $x = 4$: $(x+1)^2 = (4+1)^2 = (5)^2 = 25$

    c. Substituting $x = 4$: $x^2 + 2x + 1 = (4)^2 + 2(4) + 1 = 16 + 8 + 1 = 25$

75. The perimeter is $4(1 \text{ in.}) = 4$ in., and the area is $(1 \text{ in.})^2 = 1 \text{ in.}^2$.
77. The perimeter is $2(1.5 \text{ in.}) + 2(0.75 \text{ in.}) = 3.0 \text{ in.} + 1.5 \text{ in.} = 4.5$ in., and the area is $(1.5 \text{ in.})(0.75 \text{ in.}) = 1.125 \text{ in.}^2$.
79. The perimeter is $2.75 \text{ cm} + 4 \text{ cm} + 3.5 \text{ cm} = 10.25$ cm, and the area is $\frac{1}{2}(4 \text{ cm})(2.5 \text{ cm}) = 5.0 \text{ cm}^2$.
81. A loss of 8 yards corresponds to –8 on a number line. The total yards gained corresponds to –2 yards.
83. The temperature can be represented as –64°. The new (warmer) temperature corresponds to –54°.
85. His position corresponds to –100 feet. His new (deeper) position corresponds to –105 feet.
87. The area is given by: $8\frac{1}{2} \cdot 11 = \frac{17}{2} \cdot \frac{11}{1} = \frac{187}{2} = 93.5 \text{ in.}^2$

 The perimeter is given by: $2\left(8\frac{1}{2}\right) + 2(11) = 17 + 22 = 39$ in.
89. The calories consumed would be: $2(544) + 299 = 1,387$ calories
91. The calories consumed by the 180 lb person would be $3(653) = 1,959$ calories, while the calories consumed by the 120 lb person would be $3(435) = 1,305$ calories. Thus the 180 lb person consumed $1959 - 1306 = 654$ more calories.
93. Answers will vary.

1.3 Addition of Real Numbers

1. Adding all positive and negative combinations of 3 and 5:
 $3 + 5 = 8$ $3 + (-5) = -2$
 $-3 + 5 = 2$ $(-3) + (-5) = -8$
3. Adding all positive and negative combinations of 15 and 20:
 $15 + 20 = 35$ $15 + (-20) = -5$
 $-15 + 20 = 5$ $(-15) + (-20) = -35$
5. Adding the numbers: $6 + (-3) = 3$
7. Adding the numbers: $13 + (-20) = -7$
9. Adding the numbers: $18 + (-32) = -14$
11. Adding the numbers: $-6 + 3 = -3$
13. Adding the numbers: $-30 + 5 = -25$
15. Adding the numbers: $-6 + (-6) = -12$
17. Adding the numbers: $-9 + (-10) = -19$
19. Adding the numbers: $-10 + (-15) = -25$
21. Performing the additions: $5 + (-6) + (-7) = 5 + (-13) = -8$
23. Performing the additions: $-7 + 8 + (-5) = -12 + 8 = -4$
25. Performing the additions: $5 + [6 + (-2)] + (-3) = 5 + 4 + (-3) = 9 + (-3) = 6$
27. Performing the additions: $[6 + (-2)] + [3 + (-1)] = 4 + 2 = 6$
29. Performing the additions: $20 + (-6) + [3 + (-9)] = 20 + (-6) + (-6) = 20 + (-12) = 8$
31. Performing the additions: $-3 + (-2) + [5 + (-4)] = -3 + (-2) + 1 = -5 + 1 = -4$
33. Performing the additions: $(-9 + 2) + [5 + (-8)] + (-4) = -7 + (-3) + (-4) = -14$
35. Performing the additions: $[-6 + (-4)] + [7 + (-5)] + (-9) = -10 + 2 + (-9) = -19 + 2 = -17$
37. Performing the additions: $(-6 + 9) + (-5) + (-4 + 3) + 7 = 3 + (-5) + (-1) + 7 = 10 + (-6) = 4$
39. Using order of operations: $-5 + 2(-3 + 7) = -5 + 2(4) = -5 + 8 = 3$
41. Using order of operations: $9 + 3(-8 + 10) = 9 + 3(2) = 9 + 6 = 15$
43. Using order of operations: $-10 + 2(-6 + 8) + (-2) = -10 + 2(2) + (-2) = -10 + 4 + (-2) = -12 + 4 = -8$
45. Using order of operations: $2(-4 + 7) + 3(-6 + 8) = 2(3) + 3(2) = 6 + 6 = 12$
47. Adding the decimals: $3.9 + 7.1 = 11.0 = 11$
49. Adding the decimals: $8.1 + 2.7 = 10.8$
51. Substituting $x = 7,000$ and $y = 8,000$: $0.06x + 0.07y = 0.06(7,000) + 0.07(8,000) = 420 + 560 = 980$
53. Substituting $x = 10$ and $y = 12$: $0.05x + 0.10y = 0.05(10) + 0.10(12) = 0.5 + 1.2 = 1.7$
55. The pattern is to add 5, so the next two terms are $18 + 5 = 23$ and $23 + 5 = 28$.
57. The pattern is to add 5, so the next two terms are $25 + 5 = 30$ and $30 + 5 = 35$.
59. The pattern is to add –5, so the next two terms are $5 + (-5) = 0$ and $0 + (-5) = -5$.
61. The pattern is to add –6, so the next two terms are $-6 + (-6) = -12$ and $-12 + (-6) = -18$.
63. The pattern is to add –4, so the next two terms are $0 + (-4) = -4$ and $-4 + (-4) = -8$.

65. Yes, since each successive odd number is 2 added to the previous one.
67. The expression is: $5 + 9 = 14$
69. The expression is: $[-7 + (-5)] + 4 = -12 + 4 = -8$
71. The expression is: $[-2 + (-3)] + 10 = -5 + 10 = 5$
73. The number is 3, since $-8 + 3 = -5$.
75. The number is -3, since $-6 + (-3) = -9$
77. The expression is $-12° + 4° = -8°$.
79. The expression is $\$10 + (-\$6) + (-\$8) = \$10 + (-\$14) = -\$4$.
81. The new balance is $-\$30 + \$40 = \$10$.

1.4 Subtraction of Real Numbers

1. Subtracting the numbers: $5 - 8 = 5 + (-8) = -3$
3. Subtracting the numbers: $3 - 9 = 3 + (-9) = -6$
5. Subtracting the numbers: $5 - 5 = 5 + (-5) = 0$
7. Subtracting the numbers: $-8 - 2 = -8 + (-2) = -10$
9. Subtracting the numbers: $-4 - 12 = -4 + (-12) = -16$
11. Subtracting the numbers: $-6 - 6 = -6 + (-6) = -12$
13. Subtracting the numbers: $-8 - (-1) = -8 + 1 = -7$
15. Subtracting the numbers: $15 - (-20) = 15 + 20 = 35$
17. Subtracting the numbers: $-4 - (-4) = -4 + 4 = 0$
19. Using order of operations: $3 - 2 - 5 = 3 + (-2) + (-5) = 3 + (-7) = -4$
21. Using order of operations: $9 - 2 - 3 = 9 + (-2) + (-3) = 9 + (-5) = 4$
23. Using order of operations: $-6 - 8 - 10 = -6 + (-8) + (-10) = -24$
25. Using order of operations: $-22 + 4 - 10 = -22 + 4 + (-10) = -32 + 4 = -28$
27. Using order of operations: $10 - (-20) - 5 = 10 + 20 + (-5) = 30 + (-5) = 25$
29. Using order of operations: $8 - (2 - 3) - 5 = 8 - (-1) - 5 = 8 + 1 + (-5) = 9 + (-5) = 4$
31. Using order of operations: $7 - (3 - 9) - 6 = 7 - (-6) - 6 = 7 + 6 + (-6) = 13 + (-6) = 7$
33. Using order of operations: $5 - (-8 - 6) - 2 = 5 - (-14) - 2 = 5 + 14 + (-2) = 19 + (-2) = 17$
35. Using order of operations: $-(5 - 7) - (2 - 8) = -(-2) - (-6) = 2 + 6 = 8$
37. Using order of operations: $-(3 - 10) - (6 - 3) = -(-7) - 3 = 7 + (-3) = 4$
39. Using order of operations: $16 - [(4 - 5) - 1] = 16 - (-1 - 1) = 16 - (-2) = 16 + 2 = 18$
41. Using order of operations: $5 - [(2 - 3) - 4] = 5 - (-1 - 4) = 5 - (-5) = 5 + 5 = 10$
43. Using order of operations:
$21 - [-(3 - 4) - 2] - 5 = 21 - [-(-1) - 2] - 5 = 21 - (1 - 2) - 5 = 21 - (-1) - 5 = 21 + 1 + (-5) = 22 + (-5) = 17$
45. Using order of operations: $2 \cdot 8 - 3 \cdot 5 = 16 - 15 = 16 + (-15) = 1$
47. Using order of operations: $3 \cdot 5 - 2 \cdot 7 = 15 - 14 = 15 + (-14) = 1$
49. Using order of operations: $5 \cdot 9 - 2 \cdot 3 - 6 \cdot 2 = 45 - 6 - 12 = 45 + (-6) + (-12) = 45 + (-18) = 27$
51. Using order of operations: $3 \cdot 8 - 2 \cdot 4 - 6 \cdot 7 = 24 - 8 - 42 = 24 + (-8) + (-42) = 24 + (-50) = -26$
53. Using order of operations: $2 \cdot 3^2 - 5 \cdot 2^2 = 2 \cdot 9 - 5 \cdot 4 = 18 - 20 = 18 + (-20) = -2$
55. Using order of operations: $4 \cdot 3^3 - 5 \cdot 2^3 = 4 \cdot 27 - 5 \cdot 8 = 108 - 40 = 108 + (-40) = 68$
57. Subtracting decimals: $-3.4 - 7.9 = -3.4 + (-7.9) = -11.3$
59. Subtracting decimals: $3.3 - 6.9 = 3.3 + (-6.9) = -3.6$
61. a. Substituting $x = -3$ and $y = -2$: $x + y - 4 = -3 + (-2) - 4 = -5 - 4 = -9$
    b. Substituting $x = -9$ and $y = 3$: $x + y - 4 = -9 + 3 - 4 = -6 - 4 = -10$
    c. Substituting $x = -\frac{3}{5}$ and $y = \frac{8}{5}$: $x + y - 4 = -\frac{3}{5} + \frac{8}{5} - 4 = \frac{5}{5} - 4 = 1 + (-4) = -3$
63. Writing the expression: $-7 - 4 = -7 + (-4) = -11$
65. Writing the expression: $12 - (-8) = 12 + 8 = 20$
67. Writing the expression: $-5 - (-7) = -5 + 7 = 2$
69. Writing the expression: $[4 + (-5)] - 17 = -1 - 17 = -1 + (-17) = -18$
71. Writing the expression: $8 - 5 = 8 + (-5) = 3$
73. Writing the expression: $-8 - 5 = -8 + (-5) = -13$
75. Writing the expression: $8 - (-5) = 8 + 5 = 13$
77. The number is 10, since $8 - 10 = 8 + (-10) = -2$.
79. The number is -2, since $8 - (-2) = 8 + 2 = 10$.
81. The expression is $\$1,500 - \$730 = \$770$.
83. The expression is $-\$35 + \$15 - \$20 = -\$35 + (-\$20) + \$15 = -\$55 + \$15 = -\$40$.

6   Chapter 1   The Basics

85. The expression is $\$98 - \$65 - \$53 = \$98 + (-\$65) + (-\$53) = \$98 + (-\$118) = -\$20$.
87. The sequence of values is $4500, $3950, $3400, $2850, and $2300. This is an arithmetic sequence, since –$550 is added to each value to obtain the new value.
89. The difference is $1000 \text{ feet} - 231 \text{ feet} = 769 \text{ feet}$.
91. He is 439 feet from the starting line.
93. 2 seconds have gone by.
95. The angles add to 90°, so $x = 90° - 55° = 35°$.
97. The angles add to 180°, so $x = 180° - 120° = 60°$.
99. a. Completing the table:

Day	Plant Height (inches)
0	0
1	0.5
2	1
3	1.5
4	3
5	4
6	6
7	9
8	13
9	18
10	23

   b.   Subtracting: $13 - 1.5 = 11.5$. The grass is 11.5 inches higher after 8 days than after 3 days.

1.5 Properties of Real Numbers

1. commutative property (of addition)
3. multiplicative inverse property
5. commutative property (of addition)
7. distributive property
9. commutative and associative properties (of addition)
11. commutative and associative properties (of addition)
13. commutative property (of addition)
15. commutative and associative properties (of multiplication)
17. commutative property (of multiplication)
19. additive inverse property
21. The expression should read $3(x+2) = 3x+6$.
23. The expression should read $9(a+b) = 9a + 9b$.
25. The expression should read $3(0) = 0$.
27. The expression should read $3 + (-3) = 0$.
29. The expression should read $10(1) = 10$.
31. Simplifying the expression: $4 + (2+x) = (4+2) + x = 6+x$
33. Simplifying the expression: $(x+2) + 7 = x + (2+7) = x+9$
35. Simplifying the expression: $3(5x) = (3 \cdot 5)x = 15x$
37. Simplifying the expression: $9(6y) = (9 \cdot 6)y = 54y$
39. Simplifying the expression: $\frac{1}{2}(3a) = \left(\frac{1}{2} \cdot 3\right)a = \frac{3}{2}a$
41. Simplifying the expression: $\frac{1}{3}(3x) = \left(\frac{1}{3} \cdot 3\right)x = 1x = x$
43. Simplifying the expression: $\frac{1}{2}(2y) = \left(\frac{1}{2} \cdot 2\right)y = 1y = y$
45. Simplifying the expression: $\frac{3}{4}\left(\frac{4}{3}x\right) = \left(\frac{3}{4} \cdot \frac{4}{3}\right)x = 1x = x$
47. Simplifying the expression: $\frac{6}{5}\left(\frac{5}{6}a\right) = \left(\frac{6}{5} \cdot \frac{5}{6}\right)a = 1a = a$
49. Applying the distributive property: $8(x+2) = 8 \cdot x + 8 \cdot 2 = 8x + 16$
51. Applying the distributive property: $8(x-2) = 8 \cdot x - 8 \cdot 2 = 8x - 16$
53. Applying the distributive property: $4(y+1) = 4 \cdot y + 4 \cdot 1 = 4y + 4$
55. Applying the distributive property: $3(6x+5) = 3 \cdot 6x + 3 \cdot 5 = 18x + 15$
57. Applying the distributive property: $2(3a+7) = 2 \cdot 3a + 2 \cdot 7 = 6a + 14$
59. Applying the distributive property: $9(6y-8) = 9 \cdot 6y - 9 \cdot 8 = 54y - 72$
61. Applying the distributive property: $\frac{1}{2}(3x-6) = \frac{1}{2} \cdot 3x - \frac{1}{2} \cdot 6 = \frac{3}{2}x - 3$

63. Applying the distributive property: $\frac{1}{3}(3x+6) = \frac{1}{3} \cdot 3x + \frac{1}{3} \cdot 6 = x+2$
65. Applying the distributive property: $3(x+y) = 3x+3y$
67. Applying the distributive property: $8(a-b) = 8a-8b$
69. Applying the distributive property: $6(2x+3y) = 6 \cdot 2x + 6 \cdot 3y = 12x+18y$
71. Applying the distributive property: $4(3a-2b) = 4 \cdot 3a - 4 \cdot 2b = 12a-8b$
73. Applying the distributive property: $\frac{1}{2}(6x+4y) = \frac{1}{2} \cdot 6x + \frac{1}{2} \cdot 4y = 3x+2y$
75. Applying the distributive property: $4(a+4)+9 = 4a+16+9 = 4a+25$
77. Applying the distributive property: $2(3x+5)+2 = 6x+10+2 = 6x+12$
79. Applying the distributive property: $7(2x+4)+10 = 14x+28+10 = 14x+38$
81. Applying the distributive property: $\frac{1}{2}(4x+2) = \frac{1}{2} \cdot 4x + \frac{1}{2} \cdot 2 = 2x+1$
83. Applying the distributive property: $\frac{3}{4}(8x-4) = \frac{3}{4} \cdot 8x - \frac{3}{4} \cdot 4 = 6x-3$
85. Applying the distributive property: $\frac{5}{6}(6x+12) = \frac{5}{6} \cdot 6x + \frac{5}{6} \cdot 12 = 5x+10$
87. Applying the distributive property: $10\left(\frac{3}{5}x+\frac{1}{2}\right) = 10 \cdot \frac{3}{5}x + 10 \cdot \frac{1}{2} = 6x+5$
89. Applying the distributive property: $15\left(\frac{1}{3}x+\frac{2}{5}\right) = 15 \cdot \frac{1}{3}x + 15 \cdot \frac{2}{5} = 5x+6$
91. Applying the distributive property: $12\left(\frac{1}{2}m-\frac{5}{12}\right) = 12 \cdot \frac{1}{2}m - 12 \cdot \frac{5}{12} = 6m-5$
93. Applying the distributive property: $21\left(\frac{1}{3}+\frac{1}{7}x\right) = 21 \cdot \frac{1}{3} + 21 \cdot \frac{1}{7}x = 7+3x$
95. Applying the distributive property: $6\left(\frac{1}{2}x-\frac{1}{3}y\right) = 6 \cdot \frac{1}{2}x - 6 \cdot \frac{1}{3}y = 3x-2y$
97. Applying the distributive property: $0.09(x+2,000) = 0.09x+180$
99. Applying the distributive property: $0.12(x+500) = 0.12x+60$
101. Applying the distributive property: $a\left(1+\frac{1}{a}\right) = a \cdot 1 + a \cdot \frac{1}{a} = a+1$
103. Applying the distributive property: $a\left(\frac{1}{a}-1\right) = a \cdot \frac{1}{a} - a \cdot 1 = 1-a$
105. No. The man cannot reverse the order of putting on his socks and putting on his shoes.
107. No. The skydiver must jump out of the plane before pulling the rip cord.
109. Division is not a commutative operation. For example, $8 \div 4 = 2$ while $4 \div 8 = \frac{1}{2}$.
111. Computing the yearly take-home pay:
  $12(2400-480) = 12(1920) = \$23,040$ $12 \cdot 2400 - 12 \cdot 480 = 28,800 - 5,760 = \$23,040$
113. Computing her annual expenses:
  $3(650+225) = 3(875) = \$2,625$ $3 \cdot 650 + 3 \cdot 225 = 1950 + 675 = \$2,625$

1.6 Multiplication of Real Numbers

1. Finding the product: $7(-6) = -42$
3. Finding the product: $-8(2) = -16$
5. Finding the product: $-3(-1) = 3$
7. Finding the product: $-11(-11) = 121$
9. Using order of operations: $-3(2)(-1) = 6$
11. Using order of operations: $-3(-4)(-5) = -60$
13. Using order of operations: $-2(-4)(-3)(-1) = 24$
15. Using order of operations: $(-7)^2 = (-7)(-7) = 49$
17. Using order of operations: $(-3)^3 = (-3)(-3)(-3) = -27$
19. Using order of operations: $-2(2-5) = -2(-3) = 6$
21. Using order of operations: $-5(8-10) = -5(-2) = 10$
23. Using order of operations: $(4-7)(6-9) = (-3)(-3) = 9$
25. Using order of operations: $(-3-2)(-5-4) = (-5)(-9) = 45$
27. Using order of operations: $-3(-6)+4(-1) = 18+(-4) = 14$
29. Using order of operations: $2(3)-3(-4)+4(-5) = 6+12+(-20) = 18+(-20) = -2$

31. Using order of operations: $4(-3)^2 + 5(-6)^2 = 4(9) + 5(36) = 36 + 180 = 216$
33. Using order of operations: $7(-2)^3 - 2(-3)^3 = 7(-8) - 2(-27) = -56 + 54 = -2$
35. Using order of operations: $6 - 4(8-2) = 6 - 4(6) = 6 - 24 = 6 + (-24) = -18$
37. Using order of operations: $9 - 4(3-8) = 9 - 4(-5) = 9 + 20 = 29$
39. Using order of operations: $-4(3-8) - 6(2-5) = -4(-5) - 6(-3) = 20 + 18 = 38$
41. Using order of operations: $7 - 2[-6 - 4(-3)] = 7 - 2(-6+12) = 7 - 2(6) = 7 - 12 = 7 + (-12) = -5$
43. Using order of operations:
    $7 - 3[2(-4-4) - 3(-1-1)] = 7 - 3[2(-8) - 3(-2)] = 7 - 3(-16+6) = 7 - 3(-10) = 7 + 30 = 37$
45. a. Simplifying: $5(-4)(-3) = (-20)(-3) = 60$
    b. Simplifying: $5(-4) - 3 = -20 - 3 = -23$
    c. Simplifying: $5 - 4(-3) = 5 - (-12) = 5 + 12 = 17$
    d. Simplifying: $5 - 4 - 3 = 1 - 3 = -2$
47. Multiplying the fractions: $-\frac{2}{3} \cdot \frac{5}{7} = -\frac{2 \cdot 5}{3 \cdot 7} = -\frac{10}{21}$
49. Multiplying the fractions: $-8\left(\frac{1}{2}\right) = -\frac{8}{1} \cdot \frac{1}{2} = -\frac{8}{2} = -4$
51. Multiplying the fractions: $\left(-\frac{3}{4}\right)^2 = \left(-\frac{3}{4}\right)\left(-\frac{3}{4}\right) = \frac{9}{16}$
53. a. Simplifying: $\frac{5}{8}(24) + \frac{3}{7}(28) = 15 + 12 = 27$
    b. Simplifying: $\frac{5}{8}(24) - \frac{3}{7}(28) = 15 - 12 = 3$
    c. Simplifying: $\frac{5}{8}(-24) + \frac{3}{7}(-28) = -15 - 12 = -27$
    d. Simplifying: $-\frac{5}{8}(24) - \frac{3}{7}(28) = -15 - 12 = -27$
55. Simplifying: $\left(\frac{1}{2} \cdot 6\right)^2 = (3)^2 = (3)(3) = 9$
57. Simplifying: $\left(\frac{1}{2} \cdot 5\right)^2 = \left(\frac{5}{2}\right)^2 = \left(\frac{5}{2}\right)\left(\frac{5}{2}\right) = \frac{25}{4}$
59. Simplifying: $\left(\frac{1}{2}(-4)\right)^2 = (-2)^2 = (-2)(-2) = 4$
61. Simplifying: $\left(\frac{1}{2}(-3)\right)^2 = \left(-\frac{3}{2}\right)^2 = \left(-\frac{3}{2}\right)\left(-\frac{3}{2}\right) = \frac{9}{4}$
63. Multiplying the expressions: $-2(4x) = (-2 \cdot 4)x = -8x$
65. Multiplying the expressions: $-7(-6x) = [-7 \cdot (-6)]x = 42x$
67. Multiplying the expressions: $-\frac{1}{3}(-3x) = \left[-\frac{1}{3} \cdot (-3)\right]x = 1x = x$
69. Simplifying the expression: $-\frac{1}{2}(3x - 6) = -\frac{1}{2}(3x) - \frac{1}{2}(-6) = -\frac{3}{2}x + 3$
71. Simplifying the expression: $-3(2x - 5) - 7 = -6x + 15 - 7 = -6x + 8$
73. Simplifying the expression: $-5(3x + 4) - 10 = -15x - 20 - 10 = -15x - 30$
75. Simplifying the expression: $-4(3x + 5y) = -4(3x) - 4(5y) = -12x - 20y$
77. Simplifying the expression: $-2(3x + 5y) = -2(3x) - 2(5y) = -6x - 10y$
79. Simplifying the expression: $\frac{1}{2}(-3x + 6) = \frac{1}{2}(-3x) + \frac{1}{2}(6) = -\frac{3}{2}x + 3$
81. Simplifying the expression: $\frac{1}{3}(-2x + 6) = \frac{1}{3}(-2x) + \frac{1}{3}(6) = -\frac{2}{3}x + 2$
83. Simplifying the expression: $-\frac{1}{3}(-2x + 6) = -\frac{1}{3}(-2x) - \frac{1}{3}(6) = \frac{2}{3}x - 2$
85. Simplifying the expression: $8\left(-\frac{1}{4}x + \frac{1}{8}y\right) = 8\left(-\frac{1}{4}x\right) + 8\left(\frac{1}{8}y\right) = -2x + y$
87. a. Substituting $x = 0$: $-\frac{1}{3}x + 2 = -\frac{1}{3}(0) + 2 = 0 + 2 = 2$
    b. Substituting $x = 3$: $-\frac{1}{3}x + 2 = -\frac{1}{3}(3) + 2 = -1 + 2 = 1$
    c. Substituting $x = -3$: $-\frac{1}{3}x + 2 = -\frac{1}{3}(-3) + 2 = 1 + 2 = 3$

89. a. Substituting $x = 2$ and $y = -1$: $2x + y = 2(2) + (-1) = 4 - 1 = 3$
     b. Substituting $x = 0$ and $y = 3$: $2x + y = 2(0) + (0) = 0 + 3 = 3$
     c. Substituting $x = \frac{3}{2}$ and $y = -7$: $2x + y = 2\left(\frac{3}{2}\right) + (-7) = 3 + (-7) = -4$

91. a. Substituting $x = 4$: $2x^2 - 5x = 2(4)^2 - 5(4) = 2(16) - 5(4) = 32 - 20 = 12$
     b. Substituting $x = -\frac{3}{2}$: $2x^2 - 5x = 2\left(-\frac{3}{2}\right)^2 - 5\left(-\frac{3}{2}\right) = 2\left(\frac{9}{4}\right) - 5\left(-\frac{3}{2}\right) = \frac{9}{2} + \frac{15}{2} = \frac{24}{2} = 12$

93. a. Substituting $y = 4$: $y(2y + 3) = 4[2(4) + 3] = 4(8 + 3) = 4(11) = 44$
     b. Substituting $y = -\frac{11}{2}$: $y(2y + 3) = -\frac{11}{2}\left[2\left(-\frac{11}{2}\right) + 3\right] = -\frac{11}{2}(-11 + 3) = -\frac{11}{2}(-8) = \frac{88}{2} = 44$

95. Writing the expression: $3(-10) + 5 = -30 + 5 = -25$ 97. Writing the expression: $2(-4x) = -8x$
99. Writing the expression: $-9 \cdot 2 - 8 = -18 + (-8) = -26$
101. The pattern is to multiply by 2, so the next number is $4 \cdot 2 = 8$.
103. The pattern is to multiply by -2, so the next number is $40 \cdot (-2) = -80$.
105. The pattern is to multiply by $\frac{1}{2}$, so the next number is $\frac{1}{4} \cdot \frac{1}{2} = \frac{1}{8}$.
107. The pattern is to multiply by -2, so the next number is $12 \cdot (-2) = -24$.
109. The amount lost is: $20(\$3) = \$60$ 111. The temperature is: $25° - 4(6°) = 25° - 24° = 1°\text{F}$
113. a. The difference is: $\$2,484 - \$1,940 = \$544$
     b. The difference is: $\$1,115 - \$1,636 = -\$521$
     c. The actual costs have decreased, due to increased tax credits and student grants.

1.7 Division of Real Numbers

1. Finding the quotient: $\frac{8}{-4} = -2$ 3. Finding the quotient: $\frac{-48}{16} = -3$
5. Finding the quotient: $\frac{-7}{21} = -\frac{1}{3}$ 7. Finding the quotient: $\frac{-39}{-13} = 3$
9. Finding the quotient: $\frac{-6}{-42} = \frac{1}{7}$ 11. Finding the quotient: $\frac{0}{-32} = 0$
13. Performing the operations: $-3 + 12 = 9$ 15. Performing the operations: $-3 - 12 = -3 + (-12) = -15$
17. Performing the operations: $-3(12) = -36$ 19. Performing the operations: $-3 \div 12 = \frac{-3}{12} = -\frac{1}{4}$
21. Dividing and reducing: $\frac{4}{5} \div \frac{3}{4} = \frac{4}{5} \cdot \frac{4}{3} = \frac{16}{15}$
23. Dividing and reducing: $-\frac{5}{6} \div \left(-\frac{5}{8}\right) = -\frac{5}{6} \cdot \left(-\frac{8}{5}\right) = \frac{40}{30} = \frac{4}{3}$
25. Dividing and reducing: $\frac{10}{13} \div \left(-\frac{5}{4}\right) = \frac{10}{13} \cdot \left(-\frac{4}{5}\right) = -\frac{40}{65} = -\frac{8}{13}$
27. Dividing and reducing: $-\frac{5}{6} \div \frac{5}{6} = -\frac{5}{6} \cdot \frac{6}{5} = -\frac{30}{30} = -1$
29. Dividing and reducing: $-\frac{3}{4} \div \left(-\frac{3}{4}\right) = -\frac{3}{4} \cdot \left(-\frac{4}{3}\right) = \frac{12}{12} = 1$
31. Using order of operations: $\frac{3(-2)}{-10} = \frac{-6}{-10} = \frac{3}{5}$ 33. Using order of operations: $\frac{-5(-5)}{-15} = \frac{25}{-15} = -\frac{5}{3}$
35. Using order of operations: $\frac{-8(-7)}{-28} = \frac{56}{-28} = -2$ 37. Using order of operations: $\frac{27}{4 - 13} = \frac{27}{-9} = -3$
39. Using order of operations: $\frac{20 - 6}{5 - 5} = \frac{14}{0}$, which is undefined
41. Using order of operations: $\frac{-3 + 9}{2 \cdot 5 - 10} = \frac{6}{10 - 10} = \frac{6}{0}$, which is undefined

10 Chapter 1 The Basics

43. Using order of operations: $\dfrac{15(-5)-25}{2(-10)} = \dfrac{-75-25}{-20} = \dfrac{-100}{-20} = 5$

45. Using order of operations: $\dfrac{27-2(-4)}{-3(5)} = \dfrac{27+8}{-15} = \dfrac{35}{-15} = -\dfrac{7}{3}$

47. Using order of operations: $\dfrac{12-6(-2)}{12(-2)} = \dfrac{12+12}{-24} = \dfrac{24}{-24} = -1$

49. Using order of operations: $\dfrac{5^2-2^2}{-5+2} = \dfrac{25-4}{-3} = \dfrac{21}{-3} = -7$

51. Using order of operations: $\dfrac{8^2-2^2}{8^2+2^2} = \dfrac{64-4}{64+4} = \dfrac{60}{68} = \dfrac{15}{17}$

53. Using order of operations: $\dfrac{(5+3)^2}{-5^2-3^2} = \dfrac{8^2}{-25-9} = \dfrac{64}{-34} = -\dfrac{32}{17}$

55. Using order of operations: $\dfrac{(8-4)^2}{8^2-4^2} = \dfrac{4^2}{64-16} = \dfrac{16}{48} = \dfrac{1}{3}$

57. Using order of operations: $\dfrac{-4 \cdot 3^2 - 5 \cdot 2^2}{-8(7)} = \dfrac{-4 \cdot 9 - 5 \cdot 4}{-56} = \dfrac{-36-20}{-56} = \dfrac{-56}{-56} = 1$

59. Using order of operations: $\dfrac{3 \cdot 10^2 + 4 \cdot 10 + 5}{345} = \dfrac{300+40+5}{345} = \dfrac{345}{345} = 1$

61. Using order of operations: $\dfrac{7-[(2-3)-4]}{-1-2-3} = \dfrac{7-(-1-4)}{-6} = \dfrac{7-(-5)}{-6} = \dfrac{7+5}{-6} = \dfrac{12}{-6} = -2$

63. Using order of operations: $\dfrac{6(-4)-2(5-8)}{-6-3-5} = \dfrac{-24-2(-3)}{-14} = \dfrac{-24+6}{-14} = \dfrac{-18}{-14} = \dfrac{9}{7}$

65. Using order of operations: $\dfrac{3(-5-3)+4(7-9)}{5(-2)+3(-4)} = \dfrac{3(-8)+4(-2)}{-10+(-12)} = \dfrac{-24+(-8)}{-22} = \dfrac{-32}{-22} = \dfrac{16}{11}$

67. Using order of operations: $\dfrac{|3-9|}{3-9} = \dfrac{|-6|}{-6} = \dfrac{6}{-6} = -1$

69. Using order of operations: $\dfrac{2+0.15(10)}{10} = \dfrac{2+1.5}{10} = \dfrac{3.5}{10} = \dfrac{35}{100} = \dfrac{7}{20} = 0.35$

71. Using order of operations: $\dfrac{1-3}{3-1} = \dfrac{-2}{2} = -1$

73. **a.** Simplifying: $\dfrac{5-2}{3-1} = \dfrac{3}{2}$ **b.** Simplifying: $\dfrac{2-5}{1-3} = \dfrac{-3}{-2} = \dfrac{3}{2}$

75. **a.** Simplifying: $\dfrac{-4-1}{5-(-2)} = \dfrac{-4+(-1)}{5+2} = \dfrac{-5}{7} = -\dfrac{5}{7}$ **b.** Simplifying: $\dfrac{1-(-4)}{-2-5} = \dfrac{1+4}{-2+(-5)} = \dfrac{5}{-7} = -\dfrac{5}{7}$

77. **a.** Simplifying: $\dfrac{3+2.236}{2} = \dfrac{5.236}{2} = 2.618$ **b.** Simplifying: $\dfrac{3-2.236}{2} = \dfrac{0.764}{2} = 0.382$

   c. Simplifying: $\dfrac{3+2.236}{2} + \dfrac{3-2.236}{2} = \dfrac{5.236}{2} + \dfrac{0.764}{2} = 2.618 + 0.382 = 3$

79. **a.** Simplifying: $20 \div 4 \cdot 5 = 5 \cdot 5 = 25$ **b.** Simplifying: $-20 \div 4 \cdot 5 = -5 \cdot 5 = -25$

   c. Simplifying: $20 \div (-4) \cdot 5 = -5 \cdot 5 = -25$ **d.** Simplifying: $20 \div 4(-5) = 5(-5) = -25$

   e. Simplifying: $-20 \div 4(-5) = -5(-5) = 25$

81. a. Simplifying: $8 \div \frac{4}{5} = 8 \cdot \frac{5}{4} = 10$
    b. Simplifying: $8 \div \frac{4}{5} - 10 = 8 \cdot \frac{5}{4} - 10 = 10 - 10 = 0$
    c. Simplifying: $8 \div \frac{4}{5}(-10) = 8 \cdot \frac{5}{4}(-10) = 10(-10) = -100$
    d. Simplifying: $8 \div \left(-\frac{4}{5}\right) - 10 = 8 \cdot \left(-\frac{5}{4}\right) - 10 = -10 - 10 = -20$

83. Applying the distributive property: $10\left(\frac{x}{2} + \frac{3}{5}\right) = 10\left(\frac{x}{2}\right) + 10\left(\frac{3}{5}\right) = 5x + 6$

85. Applying the distributive property: $15\left(\frac{x}{5} + \frac{4}{3}\right) = 15\left(\frac{x}{5}\right) + 15\left(\frac{4}{3}\right) = 3x + 20$

87. Applying the distributive property: $x\left(\frac{3}{x} + 1\right) = x\left(\frac{3}{x}\right) + x(1) = 3 + x$

89. Applying the distributive property: $21\left(\frac{x}{7} - \frac{y}{3}\right) = 21\left(\frac{x}{7}\right) - 21\left(\frac{y}{3}\right) = 3x - 7y$

91. Applying the distributive property: $a\left(\frac{3}{a} - \frac{2}{a}\right) = a\left(\frac{3}{a}\right) - a\left(\frac{2}{a}\right) = 3 - 2 = 1$

93. Applying the distributive property: $2y\left(\frac{1}{y} - \frac{1}{2}\right) = 2y\left(\frac{1}{y}\right) - 2y\left(\frac{1}{2}\right) = 2 - y$

95. The quotient is $\frac{-12}{-4} = 3$.

97. The number is -10, since $\frac{-10}{-5} = 2$.

99. The number is -3, since $\frac{27}{-3} = -9$.

101. The expression is: $\frac{-20}{4} - 3 = -5 - 3 = -8$

103. Each person would lose: $\frac{13600 - 15000}{4} = \frac{-1400}{4} = -350 = \$350$ loss

105. The change per hour is: $\frac{61° - 75°}{4} = \frac{-14°}{4} = -3.5°$ per hour

107. a. Since they predict $50 revenue for every 25 people, their projected revenue is:
    $\frac{\$50}{25 \text{ people}} \cdot 10,000 \text{ people} = \$20,000$
    b. Since they predict $50 revenue for every 25 people, their projected revenue is:
    $\frac{\$50}{25 \text{ people}} \cdot 25,000 \text{ people} = \$50,000$
    c. Since they predict $50 revenue for every 25 people, their projected revenue is:
    $\frac{\$50}{25 \text{ people}} \cdot 5,000 \text{ people} = \$10,000$
    Since this projected revenue is more than the $5,000 cost for the list, it is a wise purchase.

1.8 Subsets of the Real Numbers

1. The set is $A \cup B = \{0, 1, 2, 3, 4, 5, 6\}$.
3. The set is $A \cap B = \{2, 4\}$.
5. The set is $B \cap C = \{1, 3, 5\}$.
7. The set is $A \cup (B \cap C) = \{0, 1, 2, 3, 4, 5, 6\}$.
9. The set is $\{0, 2\}$.
11. The set is $\{0, 6\}$.
13. The set is $\{0, 1, 2, 3, 4, 5, 6, 7\}$.
15. The set is $\{1, 2, 4, 5\}$.
17. The whole numbers are: 0, 1
19. The rational numbers are: $-3, -2.5, 0, 1, \frac{3}{2}$
21. The real numbers are: $-3, -2.5, 0, 1, \frac{3}{2}, \sqrt{15}$
23. The integers are: $-10, -8, -2, 9$
25. The irrational numbers are: π
27. true
29. false
31. false
33. true

12 Chapter 1 The Basics

35. This number is composite: $48 = 6 \cdot 8 = (2 \cdot 3) \cdot (2 \cdot 2 \cdot 2) = 2^4 \cdot 3$

37. This number is prime.   **39.** This number is composite: $1023 = 3 \cdot 341 = 3 \cdot 11 \cdot 31$

41. Factoring the number: $144 = 12 \cdot 12 = (3 \cdot 4) \cdot (3 \cdot 4) = (3 \cdot 2 \cdot 2) \cdot (3 \cdot 2 \cdot 2) = 2^4 \cdot 3^2$

43. Factoring the number: $38 = 2 \cdot 19$

45. Factoring the number: $105 = 5 \cdot 21 = 5 \cdot (3 \cdot 7) = 3 \cdot 5 \cdot 7$

47. Factoring the number: $180 = 10 \cdot 18 = (2 \cdot 5) \cdot (3 \cdot 6) = (2 \cdot 5) \cdot (3 \cdot 2 \cdot 3) = 2^2 \cdot 3^2 \cdot 5$

49. Factoring the number: $385 = 5 \cdot 77 = 5 \cdot (7 \cdot 11) = 5 \cdot 7 \cdot 11$

51. Factoring the number: $121 = 11 \cdot 11 = 11^2$

53. Factoring the number: $420 = 10 \cdot 42 = (2 \cdot 5) \cdot (7 \cdot 6) = (2 \cdot 5) \cdot (7 \cdot 2 \cdot 3) = 2^2 \cdot 3 \cdot 5 \cdot 7$

55. Factoring the number: $620 = 10 \cdot 62 = (2 \cdot 5) \cdot (2 \cdot 31) = 2^2 \cdot 5 \cdot 31$

57. Reducing the fraction: $\frac{105}{165} = \frac{3 \cdot 5 \cdot 7}{3 \cdot 5 \cdot 11} = \frac{7}{11}$ **59.** Reducing the fraction: $\frac{525}{735} = \frac{3 \cdot 5 \cdot 5 \cdot 7}{3 \cdot 5 \cdot 7 \cdot 7} = \frac{5}{7}$

61. Reducing the fraction: $\frac{385}{455} = \frac{5 \cdot 7 \cdot 11}{5 \cdot 7 \cdot 13} = \frac{11}{13}$ **63.** Reducing the fraction: $\frac{322}{345} = \frac{2 \cdot 7 \cdot 23}{3 \cdot 5 \cdot 23} = \frac{2 \cdot 7}{3 \cdot 5} = \frac{14}{15}$

65. Reducing the fraction: $\frac{205}{369} = \frac{5 \cdot 41}{3 \cdot 3 \cdot 41} = \frac{5}{3 \cdot 3} = \frac{5}{9}$

67. Reducing the fraction: $\frac{215}{344} = \frac{5 \cdot 43}{2 \cdot 2 \cdot 2 \cdot 43} = \frac{5}{2 \cdot 2 \cdot 2} = \frac{5}{8}$

69. **a.** Adding: $50 + (-80) = -30$ **b.** Subtracting: $50 - (-80) = 50 + 80 = 130$

   c. Multiplying: $50(-80) = -4,000$ **d.** Dividing: $\frac{50}{-80} = -\frac{5}{8}$

71. Simplifying; $\frac{6.28}{9(3.14)} = \frac{6.28}{28.26} = \frac{628}{2826} = \frac{2 \cdot 314}{9 \cdot 314} = \frac{2}{9}$ **73.** Simplifying; $\frac{9.42}{2(3.14)} = \frac{9.42}{6.28} = \frac{942}{628} = \frac{3 \cdot 314}{2 \cdot 314} = \frac{3}{2}$

75. Simplifying; $\frac{32}{0.5} = \frac{320}{5} = 64$ **77.** Simplifying; $\frac{5,599}{11} = 509$

79. **a.** Substituting $x = 10$: $\frac{2 + 0.15x}{x} = \frac{2 + 0.15(10)}{10} = \frac{2 + 1.5}{10} = \frac{3.5}{10} = 0.35$

   b. Substituting $x = 15$: $\frac{2 + 0.15x}{x} = \frac{2 + 0.15(15)}{15} = \frac{2 + 2.25}{15} = \frac{4.25}{15} \approx 0.28$

   c. Substituting $x = 20$: $\frac{2 + 0.15x}{x} = \frac{2 + 0.15(20)}{20} = \frac{2 + 3}{20} = \frac{5}{20} = 0.25$

81. Factoring into prime numbers: $6^3 = (2 \cdot 3)^3 = 2^3 \cdot 3^3$

83. Factoring into prime numbers: $9^4 \cdot 16^2 = (3 \cdot 3)^4 \cdot (2 \cdot 2 \cdot 2 \cdot 2)^2 = 3^4 \cdot 3^4 \cdot 2^2 \cdot 2^2 \cdot 2^2 \cdot 2^2 = 2^8 \cdot 3^8$

85. Simplifying and factoring: $3 \cdot 8 + 3 \cdot 7 + 3 \cdot 5 = 24 + 21 + 15 = 60 = 6 \cdot 10 = (2 \cdot 3) \cdot (2 \cdot 5) = 2^2 \cdot 3 \cdot 5$

87. They are not a subset of the irrational numbers.

89. 8, 21, and 34 are Fibonacci numbers that are composite numbers.

1.9 Addition and Subtraction with Fractions

1. Combining the fractions: $\frac{3}{6} + \frac{1}{6} = \frac{4}{6} = \frac{2}{3}$

3. Combining the fractions: $\frac{3}{8} - \frac{5}{8} = -\frac{2}{8} = -\frac{1}{4}$

5. Combining the fractions: $-\frac{1}{4} + \frac{3}{4} = \frac{2}{4} = \frac{1}{2}$

7. Combining the fractions: $\frac{x}{3} - \frac{1}{3} = \frac{x-1}{3}$

9. Combining the fractions: $\frac{1}{4} + \frac{2}{4} + \frac{3}{4} = \frac{6}{4} = \frac{3}{2}$

11. Combining the fractions: $\frac{x+7}{2} - \frac{1}{2} = \frac{x+7-1}{2} = \frac{x+6}{2}$

13. Combining the fractions: $\frac{1}{10} - \frac{3}{10} - \frac{4}{10} = -\frac{6}{10} = -\frac{3}{5}$

15. Combining the fractions: $\frac{1}{a} + \frac{4}{a} + \frac{5}{a} = \frac{10}{a}$

17. Combining the fractions: $\frac{1}{8} + \frac{3}{4} = \frac{1}{8} + \frac{3 \cdot 2}{4 \cdot 2} = \frac{1}{8} + \frac{6}{8} = \frac{7}{8}$

19. Combining the fractions: $\frac{3}{10} - \frac{1}{5} = \frac{3}{10} - \frac{1 \cdot 2}{5 \cdot 2} = \frac{3}{10} - \frac{2}{10} = \frac{1}{10}$

21. Combining the fractions: $\frac{4}{9} + \frac{1}{3} = \frac{4}{9} + \frac{1 \cdot 3}{3 \cdot 3} = \frac{4}{9} + \frac{3}{9} = \frac{7}{9}$

23. Combining the fractions: $2 + \frac{1}{3} = \frac{2 \cdot 3}{1 \cdot 3} + \frac{1}{3} = \frac{6}{3} + \frac{1}{3} = \frac{7}{3}$

25. Combining the fractions: $-\frac{3}{4} + 1 = -\frac{3}{4} + \frac{1 \cdot 4}{1 \cdot 4} = -\frac{3}{4} + \frac{4}{4} = \frac{1}{4}$

27. Combining the fractions: $\frac{1}{2} + \frac{2}{3} = \frac{1 \cdot 3}{2 \cdot 3} + \frac{2 \cdot 2}{3 \cdot 2} = \frac{3}{6} + \frac{4}{6} = \frac{7}{6}$

29. Combining the fractions: $\frac{5}{12} - \left(-\frac{3}{8}\right) = \frac{5}{12} + \frac{3}{8} = \frac{5 \cdot 2}{12 \cdot 2} + \frac{3 \cdot 3}{8 \cdot 3} = \frac{10}{24} + \frac{9}{24} = \frac{19}{24}$

31. Combining the fractions: $-\frac{1}{20} + \frac{8}{30} = -\frac{1 \cdot 3}{20 \cdot 3} + \frac{8 \cdot 2}{30 \cdot 2} = -\frac{3}{60} + \frac{16}{60} = \frac{13}{60}$

33. First factor the denominators to find the LCM:
    $30 = 2 \cdot 3 \cdot 5$
    $42 = 2 \cdot 3 \cdot 7$
    LCM $= 2 \cdot 3 \cdot 5 \cdot 7 = 210$
    Combining the fractions: $\frac{17}{30} + \frac{11}{42} = \frac{17 \cdot 7}{30 \cdot 7} + \frac{11 \cdot 5}{42 \cdot 5} = \frac{119}{210} + \frac{55}{210} = \frac{174}{210} = \frac{2 \cdot 3 \cdot 29}{2 \cdot 3 \cdot 5 \cdot 7} = \frac{29}{5 \cdot 7} = \frac{29}{35}$

35. First factor the denominators to find the LCM:
    $84 = 2 \cdot 2 \cdot 3 \cdot 7$
    $90 = 2 \cdot 3 \cdot 3 \cdot 5$
    LCM $= 2 \cdot 2 \cdot 3 \cdot 3 \cdot 5 \cdot 7 = 1260$
    Combining the fractions: $\frac{25}{84} + \frac{41}{90} = \frac{25 \cdot 15}{84 \cdot 15} + \frac{41 \cdot 14}{90 \cdot 14} = \frac{375}{1260} + \frac{574}{1260} = \frac{949}{1260}$

37. First factor the denominators to find the LCM:
    $126 = 2 \cdot 3 \cdot 3 \cdot 7$
    $180 = 2 \cdot 2 \cdot 3 \cdot 3 \cdot 5$
    LCM $= 2 \cdot 2 \cdot 3 \cdot 3 \cdot 5 \cdot 7 = 1260$
    Combining the fractions:
    $\frac{13}{126} - \frac{13}{180} = \frac{13 \cdot 10}{126 \cdot 10} - \frac{13 \cdot 7}{180 \cdot 7} = \frac{130}{1,260} - \frac{91}{1,260} = \frac{39}{1,260} = \frac{3 \cdot 13}{2 \cdot 2 \cdot 3 \cdot 3 \cdot 5 \cdot 7} = \frac{13}{2 \cdot 2 \cdot 3 \cdot 5 \cdot 7} = \frac{13}{420}$

39. Combining the fractions: $\frac{3}{4} + \frac{1}{8} + \frac{5}{6} = \frac{3 \cdot 6}{4 \cdot 6} + \frac{1 \cdot 3}{8 \cdot 3} + \frac{5 \cdot 4}{6 \cdot 4} = \frac{18}{24} + \frac{3}{24} + \frac{20}{24} = \frac{41}{24}$

41. Combining the fractions: $\frac{1}{2} + \frac{1}{3} + \frac{1}{4} + \frac{1}{6} = \frac{1 \cdot 6}{2 \cdot 6} + \frac{1 \cdot 4}{3 \cdot 4} + \frac{1 \cdot 3}{4 \cdot 3} + \frac{1 \cdot 2}{6 \cdot 2} = \frac{6}{12} + \frac{4}{12} + \frac{3}{12} + \frac{2}{12} = \frac{15}{12} = \frac{5}{4}$

43. Combining the fractions: $1 - \frac{5}{2} = 1 \cdot \frac{2}{2} - \frac{5}{2} = \frac{2}{2} - \frac{5}{2} = -\frac{3}{2}$

14 Chapter 1 The Basics

45. Combining the fractions: $1 + \frac{1}{2} = 1 \cdot \frac{2}{2} + \frac{1}{2} = \frac{2}{2} + \frac{1}{2} = \frac{3}{2}$

47. The pattern is to add $-\frac{1}{3}$, so the fourth term is: $-\frac{1}{3} + \left(-\frac{1}{3}\right) = -\frac{2}{3}$

49. The pattern is to add $\frac{2}{3}$, so the fourth term is: $\frac{5}{3} + \frac{2}{3} = \frac{7}{3}$

51. The pattern is to multiply by $\frac{1}{5}$, so the fourth term is: $\frac{1}{25} \cdot \frac{1}{5} = \frac{1}{125}$

53. Using order of operations: $9 - 3\left(\frac{5}{3}\right) = 9 - 5 = 4$

55. Using order of operations: $-\frac{1}{2} + 2\left(-\frac{3}{4}\right) = -\frac{1}{2} - \frac{3}{2} = -\frac{4}{2} = -2$

57. Using order of operations: $\frac{3}{5}(-10) + \frac{4}{7}(-21) = -6 - 12 = -18$

59. Using order of operations: $16\left(-\frac{1}{2}\right)^2 - 125\left(-\frac{2}{5}\right)^2 = 16\left(\frac{1}{4}\right) - 125\left(\frac{4}{25}\right) = 4 - 20 = -16$

61. Using order of operations: $-\frac{4}{3} \div 2 \cdot 3 = -\frac{4}{3} \cdot \frac{1}{2} \cdot 3 = -\frac{2}{3} \cdot 3 = -2$

63. Using order of operations: $-\frac{4}{3} \div 2(-3) = -\frac{4}{3} \cdot \frac{1}{2} \cdot (-3) = -\frac{2}{3} \cdot (-3) = 2$

65. Using order of operations: $-6 \div \frac{1}{2} \cdot 12 = -6 \cdot 2 \cdot 12 = -144$

67. Using order of operations: $-15 \div \frac{5}{3} \cdot 18 = -15 \cdot \frac{3}{5} \cdot 18 = -9 \cdot 18 = -162$

69. Combining the fractions: $\frac{x}{4} + \frac{1}{5} = \frac{x}{4} \cdot \frac{5}{5} + \frac{1}{5} \cdot \frac{4}{4} = \frac{5x}{20} + \frac{4}{20} = \frac{5x+4}{20}$

71. Combining the fractions: $\frac{1}{3} + \frac{a}{12} = \frac{1}{3} \cdot \frac{4}{4} + \frac{a}{12} = \frac{4}{12} + \frac{a}{12} = \frac{a+4}{12}$

73. Combining the fractions: $\frac{x}{2} + \frac{1}{3} + \frac{x}{4} = \frac{x}{2} \cdot \frac{6}{6} + \frac{1}{3} \cdot \frac{4}{4} + \frac{x}{4} \cdot \frac{3}{3} = \frac{6x}{12} + \frac{4}{12} + \frac{3x}{12} = \frac{6x+4+3x}{12} = \frac{9x+4}{12}$

75. Combining the fractions: $\frac{2}{x} + \frac{3}{5} = \frac{2}{x} \cdot \frac{5}{5} + \frac{3}{5} \cdot \frac{x}{x} = \frac{10}{5x} + \frac{3x}{5x} = \frac{3x+10}{5x}$

77. Combining the fractions: $\frac{3}{7} + \frac{4}{y} = \frac{3}{7} \cdot \frac{y}{y} + \frac{4}{y} \cdot \frac{7}{7} = \frac{3y}{7y} + \frac{28}{7y} = \frac{3y+28}{7y}$

79. Combining the fractions: $\frac{3}{a} + \frac{3}{4} + \frac{1}{5} = \frac{3}{a} \cdot \frac{20}{20} + \frac{3}{4} \cdot \frac{5a}{5a} + \frac{1}{5} \cdot \frac{4a}{4a} = \frac{60}{20a} + \frac{15a}{20a} + \frac{4a}{20a} = \frac{60+15a+4a}{20a} = \frac{19a+60}{20a}$

81. Combining the fractions: $\frac{1}{2}x + \frac{1}{6}x = \left(\frac{1}{2} + \frac{1}{6}\right)x = \left(\frac{1}{2} \cdot \frac{3}{3} + \frac{1}{6}\right)x = \left(\frac{3}{6} + \frac{1}{6}\right)x = \frac{4}{6}x = \frac{2}{3}x$

83. Combining the fractions: $\frac{1}{2}x - \frac{3}{4}x = \left(\frac{1}{2} - \frac{3}{4}\right)x = \left(\frac{1}{2} \cdot \frac{2}{2} - \frac{3}{4}\right)x = \left(\frac{2}{4} - \frac{3}{4}\right)x = -\frac{1}{4}x$

85. Combining the fractions: $\frac{1}{3}x + \frac{3}{5}x = \left(\frac{1}{3} + \frac{3}{5}\right)x = \left(\frac{1}{3} \cdot \frac{5}{5} + \frac{3}{5} \cdot \frac{3}{3}\right)x = \left(\frac{5}{15} + \frac{9}{15}\right)x = \frac{14}{15}x$

87. Combining the fractions: $\frac{3x}{4} + \frac{x}{6} = \frac{3x}{4} \cdot \frac{3}{3} + \frac{x}{6} \cdot \frac{2}{2} = \frac{9x}{12} + \frac{2x}{12} = \frac{11x}{12} = \frac{11}{12}x$

89. Combining the fractions: $\frac{2x}{5} + \frac{5x}{8} = \frac{2x}{5} \cdot \frac{8}{8} + \frac{5x}{8} \cdot \frac{5}{5} = \frac{16x}{40} + \frac{25x}{40} = \frac{41x}{40} = \frac{41}{40}x$

91. Combining the fractions: $1 - \frac{1}{x} = \frac{1}{1} \cdot \frac{x}{x} - \frac{1}{x} = \frac{x}{x} - \frac{1}{x} = \frac{x-1}{x}$

93. a. Adding the fractions: $\frac{3}{4}+\left(-\frac{1}{2}\right)=\frac{3}{4}-\frac{1}{2}\cdot\frac{2}{2}=\frac{3}{4}-\frac{2}{4}=\frac{1}{4}$

    b. Subtracting the fractions: $\frac{3}{4}-\left(-\frac{1}{2}\right)=\frac{3}{4}+\frac{1}{2}\cdot\frac{2}{2}=\frac{3}{4}+\frac{2}{4}=\frac{5}{4}$

    c. Multiplying the fractions: $\frac{3}{4}\left(-\frac{1}{2}\right)=-\frac{3}{8}$

    d. Dividing the fractions: $\frac{3}{4}\div\left(-\frac{1}{2}\right)=\frac{3}{4}\cdot\left(-\frac{2}{1}\right)=-\frac{6}{4}=-\frac{3}{2}$

95. Simplifying: $\left(1-\frac{1}{2}\right)\left(1-\frac{1}{3}\right)=\left(\frac{2}{2}-\frac{1}{2}\right)\left(\frac{3}{3}-\frac{1}{3}\right)=\left(\frac{1}{2}\right)\left(\frac{2}{3}\right)=\frac{2}{6}=\frac{1}{3}$

97. Simplifying: $\left(1+\frac{1}{2}\right)\left(1-\frac{1}{2}\right)=\left(\frac{2}{2}+\frac{1}{2}\right)\left(\frac{2}{2}-\frac{1}{2}\right)=\left(\frac{3}{2}\right)\left(\frac{1}{2}\right)=\frac{3}{4}$

99. a. Substituting $x=2$: $1+\frac{1}{x}=1+\frac{1}{2}=\frac{2}{2}+\frac{1}{2}=\frac{3}{2}$

    b. Substituting $x=3$: $1+\frac{1}{x}=1+\frac{1}{3}=\frac{3}{3}+\frac{1}{3}=\frac{4}{3}$

    c. Substituting $x=4$: $1+\frac{1}{x}=1+\frac{1}{4}=\frac{4}{4}+\frac{1}{4}=\frac{5}{4}$

101. a. Substituting $x=1$: $2x+\frac{6}{x}=2(1)+\frac{6}{1}=2+6=8$

     b. Substituting $x=2$: $2x+\frac{6}{x}=2(2)+\frac{6}{2}=4+3=7$

     c. Substituting $x=3$: $2x+\frac{6}{x}=2(3)+\frac{6}{3}=6+2=8$

Chapter 1 Review/Test

1. The expression is: $-7+(-10)=-17$
2. The expression is: $(-7+4)+5=-3+5=2$
3. The expression is: $(-3+12)+5=9+5=14$
4. The expression is: $4-9=4+(-9)=-5$
5. The expression is: $9-(-3)=9+3=12$
6. The expression is: $-7-(-9)=-7+9=2$
7. The expression is: $(-3)(-7)-6=21-6=15$
8. The expression is: $5(-6)+10=-30+10=-20$
9. The expression is: $2[-8(3x)]=2(-24x)=-48x$
10. The expression is: $\frac{-25}{-5}=5$
11. Simplifying: $|-1.8|=1.8$
12. Simplifying: $-|-10|=-(10)=-10$
13. The opposite is -6, and the reciprocal is $\frac{1}{6}$.
14. The opposite is $\frac{12}{5}$, and the reciprocal is $-\frac{5}{12}$.
15. Multiplying the fractions: $\frac{1}{2}(-10)=\frac{1}{2}\cdot\left(-\frac{10}{1}\right)=-\frac{10}{2}=-5$
16. Multiplying the fractions: $\left(-\frac{4}{5}\right)\left(\frac{25}{16}\right)=-\frac{100}{80}=-\frac{5}{4}$
17. Adding: $-9+12=3$
18. Adding: $-18+(-20)=-38$
19. Adding: $(-2)+(-8)+[-9+(-6)]=-2+(-8)+(-15)=-25$
20. Adding: $(-21)+40+(-23)+5=-44+45=1$
21. Subtracting: $6-9=6+(-9)=-3$
22. Subtracting: $14-(-8)=14+8=22$
23. Subtracting: $-12-(-8)=-12+8=-4$
24. Subtracting: $4-9-15=4+(-9)+(-15)=4+(-24)=-20$
25. Multiplying: $(-5)(6)=-30$
26. Multiplying: $4(-3)=-12$
27. Multiplying: $-2(3)(4)=-24$
28. Multiplying: $(-1)(-3)(-1)(-4)=12$
29. Finding the quotient: $\frac{12}{-3}=-4$
30. Finding the quotient: $-\frac{8}{9}\div\frac{4}{3}=-\frac{8}{9}\cdot\frac{3}{4}=-\frac{24}{36}=-\frac{2}{3}$

31. Simplifying using order of operations: $4 \cdot 5 + 3 = 20 + 3 = 23$
32. Simplifying using order of operations: $9 \cdot 3 + 4 \cdot 5 = 27 + 20 = 47$
33. Simplifying using order of operations: $2^3 - 4 \cdot 3^2 + 5^2 = 8 - 4 \cdot 9 + 25 = 8 - 36 + 25 = 33 - 36 = -3$
34. Simplifying using order of operations:
$$12 - 3(2 \cdot 5 + 7) + 4 = 12 - 3(10 + 7) + 4 = 12 - 3(17) + 4 = 12 - 51 + 4 = -39 + 4 = -35$$
35. Simplifying using order of operations: $20 + 8 \div 4 + 2 \cdot 5 = 20 + 2 + 10 = 32$
36. Simplifying using order of operations: $2(3 - 5) - (2 - 8) = 2(-2) - (-6) = -4 + 6 = 2$
37. Simplifying using order of operations: $30 \div 3 \cdot 2 = 10 \cdot 2 = 20$
38. Simplifying using order of operations: $(-2)(3) - 4(-3) - 9 = -6 - (-12) - 9 = -6 + 12 + (-9) = -15 + 12 = -3$
39. Simplifying using order of operations:
$$3(4-7)^2 - 5(3-8)^2 = 3(-3)^2 - 5(-5)^2 = 3 \cdot 9 - 5 \cdot 25 = 27 - 125 = 27 + (-125) = -98$$
40. Simplifying using order of operations: $(-5 - 2)(-3 - 7) = (-7)(-10) = 70$
41. Simplifying using order of operations: $\dfrac{4(-3)}{-6} = \dfrac{-12}{-6} = 2$
42. Simplifying using order of operations: $\dfrac{3^2 + 5^2}{(3-5)^2} = \dfrac{9 + 25}{(-2)^2} = \dfrac{34}{4} = \dfrac{17}{2}$
43. Simplifying using order of operations: $\dfrac{15 - 10}{6 - 6} = \dfrac{5}{0}$, which is undefined
44. Simplifying using order of operations: $\dfrac{2(-7) + (-11)(-4)}{7 - (-3)} = \dfrac{-14 + 44}{7 + 3} = \dfrac{30}{10} = 3$
45. associative property (of multiplication)
46. multiplicative identity property
47. commutative property (of addition)
48. additive inverse property
49. commutative and associative properties (of addition)
50. distributive property
51. Simplifying the expression: $7 + (5 + x) = (7 + 5) + x = 12 + x$
52. Simplifying the expression: $4(7a) = (4 \cdot 7)a = 28a$
53. Simplifying the expression: $\frac{1}{9}(9x) = \left(\frac{1}{9} \cdot 9\right)x = 1x = x$
54. Simplifying the expression: $\frac{4}{5}\left(\frac{5}{4}y\right) = \left(\frac{4}{5} \cdot \frac{5}{4}\right)y = 1y = y$
55. Applying the distributive property: $7(2x + 3) = 7 \cdot 2x + 7 \cdot 3 = 14x + 21$
56. Applying the distributive property: $3(2a - 4) = 3 \cdot 2a - 3 \cdot 4 = 6a - 12$
57. Applying the distributive property: $\frac{1}{2}(5x - 6) = \frac{1}{2} \cdot 5x - \frac{1}{2} \cdot 6 = \frac{5}{2}x - 3$
58. Applying the distributive property: $-\frac{1}{2}(3x - 6) = -\frac{1}{2} \cdot 3x - \left(-\frac{1}{2}\right) \cdot 6 = -\frac{3}{2}x + 3$
59. The rational numbers are: $-\frac{1}{3}, 0, 5, -4.5, \frac{2}{5}, -3$
60. The whole numbers are: $0, 5$
61. The irrational numbers are: $\sqrt{7}, \pi$
62. The integers are: $0, 5, -3$
63. Factoring the number: $90 = 9 \cdot 10 = (3 \cdot 3) \cdot (2 \cdot 5) = 2 \cdot 3^2 \cdot 5$
64. Factoring the number: $840 = 84 \cdot 10 = (21 \cdot 4) \cdot (2 \cdot 5) = (3 \cdot 7 \cdot 2 \cdot 2) \cdot (2 \cdot 5) = 2^3 \cdot 3 \cdot 5 \cdot 7$
65. First factor the denominators to find the LCM:
$35 = 5 \cdot 7$
$42 = 2 \cdot 3 \cdot 7$
$\text{LCM} = 2 \cdot 3 \cdot 5 \cdot 7 = 210$
Combining the fractions: $\dfrac{18}{35} + \dfrac{13}{42} = \dfrac{18 \cdot 6}{35 \cdot 6} + \dfrac{13 \cdot 5}{42 \cdot 5} = \dfrac{108}{210} + \dfrac{65}{210} = \dfrac{173}{210}$
66. Combining the fractions: $\dfrac{x}{6} + \dfrac{7}{12} = \dfrac{x}{6} \cdot \dfrac{2}{2} + \dfrac{7}{12} = \dfrac{2x}{12} + \dfrac{7}{12} = \dfrac{2x + 7}{12}$
67. The pattern is to add -3, so the next number is: $1 + (-3) = -2$
68. The pattern is to multiply by -3, so the next number is: $-270 \cdot (-3) = 810$

69. The pattern is to add the previous two terms, so the next number is: $3 + 5 = 8$

70. The pattern is to add 2, so the next number is: $10 + 2 = 12$

71. The pattern is to add $-\frac{1}{2}$, so the next number is: $-\frac{1}{2} + \left(-\frac{1}{2}\right) = -\frac{2}{2} = -1$

72. The pattern is to multiply by $-\frac{1}{2}$, so the next number is: $-\frac{1}{8} \cdot \left(-\frac{1}{2}\right) = \frac{1}{16}$

Chapter 2
Linear Equations and Inequalities

2.1 Simplifying Expressions

1. Simplifying the expression: $3x - 6x = (3-6)x = -3x$
3. Simplifying the expression: $-2a + a = (-2+1)a = -a$
5. Simplifying the expression: $7x + 3x + 2x = (7+3+2)x = 12x$
7. Simplifying the expression: $3a - 2a + 5a = (3-2+5)a = 6a$
9. Simplifying the expression: $4x - 3 + 2x = 4x + 2x - 3 = 6x - 3$
11. Simplifying the expression: $3a + 4a + 5 = 7a + 5$
13. Simplifying the expression: $2x - 3 + 3x - 2 = 2x + 3x - 3 - 2 = 5x - 5$
15. Simplifying the expression: $3a - 1 + a + 3 = 3a + a - 1 + 3 = 4a + 2$
17. Simplifying the expression: $-4x + 8 - 5x - 10 = -4x - 5x + 8 - 10 = -9x - 2$
19. Simplifying the expression: $7a + 3 + 2a + 3a = 7a + 2a + 3a + 3 = 12a + 3$
21. Simplifying the expression: $5(2x - 1) + 4 = 10x - 5 + 4 = 10x - 1$
23. Simplifying the expression: $7(3y + 2) - 8 = 21y + 14 - 8 = 21y + 6$
25. Simplifying the expression: $-3(2x - 1) + 5 = -6x + 3 + 5 = -6x + 8$
27. Simplifying the expression: $5 - 2(a + 1) = 5 - 2a - 2 = -2a - 2 + 5 = -2a + 3$
29. Simplifying the expression: $6 - 4(x - 5) = 6 - 4x + 20 = -4x + 20 + 6 = -4x + 26$
31. Simplifying the expression: $-9 - 4(2 - y) + 1 = -9 - 8 + 4y + 1 = 4y + 1 - 9 - 8 = 4y - 16$
33. Simplifying the expression: $-6 + 2(2 - 3x) + 1 = -6 + 4 - 6x + 1 = -6x - 6 + 4 + 1 = -6x - 1$
35. Simplifying the expression: $(4x - 7) - (2x + 5) = 4x - 7 - 2x - 5 = 4x - 2x - 7 - 5 = 2x - 12$
37. Simplifying the expression: $8(2a + 4) - (6a - 1) = 16a + 32 - 6a + 1 = 16a - 6a + 32 + 1 = 10a + 33$
39. Simplifying the expression: $3(x - 2) + (x - 3) = 3x - 6 + x - 3 = 3x + x - 6 - 3 = 4x - 9$
41. Simplifying the expression: $4(2y - 8) - (y + 7) = 8y - 32 - y - 7 = 8y - y - 32 - 7 = 7y - 39$
43. Simplifying the expression: $-9(2x + 1) - (x + 5) = -18x - 9 - x - 5 = -18x - x - 9 - 5 = -19x - 14$
45. Evaluating when $x = 2$: $3x - 1 = 3(2) - 1 = 6 - 1 = 5$
47. Evaluating when $x = 2$: $-2x - 5 = -2(2) - 5 = -4 - 5 = -9$
49. Evaluating when $x = 2$: $x^2 - 8x + 16 = (2)^2 - 8(2) + 16 = 4 - 16 + 16 = 4$
51. Evaluating when $x = 2$: $(x - 4)^2 = (2 - 4)^2 = (-2)^2 = 4$
53. Evaluating when $x = -5$: $7x - 4 - x - 3 = 7(-5) - 4 - (-5) - 3 = -35 - 4 + 5 - 3 = -42 + 5 = -37$
    Now simplifying the expression: $7x - 4 - x - 3 = 7x - x - 4 - 3 = 6x - 7$
    Evaluating when $x = -5$: $6x - 7 = 6(-5) - 7 = -30 - 7 = -37$
    Note that the two values are the same.

55. Evaluating when $x = -5$: $5(2x+1)+4 = 5[2(-5)+1]+4 = 5(-10+1)+4 = 5(-9)+4 = -45+4 = -41$
    Now simplifying the expression: $5(2x+1)+4 = 10x+5+4 = 10x+9$
    Evaluating when $x = -5$: $10x+9 = 10(-5)+9 = -50+9 = -41$
    Note that the two values are the same.

57. Evaluating when $x = -3$ and $y = 5$: $x^2 - 2xy + y^2 = (-3)^2 - 2(-3)(5) + (5)^2 = 9 + 30 + 25 = 64$

59. Evaluating when $x = -3$ and $y = 5$: $(x-y)^2 = (-3-5)^2 = (-8)^2 = 64$

61. Evaluating when $x = -3$ and $y = 5$: $x^2 + 6xy + 9y^2 = (-3)^2 + 6(-3)(5) + 9(5)^2 = 9 - 90 + 225 = 144$

63. Evaluating when $x = -3$ and $y = 5$: $(x+3y)^2 = [-3+3(5)]^2 = (-3+15)^2 = (12)^2 = 144$

65. Evaluating when $x = \frac{1}{2}$: $12x - 3 = 12\left(\frac{1}{2}\right) - 3 = 6 - 3 = 3$

67. Evaluating when $x = \frac{1}{4}$: $12x - 3 = 12\left(\frac{1}{4}\right) - 3 = 3 - 3 = 0$

69. Evaluating when $x = \frac{3}{2}$: $12x - 3 = 12\left(\frac{3}{2}\right) - 3 = 18 - 3 = 15$

71. Evaluating when $x = \frac{3}{4}$: $12x - 3 = 12\left(\frac{3}{4}\right) - 3 = 9 - 3 = 6$

73. a. Substituting the values for n:

n	1	2	3	4
$3n$	3	6	9	12

    b. Substituting the values for n:

n	1	2	3	4
n^3	1	8	27	64

75. Substituting $n = 1, 2, 3, 4$:
    $n = 1$: $3(1) - 2 = 3 - 2 = 1$
    $n = 2$: $3(2) - 2 = 6 - 2 = 4$
    $n = 3$: $3(3) - 2 = 9 - 2 = 7$
    $n = 4$: $3(4) - 2 = 12 - 2 = 10$
    The sequence is 1, 4, 7, 10, ..., which is an arithmetic sequence.

77. Substituting $n = 1, 2, 3, 4$:
    $n = 1$: $(1)^2 - 2(1) + 1 = 1 - 2 + 1 = 0$
    $n = 2$: $(2)^2 - 2(2) + 1 = 4 - 4 + 1 = 1$
    $n = 3$: $(3)^2 - 2(3) + 1 = 9 - 6 + 1 = 4$
    $n = 4$: $(4)^2 - 2(4) + 1 = 16 - 8 + 1 = 9$
    The sequence is 0, 1, 4, 9, ..., which is a sequence of squares.

79. Simplifying: $7 - 3(2y+1) = 7 - 6y - 3 = -6y + 4$

81. Simplifying: $0.08x + 0.09x = 0.17x$

83. Simplifying: $(x+y) + (x-y) = x + y + x - y = 2x$

85. Simplifying: $3x + 2(x-2) = 3x + 2x - 4 = 5x - 4$

87. Simplifying: $4(x+1) + 3(x-3) = 4x + 4 + 3x - 9 = 7x - 5$

89. Simplifying: $x + (x+3)(-3) = x - 3x - 9 = -2x - 9$

91. Simplifying: $3(4x-2) - (5x-8) = 12x - 6 - 5x + 8 = 7x + 2$

93. Simplifying: $-(3x+1) - (4x-7) = -3x - 1 - 4x + 7 = -7x + 6$

95. Simplifying: $(x+3y) + 3(2x-y) = x + 3y + 6x - 3y = 7x$

97. Simplifying: $3(2x+3y) - 2(3x+5y) = 6x + 9y - 6x - 10y = -y$

99. Simplifying: $-6\left(\frac{1}{2}x - \frac{1}{3}y\right) + 12\left(\frac{1}{4}x + \frac{2}{3}y\right) = -3x + 2y + 3x + 8y = 10y$

101. Simplifying: $0.08x + 0.09(x + 2{,}000) = 0.08x + 0.09x + 180 = 0.17x + 180$

103. Simplifying: $0.10x + 0.12(x + 500) = 0.10x + 0.12x + 60 = 0.22x + 60$

105. For the y terms to cancel, we must have $4y - ay = 0$, so $a = 4$.
 Simplifying: $(5x + 4y) + 4(2x - y) = 5x + 4y + 8x - 4y = 13x$

107. Evaluating the expression: $b^2 - 4ac = (-5)^2 - 4(1)(-6) = 25 - (-24) = 25 + 24 = 49$

109. Evaluating the expression: $b^2 - 4ac = (4)^2 - 4(2)(-3) = 16 - (-24) = 16 + 24 = 40$

111. a. Substituting $x = 8,000$: $-0.0035(8000) + 70 = 42°F$
 b. Substituting $x = 12,000$: $-0.0035(12000) + 70 = 28°F$
 c. Substituting $x = 24,000$: $-0.0035(24000) + 70 = -14°F$

113. a. Substituting $t = 10$: $35 + 0.25(10) = \$37.50$ b. Substituting $t = 20$: $35 + 0.25(20) = \$40.00$
 c. Substituting $t = 30$: $35 + 0.25(30) = \$42.50$

115. Simplifying the expression: $G - 0.21G - 0.08G = 0.71G$. Substituting $G = \$1,250$: $0.71(\$1,250) = \$887.50$

117. Simplifying: $\frac{1}{8} - \frac{1}{6} = \frac{1}{8} \cdot \frac{3}{3} - \frac{1}{6} \cdot \frac{4}{4} = \frac{3}{24} - \frac{4}{24} = -\frac{1}{24}$ 119. Simplifying: $\frac{5}{9} - \frac{4}{3} = \frac{5}{9} - \frac{4}{3} \cdot \frac{3}{3} = \frac{5}{9} - \frac{12}{9} = -\frac{7}{9}$

121. Simplifying: $-\frac{7}{30} + \frac{5}{28} = -\frac{7}{30} \cdot \frac{14}{14} + \frac{5}{28} \cdot \frac{15}{15} = -\frac{98}{420} + \frac{75}{420} = -\frac{23}{420}$

123. Simplifying: $17 - 5 = 12$ 125. Simplifying: $2 - 5 = -3$

127. Simplifying: $-2.4 + (-7.3) = -9.7$ 129. Simplifying: $-\frac{1}{2} + \left(-\frac{3}{4}\right) = -\frac{1}{2} \cdot \frac{2}{2} - \frac{3}{4} = -\frac{2}{4} - \frac{3}{4} = -\frac{5}{4}$

131. Simplifying: $4(2 \cdot 9 - 3) - 7 \cdot 9 = 4(18 - 3) - 63 = 4(15) - 63 = 60 - 63 = -3$

133. Simplifying: $4(2a - 3) - 7a = 8a - 12 - 7a = a - 12$

135. Subtracting: $-3 - \frac{1}{2} = \frac{-3}{1} - \frac{1}{2} = \frac{-3 \cdot 2}{1 \cdot 2} - \frac{1}{2} = \frac{-6}{2} - \frac{1}{2} = -\frac{7}{2}$

137. Adding: $\frac{4}{5} + \frac{1}{10} + \frac{3}{8} = \frac{4 \cdot 8}{5 \cdot 8} + \frac{1 \cdot 4}{10 \cdot 4} + \frac{3 \cdot 5}{8 \cdot 5} = \frac{32}{40} + \frac{4}{40} + \frac{15}{40} = \frac{51}{40}$

139. Evaluating when $x = 5$: $2(5) - 3 = 10 - 3 = 7$

2.2 Addition Property of Equality

1. Solving the equation:
$$x - 3 = 8$$
$$x - 3 + 3 = 8 + 3$$
$$x = 11$$

3. Solving the equation:
$$x + 2 = 6$$
$$x + 2 + (-2) = 6 + (-2)$$
$$x = 4$$

5. Solving the equation:
$$a + \frac{1}{2} = -\frac{1}{4}$$
$$a + \frac{1}{2} + \left(-\frac{1}{2}\right) = -\frac{1}{4} + \left(-\frac{1}{2}\right)$$
$$a = -\frac{1}{4} + \left(-\frac{2}{4}\right)$$
$$a = -\frac{3}{4}$$

7. Solving the equation:
$$x + 2.3 = -3.5$$
$$x + 2.3 + (-2.3) = -3.5 + (-2.3)$$
$$x = -5.8$$

9. Solving the equation:
$$y + 11 = -6$$
$$y + 11 + (-11) = -6 + (-11)$$
$$y = -17$$

11. Solving the equation:
$$x - \frac{5}{8} = -\frac{3}{4}$$
$$x - \frac{5}{8} + \frac{5}{8} = -\frac{3}{4} + \frac{5}{8}$$
$$x = -\frac{6}{8} + \frac{5}{8}$$
$$x = -\frac{1}{8}$$

13. Solving the equation:
$$m - 6 = 2m$$
$$m - 6 - m = 2m - m$$
$$m = -6$$

15. Solving the equation:
$$6.9 + x = 3.3$$
$$-6.9 + 6.9 + x = -6.9 + 3.3$$
$$x = -3.6$$

17. Solving the equation:
$$5a = 4a - 7$$
$$5a - 4a = 4a - 4a - 7$$
$$a = -7$$

19. Solving the equation:
$$-\tfrac{5}{9} = x - \tfrac{2}{5}$$
$$-\tfrac{5}{9} + \tfrac{2}{5} = x - \tfrac{2}{5} + \tfrac{2}{5}$$
$$-\tfrac{25}{45} + \tfrac{18}{45} = x$$
$$x = -\tfrac{7}{45}$$

21. Solving the equation:
$$4x + 2 - 3x = 4 + 1$$
$$x + 2 = 5$$
$$x + 2 + (-2) = 5 + (-2)$$
$$x = 3$$

23. Solving the equation:
$$8a - \tfrac{1}{2} - 7a = \tfrac{3}{4} + \tfrac{1}{8}$$
$$a - \tfrac{1}{2} = \tfrac{6}{8} + \tfrac{1}{8}$$
$$a - \tfrac{1}{2} = \tfrac{7}{8}$$
$$a - \tfrac{1}{2} + \tfrac{1}{2} = \tfrac{7}{8} + \tfrac{1}{2}$$
$$a = \tfrac{7}{8} + \tfrac{4}{8}$$
$$a = \tfrac{11}{8}$$

25. Solving the equation:
$$-3 - 4x + 5x = 18$$
$$-3 + x = 18$$
$$3 - 3 + x = 3 + 18$$
$$x = 21$$

27. Solving the equation:
$$-11x + 2 + 10x + 2x = 9$$
$$x + 2 = 9$$
$$x + 2 + (-2) = 9 + (-2)$$
$$x = 7$$

29. Solving the equation:
$$-2.5 + 4.8 = 8x - 1.2 - 7x$$
$$2.3 = x - 1.2$$
$$2.3 + 1.2 = x - 1.2 + 1.2$$
$$x = 3.5$$

31. Solving the equation:
$$2y - 10 + 3y - 4y = 18 - 6$$
$$y - 10 = 12$$
$$y - 10 + 10 = 12 + 10$$
$$y = 22$$

33. Solving the equation:
$$2(x + 3) - x = 4$$
$$2x + 6 - x = 4$$
$$x + 6 = 4$$
$$x + 6 + (-6) = 4 + (-6)$$
$$x = -2$$

35. Solving the equation:
$$-3(x - 4) + 4x = 3 - 7$$
$$-3x + 12 + 4x = -4$$
$$x + 12 = -4$$
$$x + 12 + (-12) = -4 + (-12)$$
$$x = -16$$

37. Solving the equation:
$$5(2a + 1) - 9a = 8 - 6$$
$$10a + 5 - 9a = 2$$
$$a + 5 = 2$$
$$a + 5 + (-5) = 2 + (-5)$$
$$a = -3$$

39. Solving the equation:
$$-(x + 3) + 2x - 1 = 6$$
$$-x - 3 + 2x - 1 = 6$$
$$x - 4 = 6$$
$$x - 4 + 4 = 6 + 4$$
$$x = 10$$

41. Solving the equation:
$$4y - 3(y - 6) + 2 = 8$$
$$4y - 3y + 18 + 2 = 8$$
$$y + 20 = 8$$
$$y + 20 + (-20) = 8 + (-20)$$
$$y = -12$$

43. Solving the equation:
$$-3(2m - 9) + 7(m - 4) = 12 - 9$$
$$-6m + 27 + 7m - 28 = 3$$
$$m - 1 = 3$$
$$m - 1 + 1 = 3 + 1$$
$$m = 4$$

45. Solving the equation:
$$4x = 3x + 2$$
$$4x + (-3x) = 3x + (-3x) + 2$$
$$x = 2$$

47. Solving the equation:
$$8a = 7a - 5$$
$$8a + (-7a) = 7a + (-7a) - 5$$
$$a = -5$$

49. Solving the equation:
$$2x = 3x + 1$$
$$(-2x) + 2x = (-2x) + 3x + 1$$
$$0 = x + 1$$
$$0 + (-1) = x + 1 + (-1)$$
$$x = -1$$

51. Solving the equation:
$$2y + 1 = 3y + 4$$
$$2y + (-2y) + 1 = 3y + (-2y) + 4$$
$$1 = y + 4$$
$$1 + (-4) = y + 4 + (-4)$$
$$y = -3$$

53. Solving the equation:
$$2m - 3 = m + 5$$
$$2m + (-m) - 3 = m + (-m) + 5$$
$$m - 3 = 5$$
$$m - 3 + 3 = 5 + 3$$
$$m = 8$$

55. Solving the equation:
$$4x - 7 = 5x + 1$$
$$4x + (-4x) - 7 = 5x + (-4x) + 1$$
$$-7 = x + 1$$
$$-7 + (-1) = x + 1 + (-1)$$
$$x = -8$$

57. Solving the equation:
$$4x + \tfrac{4}{3} = 5x - \tfrac{2}{3}$$
$$4x + (-4x) + \tfrac{4}{3} = 5x + (-4x) - \tfrac{2}{3}$$
$$\tfrac{4}{3} = x - \tfrac{2}{3}$$
$$\tfrac{4}{3} + \tfrac{2}{3} = x - \tfrac{2}{3} + \tfrac{2}{3}$$
$$x = \tfrac{6}{3} = 2$$

59. Solving the equation:
$$8a - 7.1 = 7a + 3.9$$
$$8a + (-7a) - 7.1 = 7a + (-7a) + 3.9$$
$$a - 7.1 = 3.9$$
$$a - 7.1 + 7.1 = 3.9 + 7.1$$
$$a = 11$$

61. a. Solving the equation:
$$2x = 3$$
$$x = \tfrac{3}{2}$$

b. Solving the equation:
$$2 + x = 3$$
$$2 + x - 2 = 3 - 2$$
$$x = 1$$

c. Solving the equation:
$$2x + 3 = 0$$
$$2x + 3 - 3 = 0 - 3$$
$$2x = -3$$
$$x = -\tfrac{3}{2}$$

d. Solving the equation:
$$2x + 3 = -5$$
$$2x + 3 - 3 = -5 - 3$$
$$2x = -8$$
$$x = -\tfrac{8}{2} = -4$$

e. Solving the equation:
$$2x + 3 = 7x - 5$$
$$2x + (-7x) + 3 = 7x + (-7x) - 5$$
$$-5x + 3 = -5$$
$$-5x + 3 + (-3) = -5 + (-3)$$
$$-5x = -8$$
$$x = \tfrac{8}{5}$$

63. a. Solving for R:
$$T + R + A = 100$$
$$88 + R + 6 = 100$$
$$94 + R = 100$$
$$R = 6\%$$

b. Solving for R:
$$T + R + A = 100$$
$$0 + R + 95 = 100$$
$$95 + R = 100$$
$$R = 5\%$$

c. Solving for A:
$$T + R + A = 100$$
$$0 + 98 + A = 100$$
$$98 + A = 100$$
$$A = 2\%$$

d. Solving for R:
$$T + R + A = 100$$
$$0 + R + 25 = 100$$
$$25 + R = 100$$
$$R = 75\%$$

65. Solving for x:
$$x + 55 + 55 = 180$$
$$x + 110 = 180$$
$$x = 70°$$

67. Applying the associative property: $3(6x) = (3 \cdot 6)x = 18x$

24 Chapter 2 Linear Equations and Inequalities

69. Applying the associative property: $\frac{1}{5}(5x) = \left(\frac{1}{5} \cdot 5\right)x = 1x = x$

71. Applying the associative property: $8\left(\frac{1}{8}y\right) = \left(8 \cdot \frac{1}{8}\right)y = 1y = y$

73. Applying the associative property: $-2\left(-\frac{1}{2}x\right) = \left[-2 \cdot \left(-\frac{1}{2}\right)\right]x = 1x = x$

75. Applying the associative property: $-\frac{4}{3}\left(-\frac{3}{4}a\right) = \left[-\frac{4}{3} \cdot \left(-\frac{3}{4}\right)\right]a = 1a = a$

77. Simplifying: $\frac{3}{2}\left(\frac{2}{3}y\right) = y$

79. Simplifying: $\frac{1}{5}(5x) = x$

81. Simplifying: $\frac{1}{5}(30) = 6$

83. Simplifying: $\frac{3}{2}(4) = \frac{12}{2} = 6$

85. Simplifying: $12\left(-\frac{3}{4}\right) = -\frac{36}{4} = -9$

87. Simplifying: $\frac{3}{2}\left(-\frac{5}{4}\right) = -\frac{15}{8}$

89. Simplifying: $-13 + (-5) = -18$

91. Simplifying: $-\frac{3}{4} + \left(-\frac{1}{2}\right) = -\frac{3}{4} - \frac{1}{2} \cdot \frac{2}{2} = -\frac{3}{4} - \frac{2}{4} = -\frac{5}{4}$

93. Simplifying: $7x + (-4x) = 3x$

2.3 Multiplication Property of Equality

1. Solving the equation:
$$5x = 10$$
$$\tfrac{1}{5}(5x) = \tfrac{1}{5}(10)$$
$$x = 2$$

3. Solving the equation:
$$7a = 28$$
$$\tfrac{1}{7}(7a) = \tfrac{1}{7}(28)$$
$$a = 4$$

5. Solving the equation:
$$-8x = 4$$
$$-\tfrac{1}{8}(-8x) = -\tfrac{1}{8}(4)$$
$$x = -\tfrac{1}{2}$$

7. Solving the equation:
$$8m = -16$$
$$\tfrac{1}{8}(8m) = \tfrac{1}{8}(-16)$$
$$m = -2$$

9. Solving the equation:
$$-3x = -9$$
$$-\tfrac{1}{3}(-3x) = -\tfrac{1}{3}(-9)$$
$$x = 3$$

11. Solving the equation:
$$-7y = -28$$
$$-\tfrac{1}{7}(-7y) = -\tfrac{1}{7}(-28)$$
$$y = 4$$

13. Solving the equation:
$$2x = 0$$
$$\tfrac{1}{2}(2x) = \tfrac{1}{2}(0)$$
$$x = 0$$

15. Solving the equation:
$$-5x = 0$$
$$-\tfrac{1}{5}(-5x) = -\tfrac{1}{5}(0)$$
$$x = 0$$

17. Solving the equation:
$$\frac{x}{3} = 2$$
$$3\left(\frac{x}{3}\right) = 3(2)$$
$$x = 6$$

19. Solving the equation:
$$-\frac{m}{5} = 10$$
$$-5\left(-\frac{m}{5}\right) = -5(10)$$
$$m = -50$$

21. Solving the equation:
$$-\frac{x}{2} = -\frac{3}{4}$$
$$-2\left(-\frac{x}{2}\right) = -2\left(-\frac{3}{4}\right)$$
$$x = \frac{3}{2}$$

23. Solving the equation:
$$\tfrac{2}{3}a = 8$$
$$\tfrac{3}{2}\left(\tfrac{2}{3}a\right) = \tfrac{3}{2}(8)$$
$$a = 12$$

25. Solving the equation:
$$-\tfrac{3}{5}x = \tfrac{9}{5}$$
$$-\tfrac{5}{3}\left(-\tfrac{3}{5}x\right) = -\tfrac{5}{3}\left(\tfrac{9}{5}\right)$$
$$x = -3$$

27. Solving the equation:
$$-\tfrac{5}{8}y = -20$$
$$-\tfrac{8}{5}\left(-\tfrac{5}{8}y\right) = -\tfrac{8}{5}(-20)$$
$$y = 32$$

29. Simplifying and then solving the equation:
$$-4x - 2x + 3x = 24$$
$$-3x = 24$$
$$-\tfrac{1}{3}(-3x) = -\tfrac{1}{3}(24)$$
$$x = -8$$

31. Simplifying and then solving the equation:
$$4x + 8x - 2x = 15 - 10$$
$$10x = 5$$
$$\tfrac{1}{10}(10x) = \tfrac{1}{10}(5)$$
$$x = \tfrac{1}{2}$$

33. Simplifying and then solving the equation:
$$-3 - 5 = 3x + 5x - 10x$$
$$-8 = -2x$$
$$-\tfrac{1}{2}(-8) = -\tfrac{1}{2}(-2x)$$
$$x = 4$$

35. Simplifying and then solving the equation:
$$18 - 13 = \tfrac{1}{2}a + \tfrac{3}{4}a - \tfrac{5}{8}a$$
$$8(5) = 8\left(\tfrac{1}{2}a + \tfrac{3}{4}a - \tfrac{5}{8}a\right)$$
$$40 = 4a + 6a - 5a$$
$$40 = 5a$$
$$\tfrac{1}{5}(40) = \tfrac{1}{5}(5a)$$
$$a = 8$$

37. Solving by multiplying both sides of the equation by -1:
$$-x = 4$$
$$-1(-x) = -1(4)$$
$$x = -4$$

39. Solving by multiplying both sides of the equation by -1:
$$-x = -4$$
$$-1(-x) = -1(-4)$$
$$x = 4$$

41. Solving by multiplying both sides of the equation by -1:
$$15 = -a$$
$$-1(15) = -1(-a)$$
$$a = -15$$

43. Solving by multiplying both sides of the equation by -1:
$$-y = \tfrac{1}{2}$$
$$-1(-y) = -1\left(\tfrac{1}{2}\right)$$
$$y = -\tfrac{1}{2}$$

45. Solving the equation:
$$3x - 2 = 7$$
$$3x - 2 + 2 = 7 + 2$$
$$3x = 9$$
$$\tfrac{1}{3}(3x) = \tfrac{1}{3}(9)$$
$$x = 3$$

47. Solving the equation:
$$2a + 1 = 3$$
$$2a + 1 + (-1) = 3 + (-1)$$
$$2a = 2$$
$$\tfrac{1}{2}(2a) = \tfrac{1}{2}(2)$$
$$a = 1$$

49. Using Method 2 to eliminate fractions:
$$\tfrac{1}{8} + \tfrac{1}{2}x = \tfrac{1}{4}$$
$$8\left(\tfrac{1}{8} + \tfrac{1}{2}x\right) = 8\left(\tfrac{1}{4}\right)$$
$$1 + 4x = 2$$
$$(-1) + 1 + 4x = (-1) + 2$$
$$4x = 1$$
$$\tfrac{1}{4}(4x) = \tfrac{1}{4}(1)$$
$$x = \tfrac{1}{4}$$

51. Solving the equation:
$$6x = 2x - 12$$
$$6x + (-2x) = 2x + (-2x) - 12$$
$$4x = -12$$
$$\tfrac{1}{4}(4x) = \tfrac{1}{4}(-12)$$
$$x = -3$$

Chapter 2 Linear Equations and Inequalities

53. Solving the equation:
$$2y = -4y + 18$$
$$2y + 4y = -4y + 4y + 18$$
$$6y = 18$$
$$\tfrac{1}{6}(6y) = \tfrac{1}{6}(18)$$
$$y = 3$$

55. Solving the equation:
$$-7x = -3x - 8$$
$$-7x + 3x = -3x + 3x - 8$$
$$-4x = -8$$
$$-\tfrac{1}{4}(-4x) = -\tfrac{1}{4}(-8)$$
$$x = 2$$

57. Solving the equation:
$$2x - 5 = 8x + 4$$
$$2x + (-8x) - 5 = 8x + (-8x) + 4$$
$$-6x - 5 = 4$$
$$-6x - 5 + 5 = 4 + 5$$
$$-6x = 9$$
$$-\tfrac{1}{6}(-6x) = -\tfrac{1}{6}(9)$$
$$x = -\tfrac{3}{2}$$

59. Using Method 2 to eliminate fractions:
$$x + \tfrac{1}{2} = \tfrac{1}{4}x - \tfrac{5}{8}$$
$$8\left(x + \tfrac{1}{2}\right) = 8\left(\tfrac{1}{4}x - \tfrac{5}{8}\right)$$
$$8x + 4 = 2x - 5$$
$$8x + (-2x) + 4 = 2x + (-2x) - 5$$
$$6x + 4 = -5$$
$$6x + 4 + (-4) = -5 + (-4)$$
$$6x = -9$$
$$\tfrac{1}{6}(6x) = \tfrac{1}{6}(-9)$$
$$x = -\tfrac{3}{2}$$

61. Solving the equation:
$$m + 2 = 6m - 3$$
$$m + (-6m) + 2 = 6m + (-6m) - 3$$
$$-5m + 2 = -3$$
$$-5m + 2 + (-2) = -3 + (-2)$$
$$-5m = -5$$
$$-\tfrac{1}{5}(-5m) = -\tfrac{1}{5}(-5)$$
$$m = 1$$

63. Using Method 2 to eliminate fractions:
$$\tfrac{1}{2}m - \tfrac{1}{4} = \tfrac{1}{12}m + \tfrac{1}{6}$$
$$12\left(\tfrac{1}{2}m - \tfrac{1}{4}\right) = 12\left(\tfrac{1}{12}m + \tfrac{1}{6}\right)$$
$$6m - 3 = m + 2$$
$$6m + (-m) - 3 = m + (-m) + 2$$
$$5m - 3 = 2$$
$$5m - 3 + 3 = 2 + 3$$
$$5m = 5$$
$$\tfrac{1}{5}(5m) = \tfrac{1}{5}(5)$$
$$m = 1$$

65. Solving the equation:
$$6y - 4 = 9y + 2$$
$$6y + (-9y) - 4 = 9y + (-9y) + 2$$
$$-3y - 4 = 2$$
$$-3y - 4 + 4 = 2 + 4$$
$$-3y = 6$$
$$-\tfrac{1}{3}(-3y) = -\tfrac{1}{3}(6)$$
$$y = -2$$

67. Using Method 2 to eliminate fractions:
$$\tfrac{3}{2}y + \tfrac{1}{3} = y - \tfrac{2}{3}$$
$$6\left(\tfrac{3}{2}y + \tfrac{1}{3}\right) = 6\left(y - \tfrac{2}{3}\right)$$
$$9y + 2 = 6y - 4$$
$$9y + (-6y) + 2 = 6y + (-6y) - 4$$
$$3y + 2 = -4$$
$$3y + 2 + (-2) = -4 + (-2)$$
$$3y = -6$$
$$\tfrac{1}{3}(3y) = \tfrac{1}{3}(-6)$$
$$y = -2$$

69. Solving the equation:
$$5x + 6 = 2$$
$$5x + 6 + (-6) = 2 + (-6)$$
$$5x = -4$$
$$x = -\tfrac{4}{5}$$

71. Solving the equation:
$$\tfrac{x}{2} = \tfrac{6}{12}$$
$$2 \cdot \tfrac{x}{2} = 2 \cdot \tfrac{6}{12}$$
$$x = \tfrac{12}{12} = 1$$

73. Solving the equation:
$$\frac{3}{x} = \frac{6}{7}$$
$$7x \cdot \frac{3}{x} = 7x \cdot \frac{6}{7}$$
$$21 = 6x$$
$$x = \frac{21}{6} = \frac{7}{2}$$

75. Solving the equation:
$$\frac{a}{3} = \frac{5}{12}$$
$$3 \cdot \frac{a}{3} = 3 \cdot \frac{5}{12}$$
$$a = \frac{15}{12} = \frac{5}{4}$$

77. Solving the equation:
$$\frac{10}{20} = \frac{20}{x}$$
$$20x \cdot \frac{10}{20} = 20x \cdot \frac{20}{x}$$
$$10x = 400$$
$$x = \frac{400}{10} = 40$$

79. Solving the equation:
$$\frac{2}{x} = \frac{6}{7}$$
$$7x \cdot \frac{2}{x} = 7x \cdot \frac{6}{7}$$
$$14 = 6x$$
$$x = \frac{14}{6} = \frac{7}{3}$$

81. Solving the equation:
$$7.5x = 1500$$
$$\frac{7.5x}{7.5} = \frac{1500}{7.5}$$
$$x = 200$$
The break-even point is 200 tickets.

83. Solving the equation:
$$G - 0.21G - 0.08G = 987.5$$
$$0.71G = 987.5$$
$$G \approx 1391$$
Your gross income is approximately $1,391.

85. Applying the distributive property: $2(3x - 5) = 2 \cdot 3x - 2 \cdot 5 = 6x - 10$

87. Applying the distributive property: $\frac{1}{2}(3x + 6) = \frac{1}{2} \cdot 3x + \frac{1}{2} \cdot 6 = \frac{3}{2}x + 3$

89. Applying the distributive property: $\frac{1}{3}(-3x + 6) = \frac{1}{3} \cdot (-3x) + \frac{1}{3} \cdot 6 = -x + 2$

91. Using the distributive property and combining like terms: $5(2x - 8) - 3 = 10x - 40 - 3 = 10x - 43$

93. Using the distributive property and combining like terms:
$-2(3x + 5) + 3(x - 1) = -6x - 10 + 3x - 3 = -6x + 3x - 10 - 3 = -3x - 13$

95. Using the distributive property and combining like terms: $7 - 3(2y + 1) = 7 - 6y - 3 = -6y + 7 - 3 = -6y + 4$

97. Solving the equation:
$$2x = 4$$
$$\tfrac{1}{2}(2x) = \tfrac{1}{2}(4)$$
$$x = 2$$

99. Solving the equation:
$$30 = 5x$$
$$5x = 30$$
$$\tfrac{1}{5}(5x) = \tfrac{1}{5}(30)$$
$$x = 6$$

101. Solving the equation:
$$0.17x = 510$$
$$x = \frac{510}{0.17} = 3{,}000$$

103. Simplifying: $3(x - 5) + 4 = 3x - 15 + 4 = 3x - 11$

105. Simplifying: $0.09(x + 2{,}000) = 0.09x + 180$

107. Simplifying: $7 - 3(2y + 1) = 7 - 6y - 3 = 4 - 6y = -6y + 4$

109. Simplifying: $3(2x - 5) - (2x - 4) = 6x - 15 - 2x + 4 = 4x - 11$

111. Simplifying: $10x + (-5x) = 5x$

113. Simplifying: $0.08x + 0.09x = 0.17x$

2.4 Solving Linear Equations

1. Solving the equation:
$$2(x+3) = 12$$
$$2x + 6 = 12$$
$$2x + 6 + (-6) = 12 + (-6)$$
$$2x = 6$$
$$\tfrac{1}{2}(2x) = \tfrac{1}{2}(6)$$
$$x = 3$$

3. Solving the equation:
$$6(x-1) = -18$$
$$6x - 6 = -18$$
$$6x - 6 + 6 = -18 + 6$$
$$6x = -12$$
$$\tfrac{1}{6}(6x) = \tfrac{1}{6}(-12)$$
$$x = -2$$

5. Solving the equation:
$$2(4a+1) = -6$$
$$8a + 2 = -6$$
$$8a + 2 + (-2) = -6 + (-2)$$
$$8a = -8$$
$$\tfrac{1}{8}(8a) = \tfrac{1}{8}(-8)$$
$$a = -1$$

7. Solving the equation:
$$14 = 2(5x-3)$$
$$14 = 10x - 6$$
$$14 + 6 = 10x - 6 + 6$$
$$20 = 10x$$
$$\tfrac{1}{10}(20) = \tfrac{1}{10}(10x)$$
$$x = 2$$

9. Solving the equation:
$$-2(3y+5) = 14$$
$$-6y - 10 = 14$$
$$-6y - 10 + 10 = 14 + 10$$
$$-6y = 24$$
$$-\tfrac{1}{6}(-6y) = -\tfrac{1}{6}(24)$$
$$y = -4$$

11. Solving the equation:
$$-5(2a+4) = 0$$
$$-10a - 20 = 0$$
$$-10a - 20 + 20 = 0 + 20$$
$$-10a = 20$$
$$-\tfrac{1}{10}(-10a) = -\tfrac{1}{10}(20)$$
$$a = -2$$

13. Solving the equation:
$$1 = \tfrac{1}{2}(4x+2)$$
$$1 = 2x + 1$$
$$1 + (-1) = 2x + 1 + (-1)$$
$$0 = 2x$$
$$\tfrac{1}{2}(0) = \tfrac{1}{2}(2x)$$
$$x = 0$$

15. Solving the equation:
$$3(t-4) + 5 = -4$$
$$3t - 12 + 5 = -4$$
$$3t - 7 = -4$$
$$3t - 7 + 7 = -4 + 7$$
$$3t = 3$$
$$\tfrac{1}{3}(3t) = \tfrac{1}{3}(3)$$
$$t = 1$$

17. Solving the equation:
$$4(2y+1) - 7 = 1$$
$$8y + 4 - 7 = 1$$
$$8y - 3 = 1$$
$$8y - 3 + 3 = 1 + 3$$
$$8y = 4$$
$$\tfrac{1}{8}(8y) = \tfrac{1}{8}(4)$$
$$y = \tfrac{1}{2}$$

19. Solving the equation:
$$\tfrac{1}{2}(x-3) = \tfrac{1}{4}(x+1)$$
$$\tfrac{1}{2}x - \tfrac{3}{2} = \tfrac{1}{4}x + \tfrac{1}{4}$$
$$4\left(\tfrac{1}{2}x - \tfrac{3}{2}\right) = 4\left(\tfrac{1}{4}x + \tfrac{1}{4}\right)$$
$$2x - 6 = x + 1$$
$$2x + (-x) - 6 = x + (-x) + 1$$
$$x - 6 = 1$$
$$x - 6 + 6 = 1 + 6$$
$$x = 7$$

21. Solving the equation:
$$-0.7(2x-7) = 0.3(11-4x)$$
$$-1.4x + 4.9 = 3.3 - 1.2x$$
$$-1.4x + 1.2x + 4.9 = 3.3 - 1.2x + 1.2x$$
$$-0.2x + 4.9 = 3.3$$
$$-0.2x + 4.9 + (-4.9) = 3.3 + (-4.9)$$
$$-0.2x = -1.6$$
$$\frac{-0.2x}{-0.2} = \frac{-1.6}{-0.2}$$
$$x = 8$$

23. Solving the equation:
$$-2(3y+1) = 3(1-6y) - 9$$
$$-6y - 2 = 3 - 18y - 9$$
$$-6y - 2 = -18y - 6$$
$$-6y + 18y - 2 = -18y + 18y - 6$$
$$12y - 2 = -6$$
$$12y - 2 + 2 = -6 + 2$$
$$12y = -4$$
$$\tfrac{1}{12}(12y) = \tfrac{1}{12}(-4)$$
$$y = -\tfrac{1}{3}$$

25. Solving the equation:
$$\tfrac{3}{4}(8x-4)+3=\tfrac{2}{5}(5x+10)-1$$
$$6x-3+3=2x+4-1$$
$$6x=2x+3$$
$$6x+(-2x)=2x+(-2x)+3$$
$$4x=3$$
$$\tfrac{1}{4}(4x)=\tfrac{1}{4}(3)$$
$$x=\tfrac{3}{4}$$

27. Solving the equation:
$$0.06x+0.08(100-x)=6.5$$
$$0.06x+8-0.08x=6.5$$
$$-0.02x+8=6.5$$
$$-0.02x+8+(-8)=6.5+(-8)$$
$$-0.02x=-1.5$$
$$\frac{-0.02x}{-0.02}=\frac{-1.5}{-0.02}$$
$$x=75$$

29. Solving the equation:
$$6-5(2a-3)=1$$
$$6-10a+15=1$$
$$-10a+21=1$$
$$-10a+21+(-21)=1+(-21)$$
$$-10a=-20$$
$$-\tfrac{1}{10}(-10a)=-\tfrac{1}{10}(-20)$$
$$a=2$$

31. Solving the equation:
$$0.2x-0.5=0.5-0.2(2x-13)$$
$$0.2x-0.5=0.5-0.4x+2.6$$
$$0.2x-0.5=-0.4x+3.1$$
$$0.2x+0.4x-0.5=-0.4x+0.4x+3.1$$
$$0.6x-0.5=3.1$$
$$0.6x-0.5+0.5=3.1+0.5$$
$$0.6x=3.6$$
$$\frac{0.6x}{0.6}=\frac{3.6}{0.6}$$
$$x=6$$

33. Solving the equation:
$$2(t-3)+3(t-2)=28$$
$$2t-6+3t-6=28$$
$$5t-12=28$$
$$5t-12+12=28+12$$
$$5t=40$$
$$\tfrac{1}{5}(5t)=\tfrac{1}{5}(40)$$
$$t=8$$

35. Solving the equation:
$$5(x-2)-(3x+4)=3(6x-8)+10$$
$$5x-10-3x-4=18x-24+10$$
$$2x-14=18x-14$$
$$2x+(-18x)-14=18x+(-18x)-14$$
$$-16x-14=-14$$
$$-16x-14+14=-14+14$$
$$-16x=0$$
$$-\tfrac{1}{16}(-16x)=-\tfrac{1}{16}(0)$$
$$x=0$$

37. Solving the equation:
$$2(5x-3)-(2x-4)=5-(6x+1)$$
$$10x-6-2x+4=5-6x-1$$
$$8x-2=-6x+4$$
$$8x+6x-2=-6x+6x+4$$
$$14x-2=4$$
$$14x-2+2=4+2$$
$$14x=6$$
$$\tfrac{1}{14}(14x)=\tfrac{1}{14}(6)$$
$$x=\tfrac{3}{7}$$

39. Solving the equation:
$$-(3x+1)-(4x-7)=4-(3x+2)$$
$$-3x-1-4x+7=4-3x-2$$
$$-7x+6=-3x+2$$
$$-7x+3x+6=-3x+3x+2$$
$$-4x+6=2$$
$$-4x+6+(-6)=2+(-6)$$
$$-4x=-4$$
$$-\tfrac{1}{4}(-4x)=-\tfrac{1}{4}(-4)$$
$$x=1$$

41. Solving the equation:
$$x+(2x-1)=2$$
$$3x-1=2$$
$$3x-1+1=2+1$$
$$3x=3$$
$$\tfrac{1}{3}(3x)=\tfrac{1}{3}(3)$$
$$x=1$$

43. Solving the equation:
$$x-(3x+5)=-3$$
$$x-3x-5=-3$$
$$-2x-5=-3$$
$$-2x-5+5=-3+5$$
$$-2x=2$$
$$-\tfrac{1}{2}(-2x)=-\tfrac{1}{2}(2)$$
$$x=-1$$

45. Solving the equation:
$$15 = 3(x-1)$$
$$15 = 3x - 3$$
$$15 + 3 = 3x - 3 + 3$$
$$18 = 3x$$
$$\tfrac{1}{3}(18) = \tfrac{1}{3}(3x)$$
$$x = 6$$

47. Solving the equation:
$$4x - (-4x + 1) = 5$$
$$4x + 4x - 1 = 5$$
$$8x - 1 = 5$$
$$8x - 1 + 1 = 5 + 1$$
$$8x = 6$$
$$\tfrac{1}{8}(8x) = \tfrac{1}{8}(6)$$
$$x = \tfrac{3}{4}$$

49. Solving the equation:
$$5x - 8(2x - 5) = 7$$
$$5x - 16x + 40 = 7$$
$$-11x + 40 = 7$$
$$-11x + 40 - 40 = 7 - 40$$
$$-11x = -33$$
$$-\tfrac{1}{11}(-11x) = -\tfrac{1}{11}(-33)$$
$$x = 3$$

51. Solving the equation:
$$7(2y - 1) - 6y = -1$$
$$14y - 7 - 6y = -1$$
$$8y - 7 = -1$$
$$8y - 7 + 7 = -1 + 7$$
$$8y = 6$$
$$\tfrac{1}{8}(8y) = \tfrac{1}{8}(6)$$
$$y = \tfrac{3}{4}$$

53. Solving the equation:
$$0.2x + 0.5(12 - x) = 3.6$$
$$0.2x + 6 - 0.5x = 3.6$$
$$-0.3x + 6 = 3.6$$
$$-0.3x + 6 - 6 = 3.6 - 6$$
$$-0.3x = -2.4$$
$$x = \frac{-2.4}{-0.3} = 8$$

55. Solving the equation:
$$0.5x + 0.2(18 - x) = 5.4$$
$$0.5x + 3.6 - 0.2x = 5.4$$
$$0.3x + 3.6 = 5.4$$
$$0.3x + 3.6 - 3.6 = 5.4 - 3.6$$
$$0.3x = 1.8$$
$$x = \frac{1.8}{0.3} = 6$$

57. Solving the equation:
$$x + (x + 3)(-3) = x - 3$$
$$x - 3x - 9 = x - 3$$
$$-2x - 9 = x - 3$$
$$-2x - x - 9 = x - x - 3$$
$$-3x - 9 = -3$$
$$-3x - 9 + 9 = -3 + 9$$
$$-3x = 6$$
$$-\tfrac{1}{3}(-3x) = -\tfrac{1}{3}(6)$$
$$x = -2$$

59. Solving the equation:
$$5(x + 2) + 3(x - 1) = -9$$
$$5x + 10 + 3x - 3 = -9$$
$$8x + 7 = -9$$
$$8x + 7 - 7 = -9 - 7$$
$$8x = -16$$
$$\tfrac{1}{8}(8x) = \tfrac{1}{8}(-16)$$
$$x = -2$$

61. Solving the equation:
$$3(x - 3) + 2(2x) = 5$$
$$3x - 9 + 4x = 5$$
$$7x - 9 = 5$$
$$7x - 9 + 9 = 5 + 9$$
$$7x = 14$$
$$\tfrac{1}{7}(7x) = \tfrac{1}{7}(14)$$
$$x = 2$$

63. Solving the equation:
$$5(y + 2) = 4(y + 1)$$
$$5y + 10 = 4y + 4$$
$$5y - 4y + 10 = 4y - 4y + 4$$
$$y + 10 = 4$$
$$y + 10 - 10 = 4 - 10$$
$$y = -6$$

65. Solving the equation:
$$3x + 2(x-2) = 6$$
$$3x + 2x - 4 = 6$$
$$5x - 4 = 6$$
$$5x - 4 + 4 = 6 + 4$$
$$5x = 10$$
$$\tfrac{1}{5}(5x) = \tfrac{1}{5}(10)$$
$$x = 2$$

67. Solving the equation:
$$50(x-5) = 30(x+5)$$
$$50x - 250 = 30x + 150$$
$$50x - 30x - 250 = 30x - 30x + 150$$
$$20x - 250 = 150$$
$$20x - 250 + 250 = 150 + 250$$
$$20x = 400$$
$$\tfrac{1}{20}(20x) = \tfrac{1}{20}(400)$$
$$x = 20$$

69. Solving the equation:
$$0.08x + 0.09(x + 2{,}000) = 860$$
$$0.08x + 0.09x + 180 = 860$$
$$0.17x + 180 = 860$$
$$0.17x + 180 - 180 = 860 - 180$$
$$0.17x = 680$$
$$x = \frac{680}{0.17} = 4{,}000$$

71. Solving the equation:
$$0.10x + 0.12(x + 500) = 214$$
$$0.10x + 0.12x + 60 = 214$$
$$0.22x + 60 = 214$$
$$0.22x + 60 - 60 = 214 - 60$$
$$0.22x = 154$$
$$x = \frac{154}{0.22} = 700$$

73. Solving the equation:
$$5x + 10(x + 8) = 245$$
$$5x + 10x + 80 = 245$$
$$15x + 80 = 245$$
$$15x + 80 - 80 = 245 - 80$$
$$15x = 165$$
$$x = \frac{165}{15} = 11$$

75. Solving the equation:
$$5x + 10(x + 3) + 25(x + 5) = 435$$
$$5x + 10x + 30 + 25x + 125 = 435$$
$$40x + 155 = 435$$
$$40x + 155 - 155 = 435 - 155$$
$$40x = 280$$
$$x = \frac{280}{40} = 7$$

77.   **a.**   Solving the equation:
$$4x - 5 = 0$$
$$4x - 5 + 5 = 0 + 5$$
$$4x = 5$$
$$\tfrac{1}{4}(4x) = \tfrac{1}{4}(5)$$
$$x = \tfrac{5}{4} = 1.25$$

b.   Solving the equation:
$$4x - 5 = 25$$
$$4x - 5 + 5 = 25 + 5$$
$$4x = 30$$
$$\tfrac{1}{4}(4x) = \tfrac{1}{4}(30)$$
$$x = \tfrac{15}{2} = 7.5$$

c.   Adding: $(4x - 5) + (2x + 25) = 6x + 20$

d.   Solving the equation:
$$4x - 5 = 2x + 25$$
$$4x - 2x - 5 = 2x - 2x + 25$$
$$2x - 5 = 25$$
$$2x - 5 + 5 = 25 + 5$$
$$2x = 30$$
$$\tfrac{1}{2}(2x) = \tfrac{1}{2}(30)$$
$$x = 15$$

e.   Multiplying: $4(x - 5) = 4x - 20$

f. Solving the equation:
$$4(x-5) = 2x + 25$$
$$4x - 20 = 2x + 25$$
$$4x - 2x - 20 = 2x - 2x + 25$$
$$2x - 20 = 25$$
$$2x - 20 + 20 = 25 + 20$$
$$2x = 45$$
$$\tfrac{1}{2}(2x) = \tfrac{1}{2}(45)$$
$$x = \tfrac{45}{2} = 22.5$$

79. Multiplying the fractions: $\tfrac{1}{2}(3) = \tfrac{1}{2} \cdot \tfrac{3}{1} = \tfrac{3}{2}$

81. Multiplying the fractions: $\tfrac{2}{3}(6) = \tfrac{2}{3} \cdot \tfrac{6}{1} = \tfrac{12}{3} = 4$

83. Multiplying the fractions: $\tfrac{5}{9} \cdot \tfrac{9}{5} = \tfrac{45}{45} = 1$

85. Completing the table:

x	$3(x+2)$	$3x+2$	$3x+6$
0	6	2	6
1	9	5	9
2	12	8	12
3	15	11	15

87. Completing the table:

a	$(2a+1)^2$	$4a^2+4a+1$
1	9	9
2	25	25
3	49	49

89. Solving the equation:
$$40 = 2x + 12$$
$$2x + 12 = 40$$
$$2x = 28$$
$$x = 14$$

91. Solving the equation:
$$12 + 2y = 6$$
$$2y = -6$$
$$y = -3$$

93. Solving the equation:
$$24x = 6$$
$$x = \tfrac{6}{24} = \tfrac{1}{4}$$

95. Solving the equation:
$$70 = x \bullet 210$$
$$x = \tfrac{70}{210} = \tfrac{1}{3}$$

97. Simplifying: $\tfrac{1}{2}(-3x+6) = \tfrac{1}{2}(-3x) + \tfrac{1}{2}(6) = -\tfrac{3}{2}x + 3$

2.5 Formulas

1. Using the perimeter formula:
$$P = 2l + 2w$$
$$300 = 2l + 2(50)$$
$$300 = 2l + 100$$
$$200 = 2l$$
$$l = 100$$
The length is 100 feet.

3. **a.** Substituting $x = 3$:
$$2(0) + 3y = 6$$
$$0 + 3y = 6$$
$$3y = 6$$
$$y = 2$$

b. Substituting $y = 1$:
$$2x + 3(1) = 6$$
$$2x + 3 = 6$$
$$2x = 3$$
$$x = \tfrac{3}{2}$$

c. Substituting $x = 3$:
$$2(3) + 3y = 6$$
$$6 + 3y = 6$$
$$6 + (-6) + 3y = 6 + (-6)$$
$$3y = 0$$
$$y = 0$$

Problem Set 2.5 33

5. **a.** Substituting $x = 0$: $y = -\frac{1}{3}(0) + 2 = 0 + 2 = 2$

   b. Substituting $y = 3$:
   $$3 = -\frac{1}{3}x + 2$$
   $$1 = -\frac{1}{3}x$$
   $$3 = -x$$
   $$x = -3$$

   c. Substituting $x = 3$: $y = -\frac{1}{3}(3) + 2 = -1 + 2 = 1$

7. Substituting $x = -2$: $y = (-2+1)^2 - 3 = (-1)^2 - 3 = 1 - 3 = -2$

9. Substituting $x = 1$: $y = (1+1)^2 - 3 = (2)^2 - 3 = 4 - 3 = 1$

11. **a.** Substituting $x = 10$: $y = \frac{20}{10} = 2$ **b.** Substituting $x = 5$: $y = \frac{20}{5} = 4$

13. **a.** Substituting $y = 15$ and $x = 3$:
    $$15 = K(3)$$
    $$K = \frac{15}{3} = 5$$

    b. Substituting $y = 72$ and $x = 4$:
    $$72 = K(4)$$
    $$K = \frac{72}{4} = 18$$

15. **a.** Substituting $x = 5$ and $y = 4$:
    $$4 = \frac{K}{5}$$
    $$K = 20$$

    b. Substituting $x = 5$ and $y = 15$:
    $$15 = \frac{K}{5}$$
    $$K = 75$$

17. Solving for l:
    $$lw = A$$
    $$\frac{lw}{w} = \frac{A}{w}$$
    $$l = \frac{A}{w}$$

19. Solving for h:
    $$lwh = V$$
    $$\frac{lwh}{lw} = \frac{V}{lw}$$
    $$h = \frac{V}{lw}$$

21. Solving for a:
    $$a + b + c = P$$
    $$a + b + c - b - c = P - b - c$$
    $$a = P - b - c$$

23. Solving for x:
    $$x - 3y = -1$$
    $$x - 3y + 3y = -1 + 3y$$
    $$x = 3y - 1$$

25. Solving for y:
    $$-3x + y = 6$$
    $$-3x + 3x + y = 6 + 3x$$
    $$y = 3x + 6$$

27. Solving for y:
    $$2x + 3y = 6$$
    $$-2x + 2x + 3y = -2x + 6$$
    $$3y = -2x + 6$$
    $$\tfrac{1}{3}(3y) = \tfrac{1}{3}(-2x + 6)$$
    $$y = -\tfrac{2}{3}x + 2$$

29. Solving for w:
    $$2l + 2w = P$$
    $$2l - 2l + 2w = P - 2l$$
    $$2w = P - 2l$$
    $$\frac{2w}{2} = \frac{P - 2l}{2}$$
    $$w = \frac{P - 2l}{2}$$

31. Solving for v:
    $$vt + 16t^2 = h$$
    $$vt + 16t^2 - 16t^2 = h - 16t^2$$
    $$vt = h - 16t^2$$
    $$\frac{vt}{t} = \frac{h - 16t^2}{t}$$
    $$v = \frac{h - 16t^2}{t}$$

34 Chapter 2 Linear Equations and Inequalities

33. Solving for h:
$$\pi r^2 + 2\pi rh = A$$
$$\pi r^2 - \pi r^2 + 2\pi rh = A - \pi r^2$$
$$2\pi rh = A - \pi r^2$$
$$\frac{2\pi rh}{2\pi r} = \frac{A - \pi r^2}{2\pi r}$$
$$h = \frac{A - \pi r^2}{2\pi r}$$

35. **a.** Solving for y:
$$y - 3 = -2(x + 4)$$
$$y - 3 = -2x - 8$$
$$y = -2x - 5$$

   b. Solving for y:
$$y - 5 = 4(x - 3)$$
$$y - 5 = 4x - 12$$
$$y = 4x - 7$$

37. **a.** Solving for y:
$$y - 1 = \tfrac{3}{4}(x - 1)$$
$$y - 1 = \tfrac{3}{4}x - \tfrac{3}{4}$$
$$y = \tfrac{3}{4}x + \tfrac{1}{4}$$

   b. Solving for y:
$$y + 2 = \tfrac{3}{4}(x - 4)$$
$$y + 2 = \tfrac{3}{4}x - 3$$
$$y = \tfrac{3}{4}x - 5$$

39. **a.** Solving for y:
$$\frac{y - 1}{x} = \frac{3}{5}$$
$$5y - 5 = 3x$$
$$5y = 3x + 5$$
$$y = \tfrac{3}{5}x + 1$$

   b. Solving for y:
$$\frac{y - 2}{x} = \frac{1}{2}$$
$$2y - 4 = x$$
$$2y = x + 4$$
$$y = \tfrac{1}{2}x + 2$$

   c. Solving for y:
$$\frac{y - 3}{x} = 4$$
$$y - 3 = 4x$$
$$y = 4x + 3$$

41. Solving for y:
$$\frac{x}{7} - \frac{y}{3} = 1$$
$$-\frac{x}{7} + \frac{x}{7} - \frac{y}{3} = -\frac{x}{7} + 1$$
$$-\frac{y}{3} = -\frac{x}{7} + 1$$
$$-3\left(-\frac{y}{3}\right) = -3\left(-\frac{x}{7} + 1\right)$$
$$y = \tfrac{3}{7}x - 3$$

43. Solving for y:
$$-\tfrac{1}{4}x + \tfrac{1}{8}y = 1$$
$$-\tfrac{1}{4}x + \tfrac{1}{4}x + \tfrac{1}{8}y = 1 + \tfrac{1}{4}x$$
$$\tfrac{1}{8}y = \tfrac{1}{4}x + 1$$
$$8\left(\tfrac{1}{8}y\right) = 8\left(\tfrac{1}{4}x + 1\right)$$
$$y = 2x + 8$$

45. **a.** Solving the equation:
$$4x + 5 = 20$$
$$4x + 5 - 5 = 20 - 5$$
$$4x = 15$$
$$\tfrac{1}{4}(4x) = \tfrac{1}{4}(15)$$
$$x = \tfrac{15}{4} = 3.75$$

   b. Evaluating when $x = 3$: $4x + 5 = 4(3) + 5 = 12 + 5 = 17$

    c. Solving for y:
$$4x + 5y = 20$$
$$4x - 4x + 5y = 20 - 4x$$
$$5y = -4x + 20$$
$$\tfrac{1}{5}(5y) = \tfrac{1}{5}(-4x + 20)$$
$$y = -\tfrac{4}{5}x + 4$$

    d. Solving for x:
$$4x + 5y = 20$$
$$4x + 5y - 5y = 20 - 5y$$
$$4x = -5y + 20$$
$$\tfrac{1}{4}(4x) = \tfrac{1}{4}(-5y + 20)$$
$$x = -\tfrac{5}{4}y + 5$$

47. The complement of 30° is 90° − 30° = 60°, and the supplement is 180° − 30° = 150°.

49. The complement of 45° is 90° − 45° = 45°, and the supplement is 180° − 45° = 135°.

51. Translating into an equation and solving:
$$x = 0.25 \cdot 40$$
$$x = 10$$
The number 10 is 25% of 40.

53. Translating into an equation and solving:
$$x = 0.12 \cdot 2000$$
$$x = 240$$
The number 240 is 12% of 2000.

55. Translating into an equation and solving:
$$x \cdot 28 = 7$$
$$28x = 7$$
$$\tfrac{1}{28}(28x) = \tfrac{1}{28}(7)$$
$$x = 0.25 = 25\%$$
The number 7 is 25% of 28.

57. Translating into an equation and solving:
$$x \cdot 40 = 14$$
$$40x = 14$$
$$\tfrac{1}{40}(40x) = \tfrac{1}{40}(14)$$
$$x = 0.35 = 35\%$$
The number 14 is 35% of 40.

59. Translating into an equation and solving:
$$0.50 \cdot x = 32$$
$$\frac{0.50x}{0.50} = \frac{32}{0.50}$$
$$x = 64$$
The number 32 is 50% of 64.

61. Translating into an equation and solving:
$$0.12 \cdot x = 240$$
$$\frac{0.12x}{0.12} = \frac{240}{0.12}$$
$$x = 2000$$
The number 240 is 12% of 2000.

63. Substituting $F = 212$: $C = \tfrac{5}{9}(212 - 32) = \tfrac{5}{9}(180) = 100°C$. This value agrees with the information in Table 1.

65. Substituting $F = 68$: $C = \tfrac{5}{9}(68 - 32) = \tfrac{5}{9}(36) = 20°C$. This value agrees with the information in Table 1.

67. Solving for C:
$$\tfrac{9}{5}C + 32 = F$$
$$\tfrac{9}{5}C + 32 - 32 = F - 32$$
$$\tfrac{9}{5}C = F - 32$$
$$\tfrac{5}{9}\left(\tfrac{9}{5}C\right) = \tfrac{5}{9}(F - 32)$$
$$C = \tfrac{5}{9}(F - 32)$$

69. We need to find what percent of 150 is 90:
$$x \cdot 150 = 90$$
$$\tfrac{1}{150}(150x) = \tfrac{1}{150}(90)$$
$$x = 0.60 = 60\%$$
So 60% of the calories in one serving of vanilla ice cream are fat calories.

71. We need to find what percent of 98 is 26:
$$x \cdot 98 = 26$$
$$\tfrac{1}{98}(98x) = \tfrac{1}{98}(26)$$
$$x \approx 0.265 = 26.5\%$$
So 26.5% of one serving of frozen yogurt are carbohydrates.

6   Chapter 2 Linear Equations and Inequalities

73. Solving for r:
$$2\pi r = C$$
$$2 \cdot \tfrac{22}{7} r = 44$$
$$\tfrac{44}{7} r = 44$$
$$\tfrac{7}{44}\left(\tfrac{44}{7} r\right) = \tfrac{7}{44}(44)$$
$$r = 7 \text{ meters}$$

75. Solving for r:
$$2\pi r = C$$
$$2 \cdot 3.14 r = 9.42$$
$$6.28 r = 9.42$$
$$\frac{6.28 r}{6.28} = \frac{9.42}{6.28}$$
$$r = 1.5 \text{ inches}$$

77. Solving for h:
$$\pi r^2 h = V$$
$$\tfrac{22}{7}\left(\tfrac{7}{22}\right)^2 h = 42$$
$$\tfrac{7}{22} h = 42$$
$$\tfrac{22}{7}\left(\tfrac{7}{22} h\right) = \tfrac{22}{7}(42)$$
$$h = 132 \text{ feet}$$

79. Solving for h:
$$\pi r^2 h = V$$
$$3.14(3)^2 h = 6.28$$
$$28.26 h = 6.28$$
$$\frac{28.26 h}{28.26} = \frac{6.28}{28.26}$$
$$h = \tfrac{2}{9} \text{ centimeters}$$

81. a. Simplifying: $27 - (-68) = 27 + 68 = 95$
    b. Simplifying: $27 + (-68) = 27 - 68 = -41$
    c. Simplifying: $-27 - 68 = -95$
    d. Simplifying: $-27 + 68 = 41$

83. a. Simplifying: $-32 - (-41) = -32 + 41 = 9$
    b. Simplifying: $-32 + (-41) = -32 - 41 = -73$
    c. Simplifying: $-32 + 41 = 9$
    d. Simplifying: $-32 - 41 = -73$

85. The sum of 4 and 1 is 5.
87. The difference of 6 and 2 is 4.
89. The difference of a number and 5 is -12.
91. The sum of a number and 3 is four times the difference of that number and 3.
93. An equivalent expression is: $2(6+3) = 2(9) = 18$
95. An equivalent expression is: $2(5) + 3 = 10 + 3 = 13$
97. An equivalent expression is: $x + 5 = 13$
99. An equivalent expression is: $5(x+7) = 30$

2.6 Applications

1. Let x represent the number. The equation is:
$$x + 5 = 13$$
$$x = 8$$
The number is 8.

3. Let x represent the number. The equation is:
$$2x + 4 = 14$$
$$2x = 10$$
$$x = 5$$
The number is 5.

5. Let x represent the number. The equation is:
$$5(x+7) = 30$$
$$5x + 35 = 30$$
$$5x = -5$$
$$x = -1$$
The number is -1.

7. Let x and $x+2$ represent the two numbers. The equation is:
$$x + x + 2 = 8$$
$$2x + 2 = 8$$
$$2x = 6$$
$$x = 3$$
$$x + 2 = 5$$
The two numbers are 3 and 5.

9. Let x and $3x - 4$ represent the two numbers. The equation is:
$$(x + 3x - 4) + 5 = 25$$
$$4x + 1 = 25$$
$$4x = 24$$
$$x = 6$$
$$3x - 4 = 3(6) - 4 = 14$$
The two numbers are 6 and 14.

11. Completing the table:

	Four Years Ago	Now
Shelly	$x + 3 - 4 = x - 1$	$x + 3$
Michele	$x - 4$	x

The equation is:
$$x - 1 + x - 4 = 67$$
$$2x - 5 = 67$$
$$2x = 72$$
$$x = 36$$
$$x + 3 = 39$$
Shelly is 39 and Michele is 36.

13. Completing the table:

	Three Years Ago	Now
Cody	$2x - 3$	$2x$
Evan	$x - 3$	x

The equation is:
$$2x - 3 + x - 3 = 27$$
$$3x - 6 = 27$$
$$3x = 33$$
$$x = 11$$
$$2x = 22$$
Evan is 11 and Cody is 22.

15. Completing the table:

	Five Years Ago	Now
Fred	$x + 4 - 5 = x - 1$	$x + 4$
Barney	$x - 5$	x

The equation is:
$$x - 1 + x - 5 = 48$$
$$2x - 6 = 48$$
$$2x = 54$$
$$x = 27$$
$$x + 4 = 31$$
Barney is 27 and Fred is 31.

17. Completing the table:

	Now	Three Years from Now
Jack	$2x$	$2x + 3$
Lacy	x	$x + 3$

The equation is:
$$2x + 3 + x + 3 = 54$$
$$3x + 6 = 54$$
$$3x = 48$$
$$x = 16$$
$$2x = 32$$
Lacy is 16 and Jack is 32.

19. Completing the table:

	Now	Two Years from Now
Pat	$x + 20$	$x + 20 + 2 = x + 22$
Patrick	x	$x + 2$

The equation is:
$$x + 22 = 2(x + 2)$$
$$x + 22 = 2x + 4$$
$$22 = x + 4$$
$$x = 18$$
$$x + 20 = 38$$
Patrick is 18 and Pat is 38.

21. Using the formula $P = 4s$, the equation is:
$$4s = 36$$
$$s = 9$$
The length of each side is 9 inches.

23. Using the formula $P = 4s$, the equation is:
$$4s = 60$$
$$s = 15$$
The length of each side is 15 feet.

38 Chapter 2 Linear Equations and Inequalities

25. Let x, $3x$, and $x + 7$ represent the sides of the triangle. The equation is:
$$x + 3x + x + 7 = 62$$
$$5x + 7 = 62$$
$$5x = 55$$
$$x = 11$$
$$x + 7 = 18$$
$$3x = 33$$
The sides are 11 feet, 18 feet, and 33 feet.

27. Let x, $2x$, and $2x - 12$ represent the sides of the triangle. The equation is:
$$x + 2x + 2x - 12 = 53$$
$$5x - 12 = 53$$
$$5x = 65$$
$$x = 13$$
$$2x = 26$$
$$2x - 12 = 14$$
The sides are 13 feet, 14 feet, and 26 feet.

29. Let w represent the width and $w + 5$ represent the length. The equation is:
$$2w + 2(w + 5) = 34$$
$$2w + 2w + 10 = 34$$
$$4w + 10 = 34$$
$$4w = 24$$
$$w = 6$$
$$w + 5 = 11$$
The length is 11 inches and the width is 6 inches.

31. Let w represent the width and $2w + 7$ represent the length. The equation is:
$$2w + 2(2w + 7) = 68$$
$$2w + 4w + 14 = 68$$
$$6w + 14 = 68$$
$$6w = 54$$
$$w = 9$$
$$2w + 7 = 2(9) + 7 = 25$$
The length is 25 inches and the width is 9 inches.

33. Let w represent the width and $3w + 6$ represent the length. The equation is:
$$2w + 2(3w + 6) = 36$$
$$2w + 6w + 12 = 36$$
$$8w + 12 = 36$$
$$8w = 24$$
$$w = 3$$
$$3w + 6 = 3(3) + 6 = 15$$
The length is 15 feet and the width is 3 feet.

35. Completing the table:

	Dimes	Quarters
Number	x	$x + 5$
Value (cents)	$10(x)$	$25(x+5)$

The equation is:
$$10(x) + 25(x + 5) = 440$$
$$10x + 25x + 125 = 440$$
$$35x + 125 = 440$$
$$35x = 315$$
$$x = 9$$
$$x + 5 = 14$$
Marissa has 9 dimes and 14 quarters.

37. Completing the table:

	Nickels	Quarters
Number	$x + 15$	x
Value (cents)	$5(x+15)$	$25(x)$

The equation is:
$$5(x + 15) + 25(x) = 435$$
$$5x + 75 + 25x = 435$$
$$30x + 75 = 435$$
$$30x = 360$$
$$x = 12$$
$$x + 15 = 27$$
Tanner has 12 quarters and 27 nickels.

39. Completing the table:

	Nickels	Dimes
Number	x	$x+9$
Value (cents)	$5(x)$	$10(x+9)$

The equation is:
$$5(x)+10(x+9)=210$$
$$5x+10x+90=210$$
$$15x+90=210$$
$$15x=120$$
$$x=8$$
$$x+9=17$$

Sue has 8 nickels and 17 dimes.

41. Completing the table:

	Nickels	Dimes	Quarters
Number	x	$x+3$	$x+5$
Value (cents)	$5(x)$	$10(x+3)$	$25(x+5)$

The equation is:
$$5(x)+10(x+3)+25(x+5)=435$$
$$5x+10x+30+25x+125=435$$
$$40x+155=435$$
$$40x=280$$
$$x=7$$
$$x+3=10$$
$$x+5=12$$

Katie has 7 nickels, 10 dimes, and 12 quarters.

43. Completing the table:

	Nickels	Dimes	Quarters
Number	x	$x+6$	$2x$
Value (cents)	$5(x)$	$10(x+6)$	$25(2x)$

The equation is:
$$5(x)+10(x+6)+25(2x)=255$$
$$5x+10x+60+50x=255$$
$$65x+60=255$$
$$65x=195$$
$$x=3$$
$$x+6=9$$
$$2x=6$$

Cory has 3 nickels, 9 dimes, and 6 quarters.

45. The statement is: 4 is less than 10

47. The statement is: 9 is greater than or equal to –5

49. The correct statement is: $12 < 20$

51. The correct statement is: $-8 < -6$

53. Simplifying: $|8-3|-|5-2|=|5|-|3|=5-3=2$

55. Simplifying: $15-|9-3(7-5)|=15-|9-3(2)|=15-|9-6|=15-|3|=15-3=12$

57. Simplifying: $x+2x+2x=5x$

59. Simplifying: $x+0.075x=1.075x$

61. Simplifying: $0.09(x+2{,}000)=0.09x+180$

63. Solving the equation:
$$0.05x+0.06(x-1{,}500)=570$$
$$0.05x+0.06x-90=570$$
$$0.11x=660$$
$$x=6{,}000$$

65. Solving the equation:
$$x+2x+3x=180$$
$$6x=180$$
$$x=30$$

40 Chapter 2 Linear Equations and Inequalities

2.7 More Applications

1. Let x and $x + 1$ represent the two numbers. The equation is:
$$x + x + 1 = 11$$
$$2x + 1 = 11$$
$$2x = 10$$
$$x = 5$$
$$x + 1 = 6$$
The numbers are 5 and 6.

3. Let x and $x + 1$ represent the two numbers. The equation is:
$$x + x + 1 = -9$$
$$2x + 1 = -9$$
$$2x = -10$$
$$x = -5$$
$$x + 1 = -4$$
The numbers are −5 and −4.

5. Let x and $x + 2$ represent the two numbers. The equation is:
$$x + x + 2 = 28$$
$$2x + 2 = 28$$
$$2x = 26$$
$$x = 13$$
$$x + 2 = 15$$
The numbers are 13 and 15.

7. Let x and $x + 2$ represent the two numbers. The equation is:
$$x + x + 2 = 106$$
$$2x + 2 = 106$$
$$2x = 104$$
$$x = 52$$
$$x + 2 = 54$$
The numbers are 52 and 54.

9. Let x and $x + 2$ represent the two numbers. The equation is:
$$x + x + 2 = -30$$
$$2x + 2 = -30$$
$$2x = -322$$
$$x = -16$$
$$x + 2 = -14$$
The numbers are −16 and −14.

11. Let x, $x + 2$, and $x + 4$ represent the three numbers. The equation is:
$$x + x + 2 + x + 4 = 57$$
$$3x + 6 = 57$$
$$3x = 51$$
$$x = 17$$
$$x + 2 = 19$$
$$x + 4 = 21$$
The numbers are 17, 19, and 21.

13. Let x, $x + 2$, and $x + 4$ represent the three numbers. The equation is:
$$x + x + 2 + x + 4 = 132$$
$$3x + 6 = 132$$
$$3x = 126$$
$$x = 42$$
$$x + 2 = 44$$
$$x + 4 = 46$$
The numbers are 42, 44, and 46.

15. Completing the table:

	Dollars Invested at 8%	Dollars Invested at 9%
Number of	x	$x + 2000$
Interest on	$0.08(x)$	$0.09(x + 2000)$

The equation is:
$$0.08(x) + 0.09(x + 2000) = 860$$
$$0.08x + 0.09x + 180 = 860$$
$$0.17x + 180 = 860$$
$$0.17x = 680$$
$$x = 4000$$
$$x + 2000 = 6000$$

You have $4,000 invested at 8% and $6,000 invested at 9%.

17. Completing the table:

	Dollars Invested at 10%	Dollars Invested at 12%
Number of	x	$x + 500$
Interest on	$0.10(x)$	$0.12(x + 500)$

The equation is:
$$0.10(x) + 0.12(x + 500) = 214$$
$$0.10x + 0.12x + 60 = 214$$
$$0.22x + 60 = 214$$
$$0.22x = 154$$
$$x = 700$$
$$x + 500 = 1200$$

Tyler has $700 invested at 10% and $1,200 invested at 12%.

19. Completing the table:

	Dollars Invested at 8%	Dollars Invested at 9%	Dollars Invested at 10%
Number of	x	$2x$	$3x$
Interest on	$0.08(x)$	$0.09(2x)$	$0.10(3x)$

The equation is:
$$0.08(x) + 0.09(2x) + 0.10(3x) = 280$$
$$0.08x + 0.18x + 0.30x = 280$$
$$0.56x = 280$$
$$x = 500$$
$$2x = 1000$$
$$3x = 1500$$

She has $500 invested at 8%, $1,000 invested at 9%, and $1,500 invested at 10%.

21. Let x represent the measure of the two equal angles, so $x + x = 2x$ represents the measure of the third angle. Since the sum of the three angles is 180°, the equation is:
$$x + x + 2x = 180°$$
$$4x = 180°$$
$$x = 45°$$
$$2x = 90°$$

The measures of the three angles are 45°, 45°, and 90°.

23. Let x represent the measure of the largest angle. Then $\frac{1}{5}x$ represents the measure of the smallest angle, and $2\left(\frac{1}{5}x\right) = \frac{2}{5}x$ represents the measure of the other angle. Since the sum of the three angles is 180°, the equation is:
$$x + \frac{1}{5}x + \frac{2}{5}x = 180°$$
$$\frac{5}{5}x + \frac{1}{5}x + \frac{2}{5}x = 180°$$
$$\frac{8}{5}x = 180°$$
$$x = 112.5°$$
$$\frac{1}{5}x = 22.5°$$
$$\frac{2}{5}x = 45°$$
The measures of the three angles are 22.5°, 45°, and 112.5°.

25. Let x represent the measure of the other acute angle, and 90° is the measure of the right angle. Since the sum of the three angles is 180°, the equation is:
$$x + 37° + 90° = 180°$$
$$x + 127° = 180°$$
$$x = 53°$$
The other two angles are 53° and 90°.

27. Let x represent the measure of the smallest angle, so $x + 20$ represents the measure of the second angle and $2x$ represents the measure of the third angle. Since the sum of the three angles is 180°, the equation is:
$$x + x + 20 + 2x = 180°$$
$$4x + 20 = 180°$$
$$4x = 160°$$
$$x = 40°$$
$$x + 20 = 60°$$
$$2x = 80°$$
The measures of the three angles are 40°, 60°, and 80°.

29. Completing the table:

	Adult	Child
Number	x	$x+6$
Income	$6(x)$	$4(x+6)$

The equation is:
$$6(x) + 4(x+6) = 184$$
$$6x + 4x + 24 = 184$$
$$10x + 24 = 184$$
$$10x = 160$$
$$x = 16$$
$$x + 6 = 22$$
Miguel sold 16 adult and 22 children's tickets.

31. Let x represent the total minutes for the call. Then $0.41 is charged for the first minute, and $0.32 is charged for the additional $x - 1$ minutes. The equation is:
$$0.41(1) + 0.32(x-1) = 5.21$$
$$0.41 + 0.32x - 0.32 = 5.21$$
$$0.32x + 0.09 = 5.21$$
$$0.32x = 5.12$$
$$x = 16$$
The call was 16 minutes long.

33. Let x represent the hours Jo Ann worked that week. Then \$12/hour is paid for the first 35 hours and \$18/hour is paid for the additional $x - 35$ hours. The equation is:
$$12(35) + 18(x - 35) = 492$$
$$420 + 18x - 630 = 492$$
$$18x - 210 = 492$$
$$18x = 702$$
$$x = 39$$
Jo Ann worked 39 hours that week.

35. Let x and $x + 2$ represent the two office numbers. The equation is:
$$x + x + 2 = 14,660$$
$$2x + 2 = 14,660$$
$$2x = 14,658$$
$$x = 7329$$
$$x + 2 = 7331$$
They are in offices 7329 and 7331.

37. Let x represent Kendra's age and $x + 2$ represent Marissa's age. The equation is:
$$x + 2 + 2x = 26$$
$$3x + 2 = 26$$
$$3x = 24$$
$$x = 8$$
$$x + 2 = 10$$
Kendra is 8 years old and Marissa is 10 years old.

39. For Jeff, the total time traveled is $\dfrac{425 \text{ miles}}{55 \text{ miles/hour}} \approx 7.72$ hours ≈ 463 minutes. Since he left at 11:00 AM, he will arrive at 6:43 PM. For Carla, the total time traveled is $\dfrac{425 \text{ miles}}{65 \text{ miles/hour}} \approx 6.54$ hours ≈ 392 minutes. Since she left at 1:00 PM, she will arrive at 7:32 PM. Thus Jeff will arrive in Lake Tahoe first.

41. Since $\frac{1}{5}$ mile $= 0.2$ mile, the taxi charge is \$1.25 for the first $\frac{1}{5}$ mile and \$0.25 per fifth mile for the remaining 7.3 miles. Since 7.3 miles $= \dfrac{7.3}{0.2} = 36.5$ fifths, the total charge is: $\$1.25 + \$0.25(36.5) \approx \$10.38$

43. Let w represent the width and $w + 2$ represent the length. The equation is:
$$2w + 2(w + 2) = 44$$
$$2w + 2w + 4 = 44$$
$$4w + 4 = 44$$
$$4w = 40$$
$$w = 10$$
$$w + 2 = 12$$
The length is 12 meters and the width is 10 meters.

45. Let x, $x + 1$, and $x + 2$ represent the measures of the three angles. Since the sum of the three angles is 180°, the equation is:
$$x + x + 1 + x + 2 = 180°$$
$$3x + 3 = 180°$$
$$3x = 177°$$
$$x = 59°$$
$$x + 1 = 60°$$
$$x + 2 = 61°$$
The measures of the three angles are 59°, 60°, and 61°.

47. If all 36 people are Elk's Lodge members (which would be the least amount), the cost of the lessons would be $\$3(36) = \$108$. Since half of the money is paid to Ike and Nancy, the least amount they could make is $\frac{1}{2}(\$108) = \$54$.

49. Yes. The total receipts were \$160, which is possible if there were 10 Elk's members and 26 nonmembers. Computing the total receipts: $10(\$3) + 26(\$5) = \$30 + \$130 = \$160$

44 Chapter 2 Linear Equations and Inequalities

51. Substituting: $36x - 12 = 36\left(\frac{1}{4}\right) - 12 = 9 - 12 = -3$ **53.** Substituting: $36x - 12 = 36\left(\frac{1}{9}\right) - 12 = 4 - 12 = -8$

55. Substituting: $36x - 12 = 36\left(\frac{1}{3}\right) - 12 = 12 - 12 = 0$ **57.** Substituting: $36x - 12 = 36\left(\frac{5}{9}\right) - 12 = 20 - 12 = 8$

59. Evaluating when $x = -4$: $3(x - 4) = 3(-4 - 4) = 3(-8) = -24$

61. Evaluating when $x = -4$: $-5x + 8 = -5(-4) + 8 = 20 + 8 = 28$

63. Evaluating when $x = -4$: $\dfrac{x - 14}{36} = \dfrac{-4 - 14}{36} = -\dfrac{18}{36} = -\dfrac{1}{2}$

65. Evaluating when $x = -4$: $\dfrac{16}{x} + 3x = \dfrac{16}{-4} + 3(-4) = -4 - 12 = -16$

67. Evaluating when $x = -4$: $7x - \dfrac{12}{x} = 7(-4) - \dfrac{12}{-4} = -28 + 3 = -25$

69. Evaluating when $x = -4$: $8\left(\dfrac{x}{2} + 5\right) = 8\left(\dfrac{-4}{2} + 5\right) = 8(-2 + 5) = 8(3) = 24$

71. a. Solving the equation:
$x - 3 = 6$
$x = 9$

  b. Solving the equation:
$x + 3 = 6$
$x = 3$

  c. Solving the equation:
$-x - 3 = 6$
$-x = 9$
$x = -9$

  d. Solving the equation:
$-x + 3 = 6$
$-x = 3$
$x = -3$

73. a. Solving the equation:
$\dfrac{x}{4} = -2$
$x = -2(4) = -8$

  b. Solving the equation:
$-\dfrac{x}{4} = -2$
$x = -2(-4) = 8$

  c. Solving the equation:
$\dfrac{x}{4} = 2$
$x = 2(4) = 8$

  d. Solving the equation:
$-\dfrac{x}{4} = 2$
$x = 2(-4) = -8$

75. Solving the equation:
$2.5x - 3.48 = 4.9x + 2.07$
$-2.4x - 3.48 = 2.07$
$-2.4x = 5.55$
$x = -2.3125$

77. Solving the equation:
$3(x - 4) = -2$
$3x - 12 = -2$
$3x = 10$
$x = \dfrac{10}{3}$

2.8 Linear Inequalities

1. Solving the inequality:
$$x - 5 < 7$$
$$x - 5 + 5 < 7 + 5$$
$$x < 12$$
Graphing the solution set:
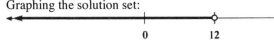

3. Solving the inequality:
$$a - 4 \leq 8$$
$$a - 4 + 4 \leq 8 + 4$$
$$a \leq 12$$
Graphing the solution set:
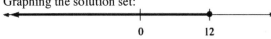

5. Solving the inequality:
$$x - 4.3 > 8.7$$
$$x - 4.3 + 4.3 > 8.7 + 4.3$$
$$x > 13$$
Graphing the solution set:
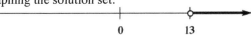

7. Solving the inequality:
$$y + 6 \geq 10$$
$$y + 6 + (-6) \geq 10 + (-6)$$
$$y \geq 4$$
Graphing the solution set:

9. Solving the inequality:
$$2 < x - 7$$
$$2 + 7 < x - 7 + 7$$
$$9 < x$$
$$x > 9$$
Graphing the solution set:
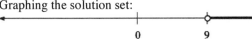

11. Solving the inequality:
$$3x < 6$$
$$\tfrac{1}{3}(3x) < \tfrac{1}{3}(6)$$
$$x < 2$$
Graphing the solution set:
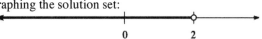

13. Solving the inequality:
$$5a \leq 25$$
$$\tfrac{1}{5}(5a) \leq \tfrac{1}{5}(25)$$
$$a \leq 5$$
Graphing the solution set:
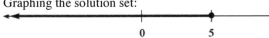

15. Solving the inequality:
$$\tfrac{x}{3} > 5$$
$$3\left(\tfrac{x}{3}\right) > 3(5)$$
$$x > 15$$
Graphing the solution set:

17. Solving the inequality:
$$-2x > 6$$
$$-\tfrac{1}{2}(-2x) < -\tfrac{1}{2}(6)$$
$$x < -3$$
Graphing the solution set:
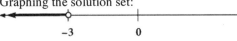

19. Solving the inequality:
$$-3x \geq -18$$
$$-\tfrac{1}{3}(-3x) \leq -\tfrac{1}{3}(-18)$$
$$x \leq 6$$
Graphing the solution set:

21. Solving the inequality:
$$-\tfrac{x}{5} \leq 10$$
$$-5\left(-\tfrac{x}{5}\right) \geq -5(10)$$
$$x \geq -50$$
Graphing the solution set:
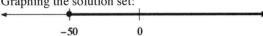

23. Solving the inequality:
$$-\tfrac{2}{3}y > 4$$
$$-\tfrac{3}{2}\left(-\tfrac{2}{3}y\right) < -\tfrac{3}{2}(4)$$
$$y < -6$$
Graphing the solution set:
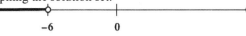

46 Chapter 2 Linear Equations and Inequalities

25. Solving the inequality:
$$2x - 3 < 9$$
$$2x - 3 + 3 < 9 + 3$$
$$2x < 12$$
$$\tfrac{1}{2}(2x) < \tfrac{1}{2}(12)$$
$$x < 6$$

Graphing the solution set:

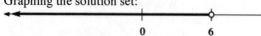

27. Solving the inequality:
$$-\tfrac{1}{5}y - \tfrac{1}{3} \leq \tfrac{2}{3}$$
$$-\tfrac{1}{5}y - \tfrac{1}{3} + \tfrac{1}{3} \leq \tfrac{2}{3} + \tfrac{1}{3}$$
$$-\tfrac{1}{5}y \leq 1$$
$$-5\left(-\tfrac{1}{5}y\right) \geq -5(1)$$
$$y \geq -5$$

Graphing the solution set:

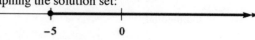

29. Solving the inequality:
$$-7.2x + 1.8 > -19.8$$
$$-7.2x + 1.8 - 1.8 > -19.8 - 1.8$$
$$-7.2x > -21.6$$
$$\frac{-7.2x}{-7.2} < \frac{-21.6}{-7.2}$$
$$x < 3$$

Graphing the solution set:

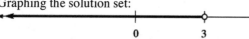

31. Solving the inequality:
$$\tfrac{2}{3}x - 5 \leq 7$$
$$\tfrac{2}{3}x - 5 + 5 \leq 7 + 5$$
$$\tfrac{2}{3}x \leq 12$$
$$\tfrac{3}{2}\left(\tfrac{2}{3}x\right) \leq \tfrac{3}{2}(12)$$
$$x \leq 18$$

Graphing the solution set:

33. Solving the inequality:
$$-\tfrac{2}{5}a - 3 > 5$$
$$-\tfrac{2}{5}a - 3 + 3 > 5 + 3$$
$$-\tfrac{2}{5}a > 8$$
$$-\tfrac{5}{2}\left(-\tfrac{2}{5}a\right) < -\tfrac{5}{2}(8)$$
$$a < -20$$

Graphing the solution set:

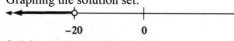

35. Solving the inequality:
$$5 - \tfrac{3}{5}y > -10$$
$$-5 + 5 - \tfrac{3}{5}y > -5 + (-10)$$
$$-\tfrac{3}{5}y > -15$$
$$-\tfrac{5}{3}\left(-\tfrac{3}{5}y\right) < -\tfrac{5}{3}(-15)$$
$$y < 25$$

Graphing the solution set:

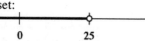

37. Solving the inequality:
$$0.3(a + 1) \leq 1.2$$
$$0.3a + 0.3 \leq 1.2$$
$$0.3a + 0.3 + (-0.3) \leq 1.2 + (-0.3)$$
$$0.3a \leq 0.9$$
$$\frac{0.3a}{0.3} \leq \frac{0.9}{0.3}$$
$$a \leq 3$$

Graphing the solution set:

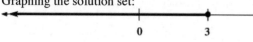

39. Solving the inequality:
$$2(5 - 2x) \leq -20$$
$$10 - 4x \leq -20$$
$$-10 + 10 - 4x \leq -10 + (-20)$$
$$-4x \leq -30$$
$$-\tfrac{1}{4}(-4x) \geq -\tfrac{1}{4}(-30)$$
$$x \geq \tfrac{15}{2}$$

Graphing the solution set:

41. Solving the inequality:

$$3x - 5 > 8x$$
$$-3x + 3x - 5 > -3x + 8x$$
$$-5 > 5x$$
$$\tfrac{1}{5}(-5) > \tfrac{1}{5}(5x)$$
$$-1 > x$$
$$x < -1$$

Graphing the solution set:

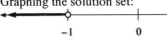

43. First multiply by 6 to clear the inequality of fractions:

$$\tfrac{1}{3}y - \tfrac{1}{2} \le \tfrac{5}{6}y + \tfrac{1}{2}$$
$$6\left(\tfrac{1}{3}y - \tfrac{1}{2}\right) \le 6\left(\tfrac{5}{6}y + \tfrac{1}{2}\right)$$
$$2y - 3 \le 5y + 3$$
$$-5y + 2y - 3 \le -5y + 5y + 3$$
$$-3y - 3 \le 3$$
$$-3y - 3 + 3 \le 3 + 3$$
$$-3y \le 6$$
$$-\tfrac{1}{3}(-3y) \ge -\tfrac{1}{3}(6)$$
$$y \ge -2$$

Graphing the solution set:

45. Solving the inequality:

$$-2.8x + 8.4 < -14x - 2.8$$
$$-2.8x + 14x + 8.4 < -14x + 14x - 2.8$$
$$11.2x + 8.4 < -2.8$$
$$11.2x + 8.4 - 8.4 < -2.8 - 8.4$$
$$11.2x < -11.2$$
$$\frac{11.2x}{11.2} < \frac{-11.2}{11.2}$$
$$x < -1$$

Graphing the solution set:

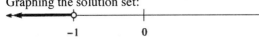

47. Solving the inequality:

$$3(m-2) - 4 \ge 7m + 14$$
$$3m - 6 - 4 \ge 7m + 14$$
$$3m - 10 \ge 7m + 14$$
$$-7m + 3m - 10 \ge -7m + 7m + 14$$
$$-4m - 10 \ge 14$$
$$-4m - 10 + 10 \ge 14 + 10$$
$$-4m \ge 24$$
$$-\tfrac{1}{4}(-4m) \le -\tfrac{1}{4}(24)$$
$$m \le -6$$

Graphing the solution set:

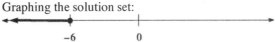

49. Solving the inequality:
$$3 - 4(x-2) \le -5x + 6$$
$$3 - 4x + 8 \le -5x + 6$$
$$-4x + 11 \le -5x + 6$$
$$-4x + 5x + 11 \le -5x + 5x + 6$$
$$x + 11 \le 6$$
$$x + 11 + (-11) \le 6 + (-11)$$
$$x \le -5$$

Graphing the solution set:

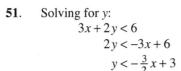

51. Solving for y:
$$3x + 2y < 6$$
$$2y < -3x + 6$$
$$y < -\tfrac{3}{2}x + 3$$

53. Solving for y:
$$2x - 5y > 10$$
$$-5y > -2x + 10$$
$$y < \tfrac{2}{5}x - 2$$

55. Solving for y:
$$-3x + 7y \le 21$$
$$7y \le 3x + 21$$
$$y \le \tfrac{3}{7}x + 3$$

57. Solving for y:
$$2x - 4y \ge -4$$
$$-4y \ge -2x - 4$$
$$y \le \tfrac{1}{2}x + 1$$

48 Chapter 2 Linear Equations and Inequalities

59. a. Evaluating when $x = 0$: $-5x + 3 = -5(0) + 3 = 0 + 3 = 3$

b. Solving the equation:
$$-5x + 3 = -7$$
$$-5x + 3 - 3 = -7 - 3$$
$$-5x = -10$$
$$-\tfrac{1}{5}(-5x) = -\tfrac{1}{5}(-10)$$
$$x = 2$$

c. Substituting $x = 0$:
$$-5(0) + 3 < -7$$
$$3 < -7 \quad \text{(false)}$$

No, $x = 0$ is not a solution to the inequality.

d. Solving the inequality:
$$-5x + 3 < -7$$
$$-5x + 3 - 3 < -7 - 3$$
$$-5x < -10$$
$$-\tfrac{1}{5}(-5x) > -\tfrac{1}{5}(-10)$$
$$x > 2$$

61. The inequality is $x < 3$.

63. The inequality is $x \geq 3$.

65. Let x and $x + 1$ represent the integers. Solving the inequality:
$$x + x + 1 \geq 583$$
$$2x + 1 \geq 583$$
$$2x \geq 582$$
$$x \geq 291$$
The two numbers are at least 291.

67. Let x represent the number. Solving the inequality:
$$2x + 6 < 10$$
$$2x < 4$$
$$x < 2$$

69. Let x represent the number. Solving the inequality:
$$4x > x - 8$$
$$3x > -8$$
$$x > -\tfrac{8}{3}$$

71. Let w represent the width, so $3w$ represents the length. Using the formula for perimeter:
$$2(w) + 2(3w) \geq 48$$
$$2w + 6w \geq 48$$
$$8w \geq 48$$
$$w \geq 6$$
The width is at least 6 meters.

73. Let x, $x + 2$, and $x + 4$ represent the sides of the triangle. The inequality is:
$$x + (x + 2) + (x + 4) > 24$$
$$3x + 6 > 24$$
$$3x > 18$$
$$x > 6$$
The shortest side is an even number greater than 6 inches (greater than or equal to 8 inches).

75. The inequality is $t \geq 100$.

77. Let n represent the number of tickets they sell. The inequality is:
$$7.5n < 1500$$
$$n < 200$$
They will lose money if they sell less than 200 tickets. They will make a profit if they sell more than 200 tickets.

79. Simplifying: $\tfrac{1}{6}(12x + 6) = \tfrac{1}{6} \cdot 12x + \tfrac{1}{6} \cdot 6 = 2x + 1$

81. Simplifying: $\tfrac{2}{3}(-3x - 6) = \tfrac{2}{3} \cdot (-3x) - \tfrac{2}{3} \cdot 6 = -2x - 4$

83. Simplifying: $3\left(\tfrac{5}{6}a + \tfrac{4}{9}\right) = 3 \cdot \tfrac{5}{6}a + 3 \cdot \tfrac{4}{9} = \tfrac{5}{2}a + \tfrac{4}{3}$

85. Simplifying: $-3\left(\tfrac{2}{3}a + \tfrac{5}{6}\right) = -3 \cdot \tfrac{2}{3}a - 3 \cdot \tfrac{5}{6} = -2a - \tfrac{5}{2}$

87. Simplifying: $\tfrac{1}{2}x + \tfrac{1}{6}x = \tfrac{3}{3} \cdot \tfrac{1}{2}x + \tfrac{1}{6}x = \tfrac{3}{6}x + \tfrac{1}{6}x = \tfrac{4}{6}x = \tfrac{2}{3}x$

89. Simplifying: $\tfrac{2}{3}x - \tfrac{5}{6}x = \tfrac{2}{2} \cdot \tfrac{2}{3}x - \tfrac{5}{6}x = \tfrac{4}{6}x - \tfrac{5}{6}x = -\tfrac{1}{6}x$

91. Simplifying: $\tfrac{3}{4}x + \tfrac{1}{6}x = \tfrac{3}{3} \cdot \tfrac{3}{4}x + \tfrac{2}{2} \cdot \tfrac{1}{6}x = \tfrac{9}{12}x + \tfrac{2}{12}x = \tfrac{11}{12}x$

93. Simplifying: $\tfrac{2}{5}x + \tfrac{5}{8}x = \tfrac{8}{8} \cdot \tfrac{2}{5}x + \tfrac{5}{5} \cdot \tfrac{5}{8}x = \tfrac{16}{40}x + \tfrac{25}{40}x = \tfrac{41}{40}x$

95. Solving the inequality:
$$2x - 1 \geq 3$$
$$2x \geq 4$$
$$x \geq 2$$

97. Solving the inequality:
$$-2x > -8$$
$$x < 4$$

99. Solving the inequality:
$$-3 \leq 4x + 1$$
$$4x \geq -4$$
$$x \geq -1$$

Chapter 2 Review/Test

1. Simplifying the expression: $5x - 8x = (5-8)x = -3x$
2. Simplifying the expression: $6x - 3 - 8x = 6x - 8x - 3 = -2x - 3$
3. Simplifying the expression: $-a + 2 + 5a - 9 = -a + 5a + 2 - 9 = 4a - 7$
4. Simplifying the expression: $5(2a-1) - 4(3a-2) = 10a - 5 - 12a + 8 = 10a - 12a - 5 + 8 = -2a + 3$
5. Simplifying the expression: $6 - 2(3y+1) - 4 = 6 - 6y - 2 - 4 = -6y + 6 - 2 - 4 = -6y$
6. Simplifying the expression: $4 - 2(3x-1) - 5 = 4 - 6x + 2 - 5 = -6x + 4 + 2 - 5 = -6x + 1$
7. Evaluating when $x = 3$: $7x - 2 = 7(3) - 2 = 21 - 2 = 19$
8. Evaluating when $x = 3$: $-4x - 5 + 2x = -2x - 5 = -2(3) - 5 = -6 - 5 = -11$
9. Evaluating when $x = 3$: $-x - 2x - 3x = -6x = -6(3) = -18$
10. Evaluating when $x = -2$: $5x - 3 = 5(-2) - 3 = -10 - 3 = -13$
11. Evaluating when $x = -2$: $-3x + 2 = -3(-2) + 2 = 6 + 2 = 8$
12. Evaluating when $x = -2$: $7 - x - 3 = 4 - x = 4 - (-2) = 4 + 2 = 6$

13. Solving the equation:
$$x + 2 = -6$$
$$x + 2 + (-2) = -6 + (-2)$$
$$x = -8$$

14. Solving the equation:
$$x - \tfrac{1}{2} = \tfrac{4}{7}$$
$$x - \tfrac{1}{2} + \tfrac{1}{2} = \tfrac{4}{7} + \tfrac{1}{2}$$
$$x = \tfrac{8}{14} + \tfrac{7}{14}$$
$$x = \tfrac{15}{14}$$

15. Solving the equation:
$$10 - 3y + 4y = 12$$
$$10 + y = 12$$
$$-10 + 10 + y = -10 + 12$$
$$y = 2$$

16. Solving the equation:
$$-3 - 4 = -y - 2 + 2y$$
$$-7 = y - 2$$
$$-7 + 2 = y - 2 + 2$$
$$y = -5$$

17. Solving the equation:
$$2x = -10$$
$$\tfrac{1}{2}(2x) = \tfrac{1}{2}(-10)$$
$$x = -5$$

18. Solving the equation:
$$3x = 0$$
$$\tfrac{1}{3}(3x) = \tfrac{1}{3}(0)$$
$$x = 0$$

19. Solving the equation:
$$\tfrac{x}{3} = 4$$
$$3\left(\tfrac{x}{3}\right) = 3(4)$$
$$x = 12$$

20. Solving the equation:
$$-\tfrac{x}{4} = 2$$
$$-4\left(-\tfrac{x}{4}\right) = -4(2)$$
$$x = -8$$

50 Chapter 2 Linear Equations and Inequalities

21. Solving the equation:
$$3a - 2 = 5a$$
$$-3a + 3a - 2 = -3a + 5a$$
$$-2 = 2a$$
$$\tfrac{1}{2}(-2) = \tfrac{1}{2}(2a)$$
$$a = -1$$

22. First multiply by 10 to clear the equation of fractions:
$$\tfrac{7}{10}a = \tfrac{1}{5}a + \tfrac{1}{2}$$
$$10\left(\tfrac{7}{10}a\right) = 10\left(\tfrac{1}{5}a + \tfrac{1}{2}\right)$$
$$7a = 2a + 5$$
$$-2a + 7a = -2a + 2a + 5$$
$$5a = 5$$
$$\tfrac{1}{5}(5a) = \tfrac{1}{5}(5)$$
$$a = 1$$

23. Solving the equation:
$$3x + 2 = 5x - 8$$
$$3x + (-5x) + 2 = 5x + (-5x) - 8$$
$$-2x + 2 = -8$$
$$-2x + 2 + (-2) = -8 + (-2)$$
$$-2x = -10$$
$$-\tfrac{1}{2}(-2x) = -\tfrac{1}{2}(-10)$$
$$x = 5$$

24. Solving the equation:
$$6x - 3 = x + 7$$
$$6x + (-x) - 3 = x + (-x) + 7$$
$$5x - 3 = 7$$
$$5x - 3 + 3 = 7 + 3$$
$$5x = 10$$
$$\tfrac{1}{5}(5x) = \tfrac{1}{5}(10)$$
$$x = 2$$

25. Solving the equation:
$$0.7x - 0.1 = 0.5x - 0.1$$
$$0.7x + (-0.5x) - 0.1 = 0.5x + (-0.5x) - 0.1$$
$$0.2x - 0.1 = -0.1$$
$$0.2x - 0.1 + 0.1 = -0.1 + 0.1$$
$$0.2x = 0$$
$$\tfrac{0.2x}{0.2} = \tfrac{0}{0.2}$$
$$x = 0$$

26. Solving the equation:
$$0.2x - 0.3 = 0.8x - 0.3$$
$$0.2x + (-0.8x) - 0.3 = 0.8x + (-0.8x) - 0.3$$
$$-0.6x - 0.3 = -0.3$$
$$-0.6x - 0.3 + 0.3 = -0.3 + 0.3$$
$$-0.6x = 0$$
$$\tfrac{-0.6x}{-0.6} = \tfrac{0}{-0.6}$$
$$x = 0$$

27. Simplifying and then solving the equation:
$$2(x - 5) = 10$$
$$2x - 10 = 10$$
$$2x - 10 + 10 = 10 + 10$$
$$2x = 20$$
$$\tfrac{1}{2}(2x) = \tfrac{1}{2}(20)$$
$$x = 10$$

28. Simplifying and then solving the equation:
$$12 = 2(5x - 4)$$
$$12 = 10x - 8$$
$$12 + 8 = 10x - 8 + 8$$
$$20 = 10x$$
$$\tfrac{1}{10}(20) = \tfrac{1}{10}(10x)$$
$$x = 2$$

29. Simplifying and then solving the equation:
$$\tfrac{1}{2}(3t - 2) + \tfrac{1}{2} = \tfrac{5}{2}$$
$$\tfrac{3}{2}t - 1 + \tfrac{1}{2} = \tfrac{5}{2}$$
$$\tfrac{3}{2}t - \tfrac{1}{2} = \tfrac{5}{2}$$
$$\tfrac{3}{2}t - \tfrac{1}{2} + \tfrac{1}{2} = \tfrac{5}{2} + \tfrac{1}{2}$$
$$\tfrac{3}{2}t = 3$$
$$\tfrac{2}{3}\left(\tfrac{3}{2}t\right) = \tfrac{2}{3}(3)$$
$$t = 2$$

30. Simplifying and then solving the equation:
$$\tfrac{3}{5}(5x - 10) = \tfrac{2}{3}(9x + 3)$$
$$3x - 6 = 6x + 2$$
$$3x + (-6x) - 6 = 6x + (-6x) + 2$$
$$-3x - 6 = 2$$
$$-3x - 6 + 6 = 2 + 6$$
$$-3x = 8$$
$$-\tfrac{1}{3}(-3x) = -\tfrac{1}{3}(8)$$
$$x = -\tfrac{8}{3}$$

31. Simplifying and then solving the equation:
$$2(3x+7) = 4(5x-1) + 18$$
$$6x + 14 = 20x - 4 + 18$$
$$6x + 14 = 20x + 14$$
$$6x + (-20x) + 14 = 20x + (-20x) + 14$$
$$-14x + 14 = 14$$
$$-14x + 14 + (-14) = 14 + (-14)$$
$$-14x = 0$$
$$-\tfrac{1}{14}(-14x) = -\tfrac{1}{14}(0)$$
$$x = 0$$

32. Simplifying and then solving the equation:
$$7 - 3(y+4) = 10$$
$$7 - 3y - 12 = 10$$
$$-3y - 5 = 10$$
$$-3y - 5 + 5 = 10 + 5$$
$$-3y = 15$$
$$-\tfrac{1}{3}(-3y) = -\tfrac{1}{3}(15)$$
$$y = -5$$

33. Substituting $x = 5$:
$$4(5) - 5y = 20$$
$$20 - 5y = 20$$
$$-5y = 0$$
$$y = 0$$

34. Substituting $x = 0$:
$$4(0) - 5y = 20$$
$$-5y = 20$$
$$y = -4$$

35. Substituting $x = -5$:
$$4(-5) - 5y = 20$$
$$-20 - 5y = 20$$
$$-5y = 40$$
$$y = -8$$

36. Substituting $x = 10$:
$$4(10) - 5y = 20$$
$$40 - 5y = 20$$
$$-5y = -20$$
$$y = 4$$

37. Solving for y:
$$2x - 5y = 10$$
$$-5y = -2x + 10$$
$$y = \tfrac{2}{5}x - 2$$

38. Solving for y:
$$5x - 2y = 10$$
$$-2y = -5x + 10$$
$$y = \tfrac{5}{2}x - 5$$

39. Solving for h:
$$\pi r^2 h = V$$
$$\frac{\pi r^2 h}{\pi r^2} = \frac{V}{\pi r^2}$$
$$h = \frac{V}{\pi r^2}$$

40. Solving for w:
$$2l + 2w = P$$
$$2w = P - 2l$$
$$w = \frac{P - 2l}{2}$$

41. Computing the amount:
$$0.86(240) = x$$
$$x = 206.4$$

So 86% of 240 is 206.4.

42. Finding the percent:
$$p \cdot 2000 = 180$$
$$p = \tfrac{180}{2000}$$
$$p = 0.09 = 9\%$$

So 9% of 2000 is 180.

43. Let x represent the number. The equation is:
$$2x + 6 = 28$$
$$2x = 22$$
$$x = 11$$

The number is 11.

44. Let w represent the width and $5w$ represent the length. Using the perimeter formula:
$$2w + 2(5w) = 60$$
$$2w + 10w = 60$$
$$12w = 60$$
$$w = 5$$
$$5w = 25$$

The width is 5 meters and the length is 25 meters.

45. Completing the table:

	Dollars Invested at 9%	Dollars Invested at 10%
Number of	x	$x + 300$
Interest on	$0.09(x)$	$0.10(x + 300)$

The equation is:
$$0.09(x) + 0.10(x + 300) = 125$$
$$0.09x + 0.10x + 30 = 125$$
$$0.19x + 30 = 125$$
$$0.19x = 95$$
$$x = 500$$
$$x + 300 = 800$$
The man invested $500 at 9% and $800 at 10%.

46. Completing the table:

	Nickels	Dimes
Number	x	$15 - x$
Value (cents)	$5(x)$	$10(15 - x)$

The equation is:
$$5(x) + 10(15 - x) = 100$$
$$5x + 150 - 10x = 100$$
$$-5x + 150 = 100$$
$$-5x = -150$$
$$x = 10$$
$$15 - x = 5$$
The collection has 10 nickels and 5 dimes.

47. Solving the inequality:
$$-2x < 4$$
$$-\tfrac{1}{2}(-2x) > -\tfrac{1}{2}(4)$$
$$x > -2$$

48. Solving the inequality:
$$-5x > -10$$
$$-\tfrac{1}{5}(-5x) < -\tfrac{1}{5}(-10)$$
$$x < 2$$

49. Solving the inequality:
$$-\frac{a}{2} \leq -3$$
$$-2\left(-\frac{a}{2}\right) \geq -2(-3)$$
$$a \geq 6$$

50. Solving the inequality:
$$-\frac{a}{3} > 5$$
$$-3\left(-\frac{a}{3}\right) < -3(5)$$
$$a < -15$$

51. Solving the inequality:
$$-4x + 5 > 37$$
$$-4x > 32$$
$$x < -8$$

Graphing the solution set:

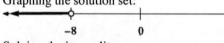

52. Solving the inequality:
$$2x + 10 < 5x - 11$$
$$-3x + 10 < -11$$
$$-3x < -21$$
$$x > 7$$

Graphing the solution set:

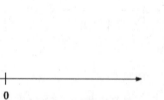

53. Solving the inequality:
$$2(3t + 1) + 6 \geq 5(2t + 4)$$
$$6t + 2 + 6 \geq 10t + 20$$
$$6t + 8 \geq 10t + 20$$
$$-4t + 8 \geq 20$$
$$-4t \geq 12$$
$$t \leq -3$$

Graphing the solution set:

Chapter 3
Linear Equations and Inequalities in Two Variables

3.1 Paired Data and Graphing Ordered Pairs

1. Graphing the ordered pair:

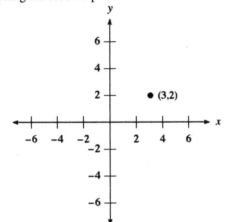

3. Graphing the ordered pair:

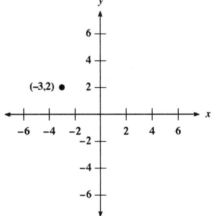

5. Graphing the ordered pair:

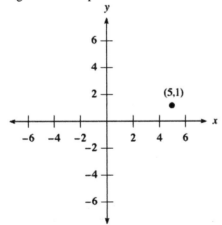

7. Graphing the ordered pair:

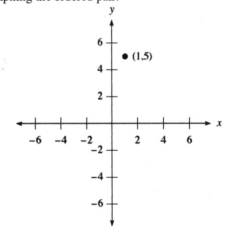

54 Chapter 3 Linear Equations and Inequalities in Two Variables

9. Graphing the ordered pair:

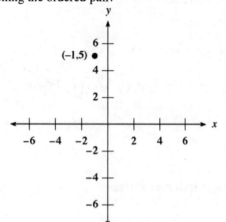

11. Graphing the ordered pair:

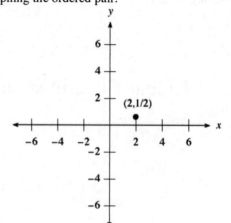

13. Graphing the ordered pair:

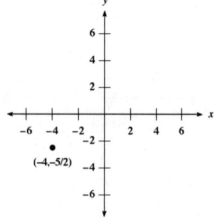

15. Graphing the ordered pair:

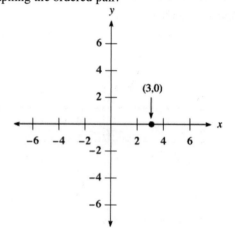

17. Graphing the ordered pair:

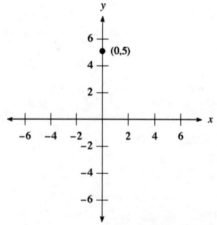

19. The coordinates are $(-4, 4)$.

21. The coordinates are $(-4, 2)$.

23. The coordinates are $(-3, 0)$.

25. The coordinates are $(2, -2)$.

27. The coordinates are $(-5, -5)$.

29. Graphing the line:

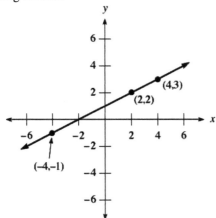

Yes, the point $(2,2)$ lies on the line.

31. Graphing the line:

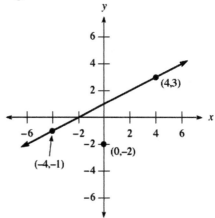

No, the point $(0,-2)$ does not lie on the line.

33. Graphing the line:

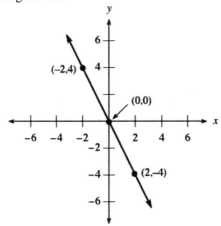

Yes, the point $(0,0)$ lies on the line.

35. Graphing the line:

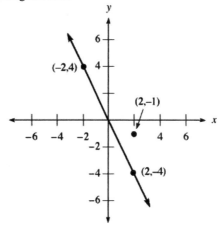

No, the point $(2,-1)$ does not lie on the line.

37. Graphing the line:

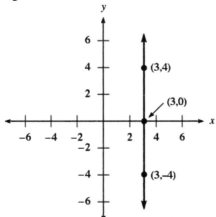

Yes, the point $(3,0)$ lies on the line.

39. No, the x-coordinate of every point on this line is 3.

56 Chapter 3 Linear Equations and Inequalities in Two Variables

41. Graphing the line:

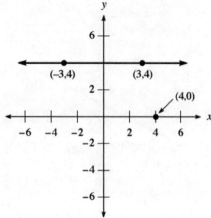

No, the point $(4,0)$ does not lie on the line.

43. No, the y-coordinate of every point on this line is 4.

45. a. Three ordered pairs on the graph are (5,40), (10,80), and (20,160).
  b. She will earn \$320 for working 40 hours.
  c. If her check is \$240, she worked 30 hours that week.
  d. No. She should be paid \$280 for working 35 hours, not \$260.

47. Constructing a line graph:

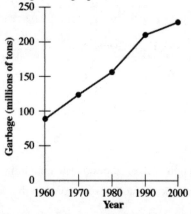

49. Constructing a line graph:

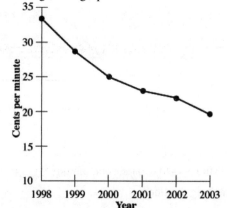

51. Six ordered pairs are (1981,5), (1985,20.2), (1990,34.4), (1995,44.8), (2000,65.4), and (2004,99.4).
53. Point A is (6 – 5,2) = (1,2), and point B is (6,2 + 5) = (6,7).
55. Point A is (7 – 5,2) = (2,2), point B is (2,2 + 3) = (2,5), and point C is (2 + 5,5) = (7,5).
57. Simplifying: $\dfrac{x}{5}+\dfrac{3}{5}=\dfrac{x+3}{5}$

59. Simplifying: $\dfrac{2}{7}-\dfrac{a}{7}=\dfrac{2-a}{7}$

61. Simplifying: $\dfrac{1}{14}-\dfrac{y}{7}=\dfrac{1}{14}-\dfrac{y}{7}\cdot\dfrac{2}{2}=\dfrac{1-2y}{14}$

63. Simplifying: $\dfrac{1}{2}+\dfrac{3}{x}=\dfrac{1}{2}\cdot\dfrac{x}{x}+\dfrac{3}{x}\cdot\dfrac{2}{2}=\dfrac{x+6}{2x}$

65. Simplifying: $\dfrac{5+x}{6}-\dfrac{5}{6}=\dfrac{5+x-5}{6}=\dfrac{x}{6}$

67. Simplifying: $\dfrac{4}{x}+\dfrac{1}{2}=\dfrac{4}{x}\cdot\dfrac{2}{2}+\dfrac{1}{2}\cdot\dfrac{x}{x}=\dfrac{8+x}{2x}$

69. a. Substituting $y = 4$:
$$2x + 3(4) = 6$$
$$2x + 12 = 6$$
$$2x = -6$$
$$x = -3$$
   c. Substituting $x = 3$:
$$2(3) + 3y = 6$$
$$6 + 3y = 6$$
$$3y = 0$$
$$y = 0$$

   b. Substituting $y = -2$:
$$2x + 3(-2) = 6$$
$$2x - 6 = 6$$
$$2x = 12$$
$$x = 6$$
   d. Substituting $x = 9$:
$$2(9) + 3y = 6$$
$$18 + 3y = 6$$
$$3y = -12$$
$$y = -4$$

71. a. Substituting $y = 7$:
$$2x - 1 = 7$$
$$2x = 8$$
$$x = 4$$
   c. Substituting $x = 0$:
$$y = 2(0) - 1 = 0 - 1 = -1$$

   b. Substituting $y = 3$:
$$2x - 1 = 3$$
$$2x = 4$$
$$x = 2$$
   d. Substituting $x = 5$:
$$y = 2(5) - 1 = 10 - 1 = 9$$

3.2 Solutions to Linear Equations in Two Variables

1. Substituting $x = 0$, $y = 0$, and $y = -6$:
$$2(0) + y = 6 \qquad 2x + 0 = 6 \qquad 2x + (-6) = 6$$
$$0 + y = 6 \qquad 2x = 6 \qquad 2x = 12$$
$$y = 6 \qquad x = 3 \qquad x = 6$$
   The ordered pairs are $(0, 6)$, $(3, 0)$, and $(6, -6)$.

3. Substituting $x = 0$, $y = 0$, and $x = -4$:
$$3(0) + 4y = 12 \qquad 3x + 4(0) = 12 \qquad 3(-4) + 4y = 12$$
$$0 + 4y = 12 \qquad 3x + 0 = 12 \qquad -12 + 4y = 12$$
$$4y = 12 \qquad 3x = 12 \qquad 4y = 24$$
$$y = 3 \qquad x = 4 \qquad y = 6$$
   The ordered pairs are $(0, 3)$, $(4, 0)$, and $(-4, 6)$.

5. Substituting $x = 1$, $y = 0$, and $x = 5$:
$$y = 4(1) - 3 \qquad 0 = 4x - 3 \qquad y = 4(5) - 3$$
$$y = 4 - 3 \qquad 3 = 4x \qquad y = 20 - 3$$
$$y = 1 \qquad x = \tfrac{3}{4} \qquad y = 17$$
   The ordered pairs are $(1, 1)$, $\left(\tfrac{3}{4}, 0\right)$, and $(5, 17)$.

7. Substituting $x = 2$, $y = 6$, and $x = 0$:
$$y = 7(2) - 1 \qquad 6 = 7x - 1 \qquad y = 7(0) - 1$$
$$y = 14 - 1 \qquad 7 = 7x \qquad y = 0 - 1$$
$$y = 13 \qquad x = 1 \qquad y = -1$$
   The ordered pairs are $(2, 13)$, $(1, 6)$, and $(0, -1)$.

9. Substituting $y = 4$, $y = -3$, and $y = 0$ results (in each case) in $x = -5$. The ordered pairs are $(-5, 4)$, $(-5, -3)$, and $(-5, 0)$.

11. Completing the table:

x	y
1	3
-3	-9
4	12
6	18

13. Completing the table:

x	y
0	0
-1/2	-2
-3	-12
3	12

58 Chapter 3 Linear Equations and Inequalities in Two Variables

15. Completing the table:

x	y
2	3
3	2
5	0
9	−4

17. Completing the table:

x	y
2	0
3	2
1	−2
−3	−10

19. Completing the table:

x	y
0	−1
−1	−7
−3	−19
3/2	8

21. Substituting each ordered pair into the equation:
$(2,3)$: $2(2)-5(3) = 4-15 = -11 \neq 10$
$(0,-2)$: $2(0)-5(-2) = 0+10 = 10$
$\left(\frac{5}{2},1\right)$: $2\left(\frac{5}{2}\right)-5(1) = 5-5 = 0 \neq 10$

Only the ordered pair $(0,-2)$ is a solution.

23. Substituting each ordered pair into the equation:
$(1,5)$: $7(1)-2 = 7-2 = 5$
$(0,-2)$: $7(0)-2 = 0-2 = -2$
$(-2,-16)$: $7(-2)-2 = -14-2 = -16$

All the ordered pairs $(1,5)$, $(0,-2)$ and $(-2,-16)$ are solutions.

25. Substituting each ordered pair into the equation:
$(1,6)$: $6(1) = 6$
$(-2,-12)$: $6(-2) = -12$
$(0,0)$: $6(0) = 0$

All the ordered pairs $(1,6)$, $(-2,-12)$ and $(0,0)$ are solutions.

27. Substituting each ordered pair into the equation:
$(1,1)$: $1+1 = 2 \neq 0$
$(2,-2)$: $2+(-2) = 0$
$(3,3)$: $3+3 = 6 \neq 0$

Only the ordered pair $(2,-2)$ is a solution.

29. Since $x = 3$, the ordered pair $(5,3)$ cannot be a solution. The ordered pairs $(3,0)$ and $(3,-3)$ are solutions.

31. Substituting $w = 3$:
$2l + 2(3) = 30$
$2l + 6 = 30$
$2l = 24$
$l = 12$

The length is 12 inches.

33.
a. This is correct, since; $y = 12(5) = \$60$
b. This is not correct, since: $y = 12(9) = \$108$. Her check should be for $108.
c. This is not correct, since: $y = 12(7) = \$84$. Her check should be for $84.
d. This is correct, since: $y = 12(14) = \$168$

35. **a.** Substituting $t = 5$: $V = -45,000(5) + 600,000 = -225,000 + 600,000 = \$375,000$

b. Solving when $V = 330,000$:
$$-45,000t + 600,000 = 330,000$$
$$-45,000t = -270,000$$
$$t = 6$$
The crane will be worth $330,000 at the end of 6 years.

c. Substituting $t = 9$: $V = -45,000(9) + 600,000 = -405,000 + 600,000 = \$195,000$

No, the crane will be worth $195,000 after 9 years.

d. The crane cost $600,000 (the value when $t = 0$).

37. Simplifying: $\dfrac{11(-5) - 17}{2(-6)} = \dfrac{-55 - 17}{-12} = \dfrac{-72}{-12} = 6$

39. Simplifying: $\dfrac{13(-6) + 18}{4(-5)} = \dfrac{-78 + 18}{-20} = \dfrac{-60}{-20} = 3$

41. Simplifying: $\dfrac{7^2 - 5^2}{(7-5)^2} = \dfrac{49 - 25}{2^2} = \dfrac{24}{4} = 6$

43. Simplifying: $\dfrac{-3 \cdot 4^2 - 3 \cdot 2^4}{-3(8)} = \dfrac{-3 \cdot 16 - 3 \cdot 16}{-3(8)} = \dfrac{-48 - 48}{-24} = \dfrac{-96}{-24} = 4$

45. Substituting $x = 4$:
$$3(4) + 2y = 6$$
$$12 + 2y = 6$$
$$2y = -6$$
$$y = -3$$

47. Substituting $x = 0$: $y = -\tfrac{1}{3}(0) + 2 = 0 + 2 = 2$

49. Substituting $x = 2$: $y = \tfrac{3}{2}(2) - 3 = 3 - 3 = 0$

51. Solving for y:
$$5x + y = 4$$
$$y = -5x + 4$$

53. Solving for y:
$$3x - 2y = 6$$
$$-2y = -3x + 6$$
$$y = \tfrac{3}{2}x - 3$$

3.3 Graphing Linear Equations in Two Variables

1. The ordered pairs are $(0,4)$, $(2,2)$, and $(4,0)$:

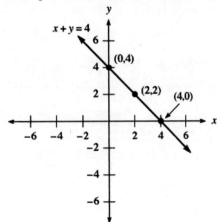

3. The ordered pairs are $(0,3)$, $(2,1)$, and $(4,-1)$:

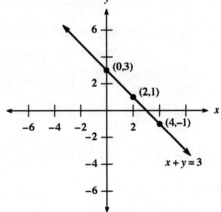

5. The ordered pairs are $(0,0)$, $(-2,-4)$, and $(2,4)$:

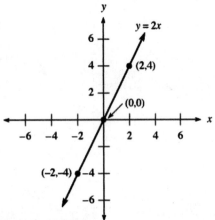

7. The ordered pairs are $(-3,-1)$, $(0,0)$, and $(3,1)$:

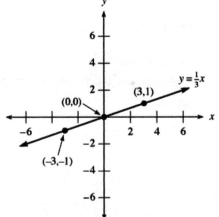

9. The ordered pairs are $(0,1)$, $(-1,-1)$, and $(1,3)$:

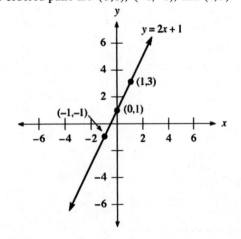

11. The ordered pairs are $(0,4)$, $(-1,4)$, and $(2,4)$:

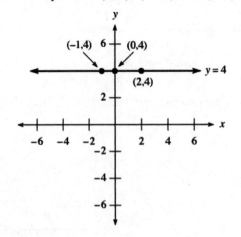

13. The ordered pairs are $(-2,2)$, $(0,3)$, and $(2,4)$:

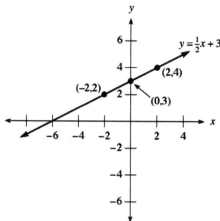

15. The ordered pairs are $(-3,3)$, $(0,1)$, and $(3,-1)$:

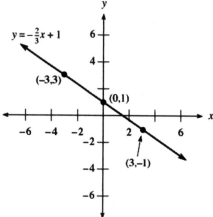

17. Solving for y:
$$2x + y = 3$$
$$y = -2x + 3$$

The ordered pairs are $(-1,5)$, $(0,3)$, and $(1,1)$:

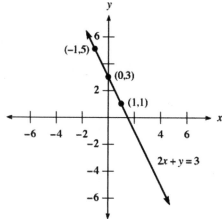

19. Solving for y:
$$3x + 2y = 6$$
$$2y = -3x + 6$$
$$y = -\frac{3}{2}x + 3$$

The ordered pairs are $(0,3)$, $(2,0)$, and $(4,-3)$:

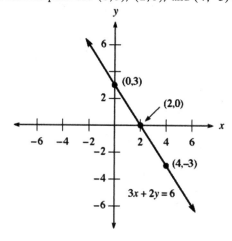

21. Solving for y:
$$-x + 2y = 6$$
$$2y = x + 6$$
$$y = \frac{1}{2}x + 3$$

The ordered pairs are $(-2,2)$, $(0,3)$, and $(2,4)$:

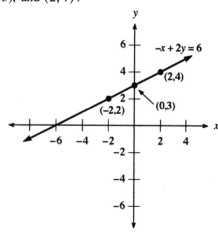

62 Chapter 3 Linear Equations and Inequalities in Two Variables

23. Three solutions are $(-4,2)$, $(0,0)$, and $(4,-2)$:

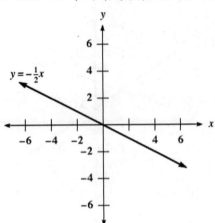

25. Three solutions are $(-1,-4)$, $(0,-1)$, and $(1,2)$:

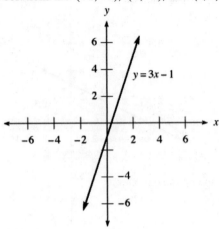

27. Solving for y:
$$-2x+y=1$$
$$y=2x+1$$

Three solutions are $(-2,-3)$, $(0,1)$, and $(2,5)$:

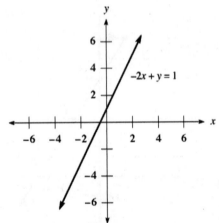

29. Solving for y:
$$3x+4y=8$$
$$4y=-3x+8$$
$$y=-\frac{3}{4}x+2$$

Three solutions are $(-4,5)$, $(0,2)$, and $(4,-1)$:

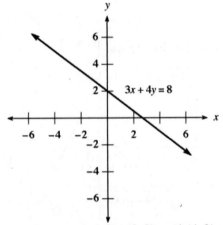

31. Three solutions are $(-2,-4)$, $(-2,0)$, and $(-2,4)$:

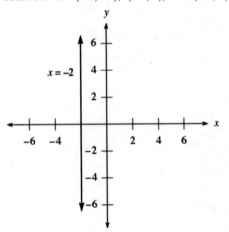

33. Three solutions are $(-4,2)$, $(0,2)$, and $(4,2)$:

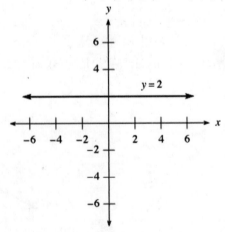

35. Graphing the equation:

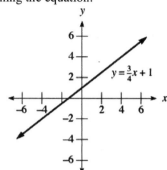

37. Graphing the equation:

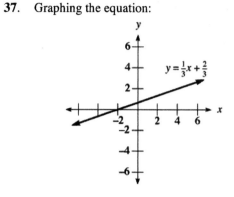

39. Graphing the equation:

41. Completing the table:

Equation	H, V, and/or O
$x = 3$	V
$y = 3$	H
$y = 3x$	O
$y = 0$	O,H

43. Completing the table:

Equation	H, V, and/or O
$x = -\frac{3}{5}$	V
$y = -\frac{3}{5}$	H
$y = -\frac{3}{5}x$	O
$x = 0$	O,V

45. **a.** Solving the equation:
$$2x + 5 = 10$$
$$2x = 5$$
$$x = \frac{5}{2}$$

b. Substituting $y = 0$:
$$2x + 5(0) = 10$$
$$2x = 10$$
$$x = 5$$

c. Substituting $x = 0$:
$$2(0) + 5y = 10$$
$$5y = 10$$
$$y = 2$$

d. Graphing the line:

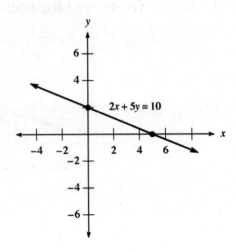

e. Solving for y:
$$2x + 5y = 10$$
$$5y = -2x + 10$$
$$y = -\tfrac{2}{5}x + 2$$

47. Simplifying: $\tfrac{1}{2}(4x+10) = \tfrac{1}{2} \cdot 4x + \tfrac{1}{2} \cdot 10 = 2x + 5$

49. Simplifying: $\tfrac{2}{3}(3x-9) = \tfrac{2}{3} \cdot 3x - \tfrac{2}{3} \cdot 9 = 2x - 6$

51. Simplifying: $\tfrac{3}{4}(4x+10) = \tfrac{3}{4} \cdot 4x + \tfrac{3}{4} \cdot 10 = 3x + \tfrac{15}{2}$

53. Simplifying: $\tfrac{3}{5}(10x+15) = \tfrac{3}{5} \cdot 10x + \tfrac{3}{5} \cdot 15 = 6x + 9$

55. Simplifying: $5\left(\tfrac{2}{5}x + 10\right) = 5 \cdot \tfrac{2}{5}x + 5 \cdot 10 = 2x + 50$

57. Simplifying: $4\left(\tfrac{3}{2}x - 7\right) = 4 \cdot \tfrac{3}{2}x - 4 \cdot 7 = 6x - 28$

59. Simplifying: $\tfrac{3}{4}(2x+12y) = \tfrac{3}{4} \cdot 2x + \tfrac{3}{4} \cdot 12y = \tfrac{3}{2}x + 9y$

61. Simplifying: $\tfrac{1}{2}(5x-10y) + 6 = \tfrac{1}{2} \cdot 5x - \tfrac{1}{2} \cdot 10y + 6 = \tfrac{5}{2}x - 5y + 6$

63. a. Substituting $y = 0$:
$$3x - 2(0) = 6$$
$$3x - 0 = 6$$
$$3x = 6$$
$$x = 2$$

b. Substituting $x = 0$:
$$3(0) - 2y = 6$$
$$0 - 2y = 6$$
$$-2y = 6$$
$$y = -3$$

65. a. Substituting $y = 0$:
$$-x + 2(0) = 4$$
$$-x + 0 = 4$$
$$-x = 4$$
$$x = -4$$

b. Substituting $x = 0$:
$$-(0) + 2y = 4$$
$$0 + 2y = 4$$
$$2y = 4$$
$$y = 2$$

67. a. Substituting $y = 0$:
$$0 = -\tfrac{1}{3}x + 2$$
$$-\tfrac{1}{3}x = -2$$
$$x = 6$$

b. Substituting $x = 0$:
$$y = -\tfrac{1}{3}(0) + 2 = 2$$

3.4 More on Graphing: Intercepts

1. To find the x-intercept, let $y = 0$:
$$2x + 0 = 4$$
$$2x = 4$$
$$x = 2$$
To find the y-intercept, let $x = 0$:
$$2(0) + y = 4$$
$$0 + y = 4$$
$$y = 4$$
Graphing the line:

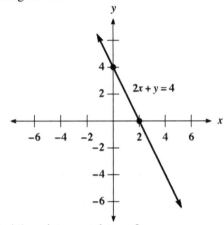

3. To find the x-intercept, let $y = 0$:
$$-x + 0 = 3$$
$$-x = 3$$
$$x = -3$$
To find the y-intercept, let $x = 0$:
$$-0 + y = 3$$
$$y = 3$$
Graphing the line:

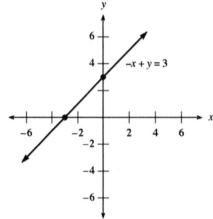

5. To find the x-intercept, let $y = 0$:
$$-x + 2(0) = 2$$
$$-x = 2$$
$$x = -2$$
To find the y-intercept, let $x = 0$:
$$-0 + 2y = 2$$
$$2y = 2$$
$$y = 1$$
Graphing the line:

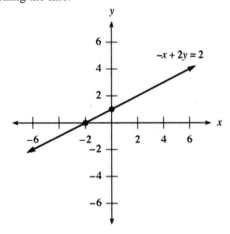

7. To find the x-intercept, let $y = 0$:
$$5x + 2(0) = 10$$
$$5x = 10$$
$$x = 2$$
To find the y-intercept, let $x = 0$:
$$5(0) + 2y = 10$$
$$2y = 10$$
$$y = 5$$
Graphing the line:

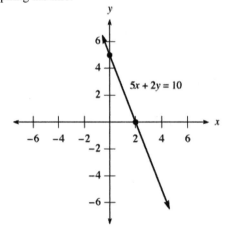

9. To find the x-intercept, let $y = 0$:
$$-4x + 5(0) = 20$$
$$-4x = 20$$
$$x = -5$$

To find the y-intercept, let $x = 0$:
$$-4(0) + 5y = 20$$
$$5y = 20$$
$$y = 4$$

Graphing the line:

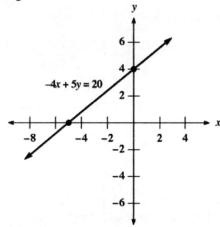

11. To find the x-intercept, let $y = 0$:
$$3x - 4(0) = -4$$
$$3x = -4$$
$$x = -\tfrac{4}{3}$$

To find the y-intercept, let $x = 0$:
$$3(0) - 4y = -4$$
$$-4y = -4$$
$$y = 1$$

Graphing the line:

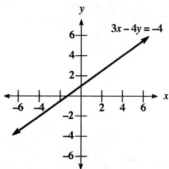

13. To find the x-intercept, let $y = 0$:
$$x - 3(0) = 2$$
$$x = 2$$

To find the y-intercept, let $x = 0$:
$$0 - 3y = 2$$
$$-3y = 2$$
$$y = -\tfrac{2}{3}$$

Graphing the line:

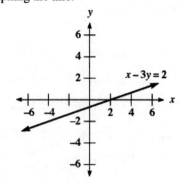

15. To find the x-intercept, let $y = 0$:
$$2x - 3(0) = -2$$
$$2x = -2$$
$$x = -1$$

To find the y-intercept, let $x = 0$:
$$2(0) - 3y = -2$$
$$-3y = -2$$
$$y = \tfrac{2}{3}$$

Graphing the line:

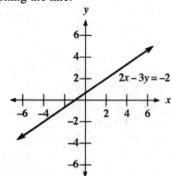

Problem Set 3.4 67

17. To find the x-intercept, let $y = 0$:
$$2x - 6 = 0$$
$$2x = 6$$
$$x = 3$$

To find the y-intercept, let $x = 0$:
$$y = 2(0) - 6$$
$$y = -6$$

Graphing the line:

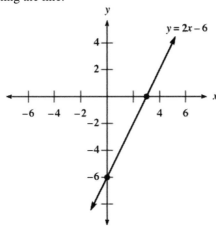

19. To find the x-intercept, let $y = 0$:
$$2x - 1 = 0$$
$$2x = 1$$
$$x = \tfrac{1}{2}$$

To find the y-intercept, let $x = 0$:
$$y = 2(0) - 1$$
$$y = -1$$

Graphing the line:

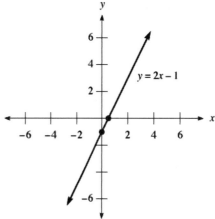

21. To find the x-intercept, let $y = 0$:
$$\tfrac{1}{2}x + 3 = 0$$
$$\tfrac{1}{2}x = -3$$
$$x = -6$$

To find the y-intercept, let $x = 0$:
$$y = \tfrac{1}{2}(0) + 3$$
$$y = 3$$

Graphing the line:

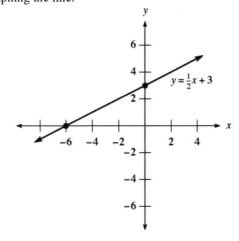

23. To find the x-intercept, let $y = 0$:
$$-\tfrac{1}{3}x - 2 = 0$$
$$-\tfrac{1}{3}x = 2$$
$$x = -6$$

To find the y-intercept, let $x = 0$:
$$y = -\tfrac{1}{3}(0) - 2$$
$$y = -2$$

Graphing the line:

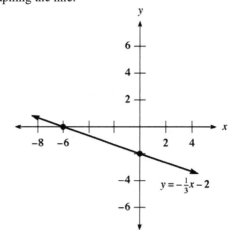

25. Another point on the line is $(2,-4)$:

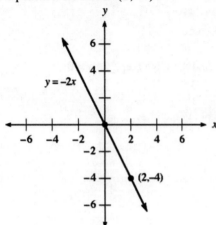

27. Another point on the line is $(3,-1)$:

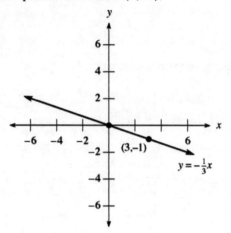

29. Another point on the line is $(3,2)$:

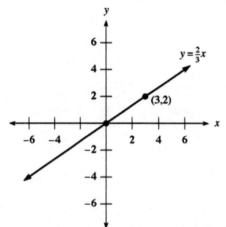

31. Completing the table:

Equation	x-intercept	y-intercept
$3x+4y=12$	4	3
$3x+4y=4$	$4/3$	1
$3x+4y=3$	1	$3/4$
$3x+4y=2$	$2/3$	$1/2$

33. Completing the table:

Equation	x-intercept	y-intercept
$x-3y=2$	2	$-2/3$
$y=\frac{1}{3}x-\frac{2}{3}$	2	$-2/3$
$x-3y=0$	0	0
$y=\frac{1}{3}x$	0	0

35. **a.** Solving the equation:

$$2x-3=-3$$
$$2x=0$$
$$x=0$$

b. Substituting $y=0$:

$$2x-3(0)=-3$$
$$2x-0=-3$$
$$2x=-3$$
$$x=-\frac{3}{2}$$

c. Substituting $x=0$:

$$2(0)-3y=-3$$
$$0-3y=-3$$
$$-3y=-3$$
$$y=1$$

d. Graphing the equation:

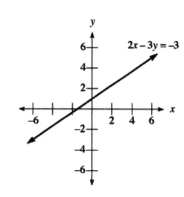

e. Solving for y:
$$2x - 3y = -3$$
$$-3y = -2x - 3$$
$$y = \tfrac{2}{3}x + 1$$

37. Graphing the line:

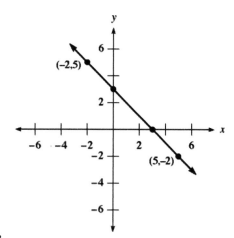

The x- and y-intercepts are both 3.

39. The x-intercept is -1 and the y-intercept is -3.

41. Completing the table:

x	y
-2	1
0	-1
-1	0
1	-2

43. Graphing the line:

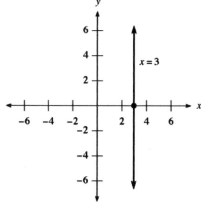

The x-intercept is 3.

45. Graphing the line:

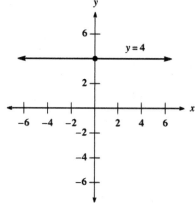

The y-intercept is 4.

47. a. The equation is $10x + 12y = 480$.
     b. Let $y = 0$:
         $10x + 12(0) = 480$
         $10x = 480$
         $x = 48$

         Let $x = 0$:
         $10(0) + 12y = 480$
         $12y = 480$
         $y = 40$

         The x-intercept is 48 and the y-intercept is 40.
     c. Graphing the equation in the first quadrant:

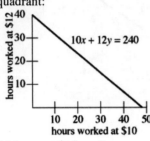

     d. When $x = 36$, $y = 10$. She worked 10 hours at $12 per hour.
     e. When $y = 25$, $x = 18$. She worked 18 hours at $10 per hour.

49. Solving the equation:
     $-12y - 4 = -148$
     $-12y = -144$
     $y = 12$

51. Solving the equation:
     $-5y - 4 = 51$
     $-5y = 55$
     $y = -11$

53. Solving the equation:
     $11x - 12 = -78$
     $11x = -66$
     $x = -6$

55. Solving the equation:
     $9x + 3 = 66$
     $9x = 63$
     $x = 7$

57. Solving the equation:
     $-9c - 6 = 12$
     $-9c = 18$
     $c = -2$

59. Solving the equation:
     $4 + 13c = -9$
     $13c = -13$
     $c = -1$

61. Solving the equation:
     $3y - 12 = 30$
     $3y = 42$
     $y = 14$

63. Solving the equation:
     $-11y + 9 = 75$
     $-11y = 66$
     $y = -6$

65. a. Evaluating: $\dfrac{5-2}{3-1} = \dfrac{3}{2}$
     b. Evaluating: $\dfrac{2-5}{1-3} = \dfrac{-3}{-2} = \dfrac{3}{2}$

67. a. Evaluating when $x = 3$ and $y = 5$: $\dfrac{y-2}{x-1} = \dfrac{5-2}{3-1} = \dfrac{3}{2}$
     b. Evaluating when $x = 3$ and $y = 5$: $\dfrac{2-y}{1-x} = \dfrac{2-5}{1-3} = \dfrac{-3}{-2} = \dfrac{3}{2}$

3.5 Slope and the Equation of a Line

1. The slope is given by: $m = \dfrac{4-1}{4-2} = \dfrac{3}{2}$

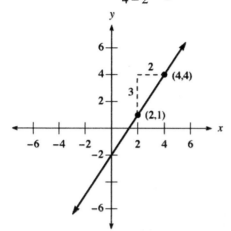

3. The slope is given by: $m = \dfrac{2-4}{5-1} = \dfrac{-2}{4} = -\dfrac{1}{2}$

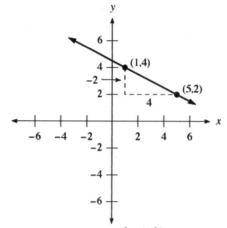

5. The slope is given by: $m = \dfrac{2-(-3)}{4-1} = \dfrac{5}{3}$

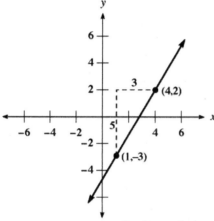

7. The slope is given by: $m = \dfrac{3-(-2)}{1-(-3)} = \dfrac{5}{4}$

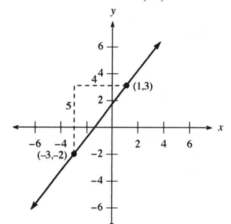

9. The slope is given by: $m = \dfrac{-2-2}{3-(-3)} = \dfrac{-4}{6} = -\dfrac{2}{3}$

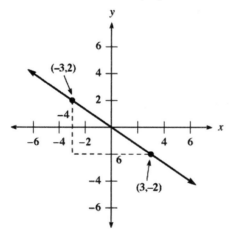

11. The slope is given by: $m = \dfrac{-2-(-5)}{3-2} = \dfrac{3}{1} = 3$

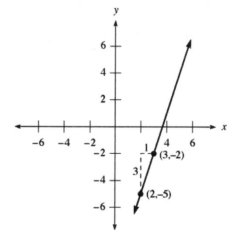

72 Chapter 3 Linear Equations and Inequalities in Two Variables

13. Graphing the line:
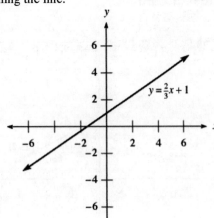

15. Graphing the line:
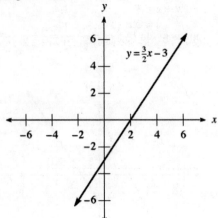

17. Graphing the line:
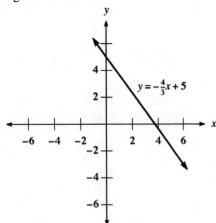

19. Graphing the line:
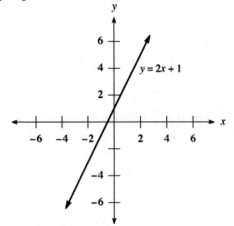

21. Graphing the line:
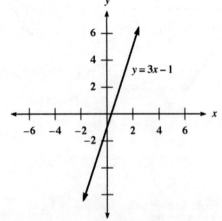

23. The y-intercept is 2, and the slope is given by: $m = \dfrac{5-(-1)}{1-(-1)} = \dfrac{6}{2} = 3$

25. The y-intercept is –2, and the slope is given by: $m = \dfrac{2-0}{2-1} = \dfrac{2}{1} = 2$

27. Solving for y:
$$-2x + y = 4$$
$$y = 2x + 4$$
The slope is 2 and the y-intercept is 4:

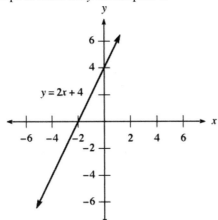

29. Solving for y:
$$3x + y = 3$$
$$y = -3x + 3$$
The slope is -3 and the y-intercept is 3:

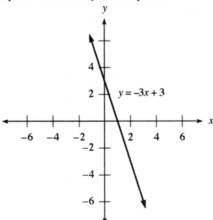

31. Solving for y:
$$3x + 2y = 6$$
$$2y = -3x + 6$$
$$y = -\tfrac{3}{2}x + 3$$
The slope is $-\tfrac{3}{2}$ and the y-intercept is 3:

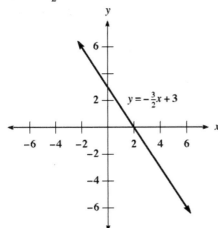

33. Solving for y:
$$4x - 5y = 20$$
$$-5y = -4x + 20$$
$$y = \tfrac{4}{5}x - 4$$
The slope is $\tfrac{4}{5}$ and the y-intercept is -4:

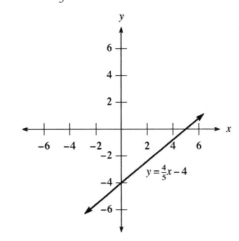

35. Solving for y:
$$-2x - 5y = 10$$
$$-5y = 2x + 10$$
$$y = -\tfrac{2}{5}x - 2$$

The slope is $-\tfrac{2}{5}$ and the y-intercept is –2:

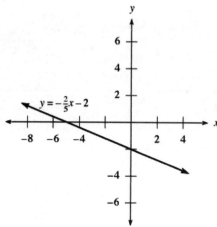

37. The slope is given by: $m = \dfrac{0-(-2)}{3-0} = \dfrac{2}{3}$

39. The slope is given by: $m = \dfrac{0-2}{4-0} = \dfrac{-2}{4} = -\dfrac{1}{2}$

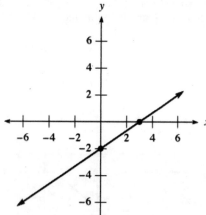

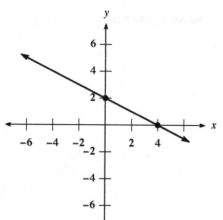

41. Graphing the line:

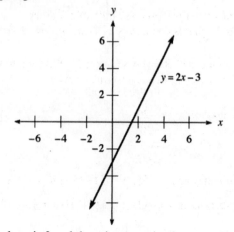

The slope is 2 and the y-intercept is –3.

43. Graphing the line:

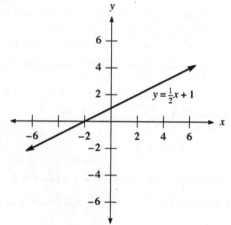

The slope is $\tfrac{1}{2}$ and the y-intercept is 1.

45. Using the slope formula:
$$\frac{y-2}{6-4} = 2$$
$$\frac{y-2}{2} = 2$$
$$y - 2 = 4$$
$$y = 6$$

47. The slopes are given in the table:

Equation	Slope
$y = 3$	0
$y = 3x$	3

49. The slopes are given in the table:

Equation	Slope
$y = -\frac{2}{3}$	0
$y = -\frac{2}{3}x$	$-\frac{2}{3}$

51.
  a. Solving the equation:
  $$-2x + 1 = 6$$
  $$-2x = 5$$
  $$x = -\frac{5}{2}$$

  b. Writing in slope-intercept form:
  $$-2x + y = 6$$
  $$y = 2x + 6$$

  c. Substituting $x = 0$:
  $$-2(0) + y = 6$$
  $$0 + y = 6$$
  $$y = 6$$

  d. Finding the slope:
  $$-2x + y = 6$$
  $$y = 2x + 6$$

  The slope is 2.

  e. Graphing the line:

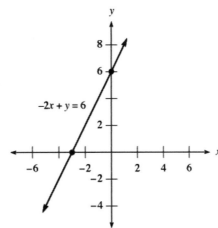

53.
  a. After 5 years, the copier is worth $6,000.
  b. The copier is worth $12,000 after 3 years.
  c. The slope of the line is –3000.
  d. The copier is decreasing in value by $3000 per year.
  e. The equation is $V = -3000t + 21,000$.

55.
  a. He will earn $3,000 for selling 1,000 shirts.
  b. He must sell 500 shirts to earn $2,000 for a month.
  c. The slope of the line is 2.
  d. Kevin earns $2 for each shirt he sells.
  e. The equation is $y = 2x + 1,000$.

57. Substituting $y = 13$:
$$3x - 5 = 13$$
$$3x = 18$$
$$x = 6$$

58. Substituting $y = -20$:
$$3x - 5 = -20$$
$$3x = -15$$
$$x = -5$$

61. Substituting $x = 0$: $y = \frac{3}{7}(0) + 4 = 0 + 4 = 4$

63. Substituting $x = -35$: $y = \frac{3}{7}(-35) + 4 = -15 + 4 = -11$

65. Substituting $x = 5$: $y = 3(5) - 5 = 15 - 5 = 10$

67. Substituting $x = -11$: $y = 3(-11) - 5 = -33 - 5 = -38$

76 Chapter 3 Linear Equations and Inequalities in Two Variables

69. Substituting $y = -9$:
$$\frac{x-6}{2} = -9$$
$$x - 6 = -18$$
$$x = -12$$

71. Substituting $y = -12$:
$$\frac{x-6}{2} = -12$$
$$x - 6 = -24$$
$$x = -18$$

73. Graphing the line:

75. Graphing the line:

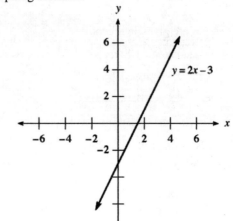

77. Graphing the line:

3.6 Solving Linear Systems by Graphing

1. Graphing both lines:

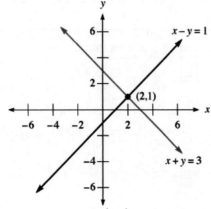

The intersection point is $(2, 1)$.

3. Graphing both lines:

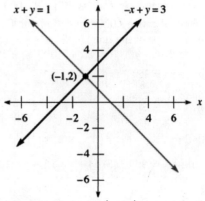

The intersection point is $(-1, 2)$.

5. Graphing both lines:

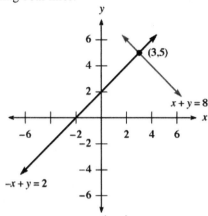

The intersection point is $(3, 5)$.

7. Graphing both lines:

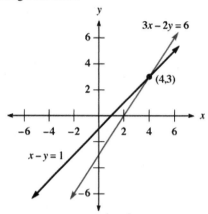

The intersection point is $(4, 3)$.

9. Graphing both lines:

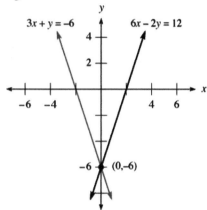

The intersection point is $(0, -6)$.

11. Graphing both lines:

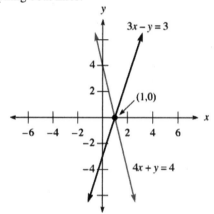

The intersection point is $(1, 0)$.

13. Graphing both lines:

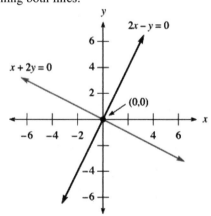

The intersection point is $(0, 0)$.

15. Graphing both lines:

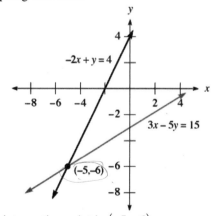

The intersection point is $(-5, -6)$.

78 Chapter 3 Linear Equations and Inequalities in Two Variables

17. Graphing both lines:

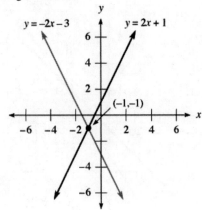

The intersection point is $(-1,-1)$.

19. Graphing both lines:

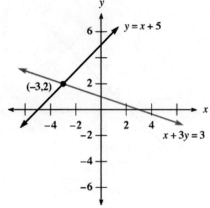

The intersection point is $(-3,2)$.

21. Graphing both lines:

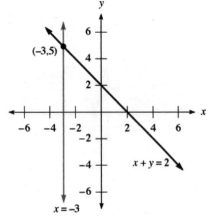

The intersection point is $(-3,5)$.

23. Graphing both lines:

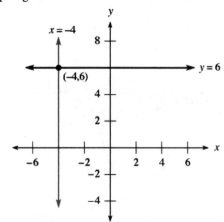

The intersection point is $(-4,6)$.

25. Graphing both lines:

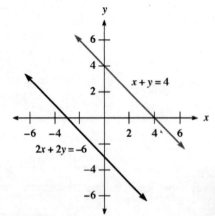

There is no intersection (the lines are parallel).

27. Graphing both lines:

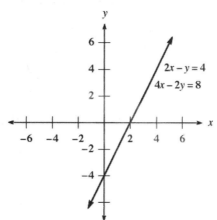

The system is dependent (both lines are the same, they coincide).

29. a. Simplifying: $(3x - 4y) + (x - y) = 4x - 5y$

b. Substituting $x = 4$:
$3(4) - 4y = 8$
$12 - 4y = 8$
$-4y = -4$
$y = 1$

c. Substituting $x = 0$:
$3(0) - 4y = 8$
$-4y = 8$
$y = -2$

d. Graphing the line:

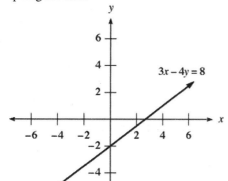

e. Graphing both lines:

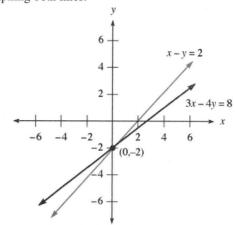

The intersection point is $(0,-2)$.

31. Graphing both lines:

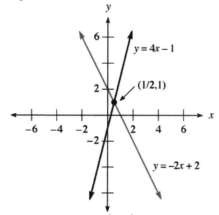

The intersection point is $\left(\frac{1}{2}, 1\right)$.

33. Graphing both lines:

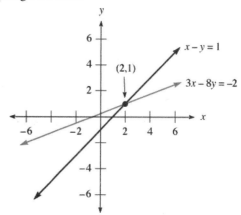

The intersection point is $(2,1)$.

80 Chapter 3 Linear Equations and Inequalities in Two Variables

35. a. If Jane worked 25 hours, she would earn the same amount at each position.
    b. If Jane worked less than 20 hours, she should choose Gigi's since she earns more in that position.
    c. If Jane worked more than 30 hours, she should choose Marcy's since she earns more in that position.
37. Simplifying: $6x + 100(0.04x + 0.75) = 6x + 4x + 75 = 10x + 75$
39. Simplifying: $13x - 1,000(0.002x + 0.035) = 13x - 2x - 35 = 11x - 35$
41. Simplifying: $16x - 10(1.7x - 5.8) = 16x - 17x + 58 = -x + 58$
43. Simplifying: $0.04x + 0.06(100 - x) = 0.04x + 6 - 0.06x = -0.02x + 6$
45. Simplifying: $0.025x - 0.028(1,000 + x) = 0.025x - 28 - 0.028x = -0.003x - 28$
47. Simplifying: $2.56x - 1.25(100 + x) = 2.56x - 125 - 1.25x = 1.31x - 125$
49. Simplifying: $(x + y) + (x - y) = x + y + x - y = 2x$
51. Simplifying: $(6x - 3y) + (x + 3y) = 6x - 3y + x + 3y = 7x$
53. Simplifying: $(-12x - 20y) + (25x + 20y) = -12x - 20y + 25x + 20y = 13x$
55. Simplifying: $-4(3x + 5y) = -12x - 20y$
57. Simplifying: $12\left(\frac{1}{4}x + \frac{2}{3}y\right) = 12 \cdot \frac{1}{4}x + 12 \cdot \frac{2}{3}y = 3x + 8y$
59. Simplifying: $-2(2x - y) = -4x + 2y$
61. Substituting $x = 3$:
    $3 + y = 4$
    $y = 1$
63. Substituting $x = 3$:
    $3 + 3y = 3$
    $3y = 0$
    $y = 0$
65. Substituting $x = 6$:
    $3(6) + 5y = -7$
    $18 + 5y = -7$
    $5y = -25$
    $y = -5$

3.7 The Elimination Method

1. Adding the two equations:
   $2x = 4$
   $x = 2$
   Substituting into the first equation:
   $2 + y = 3$
   $y = 1$
   The solution is $(2, 1)$.
3. Adding the two equations:
   $2y = 14$
   $y = 7$
   Substituting into the first equation:
   $x + 7 = 10$
   $x = 3$
   The solution is $(3, 7)$.
5. Adding the two equations:
   $-2y = 10$
   $y = -5$
   Substituting into the first equation:
   $x - (-5) = 7$
   $x + 5 = 7$
   $x = 2$
   The solution is $(2, -5)$.
7. Adding the two equations:
   $4x = -4$
   $x = -1$
   Substituting into the first equation:
   $-1 + y = -1$
   $y = 0$
   The solution is $(-1, 0)$.
9. Adding the two equations: $0 = 0$
   The lines coincide (the system is dependent).

11. Multiplying the first equation by 2:
$$6x - 2y = 8$$
$$2x + 2y = 24$$
Adding the two equations:
$$8x = 32$$
$$x = 4$$
Substituting into the first equation:
$$3(4) - y = 4$$
$$12 - y = 4$$
$$-y = -8$$
$$y = 8$$
The solution is $(4, 8)$.

13. Multiplying the second equation by -3:
$$5x - 3y = -2$$
$$-30x + 3y = -3$$
Adding the two equations:
$$-25x = -5$$
$$x = \tfrac{1}{5}$$
Substituting into the first equation:
$$5\left(\tfrac{1}{5}\right) - 3y = -2$$
$$1 - 3y = -2$$
$$-3y = -3$$
$$y = 1$$
The solution is $\left(\tfrac{1}{5}, 1\right)$.

15. Multiplying the second equation by 4:
$$11x - 4y = 11$$
$$20x + 4y = 20$$
Adding the two equations:
$$31x = 31$$
$$x = 1$$
Substituting into the second equation:
$$5(1) + y = 5$$
$$5 + y = 5$$
$$y = 0$$
The solution is $(1, 0)$.

17. Multiplying the second equation by 3:
$$3x - 5y = 7$$
$$-3x + 3y = -3$$
Adding the two equations:
$$-2y = 4$$
$$y = -2$$
Substituting into the second equation:
$$-x - 2 = -1$$
$$-x = 1$$
$$x = -1$$
The solution is $(-1, -2)$.

19. Multiplying the first equation by -2:
$$2x + 16y = 2$$
$$-2x + 4y = 13$$
Adding the two equations:
$$20y = 15$$
$$y = \tfrac{3}{4}$$
Substituting into the first equation:
$$-x - 8\left(\tfrac{3}{4}\right) = -1$$
$$-x - 6 = -1$$
$$-x = 5$$
$$x = -5$$
The solution is $\left(-5, \tfrac{3}{4}\right)$.

21. Multiplying the first equation by 2:
$$-6x - 2y = 14$$
$$6x + 7y = 11$$
Adding the two equations:
$$5y = 25$$
$$y = 5$$
Substituting into the first equation:
$$-3x - 5 = 7$$
$$-3x = 12$$
$$x = -4$$
The solution is $(-4, 5)$.

23. Adding the two equations:
$$8x = -24$$
$$x = -3$$
Substituting into the second equation:
$$2(-3) + y = -16$$
$$-6 + y = -16$$
$$y = -10$$
The solution is $(-3, -10)$.

25. Multiplying the second equation by 3:
$$x + 3y = 9$$
$$6x - 3y = 12$$
Adding the two equations:
$$7x = 21$$
$$x = 3$$
Substituting into the first equation:
$$3 + 3y = 9$$
$$3y = 6$$
$$y = 2$$
The solution is $(3, 2)$.

27. Multiplying the second equation by 2:
$$x - 6y = 3$$
$$8x + 6y = 42$$
Adding the two equations:
$$9x = 45$$
$$x = 5$$
Substituting into the second equation:
$$4(5) + 3y = 21$$
$$20 + 3y = 21$$
$$3y = 1$$
$$y = \tfrac{1}{3}$$
The solution is $\left(5, \tfrac{1}{3}\right)$.

29. Multiplying the second equation by -3:
$$2x + 9y = 2$$
$$-15x - 9y = 24$$
Adding the two equations:
$$-13x = 26$$
$$x = -2$$
Substituting into the first equation:
$$2(-2) + 9y = 2$$
$$-4 + 9y = 2$$
$$9y = 6$$
$$y = \tfrac{2}{3}$$
The solution is $\left(-2, \tfrac{2}{3}\right)$.

31. To clear each equation of fractions, multiply the first equation by 12 and the second equation by 6:
$$12\left(\tfrac{1}{3}x + \tfrac{1}{4}y\right) = 12\left(\tfrac{7}{6}\right) \qquad 6\left(\tfrac{3}{2}x - \tfrac{1}{3}y\right) = 6\left(\tfrac{7}{3}\right)$$
$$4x + 3y = 14 \qquad\qquad\qquad 9x - 2y = 14$$
The system of equations is:
$$4x + 3y = 14$$
$$9x - 2y = 14$$
Multiplying the first equation by 2 and the second equation by 3:
$$8x + 6y = 28$$
$$27x - 6y = 42$$
Adding the two equations:
$$35x = 70$$
$$x = 2$$
Substituting into $4x + 3y = 14$:
$$4(2) + 3y = 14$$
$$8 + 3y = 14$$
$$3y = 6$$
$$y = 2$$
The solution is $(2, 2)$.

33. Multiplying the first equation by -2:
$$-6x - 4y = 2$$
$$6x + 4y = 0$$
Adding the two equations:
$$0 = 2$$
Since this statement is false, the two lines are parallel, so the system has no solution.

35. Multiplying the first equation by 2 and the second equation by 3:
$$22x + 12y = 34$$
$$15x - 12y = 3$$
Adding the two equations:
$$37x = 37$$
$$x = 1$$
Substituting into the second equation:
$$5(1) - 4y = 1$$
$$5 - 4y = 1$$
$$-4y = -4$$
$$y = 1$$
The solution is $(1, 1)$.

37. To clear each equation of fractions, multiply the first equation by 6 and the second equation by 6:
$$6\left(\tfrac{1}{2}x + \tfrac{1}{6}y\right) = 6\left(\tfrac{1}{3}\right) \qquad 6\left(-x - \tfrac{1}{3}y\right) = 6\left(-\tfrac{1}{6}\right)$$
$$3x + y = 2 \qquad\qquad -6x - 2y = -1$$
The system of equations is:
$$3x + y = 2$$
$$-6x - 2y = -1$$
Multiplying the first equation by 2:
$$6x + 2y = 4$$
$$-6x - 2y = -1$$
Adding the two equations:
$$0 = 3$$
Since this statement is false, the two lines are parallel, so the system has no solution.

39. Multiplying the first equation by –5:
$$-5x - 5y = -110$$
$$5x + 10y = 170$$
Adding the two equations:
$$5y = 60$$
$$y = 12$$
Substituting into the first equation:
$$x + 12 = 22$$
$$x = 10$$
The solution is $(10, 12)$.

41. Multiplying the first equation by –5:
$$-5x - 5y = -70$$
$$5x + 25y = 230$$
Adding the two equations:
$$20y = 160$$
$$y = 8$$
Substituting into the first equation:
$$x + 8 = 14$$
$$x = 6$$
The solution is $(6, 8)$.

43. Multiplying the first equation by –6:
$$-6x - 6y = -90,000$$
$$6x + 7y = 98,000$$
Adding the two equations:
$$y = 8,000$$
Substituting into the first equation:
$$x + 8,000 = 15,000$$
$$x = 7,000$$
The solution is $(7000, 8000)$.

45. Multiplying the first equation by –4:
$$-4x - 4y = -44,000$$
$$4x + 7y = 68,000$$
Adding the two equations:
$$3y = 24,000$$
$$y = 8,000$$
Substituting into the first equation:
$$x + 8,000 = 11,000$$
$$x = 3,000$$
The solution is $(3000, 8000)$.

47. Multiplying the first equation by –5:
$$-5x - 5y = -115$$
$$5x + 10y = 175$$
Adding the two equations:
$$5y = 60$$
$$y = 12$$
Substituting into the first equation:
$$x + 12 = 23$$
$$x = 11$$
The solution is $(11, 12)$.

49. Multiplying the second equation by 100 (to eliminate decimals):
$$x + y = 22$$
$$5x + 10y = 170$$
Multiplying the first equation by –5:
$$-5x - 5y = -110$$
$$5x + 10y = 170$$
Adding the two equations:
$$5y = 60$$
$$y = 12$$
Substituting into the first equation:
$$x + 12 = 22$$
$$x = 10$$
The solution is $(10, 12)$.

51. Solving for y:
$$3x - y = 3$$
$$-y = -3x + 3$$
$$y = 3x - 3$$
The slope is 3 and the y-intercept is –3.

53. Solving for y:
$$2x - 5y = 25$$
$$-5y = -2x + 25$$
$$y = \tfrac{2}{5}x - 5$$
The slope is $\tfrac{2}{5}$ and the y-intercept is –5.

55. Finding the slope: $m = \dfrac{-5 - 3}{6 - (-2)} = \dfrac{-8}{8} = -1$

57. Finding the slope: $m = \dfrac{-3 - 3}{2 - 5} = \dfrac{-6}{-3} = 2$

59. Solving the equation:
$$x + (2x - 1) = 2$$
$$x + 2x - 1 = 2$$
$$3x - 1 = 2$$
$$3x = 3$$
$$x = 1$$

61. Solving the equation:
$$2(3y - 1) - 3y = 4$$
$$6y - 2 - 3y = 4$$
$$3y - 2 = 4$$
$$3y = 6$$
$$y = 2$$

63. Solving the equation:
$$-2x + 3(5x - 1) = 10$$
$$-2x + 15x - 3 = 10$$
$$13x - 3 = 10$$
$$13x = 13$$
$$x = 1$$

65. Solving for x:
$$x - 3y = -1$$
$$x = 3y - 1$$

67. Solving for y: $y = 2(1) - 1 = 2 - 1 = 1$

69. Solving for x: $x = 3(2) - 1 = 6 - 1 = 5$

71. Substituting $x = 13$: $y = 1.5(13) + 15 = 19.5 + 15 = 34.5$

73. Substituting $x = 12$: $y = 0.75(12) + 24.95 = 9 + 24.95 = 33.95$

3.8 The Substitution Method

1. Substituting into the first equation:
$$x + (2x - 1) = 11$$
$$3x - 1 = 11$$
$$3x = 12$$
$$x = 4$$
$$y = 2(4) - 1 = 7$$
The solution is $(4, 7)$.

3. Substituting into the first equation:
$$x + (5x + 2) = 20$$
$$6x + 2 = 20$$
$$6x = 18$$
$$x = 3$$
$$y = 5(3) + 2 = 17$$
The solution is $(3, 17)$.

5. Substituting into the first equation:
$$-2x + (-4x + 8) = -1$$
$$-6x + 8 = -1$$
$$-6x = -9$$
$$x = \tfrac{3}{2}$$
$$y = -4\left(\tfrac{3}{2}\right) + 8 = -6 + 8 = 2$$
The solution is $\left(\tfrac{3}{2}, 2\right)$.

7. Substituting into the first equation:
$$3(-y + 6) - 2y = -2$$
$$-3y + 18 - 2y = -2$$
$$-5y + 18 = -2$$
$$-5y = -20$$
$$y = 4$$
$$x = -4 + 6 = 2$$
The solution is $(2, 4)$.

9. Substituting into the first equation:
$$5x - 4(4) = -16$$
$$5x - 16 = -16$$
$$5x = 0$$
$$x = 0$$
The solution is $(0, 4)$.

11. Substituting into the first equation:
$$5x + 4(-3x) = 7$$
$$5x - 12x = 7$$
$$-7x = 7$$
$$x = -1$$
$$y = -3(-1) = 3$$
The solution is $(-1, 3)$.

13. Solving the second equation for x:
$$x - 2y = -1$$
$$x = 2y - 1$$
Substituting into the first equation:
$$(2y - 1) + 3y = 4$$
$$5y - 1 = 4$$
$$5y = 5$$
$$y = 1$$
$$x = 2(1) - 1 = 1$$
The solution is $(1, 1)$.

15. Solving the second equation for x:
$$x - 5y = 17$$
$$x = 5y + 17$$
Substituting into the first equation:
$$2(5y + 17) + y = 1$$
$$10y + 34 + y = 1$$
$$11y + 34 = 1$$
$$11y = -33$$
$$y = -3$$
$$x = 5(-3) + 17 = 2$$
The solution is $(2, -3)$.

17. Solving the second equation for x:
$$x - 5y = -5$$
$$x = 5y - 5$$
Substituting into the first equation:
$$3(5y - 5) + 5y = -3$$
$$15y - 15 + 5y = -3$$
$$20y - 15 = -3$$
$$20y = 12$$
$$y = \tfrac{3}{5}$$
$$x = 5\left(\tfrac{3}{5}\right) - 5 = 3 - 5 = -2$$
The solution is $\left(-2, \tfrac{3}{5}\right)$.

19. Solving the second equation for x:
$$x - 3y = -18$$
$$x = 3y - 18$$
Substituting into the first equation:
$$5(3y - 18) + 3y = 0$$
$$15y - 90 + 3y = 0$$
$$18y - 90 = 0$$
$$18y = 90$$
$$y = 5$$
$$x = 3(5) - 18 = -3$$
The solution is $(-3, 5)$.

21. Solving the second equation for *x*:
$$x + 3y = 12$$
$$x = -3y + 12$$
Substituting into the first equation:
$$-3(-3y + 12) - 9y = 7$$
$$9y - 36 - 9y = 7$$
$$-36 = 7$$
Since this statement is false, there is no solution to the system. The two lines are parallel.

23. Substituting into the first equation:
$$5x - 8(2x - 5) = 7$$
$$5x - 16x + 40 = 7$$
$$-11x + 40 = 7$$
$$-11x = -33$$
$$x = 3$$
$$y = 2(3) - 5 = 1$$
The solution is $(3, 1)$.

25. Substituting into the first equation:
$$7(2y - 1) - 6y = -1$$
$$14y - 7 - 6y = -1$$
$$8y - 7 = -1$$
$$8y = 6$$
$$y = \tfrac{3}{4}$$
$$x = 2\left(\tfrac{3}{4}\right) - 1 = \tfrac{3}{2} - 1 = \tfrac{1}{2}$$
The solution is $\left(\tfrac{1}{2}, \tfrac{3}{4}\right)$.

27. Substituting into the first equation:
$$-3x + 2(3x) = 6$$
$$-3x + 6x = 6$$
$$3x = 6$$
$$x = 2$$
$$y = 3(2) = 6$$
The solution is $(2, 6)$.

29. Substituting into the first equation:
$$5(y) - 6y = -4$$
$$-y = -4$$
$$y = 4$$
$$x = 4$$
The solution is $(4, 4)$.

31. Substituting into the first equation:
$$3x + 3(2x - 12) = 9$$
$$3x + 6x - 36 = 9$$
$$9x - 36 = 9$$
$$9x = 45$$
$$x = 5$$
$$y = 2(5) - 12 = -2$$
The solution is $(5, -2)$.

33. Substituting into the first equation:
$$7x - 11(10) = 16$$
$$7x - 110 = 16$$
$$7x = 126$$
$$x = 18$$
$$y = 10$$
The solution is $(18, 10)$.

35. Substituting into the first equation:
$$-4x + 4(x - 2) = -8$$
$$-4x + 4x - 8 = -8$$
$$-8 = -8$$
Since this statement is true, the system is dependent. The two lines coincide.

37. Substituting into the first equation:
$$2x + 5(12 - x) = 36$$
$$2x + 60 - 5x = 36$$
$$-3x + 60 = 36$$
$$-3x = -24$$
$$x = 8$$
$$y = 4$$
The solution is $(8, 4)$.

39. Substituting into the first equation:
$$5x + 2(18 - x) = 54$$
$$5x + 36 - 2x = 54$$
$$3x + 36 = 54$$
$$3x = 18$$
$$x = 6$$
$$y = 12$$
The solution is $(6, 12)$.

41. Substituting into the first equation:
$$2x + 2(2x) = 96$$
$$2x + 4x = 96$$
$$6x = 96$$
$$x = 16$$
$$y = 32$$

The solution is $(16, 32)$.

43. Substituting into the first equation:
$$0.05x + 0.10(22 - x) = 1.70$$
$$0.05x + 2.2 - 0.10x = 1.70$$
$$-0.05x + 2.2 = 1.7$$
$$-0.05x = -0.5$$
$$x = 10$$
$$y = 22 - 10 = 12$$

The solution is $(10, 12)$.

45. **a.** Solving the equation:
$$4x - 5 = 20$$
$$4x = 25$$
$$x = \tfrac{25}{4}$$

  b. Solving for y:
$$4x - 5y = 20$$
$$-5y = -4x + 20$$
$$y = \tfrac{4}{5}x - 4$$

  c. Solving for x:
$$x - y = 5$$
$$x = y + 5$$

  d. Substituting into the first equation:
$$4(y + 5) - 5y = 20$$
$$4y + 20 - 5y = 20$$
$$-y + 20 = 20$$
$$-y = 0$$
$$y = 0$$
$$x = 0 + 5 = 5$$

The solution is $(5, 0)$.

47. **a.** At 1,000 miles the car and truck cost the same to operate.
  b. If Daniel drives more than 1,200 miles, the car will be cheaper to operate.
  c. If Daniel drives less than 800 miles, the truck will be cheaper to operate.
  d. The graphs appear in the first quadrant only because all quantities are positive.

49. Simplifying: $6(3 + 4) + 5 = 6(7) + 5 = 42 + 5 = 47$

51. Simplifying: $1^2 + 2^2 + 3^2 = 1 + 4 + 9 = 14$

53. Simplifying: $5(6 + 3 \bullet 2) + 4 + 3 \bullet 2 = 5(6 + 6) + 4 + 3 \bullet 2 = 5(12) + 4 + 3 \bullet 2 = 60 + 4 + 6 = 70$

55. Simplifying: $(1^3 + 2^3) + [(2 \bullet 3) + (4 \bullet 5)] = (1 + 8) + [6 + 20] = 9 + 26 = 35$

57. Simplifying: $[2 \bullet 3 + 4 + 5] \div 3 = (6 + 4 + 5) \div 3 = 15 \div 3 = 5$

59. Simplifying: $6 \bullet 10^3 + 5 \bullet 10^2 + 4 \bullet 10^1 = 6,000 + 500 + 40 = 6,540$

61. Simplifying: $1 \bullet 10^3 + 7 \bullet 10^2 + 6 \bullet 10^1 + 0 = 1,000 + 700 + 60 + 0 = 1,760$

63. Simplifying: $4 \bullet 2 - 1 + 5 \bullet 3 - 2 = 8 - 1 + 15 - 2 = 20$

65. Simplifying: $(2^3 + 3^2) \bullet 4 - 5 = (8 + 9) \bullet 4 - 5 = 17 \bullet 4 - 5 = 68 - 5 = 63$

67. Simplifying: $2(2^2 + 3^2) + 3(3^2) = 2(4 + 9) + 3(9) = 2(13) + 3(9) = 26 + 27 = 53$

69. Let x and $5x + 8$ represent the two numbers. The equation is:
$$x + 5x + 8 = 26$$
$$6x + 8 = 26$$
$$6x = 18$$
$$x = 3$$
$$5x + 8 = 23$$

The two numbers are 3 and 23.

71. Let x represent the smaller number and $2x - 6$ represent the larger number. The equation is:
$$2x - 6 - x = 9$$
$$x - 6 = 9$$
$$x = 15$$
$$2x - 6 = 24$$
The two numbers are 15 and 24.

73. Let w represent the width and $3w + 5$ represent the length. Using the perimeter formula:
$$2(w) + 2(3w + 5) = 58$$
$$2w + 6w + 10 = 58$$
$$8w + 10 = 58$$
$$8w = 48$$
$$w = 6$$
$$3w + 5 = 23$$
The width is 6 inches and the length is 23 inches.

75. Completing the table:

	Nickels	Dimes
Number	$x+4$	x
Value (cents)	$5(x+4)$	$10x$

The equation is:
$$5(x+4) + 10x = 170$$
$$5x + 20 + 10x = 170$$
$$15x + 20 = 170$$
$$15x = 150$$
$$x = 10$$
$$x + 4 = 14$$
John has 14 nickels and 10 dimes.

77. Substituting into the second equation:
$$x + (5x + 2) = 20$$
$$6x + 2 = 20$$
$$6x = 18$$
$$x = 3$$
$$y = 5(3) + 2 = 17$$
The solution is $(3, 17)$.

79. Multiplying the second equation by 100 (to eliminate decimals):
$$x + y = 15000$$
$$6x + 7y = 98000$$
Multiplying the first equation by -6:
$$-6x - 6y = -90000$$
$$6x + 7y = 98000$$
Adding the two equations:
$$y = 8000$$
Substituting into the first equation:
$$x + 8000 = 15000$$
$$x = 7000$$
The solution is $(7000, 8000)$.

3.9 Applications

1. Let x and y represent the two numbers. The system of equations is:
$$x + y = 25$$
$$y = 5 + x$$
Substituting into the first equation:
$$x + (5 + x) = 25$$
$$2x + 5 = 25$$
$$2x = 20$$
$$x = 10$$
$$y = 5 + 10 = 15$$
The two numbers are 10 and 15.

3. Let x and y represent the two numbers. The system of equations is:
$$x + y = 15$$
$$y = 4x$$
Substituting into the first equation:
$$x + 4x = 15$$
$$5x = 15$$
$$x = 3$$
$$y = 4(3) = 12$$
The two numbers are 3 and 12.

5. Let x represent the larger number and y represent the smaller number. The system of equations is:
$$x - y = 5$$
$$x = 2y + 1$$
Substituting into the first equation:
$$2y + 1 - y = 5$$
$$y + 1 = 5$$
$$y = 4$$
$$x = 2(4) + 1 = 9$$
The two numbers are 4 and 9.

7. Let x and y represent the two numbers. The system of equations is:
$$y = 4x + 5$$
$$x + y = 35$$
Substituting into the second equation:
$$x + 4x + 5 = 35$$
$$5x + 5 = 35$$
$$5x = 30$$
$$x = 6$$
$$y = 4(6) + 5 = 29$$
The two numbers are 6 and 29.

90 Chapter 3 Linear Equations and Inequalities in Two Variables

9. Let x represent the amount invested at 6% and y represent the amount invested at 8%. The system of equations is:
$$x + y = 20000$$
$$0.06x + 0.08y = 1380$$
Multiplying the first equation by -0.06:
$$-0.06x - 0.06y = -1200$$
$$0.06x + 0.08y = 1380$$
Adding the two equations:
$$0.02y = 180$$
$$y = 9000$$
Substituting into the first equation:
$$x + 9000 = 20000$$
$$x = 11000$$
Mr. Wilson invested $9,000 at 8% and $11,000 at 6%.

11. Let x represent the amount invested at 5% and y represent the amount invested at 6%. The system of equations is:
$$x = 4y$$
$$0.05x + 0.06y = 520$$
Substituting into the second equation:
$$0.05(4y) + 0.06y = 520$$
$$0.20y + 0.06y = 520$$
$$0.26y = 520$$
$$y = 2000$$
$$x = 4(2000) = 8000$$
She invested $8,000 at 5% and $2,000 at 6%.

13. Let x represent the number of nickels and y represent the number of quarters. The system of equations is:
$$x + y = 14$$
$$0.05x + 0.25y = 2.30$$
Multiplying the first equation by -0.05:
$$-0.05x - 0.05y = -0.7$$
$$0.05x + 0.25y = 2.30$$
Adding the two equations:
$$0.20y = 1.6$$
$$y = 8$$
Substituting into the first equation:
$$x + 8 = 14$$
$$x = 6$$
Ron has 6 nickels and 8 quarters.

15. Let x represent the number of dimes and y represent the number of quarters. The system of equations is:
$$x + y = 21$$
$$0.10x + 0.25y = 3.45$$
Multiplying the first equation by -0.10:
$$-0.10x - 0.10y = -2.10$$
$$0.10x + 0.25y = 3.45$$
Adding the two equations:
$$0.15y = 1.35$$
$$y = 9$$
Substituting into the first equation:
$$x + 9 = 21$$
$$x = 12$$
Tom has 12 dimes and 9 quarters.

17. Let x represent the liters of 50% alcohol solution and y represent the liters of 20% alcohol solution. The system of equations is:
$$x + y = 18$$
$$0.50x + 0.20y = 0.30(18)$$
Multiplying the first equation by -0.20:
$$-0.20x - 0.20y = -3.6$$
$$0.50x + 0.20y = 5.4$$
Adding the two equations:
$$0.30x = 1.8$$
$$x = 6$$
Substituting into the first equation:
$$6 + y = 18$$
$$y = 12$$
The mixture contains 6 liters of 50% alcohol solution and 12 liters of 20% alcohol solution.

19. Let x represent the gallons of 10% disinfectant and y represent the gallons of 7% disinfectant. The system of equations is:
$$x + y = 30$$
$$0.10x + 0.07y = 0.08(30)$$
Multiplying the first equation by -0.07:
$$-0.07x - 0.07y = -2.1$$
$$0.10x + 0.07y = 2.4$$
Adding the two equations:
$$0.03x = 0.3$$
$$x = 10$$
Substituting into the first equation:
$$10 + y = 30$$
$$y = 20$$
The mixture contains 10 gallons of 10% disinfectant and 20 gallons of 7% disinfectant.

21. Let x represent the number of adult tickets and y represent the number of kids tickets. The system of equations is:
$$x + y = 70$$
$$5.50x + 4.00y = 310$$
Multiplying the first equation by -4:
$$-4.00x - 4.00y = -280$$
$$5.50x + 4.00y = 310$$
Adding the two equations:
$$1.5x = 30$$
$$x = 20$$
Substituting into the first equation:
$$20 + y = 70$$
$$y = 50$$
The matinee had 20 adult tickets sold and 50 kids tickets sold.

23. Let x represent the width and y represent the length. The system of equations is:
$$2x + 2y = 96$$
$$y = 2x$$
Substituting into the first equation:
$$2x + 2(2x) = 96$$
$$2x + 4x = 96$$
$$6x = 96$$
$$x = 16$$
$$y = 2(16) = 32$$
The width is 16 feet and the length is 32 feet.

92 Chapter 3 Linear Equations and Inequalities in Two Variables

25. Let x represent the number of $5 chips and y represent the number of $25 chips. The system of equations is:
$$x + y = 45$$
$$5x + 25y = 465$$
Multiplying the first equation by –5:
$$-5x - 5y = -225$$
$$5x + 25y = 465$$
Adding the two equations:
$$20y = 240$$
$$y = 12$$
Substituting into the first equation:
$$x + 12 = 45$$
$$x = 33$$
The gambler has 33 $5 chips and 12 $25 chips.

27. Let x represent the number of shares of $11 stock and y represent the number of shares of $20 stock. The system of equations is:
$$x + y = 150$$
$$11x + 20y = 2550$$
Multiplying the first equation by –11:
$$-11x - 11y = -1650$$
$$11x + 20y = 2550$$
Adding the two equations:
$$9y = 900$$
$$y = 100$$
Substituting into the first equation:
$$x + 100 = 150$$
$$x = 50$$
She bought 50 shares at $11 and 100 shares at $20.

29. Substituting $x = -2$, $x = 0$, and $x = 2$:
$$y = \tfrac{1}{2}(-2) + 3 = -1 + 3 = 2 \qquad y = \tfrac{1}{2}(0) + 3 = 0 + 3 = 3 \qquad y = \tfrac{1}{2}(2) + 3 = 1 + 3 = 4$$
The ordered pairs are $(-2, 2)$, $(0, 3)$, and $(2, 4)$.

31. Graphing the line:

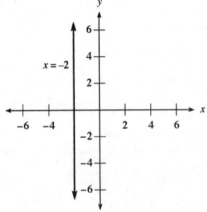

33. Computing the slope: $m = \dfrac{1-5}{0-2} = \dfrac{-4}{-2} = 2$

Chapter 3 Review/Test

1. Substituting $x = 4$, $x = 0$, $y = 3$, and $y = 0$:
$$3(4) + y = 6 \qquad 3(0) + y = 6 \qquad 3x + 3 = 6 \qquad 3x + 0 = 6$$
$$12 + y = 6 \qquad 0 + y = 6 \qquad 3x = 3 \qquad 3x = 6$$
$$y = -6 \qquad y = 6 \qquad x = 1 \qquad x = 2$$
The ordered pairs are $(4,-6), (0,6), (1,3)$, and $(2,0)$.

2. Substituting $x = 5$, $x = 0$, $y = 2$, and $y = 0$:
$$2(5) - 5y = 20 \qquad 2(0) - 5y = 20 \qquad 2x - 5(2) = 20 \qquad 2x - 5(0) = 20$$
$$10 - 5y = 20 \qquad 0 - 5y = 20 \qquad 2x - 10 = 20 \qquad 2x - 0 = 20$$
$$-5y = 10 \qquad -5y = 20 \qquad 2x = 30 \qquad 2x = 20$$
$$y = -2 \qquad y = -4 \qquad x = 15 \qquad x = 10$$
The ordered pairs are $(5,-2), (0,-4), (15,2)$, and $(10,0)$.

3. Substituting $x = 4$, $y = -2$, and $y = 3$:
$$y = 2(4) - 6 \qquad -2 = 2x - 6 \qquad 3 = 2x - 6$$
$$y = 8 - 6 \qquad 4 = 2x \qquad 9 = 2x$$
$$y = 2 \qquad x = 2 \qquad x = \tfrac{9}{2}$$
The ordered pairs are $(4,2), (2,-2)$, and $\left(\tfrac{9}{2}, 3\right)$.

4. Substituting $x = 2$, $y = 0$, and $y = -3$:
$$y = 5(2) + 3 \qquad 0 = 5x + 3 \qquad -3 = 5x + 3$$
$$y = 10 + 3 \qquad -3 = 5x \qquad -6 = 5x$$
$$y = 13 \qquad x = -\tfrac{3}{5} \qquad x = -\tfrac{6}{5}$$
The ordered pairs are $(2,13), \left(-\tfrac{3}{5}, 0\right)$, and $\left(-\tfrac{6}{5}, -3\right)$.

5. Substituting $x = 2$, $x = -1$, and $x = -3$ results (in each case) in $y = -3$. The ordered pairs are $(2,-3), (-1,-3)$, and $(-3,-3)$.

6. Substituting $y = 5$, $y = 0$, and $y = -1$ results (in each case) in $x = 6$. The ordered pairs are $(6,5), (6,0)$, and $(6,-1)$.

7. Substituting each ordered pair into the equation:
$$\left(-2, \tfrac{9}{2}\right): \quad 3(-2) - 4\left(\tfrac{9}{2}\right) = -6 - 18 = -24 \neq 12$$
$$(0, 3): \quad 3(0) - 4(3) = 0 - 12 = -12 \neq 12$$
$$\left(2, -\tfrac{3}{2}\right): \quad 3(2) - 4\left(-\tfrac{3}{2}\right) = 6 + 6 = 12$$
Only the ordered pair $\left(2, -\tfrac{3}{2}\right)$ is a solution.

8. Substituting each ordered pair into the equation:
$$\left(-\tfrac{8}{3}, -1\right): \quad 3\left(-\tfrac{8}{3}\right) + 7 = -8 + 7 = -1$$
$$\left(\tfrac{7}{3}, 0\right): \quad 3\left(\tfrac{7}{3}\right) + 7 = 7 + 7 = 14 \neq 0$$
$$(-3, -2): \quad 3(-3) + 7 = -9 + 7 = -2$$
The ordered pairs $\left(-\tfrac{8}{3}, -1\right)$ and $(-3, -2)$ are solutions.

9. Graphing the ordered pair:

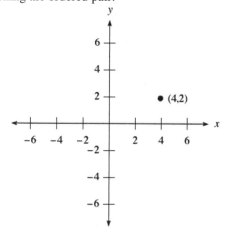

10. Graphing the ordered pair:

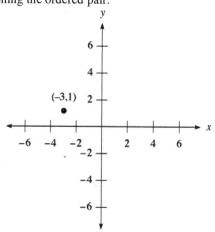

94   Chapter 3 Linear Equations and Inequalities in Two Variables

11.   Graphing the ordered pair:

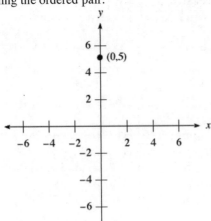

12.   Graphing the ordered pair:

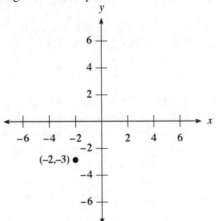

13.   Graphing the ordered pair:

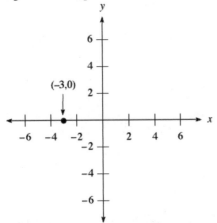

14.   Graphing the ordered pair:

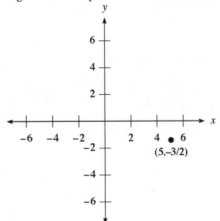

15.   The ordered pairs are $(-2,0), (0,-2)$, and $(1,-3)$:

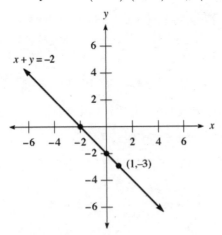

16.   The ordered pairs are $(-1,-3), (1,3)$, and $(0,0)$:

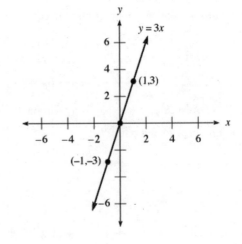

Chapter 3 Review/Test 95

17. The ordered pairs are $(1,1), (0,-1)$, and $(-1,-3)$:

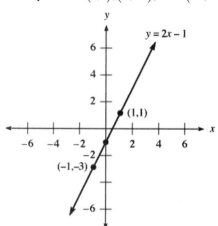

18. The ordered pairs are: $(-3,0), (-3,5)$, and $(-3,-5)$

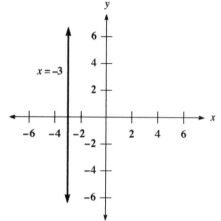

19. Graphing the equation:

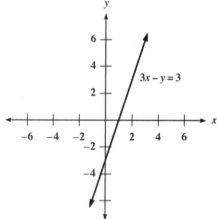

20. Graphing the equation:

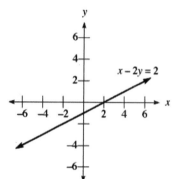

21. Graphing the equation:

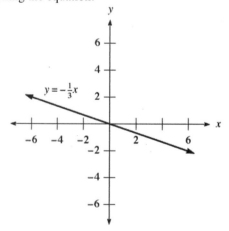

22. Graphing the equation:

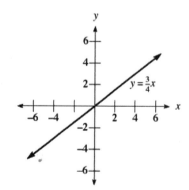

96   Chapter 3 Linear Equations and Inequalities in Two Variables

23. Graphing the equation:
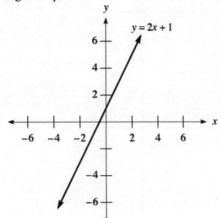

24. Graphing the equation:
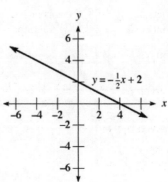

25. Graphing the equation:
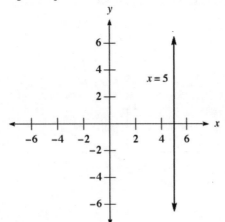

26. Graphing the equation:
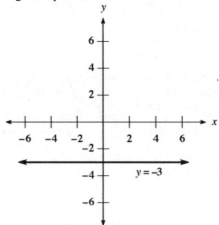

27. Graphing the equation:
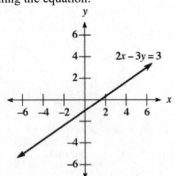

28. Graphing the equation:
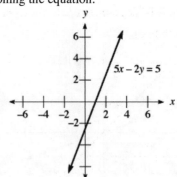

29. To find the x-intercept, let $y = 0$:
$3x - 0 = 6$
$3x = 6$
$x = 2$

To find the y-intercept, let $x = 0$:
$3(0) - y = 6$
$-y = 6$
$y = -6$

30. To find the x-intercept, let $y = 0$:
$2x - 6(0) = 24$
$2x = 24$
$x = 12$

To find the y-intercept, let $x = 0$:
$2(0) - 6y = 24$
$-6y = 24$
$y = -4$

31. To find the x-intercept, let $y = 0$:
$0 = x - 3$
$x = 3$

To find the y-intercept, let $x = 0$:
$y = 0 - 3$
$y = -3$

32. To find the *x*-intercept, let $y = 0$:
$$0 = 3x - 6$$
$$6 = 3x$$
$$x = 2$$

    To find the *y*-intercept, let $x = 0$:
$$y = 3(0) - 6$$
$$y = -6$$

33. Since the equation is $y = -5$, there is no *x*-intercept and the *y*-intercept is –5.
34. Since the equation is $x = 4$, the *x*-intercept is 4 and there is no *y*-intercept.
35. The slope is given by: $m = \dfrac{5-3}{3-2} = \dfrac{2}{1} = 2$
36. The slope is given by: $m = \dfrac{-5-3}{6-(-2)} = \dfrac{-8}{8} = -1$
37. The slope is given by: $m = \dfrac{-8-(-4)}{-3-(-1)} = \dfrac{-8+4}{-3+1} = \dfrac{-4}{-2} = 2$
38. The slope is given by: $m = \dfrac{2-4}{-\frac{1}{2}-\frac{1}{2}} = \dfrac{-2}{-1} = 2$
39. The slope-intercept form is $y = 3x + 2$.
40. The slope-intercept form is $y = -x + 6$.
41. The slope-intercept form is $y = -\frac{1}{3}x + \frac{3}{4}$.
42. The slope-intercept form is $y = 0$.
43. The equation is in slope-intercept form, so $m = 4$ and $b = -1$.
44. Solving for *y*:
$$2x + y = -5$$
$$y = -2x - 5$$
The equation is in slope-intercept form, so $m = -2$ and $b = -5$.
45. Solving for *y*:
$$6x + 3y = 9$$
$$3y = -6x + 9$$
$$y = -2x + 3$$
The equation is in slope-intercept form, so $m = -2$ and $b = 3$.
46. Solving for *y*:
$$5x + 2y = 8$$
$$2y = -5x + 8$$
$$y = -\tfrac{5}{2}x + 4$$
The equation is in slope-intercept form, so $m = -\frac{5}{2}$ and $b = 4$.
47. Graphing both lines:

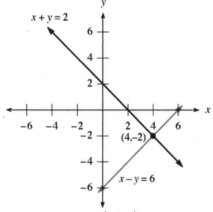

The intersection point is $(4, -2)$.

48. Graphing both lines:

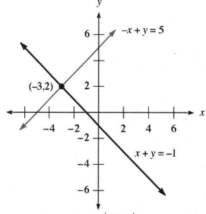

The intersection point is $(-3, 2)$.

98 Chapter 3 Linear Equations and Inequalities in Two Variables

49. Graphing both lines:

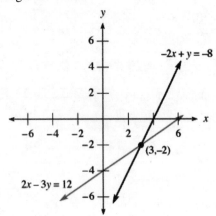

The intersection point is $(3,-2)$.

50. Graphing both lines:

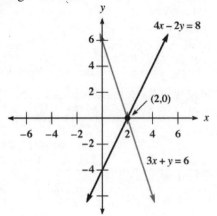

The intersection point is $(2,0)$.

51. Graphing both lines:

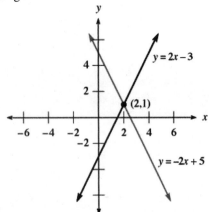

The intersection point is $(2,1)$.

52. Graphing both lines:

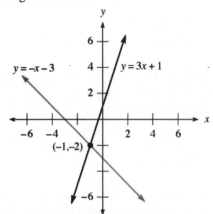

The intersection point is $(-1,-2)$.

53. Adding the two equations:
$$2x = 2$$
$$x = 1$$
Substituting into the second equation:
$$1 + y = -2$$
$$y = -3$$

The solution is $(1,-3)$.

54. Adding the two equations:
$$x = -2$$
Substituting into the second equation:
$$2(-2) + y = 1$$
$$-4 + y = 1$$
$$y = 5$$

The solution is $(-2,5)$.

55. Multiplying the first equation by 2:
$$10x - 6y = 4$$
$$-10x + 6y = -4$$
Adding the two equations:
$$0 = 0$$
Since this statement is true, the system is dependent. The two lines coincide.

56. Multiplying the first equation by 2 and the second equation by 3:
$$4x + 6y = -4$$
$$9x - 6y = 30$$
Adding the two equations:
$$13x = 26$$
$$x = 2$$
Substituting into the first equation:
$$2(2) + 3y = -2$$
$$4 + 3y = -2$$
$$3y = -6$$
$$y = -2$$
The solution is $(2, -2)$.

57. Multiplying the second equation by –4:
$$-3x + 4y = 1$$
$$16x - 4y = 12$$
Adding the two equations:
$$13x = 13$$
$$x = 1$$
Substituting into the second equation:
$$-4(1) + y = -3$$
$$-4 + y = -3$$
$$y = 1$$
The solution is $(1, 1)$.

58. Multiplying the second equation by 2:
$$-4x - 2y = 3$$
$$4x + 2y = 2$$
Adding the two equations: $0 = 5$
Since this statement is false, there is no solution to the system. The two lines are parallel.

59. Multiplying the first equation by 3 and the second equation by 5:
$$-6x + 15y = -33$$
$$35x - 15y = -25$$
Adding the two equations:
$$29x = -58$$
$$x = -2$$
Substituting into the first equation:
$$-2(-2) + 5y = -11$$
$$4 + 5y = -11$$
$$5y = -15$$
$$y = -3$$
The solution is $(-2, -3)$.

60. Multiplying the first equation by 3 and the second equation by 2:
$$-6x+15y=-45$$
$$6x-8y=38$$
Adding the two equations:
$$7y=-7$$
$$y=-1$$
Substituting into the second equation:
$$3x-4(-1)=19$$
$$3x+4=19$$
$$3x=15$$
$$x=5$$
The solution is $(5,-1)$.

61. Substituting into the first equation:
$$x+(-3x+1)=5$$
$$-2x+1=5$$
$$-2x=4$$
$$x=-2$$
Substituting into the second equation: $y=-3(-2)+1=6+1=7$. The solution is $(-2,7)$.

62. Substituting into the first equation:
$$x-(-2x-10)=-2$$
$$x+2x+10=-2$$
$$3x+10=-2$$
$$3x=-12$$
$$x=-4$$
Substituting into the second equation: $y=-2(-4)-10=8-10=-2$. The solution is $(-4,-2)$.

63. Substituting into the first equation:
$$4x-3(3x+7)=-16$$
$$4x-9x-21=-16$$
$$-5x-21=-16$$
$$-5x=5$$
$$x=-1$$
Substituting into the second equation: $y=3(-1)+7=-3+7=4$. The solution is $(-1,4)$.

64. Substituting into the first equation:
$$5x+2(-8x+10)=-2$$
$$5x-16x+20=-2$$
$$-11x+20=-2$$
$$-11x=-22$$
$$x=2$$
Substituting into the second equation: $y=-8(2)+10=-16+10=-6$. The solution is $(2,-6)$.

65. Solving the first equation for x:
$$x-4y=2$$
$$x=4y+2$$
Substituting into the second equation:
$$-3(4y+2)+12y=-8$$
$$-12y-6+12y=-8$$
$$-6=-8$$
Since this statement is false, there is no solution to the system. The two lines are parallel.

66. Solving the second equation for y:
$$3x + y = -19$$
$$y = -3x - 19$$
Substituting into the first equation:
$$4x - 2(-3x - 19) = 8$$
$$4x + 6x + 38 = 8$$
$$10x + 38 = 8$$
$$10x = -30$$
$$x = -3$$
Substituting into $y = -3x - 19$: $y = -3(-3) - 19 = 9 - 19 = -10$. The solution is $(-3, -10)$.

67. Solving the second equation for x:
$$x + 6y = -11$$
$$x = -6y - 11$$
Substituting into the first equation:
$$10(-6y - 11) - 5y = 20$$
$$-60y - 110 - 5y = 20$$
$$-65y - 110 = 20$$
$$-65y = 130$$
$$y = -2$$
Substituting into $x = -6y - 11$: $x = -6(-2) - 11 = 12 - 11 = 1$. The solution is $(1, -2)$.

68. Solving the second equation for y:
$$-6x + 2y = -4$$
$$2y = 6x - 4$$
$$y = 3x - 2$$
Substituting into the first equation:
$$3x - (3x - 2) = 2$$
$$3x - 3x + 2 = 2$$
$$2 = 2$$
Since this statement is true, the system is dependent. The two lines coincide.

69. Let x represent the smaller number and y represent the larger number. The system of equations is:
$$x + y = 18$$
$$2x = 6 + y$$
Solving the first equation for y:
$$x + y = 18$$
$$y = -x + 18$$
Substituting into the second equation:
$$2x = 6 + (-x + 18)$$
$$2x = -x + 24$$
$$3x = 24$$
$$x = 8$$
Substituting into the first equation:
$$8 + y = 18$$
$$y = 10$$
The two numbers are 8 and 10.

70. Let x and y represent the two numbers. The system of equations is:
$$x - y = 16$$
$$x = 3y$$
Substituting into the first equation:
$$3y - y = 16$$
$$2y = 16$$
$$y = 8$$
$$x = 3(8) = 24$$
The two numbers are 24 and 8.

71. Let x represent the amount invested at 4% and y represent the amount invested at 5%. The system of equations is:
$$x + y = 12000$$
$$0.04x + 0.05y = 560$$
Multiplying the first equation by –0.04:
$$-0.04x - 0.04y = -480$$
$$0.04x + 0.05y = 560$$
Adding the two equations:
$$0.01y = 80$$
$$y = 8000$$
Substituting into the first equation:
$$x + 8000 = 12000$$
$$x = 4000$$
So $4,000 was invested at 4% and $8,000 was invested at 5%.

72. Let x represent the amount invested at 6% and y represent the amount invested at 8%. The system of equations is:
$$x + y = 14000$$
$$0.06x + 0.08y = 1060$$
Multiplying the first equation by –0.06:
$$-0.06x - 0.06y = -840$$
$$0.06x + 0.08y = 1060$$
Adding the two equations:
$$0.02y = 220$$
$$y = 11000$$
Substituting into the first equation:
$$x + 11000 = 14000$$
$$x = 3000$$
So $3,000 was invested at 6% and $11,000 was invested at 8%.

73. Let x represent the number of dimes and y represent the number of nickels. The system of equations is:
$$x + y = 17$$
$$0.10x + 0.05y = 1.35$$
Multiplying the first equation by –0.05:
$$-0.05x - 0.05y = -0.85$$
$$0.10x + 0.05y = 1.35$$
Adding the two equations:
$$0.05x = 0.50$$
$$x = 10$$
Substituting into the first equation:
$$10 + y = 17$$
$$y = 7$$
Barbara has 10 dimes and 7 nickels.

74. Let x represent the number of dimes and y represent the number of quarters. The system of equations is:
$$x + y = 15$$
$$0.10x + 0.25y = 2.40$$
Multiplying the first equation by -0.10:
$$-0.10x - 0.10y = -1.50$$
$$0.10x + 0.25y = 2.40$$
Adding the two equations:
$$0.15y = 0.9$$
$$y = 6$$
Substituting into the first equation:
$$x + 6 = 15$$
$$x = 9$$
Tom has 9 dimes and 6 quarters.

75. Let x represent the liters of 20% alcohol solution and y represent the liters of 10% alcohol solution. The system of equations is:
$$x + y = 50$$
$$0.20x + 0.10y = 0.12(50)$$
Multiplying the first equation by -0.10:
$$-0.10x - 0.10y = -5$$
$$0.20x + 0.10y = 6$$
Adding the two equations:
$$0.10x = 1$$
$$x = 10$$
Substituting into the first equation:
$$10 + y = 50$$
$$y = 40$$
The solution contains 40 liters of 10% alcohol solution and 10 liters of 20% alcohol solution.

76. Let x represent the liters of 25% alcohol solution and y represent the liters of 15% alcohol solution. The system of equations is:
$$x + y = 40$$
$$0.25x + 0.15y = 0.20(40)$$
Multiplying the first equation by -0.15:
$$-0.15x - 0.15y = -6$$
$$0.25x + 0.15y = 8$$
Adding the two equations:
$$0.10x = 2$$
$$x = 20$$
Substituting into the first equation:
$$20 + y = 40$$
$$y = 20$$
The solution contains 20 liters of 25% alcohol solution and 20 liters of 15% alcohol solution.

Chapter 4
Exponents and Polynomials

4.1 Multiplication with Exponents

1. The base is 4 and the exponent is 2. Evaluating the expression: $4^2 = 4 \cdot 4 = 16$
3. The base is 0.3 and the exponent is 2. Evaluating the expression: $0.3^2 = 0.3 \cdot 0.3 = 0.09$
5. The base is 4 and the exponent is 3. Evaluating the expression: $4^3 = 4 \cdot 4 \cdot 4 = 64$
7. The base is -5 and the exponent is 2. Evaluating the expression: $(-5)^2 = (-5) \cdot (-5) = 25$
9. The base is 2 and the exponent is 3. Evaluating the expression: $-2^3 = -2 \cdot 2 \cdot 2 = -8$
11. The base is 3 and the exponent is 4. Evaluating the expression: $3^4 = 3 \cdot 3 \cdot 3 \cdot 3 = 81$
13. The base is $\frac{2}{3}$ and the exponent is 2. Evaluating the expression: $\left(\frac{2}{3}\right)^2 = \left(\frac{2}{3}\right) \cdot \left(\frac{2}{3}\right) = \frac{4}{9}$
15. The base is $\frac{1}{2}$ and the exponent is 4. Evaluating the expression: $\left(\frac{1}{2}\right)^4 = \left(\frac{1}{2}\right) \cdot \left(\frac{1}{2}\right) \cdot \left(\frac{1}{2}\right) \cdot \left(\frac{1}{2}\right) = \frac{1}{16}$

17. a. Completing the table:

Number (x)	1	2	3	4	5	6	7
Square (x^2)	1	4	9	16	25	36	49

   b. For numbers larger than 1, the square of the number is larger than the number.

19. Simplifying the expression: $x^4 \cdot x^5 = x^{4+5} = x^9$
21. Simplifying the expression: $y^{10} \cdot y^{20} = y^{10+20} = y^{30}$
23. Simplifying the expression: $2^5 \cdot 2^4 \cdot 2^3 = 2^{5+4+3} = 2^{12}$
25. Simplifying the expression: $x^4 \cdot x^6 \cdot x^8 \cdot x^{10} = x^{4+6+8+10} = x^{28}$
27. Simplifying the expression: $\left(x^2\right)^5 = x^{2 \cdot 5} = x^{10}$
29. Simplifying the expression: $\left(5^4\right)^3 = 5^{4 \cdot 3} = 5^{12}$
31. Simplifying the expression: $\left(y^3\right)^3 = y^{3 \cdot 3} = y^9$
33. Simplifying the expression: $\left(2^5\right)^{10} = 2^{5 \cdot 10} = 2^{50}$
35. Simplifying the expression: $\left(a^3\right)^x = a^{3x}$
37. Simplifying the expression: $\left(b^x\right)^y = b^{xy}$
39. Simplifying the expression: $(4x)^2 = 4^2 \cdot x^2 = 16x^2$
41. Simplifying the expression: $(2y)^5 = 2^5 \cdot y^5 = 32y^5$
43. Simplifying the expression: $(-3x)^4 = (-3)^4 \cdot x^4 = 81x^4$
45. Simplifying the expression: $(0.5ab)^2 = (0.5)^2 \cdot a^2 b^2 = 0.25 a^2 b^2$
47. Simplifying the expression: $(4xyz)^3 = 4^3 \cdot x^3 y^3 z^3 = 64 x^3 y^3 z^3$

49. Simplifying using properties of exponents: $(2x^4)^3 = 2^3 (x^4)^3 = 8x^{12}$

51. Simplifying using properties of exponents: $(4a^3)^2 = 4^2 (a^3)^2 = 16a^6$

53. Simplifying using properties of exponents: $(x^2)^3 (x^4)^2 = x^6 \cdot x^8 = x^{14}$

55. Simplifying using properties of exponents: $(a^3)^1 (a^2)^4 = a^3 \cdot a^8 = a^{11}$

57. Simplifying using properties of exponents: $(2x)^3 (2x)^4 = (2x)^7 = 2^7 x^7 = 128 x^7$

59. Simplifying using properties of exponents: $(3x^2)^3 (2x)^4 = 3^3 x^6 \cdot 2^4 x^4 = 27 x^6 \cdot 16 x^4 = 432 x^{10}$

61. Simplifying using properties of exponents: $(4x^2 y^3)^2 = 4^2 x^4 y^6 = 16 x^4 y^6$

63. Simplifying using properties of exponents: $\left(\frac{2}{3} a^4 b^5\right)^3 = \left(\frac{2}{3}\right)^3 a^{12} b^{15} = \frac{8}{27} a^{12} b^{15}$

65. Writing as a perfect square: $x^4 = (x^2)^2$

67. Writing as a perfect square: $16 x^2 = (4x)^2$

69. Writing as a perfect cube: $8 = (2)^3$

71. Writing as a perfect cube: $64 x^3 = (4x)^3$

73.
   a. Substituting $x = 2$: $x^3 x^2 = (2)^3 (2)^2 = (8)(4) = 32$
   b. Substituting $x = 2$: $(x^3)^2 = (2^3)^2 = (8)^2 = 64$
   c. Substituting $x = 2$: $x^5 = (2)^5 = 32$
   d. Substituting $x = 2$: $x^6 = (2)^6 = 64$

75. a. Completing the table:

Number (x)	Square (x^2)
−3	9
−2	4
−1	1
0	0
1	1
2	4
3	9

77. Completing the table:

Number (x)	Square (x^2)
−2.5	6.25
−1.5	2.25
−0.5	0.25
0	0
0.5	0.25
1.5	2.25
2.5	6.25

b. Constructing a line graph:

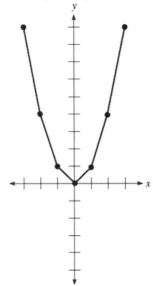

79. Writing in scientific notation: $43,200 = 4.32 \times 10^4$
81. Writing in scientific notation: $570 = 5.7 \times 10^2$
83. Writing in scientific notation: $238,000 = 2.38 \times 10^5$
85. Writing in expanded form: $2.49 \times 10^3 = 2,490$
87. Writing in expanded form: $3.52 \times 10^2 = 352$
89. Writing in expanded form: $2.8 \times 10^4 = 28,000$
91. The volume is given by: $V = (3 \text{ in.})^3 = 27 \text{ inches}^3$
93. The volume is given by: $V = (2.5 \text{ in.})^3 \approx 15.6 \text{ inches}^3$
95. The volume is given by: $V = (8 \text{ in.})(4.5 \text{ in.})(1 \text{ in.}) = 36 \text{ inches}^3$
97. Yes. If the box was 7 ft x 3 ft x 2 ft, you could fit inside of it.
99. Writing in scientific notation: $650,000,000 \text{ seconds} = 6.5 \times 10^8 \text{ seconds}$
101. Writing in expanded form: $7.4 \times 10^5 \text{ dollars} = \$740,000$
103. Writing in expanded form: $1.8 \times 10^5 \text{ dollars} = \$180,000$
105. Substitute $c = 8$, $b = 3.35$, and $s = 3.11$: $d = \pi \cdot 3.11 \cdot 8 \cdot \left(\frac{1}{2} \cdot 3.35\right)^2 \approx 219 \text{ inches}^3$
107. Substitute $c = 6$, $b = 3.59$, and $s = 2.99$: $d = \pi \cdot 2.99 \cdot 6 \cdot \left(\frac{1}{2} \cdot 3.59\right)^2 \approx 182 \text{ inches}^3$
109. Factoring into primes: $128 = 4 \cdot 32 = (2 \cdot 2) \cdot (4 \cdot 8) = (2 \cdot 2) \cdot (2 \cdot 2 \cdot 2 \cdot 2 \cdot 2) = 2^7$
111. Factoring into primes: $250 = 10 \cdot 25 = (2 \cdot 5) \cdot (5 \cdot 5) = 2 \cdot 5^3$
113. Factoring into primes: $720 = 10 \cdot 72 = (2 \cdot 5) \cdot (8 \cdot 9) = (2 \cdot 5) \cdot (2 \cdot 2 \cdot 2 \cdot 3 \cdot 3) = 2^4 \cdot 3^2 \cdot 5$
115. Factoring into primes: $820 = 10 \cdot 82 = (2 \cdot 5) \cdot (2 \cdot 41) = 2^2 \cdot 5 \cdot 41$
117. Factoring into primes: $6^3 = (2 \cdot 3)^3 = 2^3 \cdot 3^3$
119. Factoring into primes: $30^3 = (2 \cdot 3 \cdot 5)^3 = 2^3 \cdot 3^3 \cdot 5^3$
121. Factoring into primes: $25^3 = (5 \cdot 5)^3 = 5^3 \cdot 5^3 = 5^6$
123. Factoring into primes: $12^3 = (2 \cdot 2 \cdot 3)^3 = 2^3 \cdot 2^3 \cdot 3^3 = 2^6 \cdot 3^3$
125. Subtracting: $4 - 7 = 4 + (-7) = -3$
127. Subtracting: $4 - (-7) = 4 + 7 = 11$
129. Subtracting: $15 - 20 = 15 + (-20) = -5$
131. Subtracting: $-15 - (-20) = -15 + 20 = 5$
133. Simplifying: $2(3) - 4 = 6 - 4 = 2$
135. Simplifying: $4(3) - 3(2) = 12 - 6 = 6$
137. Simplifying: $2(5 - 3) = 2(2) = 4$
139. Simplifying: $5 + 4(-2) - 2(-3) = 5 - 8 + 6 = 3$

4.2 Division with Exponents

1. Writing with positive exponents: $3^{-2} = \dfrac{1}{3^2} = \dfrac{1}{9}$

3. Writing with positive exponents: $6^{-2} = \dfrac{1}{6^2} = \dfrac{1}{36}$

5. Writing with positive exponents: $8^{-2} = \dfrac{1}{8^2} = \dfrac{1}{64}$

7. Writing with positive exponents: $5^{-3} = \dfrac{1}{5^3} = \dfrac{1}{125}$

9. Writing with positive exponents: $2x^{-3} = 2 \cdot \dfrac{1}{x^3} = \dfrac{2}{x^3}$

11. Writing with positive exponents: $(2x)^{-3} = \dfrac{1}{(2x)^3} = \dfrac{1}{8x^3}$

13. Writing with positive exponents: $(5y)^{-2} = \dfrac{1}{(5y)^2} = \dfrac{1}{25y^2}$

15. Writing with positive exponents: $10^{-2} = \dfrac{1}{10^2} = \dfrac{1}{100}$

17. Completing the table:

Number (x)	Square (x^2)	Power of 2 (2^x)
-3	9	$\dfrac{1}{8}$
-2	4	$\dfrac{1}{4}$
-1	1	$\dfrac{1}{2}$
0	0	1
1	1	2
2	4	4
3	9	8

19. Simplifying: $\dfrac{5^1}{5^3} = 5^{1-3} = 5^{-2} = \dfrac{1}{5^2} = \dfrac{1}{25}$

21. Simplifying: $\dfrac{x^{10}}{x^4} = x^{10-4} = x^6$

23. Simplifying: $\dfrac{4^3}{4^0} = 4^{3-0} = 4^3 = 64$

25. Simplifying: $\dfrac{(2x)^7}{(2x)^4} = (2x)^{7-4} = (2x)^3 = 2^3 x^3 = 8x^3$

27. Simplifying: $\dfrac{6^{11}}{6} = \dfrac{6^{11}}{6^1} = 6^{11-1} = 6^{10} \ (= 60,466,176)$

29. Simplifying: $\dfrac{6}{6^{11}} = \dfrac{6^1}{6^{11}} = 6^{1-11} = 6^{-10} = \dfrac{1}{6^{10}} \ \left(= \dfrac{1}{60,466,176}\right)$

31. Simplifying: $\dfrac{2^{-5}}{2^3} = 2^{-5-3} = 2^{-8} = \dfrac{1}{2^8} = \dfrac{1}{256}$

33. Simplifying: $\dfrac{2^5}{2^{-3}} = 2^{5-(-3)} = 2^{5+3} = 2^8 = 256$

35. Simplifying: $\dfrac{(3x)^{-5}}{(3x)^{-8}} = (3x)^{-5-(-8)} = (3x)^{-5+8} = (3x)^3 = 3^3 x^3 = 27x^3$

37. Simplifying: $(3xy)^4 = 3^4 x^4 y^4 = 81 x^4 y^4$

39. Simplifying: $10^0 = 1$

41. Simplifying: $\left(2a^2 b\right)^1 = 2a^2 b$

43. Simplifying: $\left(7y^3\right)^{-2} = \dfrac{1}{\left(7y^3\right)^2} = \dfrac{1}{49 y^6}$

45. Simplifying: $x^{-3} \cdot x^{-5} = x^{-3-5} = x^{-8} = \dfrac{1}{x^8}$

47. Simplifying: $y^7 \cdot y^{-10} = y^{7-10} = y^{-3} = \dfrac{1}{y^3}$

Problem Set 4.2 109

49. Simplifying: $\dfrac{(x^2)^3}{x^4} = \dfrac{x^6}{x^4} = x^{6-4} = x^2$

51. Simplifying: $\dfrac{(a^4)^3}{(a^3)^2} = \dfrac{a^{12}}{a^6} = a^{12-6} = a^6$

53. Simplifying: $\dfrac{y^7}{(y^2)^8} = \dfrac{y^7}{y^{16}} = y^{7-16} = y^{-9} = \dfrac{1}{y^9}$

55. Simplifying: $\left(\dfrac{y^7}{y^2}\right)^8 = (y^{7-2})^8 = (y^5)^8 = y^{40}$

57. Simplifying: $\dfrac{(x^{-2})^3}{x^{-5}} = \dfrac{x^{-6}}{x^{-5}} = x^{-6-(-5)} = x^{-6+5} = x^{-1} = \dfrac{1}{x}$

59. Simplifying: $\left(\dfrac{x^{-2}}{x^{-5}}\right)^3 = (x^{-2+5})^3 = (x^3)^3 = x^9$

61. Simplifying: $\dfrac{(a^3)^2 (a^4)^5}{(a^5)^2} = \dfrac{a^6 \cdot a^{20}}{a^{10}} = \dfrac{a^{26}}{a^{10}} = a^{26-10} = a^{16}$

63. Simplifying: $\dfrac{(a^{-2})^3 (a^4)^2}{(a^{-3})^{-2}} = \dfrac{a^{-6} \cdot a^8}{a^6} = \dfrac{a^2}{a^6} = a^{2-6} = a^{-4} = \dfrac{1}{a^4}$

65. **a.** Substituting $x = 2$: $\dfrac{x^7}{x^2} = \dfrac{2^7}{2^2} = \dfrac{128}{4} = 32$ **b.** Substituting $x = 2$: $x^5 = 2^5 = 32$

     c. Substituting $x = 2$: $\dfrac{x^2}{x^7} = \dfrac{2^2}{2^7} = \dfrac{4}{128} = \dfrac{1}{32}$ **d.** Substituting $x = 2$: $x^{-5} = 2^{-5} = \dfrac{1}{2^5} = \dfrac{1}{32}$

67. **a.** Writing as a perfect square: $\dfrac{1}{25} = \left(\dfrac{1}{5}\right)^2$ **b.** Writing as a perfect square: $\dfrac{1}{64} = \left(\dfrac{1}{8}\right)^2$

     c. Writing as a perfect square: $\dfrac{1}{x^2} = \left(\dfrac{1}{x}\right)^2$ **d.** Writing as a perfect square: $\dfrac{1}{x^4} = \left(\dfrac{1}{x^2}\right)^2$

69. Completing the table:

Number (x)	Power of 2 (2^x)
−3	$\frac{1}{8}$
−2	$\frac{1}{4}$
−1	$\frac{1}{2}$
0	1
1	2
2	4
3	8

Constructing the line graph:

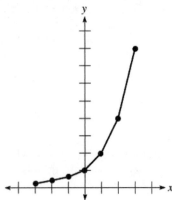

71. Writing in scientific notation: $0.0048 = 4.8 \times 10^{-3}$

73. Writing in scientific notation: $25 = 2.5 \times 10^1$

75. Writing in scientific notation: $0.000009 = 9 \times 10^{-6}$

77. Completing the table:

Expanded Form	Scientific Notation $(n \times 10^r)$
0.000357	3.57×10^{-4}
0.00357	3.57×10^{-3}
0.0357	3.57×10^{-2}
0.357	3.57×10^{-1}
3.57	3.57×10^0
35.7	3.57×10^1
357	3.57×10^2
3,570	3.57×10^3
35,700	3.57×10^4

79. Writing in expanded form: $4.23 \times 10^{-3} = 0.00423$

81. Writing in expanded form: $8 \times 10^{-5} = 0.00008$

83. Writing in expanded form: $4.2 \times 10^0 = 4.2$

85. Writing in expanded form: 2×10^{-3} seconds $= 0.002$ seconds

87. Writing each number in scientific notation:
$0.0024 = 2.4 \times 10^{-3}$, $0.0051 = 5.1 \times 10^{-3}$, $0.0096 = 9.6 \times 10^{-3}$, $0.0099 = 9.9 \times 10^{-3}$, $0.0111 = 1.11 \times 10^{-2}$

89. Writing in scientific notation: $25 \times 10^3 = 2.5 \times 10^4$

91. Writing in scientific notation: $23.5 \times 10^4 = 2.35 \times 10^5$

93. Writing in scientific notation: $0.82 \times 10^{-3} = 8.2 \times 10^{-4}$

95. The area of the smaller square is $(10 \text{ in.})^2 = 100 \text{ inches}^2$, while the area of the larger square is $(20 \text{ in.})^2 = 400 \text{ inches}^2$. It would take 4 smaller squares to cover the larger square.

97. The area of the smaller square is x^2, while the area of the larger square is $(2x)^2 = 4x^2$. It would take 4 smaller squares to cover the larger square.

99. The volume of the smaller box is $(6 \text{ in.})^3 = 216 \text{ inches}^3$, while the volume of the larger box is $(12 \text{ in.})^3 = 1,728 \text{ inches}^3$. Thus 8 smaller boxes will fit inside the larger box ($8 \cdot 216 = 1,728$).

101. The volume of the smaller box is x^3, while the volume of the larger box is $(2x)^3 = 8x^3$. Thus 8 smaller boxes will fit inside the larger box.

103. Simplifying: $4x + 3x = (4+3)x = 7x$

105. Simplifying: $5a - 3a = (5-3)a = 2a$

107. Simplifying: $4y + 5y + y = (4+5+1)y = 10y$

109. Simplifying: $3(4.5) = 13.5$

111. Simplifying: $\frac{4}{5}(10) = \frac{40}{5} = 8$

113. Simplifying: $6.8(3.9) = 26.52$

115. Simplifying: $-3 + 15 = 12$

117. Simplifying: $x^5 \cdot x^3 = x^{5+3} = x^8$

119. Simplifying: $\frac{x^3}{x^2} = x^{3-2} = x$

121. Simplifying: $\frac{y^3}{y^5} = y^{3-5} = y^{-2} = \frac{1}{y^2}$

123. Writing in expanded form: $3.4 \times 10^2 = 340$

4.3 Operations with Monomials

1. Multiplying the monomials: $(3x^4)(4x^3) = 12x^{4+3} = 12x^7$

3. Multiplying the monomials: $(-2y^4)(8y^7) = -16y^{4+7} = -16y^{11}$

5. Multiplying the monomials: $(8x)(4x) = 32x^{1+1} = 32x^2$

7. Multiplying the monomials: $(10a^3)(10a)(2a^2) = 200a^{3+1+2} = 200a^6$

9. Multiplying the monomials: $(6ab^2)(-4a^2b) = -24a^{1+2}b^{2+1} = -24a^3b^3$

11. Multiplying the monomials: $(4x^2y)(3x^3y^3)(2xy^4) = 24x^{2+3+1}y^{1+3+4} = 24x^6y^8$

13. Dividing the monomials: $\frac{15x^3}{5x^2} = \frac{15}{5} \cdot \frac{x^3}{x^2} = 3x$

15. Dividing the monomials: $\frac{18y^9}{3y^{12}} = \frac{18}{3} \cdot \frac{y^9}{y^{12}} = 6 \cdot \frac{1}{y^3} = \frac{6}{y^3}$

17. Dividing the monomials: $\frac{32a^3}{64a^4} = \frac{32}{64} \cdot \frac{a^3}{a^4} = \frac{1}{2} \cdot \frac{1}{a} = \frac{1}{2a}$

19. Dividing the monomials: $\frac{21a^2b^3}{-7ab^5} = \frac{21}{-7} \cdot \frac{a^2}{a} \cdot \frac{b^3}{b^5} = -3 \cdot a \cdot \frac{1}{b^2} = -\frac{3a}{b^2}$

21. Dividing the monomials: $\frac{3x^3y^2z}{27xy^2z^3} = \frac{3}{27} \cdot \frac{x^3}{x} \cdot \frac{y^2}{y^2} \cdot \frac{z}{z^3} = \frac{1}{9} \cdot x^2 \cdot \frac{1}{z^2} = \frac{x^2}{9z^2}$

23. Completing the table:

a	b	ab	$\dfrac{a}{b}$	$\dfrac{b}{a}$
10	$5x$	$50x$	$\dfrac{2}{x}$	$\dfrac{x}{2}$
$20x^3$	$6x^2$	$120x^5$	$\dfrac{10x}{3}$	$\dfrac{3}{10x}$
$25x^5$	$5x^4$	$125x^9$	$5x$	$\dfrac{1}{5x}$
$3x^{-2}$	$3x^2$	9	$\dfrac{1}{x^4}$	x^4
$-2y^4$	$8y^7$	$-16y^{11}$	$-\dfrac{1}{4y^3}$	$-4y^3$

25. Finding the product: $(3 \times 10^3)(2 \times 10^5) = 6 \times 10^8$

27. Finding the product: $(3.5 \times 10^4)(5 \times 10^{-6}) = 17.5 \times 10^{-2} = 1.75 \times 10^{-1}$

29. Finding the product: $(5.5 \times 10^{-3})(2.2 \times 10^{-4}) = 12.1 \times 10^{-7} = 1.21 \times 10^{-6}$

31. Finding the quotient: $\dfrac{8.4 \times 10^5}{2 \times 10^2} = 4.2 \times 10^3$

33. Finding the quotient: $\dfrac{6 \times 10^8}{2 \times 10^{-2}} = 3 \times 10^{10}$

35. Finding the quotient: $\dfrac{2.5 \times 10^{-6}}{5 \times 10^{-4}} = 0.5 \times 10^{-2} = 5.0 \times 10^{-3}$

37. Combining the monomials: $3x^2 + 5x^2 = (3+5)x^2 = 8x^2$

39. Combining the monomials: $8x^5 - 19x^5 = (8-19)x^5 = -11x^5$

41. Combining the monomials: $2a + a - 3a = (2+1-3)a = 0a = 0$

43. Combining the monomials: $10x^3 - 8x^3 + 2x^3 = (10-8+2)x^3 = 4x^3$

45. Combining the monomials: $20ab^2 - 19ab^2 + 30ab^2 = (20-19+30)ab^2 = 31ab^2$

47. Completing the table:

a	b	ab	$a+b$
$5x$	$3x$	$15x^2$	$8x$
$4x^2$	$2x^2$	$8x^4$	$6x^2$
$3x^3$	$6x^3$	$18x^6$	$9x^3$
$2x^4$	$-3x^4$	$-6x^8$	$-x^4$
x^5	$7x^5$	$7x^{10}$	$8x^5$

49. Simplifying the expression: $\dfrac{(3x^2)(8x^5)}{6x^4} = \dfrac{24x^7}{6x^4} = \dfrac{24}{6} \cdot \dfrac{x^7}{x^4} = 4x^3$

51. Simplifying the expression: $\dfrac{(9a^2b)(2a^3b^4)}{18a^5b^7} = \dfrac{18a^5b^5}{18a^5b^7} = \dfrac{18}{18} \cdot \dfrac{a^5}{a^5} \cdot \dfrac{b^5}{b^7} = 1 \cdot \dfrac{1}{b^2} = \dfrac{1}{b^2}$

53. Simplifying the expression: $\dfrac{(4x^3y^2)(9x^4y^{10})}{(3x^5y)(2x^6y)} = \dfrac{36x^7y^{12}}{6x^{11}y^2} = \dfrac{36}{6} \cdot \dfrac{x^7}{x^{11}} \cdot \dfrac{y^{12}}{y^2} = 6 \cdot \dfrac{1}{x^4} \cdot y^{10} = \dfrac{6y^{10}}{x^4}$

55. Simplifying the expression: $\dfrac{(6 \times 10^8)(3 \times 10^5)}{9 \times 10^7} = \dfrac{18 \times 10^{13}}{9 \times 10^7} = 2 \times 10^6$

57. Simplifying the expression: $\dfrac{(5\times 10^3)(4\times 10^{-5})}{2\times 10^{-2}} = \dfrac{20\times 10^{-2}}{2\times 10^{-2}} = 10 = 1\times 10^1$

59. Simplifying the expression: $\dfrac{(2.8\times 10^{-7})(3.6\times 10^4)}{2.4\times 10^3} = \dfrac{10.08\times 10^{-3}}{2.4\times 10^3} = 4.2\times 10^{-6}$

61. Simplifying the expression: $\dfrac{18x^4}{3x} + \dfrac{21x^7}{7x^4} = 6x^3 + 3x^3 = 9x^3$

63. Simplifying the expression: $\dfrac{45a^6}{9a^4} - \dfrac{50a^8}{2a^6} = 5a^2 - 25a^2 = -20a^2$

65. Simplifying the expression: $\dfrac{6x^7 y^4}{3x^2 y^2} + \dfrac{8x^5 y^8}{2y^6} = 2x^5 y^2 + 4x^5 y^2 = 6x^5 y^2$

67. Applying the distributive property: $xy\left(x + \dfrac{1}{y}\right) = xy\bullet x + xy\bullet \dfrac{1}{y} = x^2 y + x$

69. Applying the distributive property: $xy\left(\dfrac{1}{y} + \dfrac{1}{x}\right) = xy\bullet \dfrac{1}{y} + xy\bullet \dfrac{1}{x} = x + y$

71. Applying the distributive property: $x^2\left(1 - \dfrac{4}{x^2}\right) = x^2\bullet 1 - x^2\bullet \dfrac{4}{x^2} = x^2 - 4$

73. Applying the distributive property: $x^2\left(1 - \dfrac{1}{x} - \dfrac{6}{x^2}\right) = x^2\bullet 1 - x^2\bullet \dfrac{1}{x} - x^2\bullet \dfrac{6}{x^2} = x^2 - x - 6$

75. Applying the distributive property: $x^2\left(1 - \dfrac{5}{x}\right) = x^2\bullet 1 - x^2\bullet \dfrac{5}{x} = x^2 - 5x$

77. Applying the distributive property: $x^2\left(1 - \dfrac{8}{x}\right) = x^2\bullet 1 - x^2\bullet \dfrac{8}{x} = x^2 - 8x$

79. a. Dividing by $5a^2$: $\dfrac{10a^2}{5a^2} = 2$

  b. Dividing by $5a^2$: $\dfrac{-15a^2 b}{5a^2} = -3b$

  c. Dividing by $5a^2$: $\dfrac{25a^2 b^2}{5a^2} = 5b^2$

81. a. Dividing by $8x^2 y$: $\dfrac{24x^3 y^2}{8x^2 y} = 3xy$

  b. Dividing by $8x^2 y$: $\dfrac{16x^2 y^2}{8x^2 y} = 2y$

  c. Dividing by $8x^2 y$: $\dfrac{-4x^2 y^3}{8x^2 y} = -\dfrac{y^2}{2}$

83. Evaluating when $x = -2$: $4x = 4(-2) = -8$
85. Evaluating when $x = -2$: $-2x + 5 = -2(-2) + 5 = 4 + 5 = 9$
87. Evaluating when $x = -2$: $x^2 + 5x + 6 = (-2)^2 + 5(-2) + 6 = 4 - 10 + 6 = 0$

114   Chapter 4 Exponents and Polynomials

89. The ordered pairs are $(-2,-2)$, $(0,2)$, and $(2,6)$:

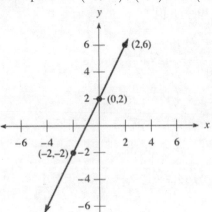

91. The ordered pairs are $(-3,0)$, $(0,1)$, and $(3,2)$:

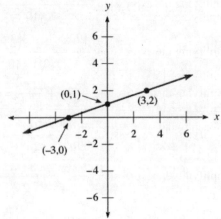

93. Simplifying: $3-8=-5$

95. Simplifying: $-1+7=6$

97. Simplifying: $3(5)^2+1=3(25)+1=75+1=76$

99. Simplifying: $2x^2+4x^2=6x^2$

101. Simplifying: $-5x+7x=2x$

103. Simplifying: $-(2x+9)=-2x-9$

105. Substituting $x=4$: $2x+3=2(4)+3=8+3=11$

4.4 Addition and Subtraction of Polynomials

1. This is a trinomial of degree 3.
3. This is a trinomial of degree 3.
5. This is a binomial of degree 1.
7. This is a binomial of degree 2.
9. This is a monomial of degree 2.
11. This is a monomial of degree 0.
13. Combining the polynomials: $(2x^2+3x+4)+(3x^2+2x+5)=(2x^2+3x^2)+(3x+2x)+(4+5)=5x^2+5x+9$
15. Combining the polynomials: $(3a^2-4a+1)+(2a^2-5a+6)=(3a^2+2a^2)+(-4a-5a)+(1+6)=5a^2-9a+7$
17. Combining the polynomials: $x^2+4x+2x+8=x^2+(4x+2x)+8=x^2+6x+8$
19. Combining the polynomials: $6x^2-3x-10x+5=6x^2+(-3x-10x)+5=6x^2-13x+5$
21. Combining the polynomials: $x^2-3x+3x-9=x^2+(-3x+3x)-9=x^2-9$
23. Combining the polynomials: $3y^2-5y-6y+10=3y^2+(-5y-6y)+10=3y^2-11y+10$
25. Combining the polynomials:
$$(6x^3-4x^2+2x)+(9x^2-6x+3)=6x^3+(-4x^2+9x^2)+(2x-6x)+3=6x^3+5x^2-4x+3$$
27. Combining the polynomials:
$$\left(\tfrac{2}{3}x^2-\tfrac{1}{5}x-\tfrac{3}{4}\right)+\left(\tfrac{4}{3}x^2-\tfrac{4}{5}x+\tfrac{7}{4}\right)=\left(\tfrac{2}{3}x^2+\tfrac{4}{3}x^2\right)+\left(-\tfrac{1}{5}x-\tfrac{4}{5}x\right)+\left(-\tfrac{3}{4}+\tfrac{7}{4}\right)=2x^2-x+1$$
29. Combining the polynomials: $(a^2-a-1)-(-a^2+a+1)=a^2-a-1+a^2-a-1=2a^2-2a-2$
31. Combining the polynomials:
$$\left(\tfrac{5}{9}x^3+\tfrac{1}{3}x^2-2x+1\right)-\left(\tfrac{2}{3}x^3+x^2+\tfrac{1}{2}x-\tfrac{3}{4}\right)=\tfrac{5}{9}x^3+\tfrac{1}{3}x^2-2x+1-\tfrac{2}{3}x^3-x^2-\tfrac{1}{2}x+\tfrac{3}{4}$$
$$=-\tfrac{1}{9}x^3-\tfrac{2}{3}x^2-\tfrac{5}{2}x+\tfrac{7}{4}$$
33. Combining the polynomials:
$$(4y^2-3y+2)+(5y^2+12y-4)-(13y^2-6y+20)=4y^2-3y+2+5y^2+12y-4-13y^2+6y-20$$
$$=(4y^2+5y^2-13y^2)+(-3y+12y+6y)+(2-4-20)$$
$$=-4y^2+15y-22$$

35. Simplifying: $(x^2 - 5x) - (x^2 - 3x) = x^2 - 5x - x^2 + 3x = -2x$

37. Simplifying: $(6x^2 - 11x) - (6x^2 - 15x) = 6x^2 - 11x - 6x^2 + 15x = 4x$

39. Simplifying: $(x^3 + 3x^2 + 9x) - (3x^2 + 9x + 27) = x^3 + 3x^2 + 9x - 3x^2 - 9x - 27 = x^3 - 27$

41. Simplifying: $(x^3 + 4x^2 + 4x) + (2x^2 + 8x + 8) = x^3 + 4x^2 + 4x + 2x^2 + 8x + 8 = x^3 + 6x^2 + 12x + 8$

43. Simplifying: $(x^2 - 4) - (x^2 - 4x + 4) = x^2 - 4 - x^2 + 4x - 4 = 4x - 8$

45. Performing the subtraction:
$$(11x^2 - 10x + 13) - (10x^2 + 23x - 50) = 11x^2 - 10x + 13 - 10x^2 - 23x + 50$$
$$= (11x^2 - 10x^2) + (-10x - 23x) + (13 + 50)$$
$$= x^2 - 33x + 63$$

47. Performing the subtraction:
$$(11y^2 + 11y + 11) - (3y^2 + 7y - 15) = 11y^2 + 11y + 11 - 3y^2 - 7y + 15$$
$$= (11y^2 - 3y^2) + (11y - 7y) + (11 + 15)$$
$$= 8y^2 + 4y + 26$$

49. Performing the addition:
$$(25x^2 - 50x + 75) + (50x^2 - 100x - 150) = (25x^2 + 50x^2) + (-50x - 100x) + (75 - 150) = 75x^2 - 150x - 75$$

51. Performing the operations:
$$(3x - 2) + (11x + 5) - (2x + 1) = 3x - 2 + 11x + 5 - 2x - 1 = (3x + 11x - 2x) + (-2 + 5 - 1) = 12x + 2$$

53. Evaluating when $x = 3$: $x^2 - 2x + 1 = (3)^2 - 2(3) + 1 = 9 - 6 + 1 = 4$

55. a. Substituting $p = 5$: $100p^2 - 1,300p + 4,000 = 100(5)^2 - 1,300(5) + 4,000 = 2,500 - 6,500 + 4,000 = 0$

    b. Substituting $p = 8$: $100p^2 - 1,300p + 4,000 = 100(8)^2 - 1,300(8) + 4,000 = 6,400 - 10,400 + 4,000 = 0$

57. a. Substituting $x = 8$: $600 + 1,000x - 100x^2 = 600 + 1,000(8) - 100(8)^2 = 600 + 8,000 - 6,400 = 2,200$

    b. Substituting $x = -2$: $600 + 1,000x - 100x^2 = 600 + 1,000(-2) - 100(-2)^2 = 600 - 2,000 - 400 = -1,800$

59. Finding the volume of the cylinder and sphere:
$$V_{cylinder} = \pi(3^2)(6) = 54\pi \qquad V_{sphere} = \tfrac{4}{3}\pi(3^3) = 36\pi$$
Subtracting to find the amount of space to pack: $V = 54\pi - 36\pi = 18\pi$ inches^3

61. Multiplying: $3x(-5x) = -15x^2$

63. Multiplying: $2x(3x^2) = 6x^{1+2} = 6x^3$

65. Multiplying: $3x^2(2x^2) = 6x^{2+2} = 6x^4$

67. Simplifying: $(-5)(-1) = 5$

69. Simplifying: $(-1)(6) = -6$

71. Simplifying: $(5x)(-4x) = -20x^2$

73. Simplifying: $3x(-7) = -21x$

75. Simplifying: $5x + (-3x) = 2x$

77. Multiplying: $3(2x - 6) = 6x - 18$

4.5 Multiplication with Polynomials

1. Using the distributive property: $2x(3x+1) = 2x(3x) + 2x(1) = 6x^2 + 2x$
3. Using the distributive property: $2x^2(3x^2 - 2x + 1) = 2x^2(3x^2) - 2x^2(2x) + 2x^2(1) = 6x^4 - 4x^3 + 2x^2$
5. Using the distributive property: $2ab(a^2 - ab + 1) = 2ab(a^2) - 2ab(ab) + 2ab(1) = 2a^3b - 2a^2b^2 + 2ab$
7. Using the distributive property: $y^2(3y^2 + 9y + 12) = y^2(3y^2) + y^2(9y) + y^2(12) = 3y^4 + 9y^3 + 12y^2$
9. Using the distributive property:
$$4x^2y(2x^3y + 3x^2y^2 + 8y^3) = 4x^2y(2x^3y) + 4x^2y(3x^2y^2) + 4x^2y(8y^3) = 8x^5y^2 + 12x^4y^3 + 32x^2y^4$$
11. Multiplying using the FOIL method: $(x+3)(x+4) = x^2 + 3x + 4x + 12 = x^2 + 7x + 12$
13. Multiplying using the FOIL method: $(x+6)(x+1) = x^2 + 6x + 1x + 6 = x^2 + 7x + 6$
15. Multiplying using the FOIL method: $\left(x + \frac{1}{2}\right)\left(x + \frac{3}{2}\right) = x^2 + \frac{1}{2}x + \frac{3}{2}x + \frac{3}{4} = x^2 + 2x + \frac{3}{4}$
17. Multiplying using the FOIL method: $(a+5)(a-3) = a^2 + 5a - 3a - 15 = a^2 + 2a - 15$
19. Multiplying using the FOIL method: $(x-a)(y+b) = xy + bx - ay - ab$
21. Multiplying using the FOIL method: $(x+6)(x-6) = x^2 + 6x - 6x - 36 = x^2 - 36$
23. Multiplying using the FOIL method: $\left(y + \frac{5}{6}\right)\left(y - \frac{5}{6}\right) = y^2 + \frac{5}{6}y - \frac{5}{6}y - \frac{25}{36} = y^2 - \frac{25}{36}$
25. Multiplying using the FOIL method: $(2x-3)(x-4) = 2x^2 - 3x - 8x + 12 = 2x^2 - 11x + 12$
27. Multiplying using the FOIL method: $(a+2)(2a-1) = 2a^2 + 4a - a - 2 = 2a^2 + 3a - 2$
29. Multiplying using the FOIL method: $(2x-5)(3x-2) = 6x^2 - 15x - 4x + 10 = 6x^2 - 19x + 10$
31. Multiplying using the FOIL method: $(2x+3)(a+4) = 2ax + 3a + 8x + 12$
33. Multiplying using the FOIL method: $(5x-4)(5x+4) = 25x^2 - 20x + 20x - 16 = 25x^2 - 16$
35. Multiplying using the FOIL method: $\left(2x - \frac{1}{2}\right)\left(x + \frac{3}{2}\right) = 2x^2 - \frac{1}{2}x + 3x - \frac{3}{4} = 2x^2 + \frac{5}{2}x - \frac{3}{4}$
37. Multiplying using the FOIL method: $(1-2a)(3-4a) = 3 - 6a - 4a + 8a^2 = 3 - 10a + 8a^2$
39. The product is $(x+2)(x+3) = x^2 + 5x + 6$:

	x	3
x	x^2	$3x$
2	$2x$	6

41. The product is $(x+1)(2x+2) = 2x^2 + 4x + 2$:

	x	x	2
x	x^2	x^2	$2x$
1	x	x	2

43. Multiplying using the column method:

    $$\begin{array}{r} a^2 - 3a + 2 \\ a - 3 \\ \hline a^3 - 3a^2 + 2a \\ -3a^2 + 9a - 6 \\ \hline a^3 - 6a^2 + 11a - 6 \end{array}$$

45. Multiplying using the column method:

    $$\begin{array}{r} x^2 - 2x + 4 \\ x + 2 \\ \hline x^3 - 2x^2 + 4x \\ 2x^2 - 4x + 8 \\ \hline x^3 + 8 \end{array}$$

47. Multiplying using the column method:

$$\begin{array}{r} x^2 + 8x + 9 \\ 2x + 1 \\ \hline 2x^3 + 16x^2 + 18x \\ x^2 + 8x + 9 \\ \hline 2x^3 + 17x^2 + 26x + 9 \end{array}$$

49. Multiplying using the column method:

$$\begin{array}{r} 5x^2 + 2x + 1 \\ x^2 - 3x + 5 \\ \hline 5x^4 + 2x^3 + x^2 \\ -15x^3 - 6x^2 - 3x \\ 25x^2 + 10x + 5 \\ \hline 5x^4 - 13x^3 + 20x^2 + 7x + 5 \end{array}$$

51. Multiplying using the FOIL method: $(x^2 + 3)(2x^2 - 5) = 2x^4 - 5x^2 + 6x^2 - 15 = 2x^4 + x^2 - 15$

53. Multiplying using the FOIL method: $(3a^4 + 2)(2a^2 + 5) = 6a^6 + 15a^4 + 4a^2 + 10$

55. First multiply two polynomials using the FOIL method: $(x + 3)(x + 4) = x^2 + 3x + 4x + 12 = x^2 + 7x + 12$
    Now using the column method:

$$\begin{array}{r} x^2 + 7x + 12 \\ x + 5 \\ \hline x^3 + 7x^2 + 12x \\ 5x^2 + 35x + 60 \\ \hline x^3 + 12x^2 + 47x + 60 \end{array}$$

57. Simplifying: $(x - 3)(x - 2) + 2 = x^2 - 3x - 2x + 6 + 2 = x^2 - 5x + 8$

59. Simplifying: $(2x - 3)(4x + 3) + 4 = 8x^2 + 6x - 12x - 9 + 4 = 8x^2 - 6x - 5$

61. Simplifying: $(x + 4)(x - 5) + (-5)(2) = x^2 - 5x + 4x - 20 - 10 = x^2 - x - 30$

63. Simplifying: $2(x - 3) + x(x + 2) = 2x - 6 + x^2 + 2x = x^2 + 4x - 6$

65. Simplifying: $3x(x + 1) - 2x(x - 5) = 3x^2 + 3x - 2x^2 + 10x = x^2 + 13x$

67. Simplifying: $x(x + 2) - 3 = x^2 + 2x - 3$ 69. Simplifying: $a(a - 3) + 6 = a^2 - 3a + 6$

71. a. Finding the product: $(x + 1)(x - 1) = x^2 - x + x - 1 = x^2 - 1$
    b. Finding the product: $(x + 1)(x + 1) = x^2 + x + x + 1 = x^2 + 2x + 1$
    c. Finding the product: $(x + 1)(x^2 + 2x + 1) = x^3 + 2x^2 + x + x^2 + 2x + 1 = x^3 + 3x^2 + 3x + 1$
    d. Finding the product:
       $(x + 1)(x^3 + 3x^2 + 3x + 1) = x^4 + 3x^3 + 3x^2 + x + x^3 + 3x^2 + 3x + 1 = x^4 + 4x^3 + 6x^2 + 4x + 1$

73. a. Finding the product: $(x + 1)(x - 1) = x^2 - x + x - 1 = x^2 - 1$
    b. Finding the product: $(x + 1)(x - 2) = x^2 - 2x + x - 2 = x^2 - x - 2$
    c. Finding the product: $(x + 1)(x - 3) = x^2 - 3x + x - 3 = x^2 - 2x - 3$
    d. Finding the product: $(x + 1)(x - 4) = x^2 - 4x + x - 4 = x^2 - 3x - 4$

75. The expression is equal to 0 if $x = 0$.   77. The expression is equal to 0 if $x = -5$.
79. The expression is equal to 0 if $x = 3$ or $x = -2$.   81. The expression is never equal to 0.
83. The expression is equal to 0 if $x = a$ or $x = b$.

85. Let x represent the width and $2x + 5$ represent the length. The area is given by: $A = x(2x + 5) = 2x^2 + 5x$

87. Let x and $x + 1$ represent the width and length, respectively. The area is given by: $A = x(x + 1) = x^2 + x$

89. The revenue is: $R = xp = (1,200 - 100p)p = 1,200p - 100p^2$

91. Graphing each line:

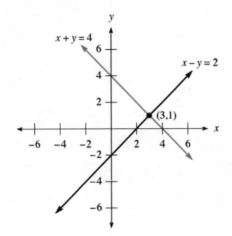

The intersection point is $(3,1)$.

93. Multiplying the second equation by –2:
$$3x+2y=1$$
$$-4x-2y=-6$$
Adding the two equations:
$$-x=-5$$
$$x=5$$
Substituting into the second equation:
$$2(5)+y=3$$
$$10+y=3$$
$$y=-7$$
The solution is $(5,-7)$.

95. Substituting into the first equation:
$$x+5x+2=20$$
$$6x+2=20$$
$$6x=18$$
$$x=3$$
Substituting into the second equation: $y=5(3)+2=15+2=17$. The solution is $(3,17)$.

97. Let x represent the amount invested at 8% and y represent the amount invested at 10%. The system of equations is:
$$x+y=1200$$
$$0.08x+0.10y=104$$
Multiplying the first equation by –0.08:
$$-0.08x-0.08y=-96$$
$$0.08x+0.10y=104$$
Adding the two equations:
$$0.02y=8$$
$$y=400$$
Substituting into the first equation:
$$x+400=1200$$
$$x=800$$
So $800 is invested at 8% and $400 is invested at 10%.

99. Simplifying: $13 \cdot 13 = 169$

101. Simplifying: $2(x)(-5) = -10x$

103. Simplifying: $6x+(-6x)=0$

105. Simplifying: $(2x)(-3)+(2x)(3)=-6x+6x=0$

107. Multiplying: $-4(3x-4) = -12x+16$

109. Multiplying: $(x-1)(x+2) = x^2 + 2x - x - 2 = x^2 + x - 2$

111. Multiplying: $(x+3)(x+3) = x^2 + 3x + 3x + 9 = x^2 + 6x + 9$

4.6 Binomial Squares and Other Special Products

1. Multiplying using the FOIL method: $(x-2)^2 = (x-2)(x-2) = x^2 - 2x - 2x + 4 = x^2 - 4x + 4$

3. Multiplying using the FOIL method: $(a+3)^2 = (a+3)(a+3) = a^2 + 3a + 3a + 9 = a^2 + 6a + 9$

5. Multiplying using the FOIL method: $(x-5)^2 = (x-5)(x-5) = x^2 - 5x - 5x + 25 = x^2 - 10x + 25$

7. Multiplying using the FOIL method: $\left(a - \frac{1}{2}\right)^2 = \left(a - \frac{1}{2}\right)\left(a - \frac{1}{2}\right) = a^2 - \frac{1}{2}a - \frac{1}{2}a + \frac{1}{4} = a^2 - a + \frac{1}{4}$

9. Multiplying using the FOIL method: $(x+10)^2 = (x+10)(x+10) = x^2 + 10x + 10x + 100 = x^2 + 20x + 100$

11. Multiplying using the square of binomial formula: $(a+0.8)^2 = a^2 + 2(a)(0.8) + (0.8)^2 = a^2 + 1.6a + 0.64$

13. Multiplying using the square of binomial formula: $(2x-1)^2 = (2x)^2 - 2(2x)(1) + (1)^2 = 4x^2 - 4x + 1$

15. Multiplying using the square of binomial formula: $(4a+5)^2 = (4a)^2 + 2(4a)(5) + (5)^2 = 16a^2 + 40a + 25$

17. Multiplying using the square of binomial formula: $(3x-2)^2 = (3x)^2 - 2(3x)(2) + (2)^2 = 9x^2 - 12x + 4$

19. Multiplying using the square of binomial formula: $(3a+5b)^2 = (3a)^2 + 2(3a)(5b) + (5b)^2 = 9a^2 + 30ab + 25b^2$

21. Multiplying using the square of binomial formula: $(4x-5y)^2 = (4x)^2 - 2(4x)(5y) + (5y)^2 = 16x^2 - 40xy + 25y^2$

23. Multiplying using the square of binomial formula: $(7m+2n)^2 = (7m)^2 + 2(7m)(2n) + (2n)^2 = 49m^2 + 28mn + 4n^2$

25. Multiplying using the square of binomial formula:
$(6x - 10y)^2 = (6x)^2 - 2(6x)(10y) + (10y)^2 = 36x^2 - 120xy + 100y^2$

27. Multiplying using the square of binomial formula: $\left(x^2 + 5\right)^2 = \left(x^2\right)^2 + 2\left(x^2\right)(5) + (5)^2 = x^4 + 10x^2 + 25$

29. Multiplying using the square of binomial formula: $\left(a^2 + 1\right)^2 = \left(a^2\right)^2 + 2\left(a^2\right)(1) + (1)^2 = a^4 + 2a^2 + 1$

31. Multiplying using the square of binomial formula: $\left(y + \frac{3}{2}\right)^2 = (y)^2 + 2(y)\left(\frac{3}{2}\right) + \left(\frac{3}{2}\right)^2 = y^2 + 3y + \frac{9}{4}$

33. Multiplying using the square of binomial formula: $\left(a + \frac{1}{2}\right)^2 = (a)^2 + 2(a)\left(\frac{1}{2}\right) + \left(\frac{1}{2}\right)^2 = a^2 + a + \frac{1}{4}$

35. Multiplying using the square of binomial formula: $\left(x + \frac{3}{4}\right)^2 = (x)^2 + 2(x)\left(\frac{3}{4}\right) + \left(\frac{3}{4}\right)^2 = x^2 + \frac{3}{2}x + \frac{9}{16}$

37. Multiplying using the square of binomial formula: $\left(t + \frac{1}{5}\right)^2 = (t)^2 + 2(t)\left(\frac{1}{5}\right) + \left(\frac{1}{5}\right)^2 = t^2 + \frac{2}{5}t + \frac{1}{25}$

39. Completing the table:

x	$(x+3)^2$	$x^2 + 9$	$x^2 + 6x + 9$
1	16	10	16
2	25	13	25
3	36	18	36
4	49	25	49

41. Completing the table:

a	b	$(a+b)^2$	a^2+b^2	a^2+ab+b^2	$a^2+2ab+b^2$
1	1	4	2	3	4
3	5	64	34	49	64
3	4	49	25	37	49
4	5	81	41	61	81

43. Multiplying using the FOIL method: $(a+5)(a-5) = a^2 + 5a - 5a - 25 = a^2 - 25$

45. Multiplying using the FOIL method: $(y-1)(y+1) = y^2 - y + y - 1 = y^2 - 1$

47. Multiplying using the difference of squares formula: $(9+x)(9-x) = (9)^2 - (x)^2 = 81 - x^2$

49. Multiplying using the difference of squares formula: $(2x+5)(2x-5) = (2x)^2 - (5)^2 = 4x^2 - 25$

51. Multiplying using the difference of squares formula: $\left(4x+\frac{1}{3}\right)\left(4x-\frac{1}{3}\right) = (4x)^2 - \left(\frac{1}{3}\right)^2 = 16x^2 - \frac{1}{9}$

53. Multiplying using the difference of squares formula: $(2a+7)(2a-7) = (2a)^2 - (7)^2 = 4a^2 - 49$

55. Multiplying using the difference of squares formula: $(6-7x)(6+7x) = (6)^2 - (7x)^2 = 36 - 49x^2$

57. Multiplying using the difference of squares formula: $(x^2+3)(x^2-3) = (x^2)^2 - (3)^2 = x^4 - 9$

59. Multiplying using the difference of squares formula: $(a^2+4)(a^2-4) = (a^2)^2 - (4)^2 = a^4 - 16$

61. Multiplying using the difference of squares formula: $(5y^4-8)(5y^4+8) = (5y^4)^2 - (8)^2 = 25y^8 - 64$

63. Multiplying and simplifying: $(x+3)(x-3) + (x+5)(x-5) = (x^2-9) + (x^2-25) = 2x^2 - 34$

65. Multiplying and simplifying:
$(2x+3)^2 - (4x-1)^2 = (4x^2+12x+9) - (16x^2-8x+1) = 4x^2+12x+9-16x^2+8x-1 = -12x^2+20x+8$

67. Multiplying and simplifying:
$(a+1)^2 - (a+2)^2 + (a+3)^2 = (a^2+2a+1) - (a^2+4a+4) + (a^2+6a+9)$
$= a^2+2a+1-a^2-4a-4+a^2+6a+9$
$= a^2+4a+6$

69. Multiplying and simplifying:
$(2x+3)^3 = (2x+3)(2x+3)^2$
$= (2x+3)(4x^2+12x+9)$
$= 8x^3+24x^2+18x+12x^2+36x+27$
$= 8x^3+36x^2+54x+27$

71. a. Evaluating when $x=6$: $x^2-25 = (6)^2-25 = 36-25 = 11$

    b. Evaluating when $x=6$: $(x-5)^2 = (6-5)^2 = (1)^2 = 1$

    c. Evaluating when $x=6$: $(x+5)(x-5) = (6+5)(6-5) = (11)(1) = 11$

73. a. Evaluating when $x=-2$: $(x+3)^2 = (-2+3)^2 = (1)^2 = 1$

    b. Evaluating when $x=-2$: $x^2+9 = (-2)^2+9 = 4+9 = 13$

    c. Evaluating when $x=-2$: $x^2+6x+9 = (-2)^2+6(-2)+9 = 4-12+9 = 1$

75. Let x and $x+1$ represent the two integers. The expression can be written as:
$(x)^2 + (x+1)^2 = x^2 + (x^2+2x+1) = 2x^2+2x+1$

77. Let x, $x+1$, and $x+2$ represent the three integers. The expression can be written as:
$$(x)^2 + (x+1)^2 + (x+2)^2 = x^2 + (x^2+2x+1) + (x^2+4x+4) = 3x^2 + 6x + 5$$

79. Verifying the areas: $(a+b)^2 = a^2 + ab + ab + b^2 = a^2 + 2ab + b^2$

81. Graphing both lines:

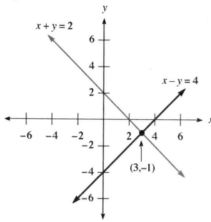

The intersection point is $(3,-1)$.

83. Graphing both lines:

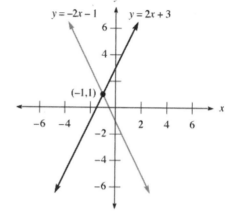

The intersection point is $(-1,1)$.

85. Simplifying: $\dfrac{10x^3}{5x} = 2x^{3-1} = 2x^2$

87. Simplifying: $\dfrac{3x^2}{3} = x^2$

89. Simplifying: $\dfrac{9x^2}{3x} = 3x^{2-1} = 3x$

91. Simplifying: $\dfrac{24x^3y^2}{8x^2y} = 3x^{3-2}y^{2-1} = 3xy$

93. Simplifying: $\dfrac{15x^2y}{3xy} = \dfrac{15}{3} \cdot \dfrac{x^2}{x} \cdot \dfrac{y}{y} = 5x$

95. Simplifying: $\dfrac{35a^6b^8}{70a^2b^{10}} = \dfrac{35}{70} \cdot \dfrac{a^6}{a^2} \cdot \dfrac{b^8}{b^{10}} = \dfrac{1}{2} \cdot a^4 \cdot \dfrac{1}{b^2} = \dfrac{a^4}{2b^2}$

4.7 Dividing a Polynomial by a Monomial

1. Performing the division: $\dfrac{5x^2 - 10x}{5x} = \dfrac{5x^2}{5x} - \dfrac{10x}{5x} = x - 2$

3. Performing the division: $\dfrac{15x - 10x^3}{5x} = \dfrac{15x}{5x} - \dfrac{10x^3}{5x} = 3 - 2x^2$

5. Performing the division: $\dfrac{25x^2y - 10xy}{5x} = \dfrac{25x^2y}{5x} - \dfrac{10xy}{5x} = 5xy - 2y$

7. Performing the division: $\dfrac{35x^5 - 30x^4 + 25x^3}{5x} = \dfrac{35x^5}{5x} - \dfrac{30x^4}{5x} + \dfrac{25x^3}{5x} = 7x^4 - 6x^3 + 5x^2$

9. Performing the division: $\dfrac{50x^5 - 25x^3 + 5x}{5x} = \dfrac{50x^5}{5x} - \dfrac{25x^3}{5x} + \dfrac{5x}{5x} = 10x^4 - 5x^2 + 1$

11. Performing the division: $\dfrac{8a^2 - 4a}{-2a} = \dfrac{8a^2}{-2a} + \dfrac{-4a}{-2a} = -4a + 2$

13. Performing the division: $\dfrac{16a^5 + 24a^4}{-2a} = \dfrac{16a^5}{-2a} + \dfrac{24a^4}{-2a} = -8a^4 - 12a^3$

15. Performing the division: $\dfrac{8ab + 10a^2}{-2a} = \dfrac{8ab}{-2a} + \dfrac{10a^2}{-2a} = -4b - 5a$

122 Chapter 4 Exponents and Polynomials

17. Performing the division: $\dfrac{12a^3b-6a^2b^2+14ab^3}{-2a} = \dfrac{12a^3b}{-2a}+\dfrac{-6a^2b^2}{-2a}+\dfrac{14ab^3}{-2a} = -6a^2b+3ab^2-7b^3$

19. Performing the division: $\dfrac{a^2+2ab+b^2}{-2a} = \dfrac{a^2}{-2a}+\dfrac{2ab}{-2a}+\dfrac{b^2}{-2a} = -\dfrac{a}{2}-b-\dfrac{b^2}{2a}$

21. Performing the division: $\dfrac{6x+8y}{2} = \dfrac{6x}{2}+\dfrac{8y}{2} = 3x+4y$

23. Performing the division: $\dfrac{7y-21}{-7} = \dfrac{7y}{-7}+\dfrac{-21}{-7} = -y+3$

25. Performing the division: $\dfrac{2x^2+16x-18}{2} = \dfrac{2x^2}{2}+\dfrac{16x}{2}-\dfrac{18}{2} = x^2+8x-9$

27. Performing the division: $\dfrac{3y^2-9y+3}{3} = \dfrac{3y^2}{3}-\dfrac{9y}{3}+\dfrac{3}{3} = y^2-3y+1$

29. Performing the division: $\dfrac{x^2y-x^3y^2}{-x^2y} = \dfrac{x^2y}{-x^2y}+\dfrac{-x^3y^2}{-x^2y} = -1+xy$

31. Performing the division: $\dfrac{a^2b^2-ab^2}{-ab^2} = \dfrac{a^2b^2}{-ab^2}+\dfrac{-ab^2}{-ab^2} = -a+1$

33. Performing the division: $\dfrac{x^3-3x^2y+xy^2}{x} = \dfrac{x^3}{x}-\dfrac{3x^2y}{x}+\dfrac{xy^2}{x} = x^2-3xy+y^2$

35. Performing the division: $\dfrac{10a^2-15a^2b+25a^2b^2}{5a^2} = \dfrac{10a^2}{5a^2}-\dfrac{15a^2b}{5a^2}+\dfrac{25a^2b^2}{5a^2} = 2-3b+5b^2$

37. Performing the division: $\dfrac{26x^2y^2-13xy}{-13xy} = \dfrac{26x^2y^2}{-13xy}+\dfrac{-13xy}{-13xy} = -2xy+1$

39. Performing the division: $\dfrac{4x^2y^2-2xy}{4xy} = \dfrac{4x^2y^2}{4xy}-\dfrac{2xy}{4xy} = xy-\dfrac{1}{2}$

41. Performing the division: $\dfrac{5a^2x-10ax^2+15a^2x^2}{20a^2x^2} = \dfrac{5a^2x}{20a^2x^2}-\dfrac{10ax^2}{20a^2x^2}+\dfrac{15a^2x^2}{20a^2x^2} = \dfrac{1}{4x}-\dfrac{1}{2a}+\dfrac{3}{4}$

43. Performing the division: $\dfrac{16x^5+8x^2+12x}{12x^3} = \dfrac{16x^5}{12x^3}+\dfrac{8x^2}{12x^3}+\dfrac{12x}{12x^3} = \dfrac{4x^2}{3}+\dfrac{2}{3x}+\dfrac{1}{x^2}$

45. Performing the division: $\dfrac{9a^{5m}-27a^{3m}}{3a^{2m}} = \dfrac{9a^{5m}}{3a^{2m}}-\dfrac{27a^{3m}}{3a^{2m}} = 3a^{5m-2m}-9a^{3m-2m} = 3a^{3m}-9a^m$

47. Performing the division:
$\dfrac{10x^{5m}-25x^{3m}+35x^m}{5x^m} = \dfrac{10x^{5m}}{5x^m}-\dfrac{25x^{3m}}{5x^m}+\dfrac{35x^m}{5x^m} = 2x^{5m-m}-5x^{3m-m}+7x^{m-m} = 2x^{4m}-5x^{2m}+7$

49. Simplifying and then dividing:
$\dfrac{2x^3(3x+2)-3x^2(2x-4)}{2x^2} = \dfrac{6x^4+4x^3-6x^3+12x^2}{2x^2}$
$= \dfrac{6x^4-2x^3+12x^2}{2x^2}$
$= \dfrac{6x^4}{2x^2}-\dfrac{2x^3}{2x^2}+\dfrac{12x^2}{2x^2}$
$= 3x^2-x+6$

51. Simplifying and then dividing:
$\dfrac{(x+2)^2-(x-2)^2}{2x} = \dfrac{\left(x^2+4x+4\right)-\left(x^2-4x+4\right)}{2x} = \dfrac{x^2+4x+4-x^2+4x-4}{2x} = \dfrac{8x}{2x} = 4$

53. Simplifying and then dividing:
$$\frac{(x+5)^2+(x+5)(x-5)}{2x} = \frac{(x^2+10x+25)+(x^2-25)}{2x} = \frac{2x^2+10x}{2x} = \frac{2x^2}{2x}+\frac{10x}{2x} = x+5$$

55. a. Evaluating when $x = 2$: $2x+3 = 2(2)+3 = 4+3 = 7$

   b. Evaluating when $x = 2$: $\frac{10x+15}{5} = \frac{10(2)+15}{5} = \frac{20+15}{5} = \frac{35}{5} = 7$

   c. Evaluating when $x = 2$: $10x+3 = 10(2)+3 = 20+3 = 23$

57. a. Evaluating when $x = 10$: $\frac{3x+8}{2} = \frac{3(10)+8}{2} = \frac{30+8}{2} = \frac{38}{2} = 19$

   b. Evaluating when $x = 10$: $3x+4 = 3(10)+4 = 30+4 = 34$

   c. Evaluating when $x = 10$: $\frac{3}{2}x+4 = \frac{3}{2}(10)+4 = 15+4 = 19$

59. a. Evaluating when $x = 2$: $2x^2-3x = 2(2)^2-3(2) = 2(4)-3(2) = 8-6 = 2$

   b. Evaluating when $x = 2$: $\frac{10x^3-15x^2}{5x} = \frac{10(2)^3-15(2)^2}{5(2)} = \frac{10(8)-15(4)}{5(2)} = \frac{80-60}{10} = \frac{20}{10} = 2$

61. Adding the two equations:
   $2x = 14$
   $x = 7$
   Substituting into the first equation:
   $7+y = 6$
   $y = -1$
   The solution is $(7,-1)$.

63. Multiplying the second equation by 3:
   $2x-3y = -5$
   $3x+3y = 15$
   Adding the two equations:
   $5x = 10$
   $x = 2$
   Substituting into the second equation:
   $2+y = 5$
   $y = 3$
   The solution is $(2,3)$.

65. Substituting into the first equation:
   $x+2x-1 = 2$
   $3x-1 = 2$
   $3x = 3$
   $x = 1$
   Substituting into the second equation: $y = 2(1)-1 = 2-1 = 1$. The solution is $(1,1)$.

67. Substituting into the first equation:
   $4x+2(-2x+4) = 8$
   $4x-4x+8 = 8$
   $8 = 8$
   Since this statement is true, the system is dependent. The two lines coincide.

124 Chapter 4 Exponents and Polynomials

69. Dividing:
$$\begin{array}{r} 146 \\ 27\overline{)3962} \\ \underline{27} \\ 126 \\ \underline{108} \\ 182 \\ \underline{162} \\ 20 \end{array}$$

The quotient is $146\frac{20}{27}$.

71. Dividing: $\dfrac{2x^2+5x}{x} = \dfrac{2x^2}{x} + \dfrac{5x}{x} = 2x+5$

73. Multiplying: $(x-3)x = x^2 - 3x$

75. Multiplying: $2x^2(x-5) = 2x^3 - 10x^2$

77. Subtracting: $(x^2-5x)-(x^2-3x) = x^2-5x-x^2+3x = -2x$

79. Subtracting: $(-2x+8)-(-2x+6) = -2x+8+2x-6 = 2$

4.8 Dividing a Polynomial by a Polynomial

1. Using long division:
$$\begin{array}{r} x-2 \\ x-3\overline{)x^2-5x+6} \\ \underline{x^2-3x} \\ -2x+6 \\ \underline{-2x+6} \\ 0 \end{array}$$

The quotient is $x-2$.

3. Using long division:
$$\begin{array}{r} a+4 \\ a+5\overline{)a^2+9a+20} \\ \underline{a^2+5a} \\ 4a+20 \\ \underline{4a+20} \\ 0 \end{array}$$

The quotient is $a+4$.

5. Using long division:
$$\begin{array}{r} x-3 \\ x-3\overline{)x^2-6x+9} \\ \underline{x^2-3x} \\ -3x+9 \\ \underline{-3x+9} \\ 0 \end{array}$$

The quotient is $x-3$.

7. Using long division:
$$\begin{array}{r} x+3 \\ 2x-1\overline{)2x^2+5x-3} \\ \underline{2x^2-x} \\ 6x-3 \\ \underline{6x-3} \\ 0 \end{array}$$

The quotient is $x+3$.

9. Using long division:
$$\begin{array}{r} a-5 \\ 2a+1\overline{)2a^2-9a-5} \\ \underline{2a^2+a} \\ -10a-5 \\ \underline{-10a-5} \\ 0 \end{array}$$

The quotient is $a-5$.

11. Using long division:
$$\begin{array}{r} x+2 \\ x+3\overline{)x^2+5x+8} \\ \underline{x^2+3x} \\ 2x+8 \\ \underline{2x+6} \\ 2 \end{array}$$

The quotient is $x+2+\dfrac{2}{x+3}$.

13. Using long division:

$$\require{enclose}\begin{array}{r}a-2\\a+5\enclose{longdiv}{a^2+3a+2}\\\underline{a^2+5a}\\-2a+2\\\underline{-2a-10}\\12\end{array}$$

The quotient is $a-2+\dfrac{12}{a+5}$.

15. Using long division:

$$\begin{array}{r}x+4\\x-2\enclose{longdiv}{x^2+2x+1}\\\underline{x^2-2x}\\4x+1\\\underline{4x-8}\\9\end{array}$$

The quotient is $x+4+\dfrac{9}{x-2}$.

17. Using long division:

$$\begin{array}{r}x+4\\x+1\enclose{longdiv}{x^2+5x-6}\\\underline{x^2+x}\\4x-6\\\underline{4x+4}\\-10\end{array}$$

The quotient is $x+4+\dfrac{-10}{x+1}$.

19. Using long division:

$$\begin{array}{r}a+1\\a+2\enclose{longdiv}{a^2+3a+1}\\\underline{a^2+2a}\\a+1\\\underline{a+2}\\-1\end{array}$$

The quotient is $a+1+\dfrac{-1}{a+2}$.

21. Using long division:

$$\begin{array}{r}x-3\\2x+4\enclose{longdiv}{2x^2-2x+5}\\\underline{2x^2+4x}\\-6x+5\\\underline{-6x-12}\\17\end{array}$$

The quotient is $x-3+\dfrac{17}{2x+4}$.

23. Using long division:

$$\begin{array}{r}3a-2\\2a+3\enclose{longdiv}{6a^2+5a+1}\\\underline{6a^2+9a}\\-4a+1\\\underline{-4a-6}\\7\end{array}$$

The quotient is $3a-2+\dfrac{7}{2a+3}$.

25. Using long division:

$$\begin{array}{r}2a^2-a-3\\3a-5\enclose{longdiv}{6a^3-13a^2-4a+15}\\\underline{6a^3-10a^2}\\-3a^2-4a\\\underline{-3a^2+5a}\\-9a+15\\\underline{-9a+15}\\0\end{array}$$

The quotient is $2a^2-a-3$.

27. Using long division:

$$\begin{array}{r}x^2-x+5\\x+1\enclose{longdiv}{x^3+0x^2+4x+5}\\\underline{x^3+x^2}\\-x^2+4x\\\underline{-x^2-x}\\5x+5\\\underline{5x+5}\\0\end{array}$$

The quotient is x^2-x+5.

29. Using long division:

$$\begin{array}{r} x^2+x+1 \\ x-1 \overline{\smash{)}x^3+0x^2+0x-1} \\ \underline{x^3-x^2} \\ x^2+0x \\ \underline{x^2-x} \\ x-1 \\ \underline{x-1} \\ 0 \end{array}$$

The quotient is x^2+x+1.

31. Using long division:

$$\begin{array}{r} x^2+2x+4 \\ x-2 \overline{\smash{)}x^3+0x^2+0x-8} \\ \underline{x^3-2x^2} \\ 2x^2+0x \\ \underline{2x^2-4x} \\ 4x-8 \\ \underline{4x-8} \\ 0 \end{array}$$

The quotient is x^2+2x+4.

33. a. Evaluating when $x=3$: $x^2+2x+4=(3)^2+2(3)+4=9+6+4=19$

　　b. Evaluating when $x=3$: $\dfrac{x^3-8}{x-2}=\dfrac{(3)^3-8}{3-2}=\dfrac{27-8}{3-2}=\dfrac{19}{1}=19$

　　c. Evaluating when $x=3$: $x^2-4=(3)^2-4=9-4=5$

35. a. Evaluating when $x=4$: $x+3=4+3=7$

　　b. Evaluating when $x=4$: $\dfrac{x^2+9}{x+3}=\dfrac{(4)^2+9}{4+3}=\dfrac{16+9}{4+3}=\dfrac{25}{7}$

　　c. Evaluating when $x=4$: $x-3+\dfrac{18}{x+3}=4-3+\dfrac{18}{4+3}=1+\dfrac{18}{7}=\dfrac{7}{7}+\dfrac{18}{7}=\dfrac{25}{7}$

37. The monthly payment is: $\dfrac{\$5,162}{12} \approx \$430.17$

39. The monthly payment is: $\dfrac{\$3,482}{12} \approx \$290.17$

41. Let x and y represent the two numbers. The system of equations is:
$$x+y=25$$
$$y=4x$$
Substituting into the first equation:
$$x+4x=25$$
$$5x=25$$
$$x=5$$
$$y=4(5)=20$$
The two numbers are 5 and 20.

43. Let x represent the amount invested at 8% and y represent the amount invested at 9%. The system of equations is:
$$x+y=1200$$
$$0.08x+0.09y=100$$
Multiplying the first equation by -0.08:
$$-0.08x-0.08y=-96$$
$$0.08x+0.09y=100$$
Adding the two equations:
$$0.01y=4$$
$$y=400$$
Substituting into the first equation:
$$x+400=1200$$
$$x=800$$
You have $800 invested at 8% and $400 invested at 9%.

45. Let x represent the number of $5 bills and $x + 4$ represent the number of $10 bills. The equation is:
$$5(x) + 10(x + 4) = 160$$
$$5x + 10x + 40 = 160$$
$$15x + 40 = 160$$
$$15x = 120$$
$$x = 8$$
$$x + 4 = 12$$
You have 8 $5 bills and 12 $10 bills.

47. Let x represent the gallons of 20% antifreeze and y represent the gallons of 60% antifreeze. The system of equations is:
$$x + y = 16$$
$$0.20x + 0.60y = 0.35(16)$$
Multiplying the first equation by -0.20:
$$-0.20x - 0.20y = -3.2$$
$$0.20x + 0.60y = 5.6$$
Adding the two equations:
$$0.40y = 2.4$$
$$y = 6$$
Substituting into the first equation:
$$x + 6 = 16$$
$$x = 10$$
The mixture contains 10 gallons of 20% antifreeze and 6 gallons of 60% antifreeze.

Chapter 4 Review/Test

1. Simplifying the expression: $(-1)^3 = (-1)(-1)(-1) = -1$
2. Simplifying the expression: $-8^2 = -8 \cdot 8 = -64$
3. Simplifying the expression: $\left(\frac{3}{7}\right)^2 = \left(\frac{3}{7}\right) \cdot \left(\frac{3}{7}\right) = \frac{9}{49}$
4. Simplifying the expression: $y^3 \cdot y^9 = y^{3+9} = y^{12}$
5. Simplifying the expression: $x^{15} \cdot x^7 \cdot x^5 \cdot x^3 = x^{15+7+5+3} = x^{30}$
6. Simplifying the expression: $\left(x^7\right)^5 = x^{7 \cdot 5} = x^{35}$
7. Simplifying the expression: $\left(2^6\right)^4 = 2^{6 \cdot 4} = 2^{24}$
8. Simplifying the expression: $(3y)^3 = 3^3 \cdot y^3 = 27y^3$
9. Simplifying the expression: $(-2xyz)^3 = (-2)^3 \cdot x^3 y^3 z^3 = -8x^3 y^3 z^3$
10. Writing with positive exponents: $7^{-2} = \frac{1}{7^2} = \frac{1}{49}$
11. Writing with positive exponents: $4x^{-5} = 4 \cdot \frac{1}{x^5} = \frac{4}{x^5}$
12. Writing with positive exponents: $(3y)^{-3} = \frac{1}{(3y)^3} = \frac{1}{3^3 y^3} = \frac{1}{27y^3}$
13. Simplifying the expression: $\frac{a^9}{a^3} = a^{9-3} = a^6$
14. Simplifying the expression: $\left(\frac{x^3}{x^5}\right)^2 = \left(x^{3-5}\right)^2 = \left(x^{-2}\right)^2 = x^{-4} = \frac{1}{x^4}$
15. Simplifying the expression: $\frac{x^9}{x^{-6}} = x^{9-(-6)} = x^{9+6} = x^{15}$
16. Simplifying the expression: $\frac{x^{-7}}{x^{-2}} = x^{-7-(-2)} = x^{-7+2} = x^{-5} = \frac{1}{x^5}$
17. Simplifying the expression: $(-3xy)^0 = 1$
18. Simplifying the expression: $3^0 - 5^1 + 5^0 = 1 - 5 + 1 = -3$

19. Simplifying the expression: $(3x^3y^2)^2 = 3^2 x^6 y^4 = 9x^6 y^4$

20. Simplifying the expression: $(2a^3b^2)^4 (2a^5b^6)^2 = 2^4 a^{12} b^8 \cdot 2^2 a^{10} b^{12} = 2^6 a^{22} b^{20} = 64 a^{22} b^{20}$

21. Simplifying the expression: $(-3xy^2)^{-3} = \dfrac{1}{(-3xy^2)^3} = \dfrac{1}{(-3)^3 x^3 y^6} = -\dfrac{1}{27 x^3 y^6}$

22. Simplifying the expression: $\dfrac{(b^3)^4 (b^2)^5}{(b^7)^3} = \dfrac{b^{12} \cdot b^{10}}{b^{21}} = \dfrac{b^{22}}{b^{21}} = b$

23. Simplifying the expression: $\dfrac{(x^{-3})^3 (x^6)^{-1}}{(x^{-5})^{-4}} = \dfrac{x^{-9} \cdot x^{-6}}{x^{20}} = \dfrac{x^{-15}}{x^{20}} = x^{-15-20} = x^{-35} = \dfrac{1}{x^{35}}$

24. Simplifying the expression: $\dfrac{(2x^4)(15x^9)}{6x^6} = \dfrac{30x^{13}}{6x^6} = \dfrac{30}{6} \cdot \dfrac{x^{13}}{x^6} = 5x^7$

25. Simplifying the expression: $\dfrac{(10x^3 y^5)(21x^2 y^6)}{(7xy^3)(5x^9 y)} = \dfrac{210 x^5 y^{11}}{35 x^{10} y^4} = \dfrac{210}{35} \cdot \dfrac{x^5}{x^{10}} \cdot \dfrac{y^{11}}{y^4} = 6 \cdot \dfrac{1}{x^5} \cdot y^7 = \dfrac{6y^7}{x^5}$

26. Simplifying the expression: $\dfrac{21a^{10}}{3a^4} - \dfrac{18a^{17}}{6a^{11}} = 7a^6 - 3a^6 = 4a^6$

27. Simplifying the expression: $\dfrac{8x^8 y^3}{2x^3 y} - \dfrac{10x^6 y^9}{5xy^7} = 4x^5 y^2 - 2x^5 y^2 = 2x^5 y^2$

28. Finding the product: $(3.2 \times 10^3)(2 \times 10^4) = 6.4 \times 10^7$

29. Finding the quotient: $\dfrac{4.6 \times 10^5}{2 \times 10^{-3}} = 2.3 \times 10^8$

30. Finding the value: $\dfrac{(4 \times 10^6)(6 \times 10^5)}{3 \times 10^8} = \dfrac{24 \times 10^{11}}{3 \times 10^8} = 8 \times 10^3$

31. Performing the operations: $(3a^2 - 5a + 5) + (5a^2 - 7a - 8) = (3a^2 + 5a^2) + (-5a - 7a) + (5 - 8) = 8a^2 - 12a - 3$

32. Performing the operations:
$$(-7x^2 + 3x - 6) - (8x^2 - 4x + 7) + (3x^2 - 2x - 1) = -7x^2 + 3x - 6 - 8x^2 + 4x - 7 + 3x^2 - 2x - 1$$
$$= (-7x^2 - 8x^2 + 3x^2) + (3x + 4x - 2x) + (-6 - 7 - 1)$$
$$= -12x^2 + 5x - 14$$

33. Performing the operations:
$$(4x^2 - 3x - 2) - (8x^2 + 3x - 2) = 4x^2 - 3x - 2 - 8x^2 - 3x + 2$$
$$= (4x^2 - 8x^2) + (-3x - 3x) + (-2 + 2)$$
$$= -4x^2 - 6x$$

34. Evaluating when $x = 3$: $2x^2 - 3x + 5 = 2(3)^2 - 3(3) + 5 = 2(9) - 9 + 5 = 18 - 9 + 5 = 14$

35. Multiplying: $3x(4x - 7) = 3x(4x) - 3x(7) = 12x^2 - 21x$

36. Multiplying: $8x^3 y(3x^2 y - 5xy^2 + 4y^3) = 8x^3 y(3x^2 y) - 8x^3 y(5xy^2) + 8x^3 y(4y^3) = 24x^5 y^2 - 40x^4 y^3 + 32x^3 y^4$

37. Multiplying using the column method:

$$\begin{array}{r} a^2 + 5a - 4 \\ a + 1 \\ \hline a^3 + 5a^2 - 4a \\ a^2 + 5a - 4 \\ \hline a^3 + 6a^2 + a - 4 \end{array}$$

38. Multiplying using the column method:

$$\begin{array}{r} x^2 - 5x + 25 \\ x + 5 \\ \hline x^3 - 5x^2 + 25x \\ 5x^2 - 25x + 125 \\ \hline x^3 + 125 \end{array}$$

39. Multiplying using the FOIL method: $(3x - 7)(2x - 5) = 6x^2 - 14x - 15x + 35 = 6x^2 - 29x + 35$

40. Multiplying using the difference of squares formula: $\left(5y + \frac{1}{5}\right)\left(5y - \frac{1}{5}\right) = (5y)^2 - \left(\frac{1}{5}\right)^2 = 25y^2 - \frac{1}{25}$

41. Multiplying using the difference of squares formula: $(a^2 - 3)(a^2 + 3) = (a^2)^2 - (3)^2 = a^4 - 9$

42. Multiplying using the square of binomial formula: $(a - 5)^2 = a^2 - 2(a)(5) + (5)^2 = a^2 - 10a + 25$

43. Multiplying using the square of binomial formula: $(3x + 4)^2 = (3x)^2 + 2(3x)(4) + (4)^2 = 9x^2 + 24x + 16$

44. Multiplying using the square of binomial formula: $(y^2 + 3)^2 = (y^2)^2 + 2(y^2)(3) + (3)^2 = y^4 + 6y^2 + 9$

45. Performing the division: $\dfrac{10ab + 20a^2}{-5a} = \dfrac{10ab}{-5a} + \dfrac{20a^2}{-5a} = -2b - 4a$

46. Performing the division: $\dfrac{40x^5y^4 - 32x^3y^3 - 16x^2y}{-8xy} = \dfrac{40x^5y^4}{-8xy} + \dfrac{-32x^3y^3}{-8xy} + \dfrac{-16x^2y}{-8xy} = -5x^4y^3 + 4x^2y^2 + 2x$

47. Using long division:

$$\begin{array}{r} x + 9 \\ x + 6 \overline{\smash{)}x^2 + 15x + 54} \\ \underline{x^2 + 6x } \\ 9x + 54 \\ \underline{9x + 54} \\ 0 \end{array}$$

The quotient is $x + 9$.

48. Using long division:

$$\begin{array}{r} 2x + 5 \\ 3x - 1 \overline{\smash{)}6x^2 + 13x - 5} \\ \underline{6x^2 - 2x } \\ 15x - 5 \\ \underline{15x - 5} \\ 0 \end{array}$$

The quotient is $2x + 5$.

49. Using long division:

$$\begin{array}{r} x^2 - 4x + 16 \\ x + 4 \overline{\smash{)}x^3 + 0x^2 + 0x + 64} \\ \underline{x^3 + 4x^2 } \\ -4x^2 + 0x \\ \underline{-4x^2 - 16x } \\ 16x + 64 \\ \underline{16x + 64} \\ 0 \end{array}$$

The quotient is $x^2 - 4x + 16$.

50. Using long division:

$$\begin{array}{r} x - 3 \\ 3x + 2 \overline{\smash{)}3x^2 - 7x + 10} \\ \underline{3x^2 + 2x } \\ -9x + 10 \\ \underline{-9x - 6} \\ 16 \end{array}$$

The quotient is $x - 3 + \dfrac{16}{3x + 2}$.

51. Using long division:

$$\begin{array}{r} x^2 - 4x + 5 \\ 2x+1 \overline{\smash{\big)}\, 2x^3 - 7x^2 + 6x + 10} \\ \underline{2x^3 + x^2 } \\ -8x^2 + 6x \\ \underline{-8x^2 - 4x } \\ 10x + 10 \\ \underline{10x + 5} \\ 5 \end{array}$$

The quotient is $x^2 - 4x + 5 + \dfrac{5}{2x+1}$.

Chapter 5
Factoring

5.1 The Greatest Common Factor and Factoring by Grouping

1. Factoring out the greatest common factor: $15x + 25 = 5(3x + 5)$
3. Factoring out the greatest common factor: $6a + 9 = 3(2a + 3)$
5. Factoring out the greatest common factor: $4x - 8y = 4(x - 2y)$
7. Factoring out the greatest common factor: $3x^2 - 6x - 9 = 3(x^2 - 2x - 3)$
9. Factoring out the greatest common factor: $3a^2 - 3a - 60 = 3(a^2 - a - 20)$
11. Factoring out the greatest common factor: $24y^2 - 52y + 24 = 4(6y^2 - 13y + 6)$
13. Factoring out the greatest common factor: $9x^2 - 8x^3 = x^2(9 - 8x)$
15. Factoring out the greatest common factor: $13a^2 - 26a^3 = 13a^2(1 - 2a)$
17. Factoring out the greatest common factor: $21x^2y - 28xy^2 = 7xy(3x - 4y)$
19. Factoring out the greatest common factor: $22a^2b^2 - 11ab^2 = 11ab^2(2a - 1)$
21. Factoring out the greatest common factor: $7x^3 + 21x^2 - 28x = 7x(x^2 + 3x - 4)$
23. Factoring out the greatest common factor: $121y^4 - 11x^4 = 11(11y^4 - x^4)$
25. Factoring out the greatest common factor: $100x^4 - 50x^3 + 25x^2 = 25x^2(4x^2 - 2x + 1)$
27. Factoring out the greatest common factor: $8a^2 + 16b^2 + 32c^2 = 8(a^2 + 2b^2 + 4c^2)$
29. Factoring out the greatest common factor: $4a^2b - 16ab^2 + 32a^2b^2 = 4ab(a - 4b + 8ab)$
31. Factoring out the greatest common factor: $121a^3b^2 - 22a^2b^3 + 33a^3b^3 = 11a^2b^2(11a - 2b + 3ab)$
33. Factoring out the greatest common factor: $12x^2y^3 - 72x^5y^3 - 36x^4y^4 = 12x^2y^3(1 - 6x^3 - 3x^2y)$
35. Factoring by grouping: $xy + 5x + 3y + 15 = x(y + 5) + 3(y + 5) = (y + 5)(x + 3)$
37. Factoring by grouping: $xy + 6x + 2y + 12 = x(y + 6) + 2(y + 6) = (y + 6)(x + 2)$
39. Factoring by grouping: $ab + 7a - 3b - 21 = a(b + 7) - 3(b + 7) = (b + 7)(a - 3)$

132 Chapter 5 Factoring

41. Factoring by grouping: $ax - bx + ay - by = x(a-b) + y(a-b) = (a-b)(x+y)$

43. Factoring by grouping: $2ax + 6x - 5a - 15 = 2x(a+3) - 5(a+3) = (a+3)(2x-5)$

45. Factoring by grouping: $3xb - 4b - 6x + 8 = b(3x-4) - 2(3x-4) = (3x-4)(b-2)$

47. Factoring by grouping: $x^2 + ax + 2x + 2a = x(x+a) + 2(x+a) = (x+a)(x+2)$

49. Factoring by grouping: $x^2 - ax - bx + ab = x(x-a) - b(x-a) = (x-a)(x-b)$

51. Factoring by grouping: $ax + ay + bx + by + cx + cy = a(x+y) + b(x+y) + c(x+y) = (x+y)(a+b+c)$

53. Factoring by grouping: $6x^2 + 9x + 4x + 6 = 3x(2x+3) + 2(2x+3) = (2x+3)(3x+2)$

55. Factoring by grouping: $20x^2 - 2x + 50x - 5 = 2x(10x-1) + 5(10x-1) = (10x-1)(2x+5)$

57. Factoring by grouping: $20x^2 + 4x + 25x + 5 = 4x(5x+1) + 5(5x+1) = (5x+1)(4x+5)$

59. Factoring by grouping: $x^3 + 2x^2 + 3x + 6 = x^2(x+2) + 3(x+2) = (x+2)(x^2+3)$

61. Factoring by grouping: $6x^3 - 4x^2 + 15x - 10 = 2x^2(3x-2) + 5(3x-2) = (3x-2)(2x^2+5)$

63. Its greatest common factor is $3 \cdot 2 = 6$.

65. The correct factoring is: $12x^2 + 6x + 3 = 3(4x^2 + 2x + 1)$

67. Dividing: $\dfrac{y^3 - 16y^2 + 64y}{y} = \dfrac{y^3}{y} - \dfrac{16y^2}{y} + \dfrac{64y}{y} = y^2 - 16y + 64$

69. Dividing: $\dfrac{-12x^4 + 48x^3 + 144x^2}{-12x^2} = \dfrac{-12x^4}{-12x^2} + \dfrac{48x^3}{-12x^2} + \dfrac{144x^2}{-12x^2} = x^2 - 4x - 12$

71. Dividing: $\dfrac{-18y^5 + 63y^4 - 108y^3}{-9y^3} = \dfrac{-18y^5}{-9y^3} + \dfrac{63y^4}{-9y^3} - \dfrac{108y^3}{-9y^3} = 2y^2 - 7y + 12$

73. Subtracting the polynomials: $(5x^2 + 5x - 4) - (3x^2 - 2x + 7) = 5x^2 + 5x - 4 - 3x^2 + 2x - 7 = 2x^2 + 7x - 11$

75. Subtracting the polynomials: $(7x+3) - (4x-5) = 7x + 3 - 4x + 5 = 3x + 8$

77. Subtracting the polynomials: $(5x^2 - 5) - (2x^2 - 4x) = 5x^2 - 5 - 2x^2 + 4x = 3x^2 + 4x - 5$

79. Multiplying using the FOIL method: $(x-7)(x+2) = x^2 - 7x + 2x - 14 = x^2 - 5x - 14$

81. Multiplying using the FOIL method: $(x-3)(x+2) = x^2 - 3x + 2x - 6 = x^2 - x - 6$

83. Multiplying using the column method:

$$\begin{array}{r} x^2 - 3x + 9 \\ x + 3 \\ \hline x^3 - 3x^2 + 9x \\ 3x^2 - 9x + 27 \\ \hline x^3 + 27 \end{array}$$

85. Multiplying using the column method:

$$\begin{array}{r} x^2 + 4x - 3 \\ 2x + 1 \\ \hline 2x^3 + 8x^2 - 6x \\ x^2 + 4x - 3 \\ \hline 2x^3 + 9x^2 - 2x - 3 \end{array}$$

87. Multiplying: $3x^4(6x^3 - 4x^2 + 2x) = 3x^4 \cdot 6x^3 - 3x^4 \cdot 4x^2 + 3x^4 \cdot 2x = 18x^7 - 12x^6 + 6x^5$

89. Multiplying: $\left(x + \tfrac{1}{3}\right)\left(x + \tfrac{2}{3}\right) = x^2 + \tfrac{2}{3}x + \tfrac{1}{3}x + \tfrac{2}{9} = x^2 + x + \tfrac{2}{9}$

91. Multiplying: $(6x + 4y)(2x - 3y) = 12x^2 - 18xy + 8xy - 12y^2 = 12x^2 - 10xy - 12y^2$

93. Multiplying: $(9a+1)(9a-1) = 81a^2 - 9a + 9a - 1 = 81a^2 - 1$

95. Multiplying: $(x-9)(x-9) = x^2 - 9x - 9x + 81 = x^2 - 18x + 81$

97. Multiplying: $(x+2)(x^2 - 2x + 4) = x(x^2 - 2x + 4) + 2(x^2 - 2x + 4) = x^3 - 2x^2 + 4x + 2x^2 - 4x + 8 = x^3 + 8$

5.2 Factoring Trinomials

1. Factoring the trinomial: $x^2 + 7x + 12 = (x+3)(x+4)$
3. Factoring the trinomial: $x^2 + 3x + 2 = (x+2)(x+1)$
5. Factoring the trinomial: $a^2 + 10a + 21 = (a+7)(a+3)$
7. Factoring the trinomial: $x^2 - 7x + 10 = (x-5)(x-2)$
9. Factoring the trinomial: $y^2 - 10y + 21 = (y-7)(y-3)$
11. Factoring the trinomial: $x^2 - x - 12 = (x-4)(x+3)$
13. Factoring the trinomial: $y^2 + y - 12 = (y+4)(y-3)$
15. Factoring the trinomial: $x^2 + 5x - 14 = (x+7)(x-2)$
17. Factoring the trinomial: $r^2 - 8r - 9 = (r-9)(r+1)$
19. Factoring the trinomial: $x^2 - x - 30 = (x-6)(x+5)$
21. Factoring the trinomial: $a^2 + 15a + 56 = (a+7)(a+8)$
23. Factoring the trinomial: $y^2 - y - 42 = (y-7)(y+6)$
25. Factoring the trinomial: $x^2 + 13x + 42 = (x+7)(x+6)$
27. Factoring the trinomial: $2x^2 + 6x + 4 = 2(x^2 + 3x + 2) = 2(x+2)(x+1)$
29. Factoring the trinomial: $3a^2 - 3a - 60 = 3(a^2 - a - 20) = 3(a-5)(a+4)$
31. Factoring the trinomial: $100x^2 - 500x + 600 = 100(x^2 - 5x + 6) = 100(x-3)(x-2)$
33. Factoring the trinomial: $100p^2 - 1300p + 4000 = 100(p^2 - 13p + 40) = 100(p-8)(p-5)$
35. Factoring the trinomial: $x^4 - x^3 - 12x^2 = x^2(x^2 - x - 12) = x^2(x-4)(x+3)$
37. Factoring the trinomial: $2r^3 + 4r^2 - 30r = 2r(r^2 + 2r - 15) = 2r(r+5)(r-3)$
39. Factoring the trinomial: $2y^4 - 6y^3 - 8y^2 = 2y^2(y^2 - 3y - 4) = 2y^2(y-4)(y+1)$
41. Factoring the trinomial: $x^5 + 4x^4 + 4x^3 = x^3(x^2 + 4x + 4) = x^3(x+2)(x+2) = x^3(x+2)^2$
43. Factoring the trinomial: $3y^4 - 12y^3 - 15y^2 = 3y^2(y^2 - 4y - 5) = 3y^2(y-5)(y+1)$
45. Factoring the trinomial: $4x^4 - 52x^3 + 144x^2 = 4x^2(x^2 - 13x + 36) = 4x^2(x-9)(x-4)$
47. Factoring the trinomial: $x^2 + 5xy + 6y^2 = (x+2y)(x+3y)$
49. Factoring the trinomial: $x^2 - 9xy + 20y^2 = (x-4y)(x-5y)$
51. Factoring the trinomial: $a^2 + 2ab - 8b^2 = (a+4b)(a-2b)$
53. Factoring the trinomial: $a^2 - 10ab + 25b^2 = (a-5b)(a-5b) = (a-5b)^2$
55. Factoring the trinomial: $a^2 + 10ab + 25b^2 = (a+5b)(a+5b) = (a+5b)^2$
57. Factoring the trinomial: $x^2 + 2xa - 48a^2 = (x+8a)(x-6a)$
59. Factoring the trinomial: $x^2 - 5xb - 36b^2 = (x-9b)(x+4b)$
61. Factoring the trinomial: $x^4 - 5x^2 + 6 = (x^2 - 2)(x^2 - 3)$
63. Factoring the trinomial: $x^2 - 80x - 2000 = (x-100)(x+20)$
65. Factoring the trinomial: $x^2 - x + \frac{1}{4} = \left(x - \frac{1}{2}\right)\left(x - \frac{1}{2}\right) = \left(x - \frac{1}{2}\right)^2$
67. Factoring the trinomial: $x^2 + 0.6x + 0.08 = (x+0.4)(x+0.2)$

69. We can use long division to find the other factor:

$$\begin{array}{r} x+16 \\ x+8 \overline{\smash{\big)}\, x^2+24x+128} \\ \underline{x^2+8x} \\ 16x+128 \\ \underline{16x+128} \\ 0 \end{array}$$

The other factor is $x+16$.

71. Using FOIL to multiply out the factors: $(4x+3)(x-1) = 4x^2 + 3x - 4x - 3 = 4x^2 - x - 3$

73. Simplifying; $\left(-\frac{2}{5}\right)^2 = \left(-\frac{2}{5}\right)\left(-\frac{2}{5}\right) = \frac{4}{25}$

75. Simplifying; $(3a^3)^2(2a^2)^3 = (9a^6)(8a^6) = 72a^{12}$

77. Simplifying; $\dfrac{(4x)^{-7}}{(4x)^{-5}} = (4x)^{-2} = \dfrac{1}{(4x)^2} = \dfrac{1}{16x^2}$

79. Simplifying; $\dfrac{12a^5b^3}{72a^2b^5} = \dfrac{1}{6}a^{5-2}b^{3-5} = \dfrac{1}{6}a^3b^{-2} = \dfrac{a^3}{6b^2}$

81. Simplifying; $\dfrac{15x^{-5}y^3}{45x^2y^5} = \dfrac{1}{3}x^{-5-2}y^{3-5} = \dfrac{1}{3}x^{-7}y^{-2} = \dfrac{1}{3x^7y^2}$

83. Simplifying; $(-7x^3y)(3xy^4) = -21x^{3+1}y^{1+4} = -21x^4y^5$

85. Simplifying; $(-5a^3b^{-1})(4a^{-2}b^4) = -20a^{3-2}b^{-1+4} = -20ab^3$

87. Simplifying; $(9a^2b^3)(-3a^3b^5) = -27a^{2+3}b^{3+5} = -27a^5b^8$

89. Multiplying using the FOIL method: $(6a+1)(a+2) = 6a^2 + a + 12a + 2 = 6a^2 + 13a + 2$

91. Multiplying using the FOIL method: $(3a+2)(2a+1) = 6a^2 + 4a + 3a + 2 = 6a^2 + 7a + 2$

93. Multiplying using the FOIL method: $(6a+2)(a+1) = 6a^2 + 2a + 6a + 2 = 6a^2 + 8a + 2$

5.3 More Trinomials to Factor

1. Factoring the trinomial: $2x^2 + 7x + 3 = (2x+1)(x+3)$

3. Factoring the trinomial: $2a^2 - a - 3 = (2a-3)(a+1)$

5. Factoring the trinomial: $3x^2 + 2x - 5 = (3x+5)(x-1)$

7. Factoring the trinomial: $3y^2 - 14y - 5 = (3y+1)(y-5)$

9. Factoring the trinomial: $6x^2 + 13x + 6 = (3x+2)(2x+3)$

11. Factoring the trinomial: $4x^2 - 12xy + 9y^2 = (2x-3y)(2x-3y) = (2x-3y)^2$

13. Factoring the trinomial: $4y^2 - 11y - 3 = (4y+1)(y-3)$

15. Factoring the trinomial: $20x^2 - 41x + 20 = (4x-5)(5x-4)$

17. Factoring the trinomial: $20a^2 + 48ab - 5b^2 = (10a-b)(2a+5b)$

19. Factoring the trinomial: $20x^2 - 21x - 5 = (4x-5)(5x+1)$

21. Factoring the trinomial: $12m^2 + 16m - 3 = (6m-1)(2m+3)$

23. Factoring the trinomial: $20x^2 + 37x + 15 = (4x+5)(5x+3)$

25. Factoring the trinomial: $12a^2 - 25ab + 12b^2 = (3a-4b)(4a-3b)$

27. Factoring the trinomial: $3x^2 - xy - 14y^2 = (3x-7y)(x+2y)$

29. Factoring the trinomial: $14x^2 + 29x - 15 = (2x+5)(7x-3)$
31. Factoring the trinomial: $6x^2 - 43x + 55 = (3x-5)(2x-11)$
33. Factoring the trinomial: $15t^2 - 67t + 38 = (5t-19)(3t-2)$
35. Factoring the trinomial: $4x^2 + 2x - 6 = 2(2x^2 + x - 3) = 2(2x+3)(x-1)$
37. Factoring the trinomial: $24a^2 - 50a + 24 = 2(12a^2 - 25a + 12) = 2(4a-3)(3a-4)$
39. Factoring the trinomial: $10x^3 - 23x^2 + 12x = x(10x^2 - 23x + 12) = x(5x-4)(2x-3)$
41. Factoring the trinomial: $6x^4 - 11x^3 - 10x^2 = x^2(6x^2 - 11x - 10) = x^2(3x+2)(2x-5)$
43. Factoring the trinomial: $10a^3 - 6a^2 - 4a = 2a(5a^2 - 3a - 2) = 2a(5a+2)(a-1)$
45. Factoring the trinomial: $15x^3 - 102x^2 - 21x = 3x(5x^2 - 34x - 7) = 3x(5x+1)(x-7)$
47. Factoring the trinomial: $35y^3 - 60y^2 - 20y = 5y(7y^2 - 12y - 4) = 5y(7y+2)(y-2)$
49. Factoring the trinomial: $15a^4 - 2a^3 - a^2 = a^2(15a^2 - 2a - 1) = a^2(5a+1)(3a-1)$
51. Factoring the trinomial: $24x^2y - 6xy - 45y = 3y(8x^2 - 2x - 15) = 3y(4x+5)(2x-3)$
53. Factoring the trinomial: $12x^2y - 34xy^2 + 14y^3 = 2y(6x^2 - 17xy + 7y^2) = 2y(2x-y)(3x-7y)$
55. Evaluating each expression when $x = 2$:
$2x^2 + 7x + 3 = 2(2)^2 + 7(2) + 3 = 8 + 14 + 3 = 25$
$(2x+1)(x+3) = (2 \cdot 2 + 1)(2+3) = (5)(5) = 25$
57. Multiplying using the difference of squares formula: $(2x+3)(2x-3) = (2x)^2 - (3)^2 = 4x^2 - 9$
59. Multiplying using the difference of squares formula:
$(x+3)(x-3)(x^2+9) = (x^2-9)(x^2+9) = (x^2)^2 - (9)^2 = x^4 - 81$
61. Factoring: $12x^2 - 71x + 105 = (x-3)(12x-35)$. The other factor is $12x - 35$.
63. Factoring: $54x^2 + 111x + 56 = (6x+7)(9x+8)$. The other factor is $9x+8$.
65. Factoring: $16t^2 - 64t + 48 = 16(t^2 - 4t + 3) = 16(t-1)(t-3)$
67. Factoring: $h = 8 + 62t - 16t^2 = 2(4 + 31t - 8t^2) = 2(4-t)(1+8t)$. Now completing the table:

Time t (seconds)	Height h (feet)
0	8
1	54
2	68
3	50
4	0

69. Simplifying: $(6x^3 - 4x^2 + 2x) + (9x^2 - 6x + 3) = 6x^3 - 4x^2 + 2x + 9x^2 - 6x + 3 = 6x^3 + 5x^2 - 4x + 3$
71. Simplifying: $(-7x^4 + 4x^3 - 6x) + (8x^4 + 7x^3 - 9) = -7x^4 + 4x^3 - 6x + 8x^4 + 7x^3 - 9 = x^4 + 11x^3 - 6x - 9$
73. Simplifying: $(2x^5 + 3x^3 + 4x) + (5x^3 - 6x - 7) = 2x^5 + 3x^3 + 4x + 5x^3 - 6x - 7 = 2x^5 + 8x^3 - 2x - 7$
75. Simplifying: $(-8x^5 - 5x^4 + 7) + (7x^4 + 2x^2 + 5) = -8x^5 - 5x^4 + 7 + 7x^4 + 2x^2 + 5 = -8x^5 + 2x^4 + 2x^2 + 12$

136 Chapter 5 Factoring

77. Simplifying: $\dfrac{24x^3y^7}{6x^{-2}y^4} + \dfrac{27x^{-2}y^{10}}{9x^{-7}y^7} = 4x^5y^3 + 3x^5y^3 = 7x^5y^3$

79. Simplifying: $\dfrac{18a^5b^9}{3a^3b^6} - \dfrac{48a^{-3}b^{-1}}{16a^{-5}b^{-4}} = 6a^2b^3 - 3a^2b^3 = 3a^2b^3$

81. Multiplying: $(x+3)(x-3) = x^2 - (3)^2 = x^2 - 9$ 83. Multiplying: $(x+5)(x-5) = x^2 - (5)^2 = x^2 - 25$

85. Multiplying: $(x+7)(x-7) = x^2 - (7)^2 = x^2 - 49$ 87. Multiplying: $(x+9)(x-9) = x^2 - (9)^2 = x^2 - 81$

89. Multiplying: $(2x-3y)(2x+3y) = (2x)^2 - (3y)^2 = 4x^2 - 9y^2$

91. Multiplying: $(x^2+4)(x+2)(x-2) = (x^2+4)(x^2-4) = (x^2)^2 - (4)^2 = x^4 - 16$

93. Multiplying: $(x+3)^2 = x^2 + 2(x)(3) + (3)^2 = x^2 + 6x + 9$

95. Multiplying: $(x+5)^2 = x^2 + 2(x)(5) + (5)^2 = x^2 + 10x + 25$

97. Multiplying: $(x+7)^2 = x^2 + 2(x)(7) + (7)^2 = x^2 + 14x + 49$

99. Multiplying: $(x+9)^2 = x^2 + 2(x)(9) + (9)^2 = x^2 + 18x + 81$

101. Multiplying: $(2x+3)^2 = (2x)^2 + 2(2x)(3) + (3)^2 = 4x^2 + 12x + 9$

103. Multiplying: $(4x-2y)^2 = (4x)^2 - 2(4x)(2y) + (2y)^2 = 16x^2 - 16xy + 4y^2$

5.4 The Difference of Two Squares

1. Factoring the binomial: $x^2 - 9 = (x+3)(x-3)$ 3. Factoring the binomial: $a^2 - 36 = (a+6)(a-6)$

5. Factoring the binomial: $x^2 - 49 = (x+7)(x-7)$

7. Factoring the binomial: $4a^2 - 16 = 4(a^2 - 4) = 4(a+2)(a-2)$

9. The expression $9x^2 + 25$ cannot be factored.

11. Factoring the binomial: $25x^2 - 169 = (5x+13)(5x-13)$

13. Factoring the binomial: $9a^2 - 16b^2 = (3a+4b)(3a-4b)$

15. Factoring the binomial: $9 - m^2 = (3+m)(3-m)$ 17. Factoring the binomial: $25 - 4x^2 = (5+2x)(5-2x)$

19. Factoring the binomial: $2x^2 - 18 = 2(x^2 - 9) = 2(x+3)(x-3)$

21. Factoring the binomial: $32a^2 - 128 = 32(a^2 - 4) = 32(a+2)(a-2)$

23. Factoring the binomial: $8x^2y - 18y = 2y(4x^2 - 9) = 2y(2x+3)(2x-3)$

25. Factoring the binomial: $a^4 - b^4 = (a^2+b^2)(a^2-b^2) = (a^2+b^2)(a+b)(a-b)$

27. Factoring the binomial: $16m^4 - 81 = (4m^2+9)(4m^2-9) = (4m^2+9)(2m+3)(2m-3)$

29. Factoring the binomial: $3x^3y - 75xy^3 = 3xy(x^2 - 25y^2) = 3xy(x+5y)(x-5y)$

31. Factoring the trinomial: $x^2 - 2x + 1 = (x-1)(x-1) = (x-1)^2$

33. Factoring the trinomial: $x^2 + 2x + 1 = (x+1)(x+1) = (x+1)^2$

35. Factoring the trinomial: $a^2 - 10a + 25 = (a-5)(a-5) = (a-5)^2$

37. Factoring the trinomial: $y^2 + 4y + 4 = (y+2)(y+2) = (y+2)^2$

39. Factoring the trinomial: $x^2 - 4x + 4 = (x-2)(x-2) = (x-2)^2$

41. Factoring the trinomial: $m^2 - 12m + 36 = (m-6)(m-6) = (m-6)^2$

43. Factoring the trinomial: $4a^2 + 12a + 9 = (2a+3)(2a+3) = (2a+3)^2$

45. Factoring the trinomial: $49x^2 - 14x + 1 = (7x-1)(7x-1) = (7x-1)^2$

47. Factoring the trinomial: $9y^2 - 30y + 25 = (3y-5)(3y-5) = (3y-5)^2$

49. Factoring the trinomial: $x^2 + 10xy + 25y^2 = (x+5y)(x+5y) = (x+5y)^2$

51. Factoring the trinomial: $9a^2 + 6ab + b^2 = (3a+b)(3a+b) = (3a+b)^2$

53. Factoring the trinomial: $y^2 - 3y + \frac{9}{4} = \left(y - \frac{3}{2}\right)\left(y - \frac{3}{2}\right) = \left(y - \frac{3}{2}\right)^2$

55. Factoring the trinomial: $a^2 + a + \frac{1}{4} = \left(a + \frac{1}{2}\right)\left(a + \frac{1}{2}\right) = \left(a + \frac{1}{2}\right)^2$

57. Factoring the trinomial: $x^2 - 7x + \frac{49}{4} = \left(x - \frac{7}{2}\right)\left(x - \frac{7}{2}\right) = \left(x - \frac{7}{2}\right)^2$

59. Factoring the trinomial: $x^2 - \frac{3}{4}x + \frac{9}{64} = \left(x - \frac{3}{8}\right)\left(x - \frac{3}{8}\right) = \left(x - \frac{3}{8}\right)^2$

61. Factoring the trinomial: $t^2 - \frac{2}{5}t + \frac{1}{25} = \left(t - \frac{1}{5}\right)\left(t - \frac{1}{5}\right) = \left(t - \frac{1}{5}\right)^2$

63. Factoring the trinomial: $3a^2 + 18a + 27 = 3(a^2 + 6a + 9) = 3(a+3)(a+3) = 3(a+3)^2$

65. Factoring the trinomial: $2x^2 + 20xy + 50y^2 = 2(x^2 + 10xy + 25y^2) = 2(x+5y)(x+5y) = 2(x+5y)^2$

67. Factoring the trinomial: $5x^3 + 30x^2y + 45xy^2 = 5x(x^2 + 6xy + 9y^2) = 5x(x+3y)(x+3y) = 5x(x+3y)^2$

69. Factoring by grouping: $x^2 + 6x + 9 - y^2 = (x+3)^2 - y^2 = (x+3+y)(x+3-y)$

71. Factoring by grouping: $x^2 + 2xy + y^2 - 9 = (x+y)^2 - 9 = (x+y+3)(x+y-3)$

73. Since $(x+7)^2 = x^2 + 14x + 49$, the value is $b = 14$. 75. Since $(x+5)^2 = x^2 + 10x + 25$, the value is $c = 25$.

77. Dividing: $\dfrac{24y^3 - 36y^2 - 18y}{6y} = \dfrac{24y^3}{6y} - \dfrac{36y^2}{6y} - \dfrac{18y}{6y} = 4y^2 - 6y - 3$

79. Dividing: $\dfrac{48x^7 - 36x^5 + 12x^2}{4x^2} = \dfrac{48x^7}{4x^2} - \dfrac{36x^5}{4x^2} + \dfrac{12x^2}{4x^2} = 12x^5 - 9x^3 + 3$

81. Dividing: $\dfrac{18x^7 + 12x^6 - 6x^5}{-3x^4} = \dfrac{18x^7}{-3x^4} + \dfrac{12x^6}{-3x^4} - \dfrac{6x^5}{-3x^4} = -6x^3 - 4x^2 + 2x$

83. Dividing: $\dfrac{-42x^5 + 24x^4 - 66x^2}{6x^2} = \dfrac{-42x^5}{6x^2} + \dfrac{24x^4}{6x^2} - \dfrac{66x^2}{6x^2} = -7x^3 + 4x^2 - 11$

85. Using long division:

$$\begin{array}{r}
x - 2 \\
x-3\overline{\smash{)}x^2 - 5x + 8} \\
\underline{x^2 - 3x} \\
-2x + 8 \\
\underline{-2x + 6} \\
2
\end{array}$$

The quotient is $x - 2 + \dfrac{2}{x-3}$.

87. Using long division:

$$\begin{array}{r}
3x - 2 \\
2x+3\overline{\smash{)}6x^2 + 5x + 3} \\
\underline{6x^2 + 9x} \\
-4x + 3 \\
\underline{-4x - 6} \\
9
\end{array}$$

The quotient is $3x - 2 + \dfrac{9}{2x+3}$.

138 Chapter 5 Factoring

89. a. Multiplying: $1^3 = 1$ b. Multiplying: $2^3 = 8$
     c. Multiplying: $3^3 = 27$ d. Multiplying: $4^3 = 64$
     e. Multiplying: $5^3 = 125$

91. a. Multiplying: $x(x^2 - x + 1) = x^3 - x^2 + x$ b. Multiplying: $1(x^2 - x + 1) = x^2 - x + 1$
     c. Multiplying: $(x+1)(x^2 - x + 1) = x(x^2 - x + 1) + 1(x^2 - x + 1) = x^3 - x^2 + x + x^2 - x + 1 = x^3 + 1$

93. a. Multiplying: $x(x^2 - 2x + 4) = x^3 - 2x^2 + 4x$ b. Multiplying: $2(x^2 - 2x + 4) = 2x^2 - 4x + 8$
     c. Multiplying: $(x+2)(x^2 - 2x + 4) = x(x^2 - 2x + 4) + 2(x^2 - 2x + 4) = x^3 - 2x^2 + 4x + 2x^2 - 4x + 8 = x^3 + 8$

95. a. Multiplying: $x(x^2 - 3x + 9) = x^3 - 3x^2 + 9x$ b. Multiplying: $3(x^2 - 3x + 9) = 3x^2 - 9x + 27$
     c. Multiplying:
$$(x+3)(x^2 - 3x + 9) = x(x^2 - 3x + 9) + 3(x^2 - 3x + 9) = x^3 - 3x^2 + 9x + 3x^2 - 9x + 27 = x^3 + 27$$

97. a. Multiplying: $x(x^2 - 4x + 16) = x^3 - 4x^2 + 16x$ b. Multiplying: $4(x^2 - 4x + 16) = 4x^2 - 16x + 64$
     c. Multiplying:
$$(x+4)(x^2 - 4x + 16) = x(x^2 - 4x + 16) + 4(x^2 - 4x + 16) = x^3 - 4x^2 + 16x + 4x^2 - 16x + 64 = x^3 + 64$$

99. a. Multiplying: $x(x^2 - 5x + 25) = x^3 - 5x^2 + 25x$ b. Multiplying: $5(x^2 - 5x + 25) = 5x^2 - 25x + 125$
     c. Multiplying:
$$(x+5)(x^2 - 5x + 25) = x(x^2 - 5x + 25) + 5(x^2 - 5x + 25) = x^3 - 5x^2 + 25x + 5x^2 - 25x + 125 = x^3 + 125$$

5.5 The Sum and Difference of Two Cubes

1. Factoring: $x^3 - y^3 = (x - y)(x^2 + xy + y^2)$ 3. Factoring: $a^3 + 8 = (a + 2)(a^2 - 2a + 4)$
5. Factoring: $27 + x^3 = (3 + x)(9 - 3x + x^2)$ 7. Factoring: $y^3 - 1 = (y - 1)(y^2 + y + 1)$
9. Factoring: $y^3 - 64 = (y - 4)(y^2 + 4y + 16)$ 11. Factoring: $125h^3 - t^3 = (5h - t)(25h^2 + 5ht + t^2)$
13. Factoring: $x^3 - 216 = (x - 6)(x^2 + 6x + 36)$
15. Factoring: $2y^3 - 54 = 2(y^3 - 27) = 2(y - 3)(y^2 + 3y + 9)$
17. Factoring: $2a^3 - 128b^3 = 2(a^3 - 64b^3) = 2(a - 4b)(a^2 + 4ab + 16b^2)$
19. Factoring: $2x^3 + 432y^3 = 2(x^3 + 216y^3) = 2(x + 6y)(x^2 - 6xy + 36y^2)$
21. Factoring: $10a^3 - 640b^3 = 10(a^3 - 64b^3) = 10(a - 4b)(a^2 + 4ab + 16b^2)$
23. Factoring: $10r^3 - 1250 = 10(r^3 - 125) = 10(r - 5)(r^2 + 5r + 25)$
25. Factoring: $64 + 27a^3 = (4 + 3a)(16 - 12a + 9a^2)$ 27. Factoring: $8x^3 - 27y^3 = (2x - 3y)(4x^2 + 6xy + 9y^2)$
29. Factoring: $t^3 + \frac{1}{27} = \left(t + \frac{1}{3}\right)\left(t^2 - \frac{1}{3}t + \frac{1}{9}\right)$ 31. Factoring: $27x^3 - \frac{1}{27} = \left(3x - \frac{1}{3}\right)\left(9x^2 + x + \frac{1}{9}\right)$
33. Factoring: $64a^3 + 125b^3 = (4a + 5b)(16a^2 - 20ab + 25b^2)$
35. Factoring: $\frac{1}{8}x^3 - \frac{1}{27}y^3 = \left(\frac{1}{2}x - \frac{1}{3}y\right)\left(\frac{1}{4}x^2 + \frac{1}{6}xy + \frac{1}{9}y^2\right)$
37. Factoring: $a^6 - b^6 = (a^3 + b^3)(a^3 - b^3) = (a + b)(a^2 - ab + b^2)(a - b)(a^2 + ab + b^2)$

39. Factoring: $64x^6 - y^6 = (8x^3 + y^3)(8x^3 - y^3) = (2x+y)(4x^2 - 2xy + y^2)(2x-y)(4x^2 + 2xy + y^2)$

41. Factoring: $x^6 - (5y)^6 = (x^3 + (5y)^3)(x^3 - (5y)^3) = (x+5y)(x^2 - 5xy + 25y^2)(x-5y)(x^2 + 5xy + 25y^2)$

43. Solving for x:
$$2x - 6y = 8$$
$$2x = 6y + 8$$
$$x = 3y + 4$$

45. Solving for x:
$$4x - 6y = 8$$
$$4x = 6y + 8$$
$$x = \tfrac{3}{2}y + 2$$

47. Solving for y:
$$3x - 6y = -18$$
$$-6y = -3x - 18$$
$$y = \tfrac{1}{2}x + 3$$

49. Solving for y:
$$4x - 6y = 24$$
$$-6y = -4x + 24$$
$$y = \tfrac{2}{3}x - 4$$

51. Multiplying: $2x^3(x+2)(x-2) = 2x^3(x^2 - 4) = 2x^5 - 8x^3$

53. Multiplying: $3x^2(x-3)^2 = 3x^2(x^2 - 6x + 9) = 3x^4 - 18x^3 + 27x^2$

55. Multiplying: $y(y^2 + 25) = y^3 + 25y$

57. Multiplying: $(5a-2)(3a+1) = 15a^2 + 5a - 6a - 2 = 15a^2 - a - 2$

59. Multiplying: $4x^2(x-5)(x+2) = 4x^2(x^2 - 3x - 10) = 4x^4 - 12x^3 - 40x^2$

61. Multiplying: $2ab^3(b^2 - 4b + 1) = 2ab^5 - 8ab^4 + 2ab^3$

5.6 Factoring: A General Review

1. Factoring the polynomial: $x^2 - 81 = (x+9)(x-9)$

3. Factoring the polynomial: $x^2 + 2x - 15 = (x+5)(x-3)$

5. Factoring the polynomial: $x^2 + 6x + 9 = (x+3)(x+3) = (x+3)^2$

7. Factoring the polynomial: $y^2 - 10y + 25 = (y-5)(y-5) = (y-5)^2$

9. Factoring the polynomial: $2a^3b + 6a^2b + 2ab = 2ab(a^2 + 3a + 1)$

11. The polynomial $x^2 + x + 1$ cannot be factored.

13. Factoring the polynomial: $12a^2 - 75 = 3(4a^2 - 25) = 3(2a+5)(2a-5)$

15. Factoring the polynomial: $9x^2 - 12xy + 4y^2 = (3x-2y)(3x-2y) = (3x-2y)^2$

17. Factoring the polynomial: $4x^3 + 16xy^2 = 4x(x^2 + 4y^2)$

19. Factoring the polynomial: $2y^3 + 20y^2 + 50y = 2y(y^2 + 10y + 25) = 2y(y+5)(y+5) = 2y(y+5)^2$

21. Factoring the polynomial: $a^6 + 4a^4b^2 = a^4(a^2 + 4b^2)$

23. Factoring the polynomial: $xy + 3x + 4y + 12 = x(y+3) + 4(y+3) = (y+3)(x+4)$

25. Factoring the polynomial: $x^3 - 27 = (x-3)(x^2 + 3x + 9)$

27. Factoring the polynomial: $xy - 5x + 2y - 10 = x(y-5) + 2(y-5) = (y-5)(x+2)$

29. Factoring the polynomial: $5a^2 + 10ab + 5b^2 = 5(a^2 + 2ab + b^2) = 5(a+b)(a+b) = 5(a+b)^2$

31. The polynomial $x^2 + 49$ cannot be factored.

33. Factoring the polynomial: $3x^2 + 15xy + 18y^2 = 3(x^2 + 5xy + 6y^2) = 3(x+2y)(x+3y)$
35. Factoring the polynomial: $2x^2 + 15x - 38 = (2x+19)(x-2)$
37. Factoring the polynomial: $100x^2 - 300x + 200 = 100(x^2 - 3x + 2) = 100(x-2)(x-1)$
39. Factoring the polynomial: $x^2 - 64 = (x+8)(x-8)$
41. Factoring the polynomial: $x^2 + 3x + ax + 3a = x(x+3) + a(x+3) = (x+3)(x+a)$
43. Factoring the polynomial: $49a^7 - 9a^5 = a^5(49a^2 - 9) = a^5(7a+3)(7a-3)$
45. The polynomial $49x^2 + 9y^2$ cannot be factored.
47. Factoring the polynomial: $25a^3 + 20a^2 + 3a = a(25a^2 + 20a + 3) = a(5a+3)(5a+1)$
49. Factoring the polynomial: $xa - xb + ay - by = x(a-b) + y(a-b) = (a-b)(x+y)$
51. Factoring the polynomial: $48a^4b - 3a^2b = 3a^2b(16a^2 - 1) = 3a^2b(4a+1)(4a-1)$
53. Factoring the polynomial: $20x^4 - 45x^2 = 5x^2(4x^2 - 9) = 5x^2(2x+3)(2x-3)$
55. Factoring the polynomial: $3x^2 + 35xy - 82y^2 = (3x+41y)(x-2y)$
57. Factoring the polynomial: $16x^5 - 44x^4 + 30x^3 = 2x^3(8x^2 - 22x + 15) = 2x^3(2x-3)(4x-5)$
59. Factoring the polynomial: $2x^2 + 2ax + 3x + 3a = 2x(x+a) + 3(x+a) = (x+a)(2x+3)$
61. Factoring the polynomial: $y^4 - 1 = (y^2+1)(y^2-1) = (y^2+1)(y+1)(y-1)$
63. Factoring the polynomial:
$$12x^4y^2 + 36x^3y^3 + 27x^2y^4 = 3x^2y^2(4x^2 + 12xy + 9y^2) = 3x^2y^2(2x+3y)(2x+3y) = 3x^2y^2(2x+3y)^2$$
65. Factoring the polynomial: $16t^2 - 64t + 48 = 16(t^2 - 4t + 3) = 16(t-1)(t-3)$
67. Factoring the polynomial: $54x^2 + 111x + 56 = (6x+7)(9x+8)$

69. Solving the equation:
$-2(x+4) = -10$
$-2x - 8 = -10$
$-2x = -2$
$x = 1$

71. Solving the equation:
$\frac{3}{5}x + 4 = 22$
$\frac{3}{5}x = 18$
$x = \frac{5}{3}(18) = 30$

73. Solving the equation:
$6x - 4(9-x) = -96$
$6x - 36 + 4x = -96$
$10x - 36 = -96$
$10x = -60$
$x = -6$

75. Solving the equation:
$2x - 3(4x - 7) = -3x$
$2x - 12x + 21 = -3x$
$-10x + 21 = -3x$
$21 = 7x$
$x = 3$

77. Solving the equation:
$\frac{1}{2}x - \frac{5}{12} = \frac{1}{12}x + \frac{5}{12}$
$12\left(\frac{1}{2}x - \frac{5}{12}\right) = 12\left(\frac{1}{12}x + \frac{5}{12}\right)$
$6x - 5 = x + 5$
$5x - 5 = 5$
$5x = 10$
$x = 2$

79. Solving the equation:
$3x - 6 = 9$
$3x = 15$
$x = 5$

81. Solving the equation:
$$2x+3=0$$
$$2x=-3$$
$$x=-\tfrac{3}{2}$$

83. Solving the equation:
$$4x+3=0$$
$$4x=-3$$
$$x=-\tfrac{3}{4}$$

5.7 Solving Equations by Factoring

1. Setting each factor equal to 0:
$$x+2=0 \qquad x-1=0$$
$$x=-2 \qquad x=1$$
The solutions are −2 and 1.

3. Setting each factor equal to 0:
$$a-4=0 \qquad a-5=0$$
$$a=4 \qquad a=5$$
The solutions are 4 and 5.

5. Setting each factor equal to 0:
$$x=0 \qquad x+1=0 \qquad x-3=0$$
$$x=-1 \qquad x=3$$
The solutions are 0, −1 and 3.

7. Setting each factor equal to 0:
$$3x+2=0 \qquad 2x+3=0$$
$$3x=-2 \qquad 2x=-3$$
$$x=-\tfrac{2}{3} \qquad x=-\tfrac{3}{2}$$
The solutions are $-\tfrac{2}{3}$ and $-\tfrac{3}{2}$.

9. Setting each factor equal to 0:
$$m=0 \qquad 3m+4=0 \qquad 3m-4=0$$
$$3m=-4 \qquad 3m=4$$
$$m=-\tfrac{4}{3} \qquad m=\tfrac{4}{3}$$
The solutions are 0, $-\tfrac{4}{3}$ and $\tfrac{4}{3}$.

11. Setting each factor equal to 0:
$$2y=0 \qquad 3y+1=0 \qquad 5y+3=0$$
$$y=0 \qquad 3y=-1 \qquad 5y=-3$$
$$y=-\tfrac{1}{3} \qquad y=-\tfrac{3}{5}$$
The solutions are 0, $-\tfrac{1}{3}$ and $-\tfrac{3}{5}$.

13. Solving by factoring:
$$x^2+3x+2=0$$
$$(x+2)(x+1)=0$$
$$x=-2,-1$$

15. Solving by factoring:
$$x^2-9x+20=0$$
$$(x-4)(x-5)=0$$
$$x=4,5$$

17. Solving by factoring:
$$a^2-2a-24=0$$
$$(a-6)(a+4)=0$$
$$a=6,-4$$

19. Solving by factoring:
$$100x^2-500x+600=0$$
$$100(x^2-5x+6)=0$$
$$100(x-2)(x-3)=0$$
$$x=2,3$$

21. Solving by factoring:
$$x^2 = -6x - 9$$
$$x^2 + 6x + 9 = 0$$
$$(x+3)^2 = 0$$
$$x + 3 = 0$$
$$x = -3$$

23. Solving by factoring:
$$a^2 - 16 = 0$$
$$(a+4)(a-4) = 0$$
$$a = -4, 4$$

25. Solving by factoring:
$$2x^2 + 5x - 12 = 0$$
$$(2x-3)(x+4) = 0$$
$$x = \tfrac{3}{2}, -4$$

27. Solving by factoring:
$$9x^2 + 12x + 4 = 0$$
$$(3x+2)^2 = 0$$
$$3x + 2 = 0$$
$$x = -\tfrac{2}{3}$$

29. Solving by factoring:
$$a^2 + 25 = 10a$$
$$a^2 - 10a + 25 = 0$$
$$(a-5)^2 = 0$$
$$a - 5 = 0$$
$$a = 5$$

31. Solving by factoring:
$$2x^2 = 3x + 20$$
$$2x^2 - 3x - 20 = 0$$
$$(2x+5)(x-4) = 0$$
$$x = -\tfrac{5}{2}, 4$$

33. Solving by factoring:
$$3m^2 = 20 - 7m$$
$$3m^2 + 7m - 20 = 0$$
$$(3m-5)(m+4) = 0$$
$$m = \tfrac{5}{3}, -4$$

35. Solving by factoring:
$$4x^2 - 49 = 0$$
$$(2x+7)(2x-7) = 0$$
$$x = -\tfrac{7}{2}, \tfrac{7}{2}$$

37. Solving by factoring:
$$x^2 + 6x = 0$$
$$x(x+6) = 0$$
$$x = 0, -6$$

39. Solving by factoring:
$$x^2 - 3x = 0$$
$$x(x-3) = 0$$
$$x = 0, 3$$

41. Solving by factoring:
$$2x^2 = 8x$$
$$2x^2 - 8x = 0$$
$$2x(x-4) = 0$$
$$x = 0, 4$$

43. Solving by factoring:
$$3x^2 = 15x$$
$$3x^2 - 15x = 0$$
$$3x(x-5) = 0$$
$$x = 0, 5$$

45. Solving by factoring:
$$1,400 = 400 + 700x - 100x^2$$
$$100x^2 - 700x + 1,000 = 0$$
$$100(x^2 - 7x + 10) = 0$$
$$100(x-5)(x-2) = 0$$
$$x = 2, 5$$

47. Solving by factoring:
$$6x^2 = -5x + 4$$
$$6x^2 + 5x - 4 = 0$$
$$(3x+4)(2x-1) = 0$$
$$x = -\tfrac{4}{3}, \tfrac{1}{2}$$

49. Solving by factoring:
$$x(2x-3) = 20$$
$$2x^2 - 3x = 20$$
$$2x^2 - 3x - 20 = 0$$
$$(2x+5)(x-4) = 0$$
$$x = -\tfrac{5}{2}, 4$$

51. Solving by factoring:
$$t(t+2) = 80$$
$$t^2 + 2t = 80$$
$$t^2 + 2t - 80 = 0$$
$$(t+10)(t-8) = 0$$
$$t = -10, 8$$

53. Solving by factoring:
$$4000 = (1300 - 100p)p$$
$$4000 = 1300p - 100p^2$$
$$100p^2 - 1300p + 4000 = 0$$
$$100(p^2 - 13p + 40) = 0$$
$$100(p-8)(p-5) = 0$$
$$p = 5, 8$$

55. Solving by factoring:
$$x(14-x) = 48$$
$$14x - x^2 = 48$$
$$-x^2 + 14x - 48 = 0$$
$$x^2 - 14x + 48 = 0$$
$$(x-6)(x-8) = 0$$
$$x = 6, 8$$

57. Solving by factoring:
$$(x+5)^2 = 2x + 9$$
$$x^2 + 10x + 25 = 2x + 9$$
$$x^2 + 8x + 16 = 0$$
$$(x+4)^2 = 0$$
$$x + 4 = 0$$
$$x = -4$$

59. Solving by factoring:
$$(y-6)^2 = y - 4$$
$$y^2 - 12y + 36 = y - 4$$
$$y^2 - 13y + 40 = 0$$
$$(y-5)(y-8) = 0$$
$$y = 5, 8$$

61. Solving by factoring:
$$10^2 = (x+2)^2 + x^2$$
$$100 = x^2 + 4x + 4 + x^2$$
$$100 = 2x^2 + 4x + 4$$
$$0 = 2x^2 + 4x - 96$$
$$0 = 2(x^2 + 2x - 48)$$
$$0 = 2(x+8)(x-6)$$
$$x = -8, 6$$

63. Solving by factoring:
$$2x^3 + 11x^2 + 12x = 0$$
$$x(2x^2 + 11x + 12) = 0$$
$$x(2x+3)(x+4) = 0$$
$$x = 0, -\tfrac{3}{2}, -4$$

65. Solving by factoring:
$$4y^3 - 2y^2 - 30y = 0$$
$$2y(2y^2 - y - 15) = 0$$
$$2y(2y+5)(y-3) = 0$$
$$y = 0, -\tfrac{5}{2}, 3$$

67. Solving by factoring:
$$8x^3 + 16x^2 = 10x$$
$$8x^3 + 16x^2 - 10x = 0$$
$$2x(4x^2 + 8x - 5) = 0$$
$$2x(2x-1)(2x+5) = 0$$
$$x = 0, \tfrac{1}{2}, -\tfrac{5}{2}$$

69. Solving by factoring:
$$20a^3 = -18a^2 + 18a$$
$$20a^3 + 18a^2 - 18a = 0$$
$$2a(10a^2 + 9a - 9) = 0$$
$$2a(5a-3)(2a+3) = 0$$
$$a = 0, \tfrac{3}{5}, -\tfrac{3}{2}$$

71. Solving by factoring:
$$16t^2 - 32t + 12 = 0$$
$$4(4t^2 - 8t + 3) = 0$$
$$4(2t-1)(2t-3) = 0$$
$$t = \tfrac{1}{2}, \tfrac{3}{2}$$

73. Solving the equation:
$$(a-5)(a+4) = -2a$$
$$a^2 - a - 20 = -2a$$
$$a^2 + a - 20 = 0$$
$$(a+5)(a-4) = 0$$
$$a = -5, 4$$

75. Solving the equation:
$$3x(x+1) - 2x(x-5) = -42$$
$$3x^2 + 3x - 2x^2 + 10x = -42$$
$$x^2 + 13x + 42 = 0$$
$$(x+7)(x+6) = 0$$
$$x = -7, -6$$

77. Solving the equation:
$$2x(x+3) = x(x+2) - 3$$
$$2x^2 + 6x = x^2 + 2x - 3$$
$$x^2 + 4x + 3 = 0$$
$$(x+3)(x+1) = 0$$
$$x = -3, -1$$

79. Solving the equation:
$$a(a-3) + 6 = 2a$$
$$a^2 - 3a + 6 = 2a$$
$$a^2 - 5a + 6 = 0$$
$$(a-2)(a-3) = 0$$
$$a = 2, 3$$

81. Solving the equation:
$$15(x+20) + 15x = 2x(x+20)$$
$$15x + 300 + 15x = 2x^2 + 40x$$
$$30x + 300 = 2x^2 + 40x$$
$$0 = 2x^2 + 10x - 300$$
$$0 = 2(x^2 + 5x - 150)$$
$$0 = 2(x+15)(x-10)$$
$$x = -15, 10$$

83. Solving the equation:
$$15 = a(a+2)$$
$$15 = a^2 + 2a$$
$$0 = a^2 + 2a - 15$$
$$0 = (a+5)(a-3)$$
$$a = -5, 3$$

85. Solving by factoring:
$$x^3 + 3x^2 - 4x - 12 = 0$$
$$x^2(x+3) - 4(x+3) = 0$$
$$(x+3)(x^2 - 4) = 0$$
$$(x+3)(x+2)(x-2) = 0$$
$$x = -3, -2, 2$$

87. Solving by factoring:
$$x^3 + x^2 - 16x - 16 = 0$$
$$x^2(x+1) - 16(x+1) = 0$$
$$(x+1)(x^2 - 16) = 0$$
$$(x+1)(x+4)(x-4) = 0$$
$$x = -1, -4, 4$$

89. Writing the equation:
$$(x-3)(x-5) = 0$$
$$x^2 - 8x + 15 = 0$$

91. a. Writing the equation:
$$(x-3)(x-2) = 0$$
$$x^2 - 5x + 6 = 0$$

  b. Writing the equation:
$$(x-1)(x-6) = 0$$
$$x^2 - 7x + 6 = 0$$

  c. Writing the equation:
$$(x-3)(x+2) = 0$$
$$x^2 - x - 6 = 0$$

93. Simplifying using the properties of exponents: $2^{-3} = \dfrac{1}{2^3} = \dfrac{1}{8}$

95. Simplifying using the properties of exponents: $\dfrac{x^5}{x^{-3}} = x^{5-(-3)} = x^{5+3} = x^8$

97. Simplifying using the properties of exponents: $\dfrac{(x^2)^3}{(x^{-3})^4} = \dfrac{x^6}{x^{-12}} = x^{6-(-12)} = x^{6+12} = x^{18}$

99. Writing in scientific notation: $0.0056 = 5.6 \times 10^{-3}$

101. Writing in scientific notation: $5{,}670{,}000{,}000 = 5.67 \times 10^9$

103. Let x and $x+1$ represent the two consecutive integers. The equation is: $x(x+1) = 72$

105. Let x and $x+2$ represent the two consecutive odd integers. The equation is: $x(x+2) = 99$

107. Let x and $x+2$ represent the two consecutive even integers. The equation is: $x(x+2) = 5[x+(x+2)] - 10$

109. Let x represent the cost of the suit and $5x$ represent the cost of the bicycle. The equation is:
$$x + 5x = 90$$
$$6x = 90$$
$$x = 15$$
$$5x = 75$$
The suit costs $15 and the bicycle costs $75.

111. Let x represent the cost of the lot and $4x$ represent the cost of the house. The equation is:
$$x + 4x = 3000$$
$$5x = 3000$$
$$x = 600$$
$$4x = 2400$$
The lot cost $600 and the house cost $2,400.

5.8 Applications

1. Let x and $x + 2$ represent the two integers. The equation is:
$$x(x+2) = 80$$
$$x^2 + 2x = 80$$
$$x^2 + 2x - 80 = 0$$
$$(x+10)(x-8) = 0$$
$$x = -10, 8$$
$$x + 2 = -8, 10$$
The two numbers are either -10 and -8, or 8 and 10.

3. Let x and $x + 2$ represent the two integers. The equation is:
$$x(x+2) = 99$$
$$x^2 + 2x = 99$$
$$x^2 + 2x - 99 = 0$$
$$(x+11)(x-9) = 0$$
$$x = -11, 9$$
$$x + 2 = -9, 11$$
The two numbers are either -11 and -9, or 9 and 11.

5. Let x and $x + 2$ represent the two integers. The equation is:
$$x(x+2) = 5(x + x + 2) - 10$$
$$x^2 + 2x = 5(2x + 2) - 10$$
$$x^2 + 2x = 10x + 10 - 10$$
$$x^2 + 2x = 10x$$
$$x^2 - 8x = 0$$
$$x(x-8) = 0$$
$$x = 0, 8$$
$$x + 2 = 2, 10$$
The two numbers are either 0 and 2, or 8 and 10.

7. Let x and $14 - x$ represent the two numbers. The equation is:
$$x(14-x) = 48$$
$$14x - x^2 = 48$$
$$0 = x^2 - 14x + 48$$
$$0 = (x-8)(x-6)$$
$$x = 8, 6$$
$$14 - x = 6, 8$$
The two numbers are 6 and 8.

9. Let x and $5x + 2$ represent the two numbers. The equation is:
$$x(5x+2) = 24$$
$$5x^2 + 2x = 24$$
$$5x^2 + 2x - 24 = 0$$
$$(5x+12)(x-2) = 0$$
$$x = -\tfrac{12}{5}, 2$$
$$5x + 2 = -10, 12$$
The two numbers are either $-\tfrac{12}{5}$ and -10, or 2 and 12.

11. Let x and $4x$ represent the two numbers. The equation is:
$$x(4x) = 4(x + 4x)$$
$$4x^2 = 4(5x)$$
$$4x^2 = 20x$$
$$4x^2 - 20x = 0$$
$$4x(x-5) = 0$$
$$x = 0, 5$$
$$4x = 0, 20$$
The two numbers are either 0 and 0, or 5 and 20.

13. Let w represent the width and $w + 1$ represent the length. The equation is:
$$w(w+1) = 12$$
$$w^2 + w = 12$$
$$w^2 + w - 12 = 0$$
$$(w+4)(w-3) = 0$$
$$w = 3 \quad (w = -4 \text{ is impossible})$$
$$w + 1 = 4$$
The width is 3 inches and the length is 4 inches.

15. Let b represent the base and $2b$ represent the height. The equation is:
$$\tfrac{1}{2}b(2b) = 9$$
$$b^2 = 9$$
$$b^2 - 9 = 0$$
$$(b+3)(b-3) = 0$$
$$b = 3 \quad (b = -3 \text{ is impossible})$$
The base is 3 inches.

17. Let x and $x + 2$ represent the two legs. The equation is:
$$x^2 + (x+2)^2 = 10^2$$
$$x^2 + x^2 + 4x + 4 = 100$$
$$2x^2 + 4x + 4 = 100$$
$$2x^2 + 4x - 96 = 0$$
$$2(x^2 + 2x - 48) = 0$$
$$2(x+8)(x-6) = 0$$
$$x = 6 \quad (x = -8 \text{ is impossible})$$
$$x + 2 = 8$$
The legs are 6 inches and 8 inches.

19. Let x represent the longer leg and $x + 1$ represent the hypotenuse. The equation is:
$$5^2 + x^2 = (x+1)^2$$
$$25 + x^2 = x^2 + 2x + 1$$
$$25 = 2x + 1$$
$$24 = 2x$$
$$x = 12$$
The longer leg is 12 meters.

21. Setting $C = \$1,400$:
$$1400 = 400 + 700x - 100x^2$$
$$100x^2 - 700x + 1000 = 0$$
$$100(x^2 - 7x + 10) = 0$$
$$100(x - 5)(x - 2) = 0$$
$$x = 2, 5$$
The company can manufacture either 200 items or 500 items.

23. Setting $C = \$2,200$:
$$2200 = 600 + 1000x - 100x^2$$
$$100x^2 - 1000x + 1600 = 0$$
$$100(x^2 - 10x + 16) = 0$$
$$100(x - 2)(x - 8) = 0$$
$$x = 2, 8$$
The company can manufacture either 200 videotapes or 800 videotapes.

25. The revenue is given by: $R = xp = (1200 - 100p)p$. Setting $R = \$3,200$:
$$3200 = (1200 - 100p)p$$
$$3200 = 1200p - 100p^2$$
$$100p^2 - 1200p + 3200 = 0$$
$$100(p^2 - 12p + 32) = 0$$
$$100(p - 4)(p - 8) = 0$$
$$p = 4, 8$$
The company should sell the ribbons for either $4 or $8.

27. The revenue is given by: $R = xp = (1700 - 100p)p$. Setting $R = \$7,000$:
$$7000 = (1700 - 100p)p$$
$$7000 = 1700p - 100p^2$$
$$100p^2 - 1700p + 7000 = 0$$
$$100(p^2 - 17p + 70) = 0$$
$$100(p - 7)(p - 10) = 0$$
$$p = 7, 10$$
The calculators should be sold for either $7 or $10.

29. **a.** Let x represent the distance from the base to the wall, and $2x + 2$ represent the height on the wall. Using the Pythagorean theorem:
$$x^2 + (2x + 2)^2 = 13^2$$
$$x^2 + 4x^2 + 8x + 4 = 169$$
$$5x^2 + 8x - 165 = 0$$
$$(5x + 33)(x - 5) = 0$$
$$x = 5 \quad \left(x = -\tfrac{33}{5} \text{ is impossible}\right)$$
The base of the ladder is 5 feet from the wall.

  b. Since $2x + 2 = 2 \cdot 5 + 2 = 12$, the ladder reaches a height of 12 feet.

31. a. Finding when $h = 0$:
$$0 = -16t^2 + 396t + 100$$
$$16t^2 - 396t - 100 = 0$$
$$4t^2 - 99t - 25 = 0$$
$$(4t+1)(t-25) = 0$$
$$t = 25 \quad \left(t = -\tfrac{1}{4} \text{ is impossible}\right)$$

The bullet will land on the ground after 25 seconds.

b. Completing the table:

t (seconds)	h (feet)
0	100
5	1,680
10	2,460
15	2,440
20	1,620
25	0

33. Simplifying the expression: $(5x^3)^2 (2x^6)^3 = 25x^6 \cdot 8x^{18} = 200x^{24}$

35. Simplifying the expression: $\dfrac{x^4}{x^{-3}} = x^{4-(-3)} = x^{4+3} = x^7$

37. Simplifying the expression: $(2 \times 10^{-4})(4 \times 10^5) = 8 \times 10^1 = 80$

39. Simplifying the expression: $20ab^2 - 16ab^2 + 6ab^2 = 10ab^2$

41. Multiplying using the distributive property:
$$2x^2(3x^2 + 3x - 1) = 2x^2(3x^2) + 2x^2(3x) - 2x^2(1) = 6x^4 + 6x^3 - 2x^2$$

43. Multiplying using the square of binomial formula: $(3y-5)^2 = (3y)^2 - 2(3y)(5) + (5)^2 = 9y^2 - 30y + 25$

45. Multiplying using the difference of squares formula: $(2a^2 + 7)(2a^2 - 7) = (2a^2)^2 - (7)^2 = 4a^4 - 49$

Chapter 5 Review/Test

1. Factoring the polynomial: $10x - 20 = 10(x - 2)$
2. Factoring the polynomial: $4x^3 - 9x^2 = x^2(4x - 9)$
3. Factoring the polynomial: $5x - 5y = 5(x - y)$
4. Factoring the polynomial: $7x^3 + 2x = x(7x^2 + 2)$
5. Factoring the polynomial: $8x + 4 = 4(2x + 1)$
6. Factoring the polynomial: $2x^2 + 14x + 6 = 2(x^2 + 7x + 3)$
7. Factoring the polynomial: $24y^2 - 40y + 48 = 8(3y^2 - 5y + 6)$
8. Factoring the polynomial: $30xy^3 - 45x^3y^2 = 15xy^2(2y - 3x^2)$
9. Factoring the polynomial: $49a^3 - 14b^3 = 7(7a^3 - 2b^3)$
10. Factoring the polynomial: $6ab^2 + 18a^3b^3 - 24a^2b = 6ab(b + 3a^2b^2 - 4a)$
11. Factoring the polynomial: $xy + bx + ay + ab = x(y+b) + a(y+b) = (y+b)(x+a)$
12. Factoring the polynomial: $xy + 4x - 5y - 20 = x(y+4) - 5(y+4) = (y+4)(x-5)$
13. Factoring the polynomial: $2xy + 10x - 3y - 15 = 2x(y+5) - 3(y+5) = (y+5)(2x-3)$

14. Factoring the polynomial: $5x^2 - 4ax - 10bx + 8ab = x(5x-4a) - 2b(5x-4a) = (5x-4a)(x-2b)$
15. Factoring the polynomial: $y^2 + 9y + 14 = (y+7)(y+2)$
16. Factoring the polynomial: $w^2 + 15w + 50 = (w+10)(w+5)$
17. Factoring the polynomial: $a^2 - 14a + 48 = (a-8)(a-6)$
18. Factoring the polynomial: $r^2 - 18r + 72 = (r-12)(r-6)$
19. Factoring the polynomial: $y^2 + 20y + 99 = (y+9)(y+11)$
20. Factoring the polynomial: $y^2 + 8y + 12 = (y+6)(y+2)$
21. Factoring the polynomial: $2x^2 + 13x + 15 = (2x+3)(x+5)$
22. Factoring the polynomial: $4y^2 - 12y + 5 = (2y-5)(2y-1)$
23. Factoring the polynomial: $5y^2 + 11y + 6 = (5y+6)(y+1)$
24. Factoring the polynomial: $20a^2 - 27a + 9 = (5a-3)(4a-3)$
25. Factoring the polynomial: $6r^2 + 5rt - 6t^2 = (3r-2t)(2r+3t)$
26. Factoring the polynomial: $10x^2 - 29x - 21 = (5x+3)(2x-7)$
27. Factoring the polynomial: $n^2 - 81 = (n+9)(n-9)$
28. Factoring the polynomial: $4y^2 - 9 = (2y+3)(2y-3)$
29. The expression $x^2 + 49$ cannot be factored.
30. Factoring the polynomial: $36y^2 - 121x^2 = (6y+11x)(6y-11x)$
31. Factoring the polynomial: $64a^2 - 121b^2 = (8a+11b)(8a-11b)$
32. Factoring the polynomial: $64 - 9m^2 = (8+3m)(8-3m)$
33. Factoring the polynomial: $y^2 + 20y + 100 = (y+10)(y+10) = (y+10)^2$
34. Factoring the polynomial: $m^2 - 16m + 64 = (m-8)(m-8) = (m-8)^2$
35. Factoring the polynomial: $64t^2 + 16t + 1 = (8t+1)(8t+1) = (8t+1)^2$
36. Factoring the polynomial: $16n^2 - 24n + 9 = (4n-3)(4n-3) = (4n-3)^2$
37. Factoring the polynomial: $4r^2 - 12rt + 9t^2 = (2r-3t)(2r-3t) = (2r-3t)^2$
38. Factoring the polynomial: $9m^2 + 30mn + 25n^2 = (3m+5n)(3m+5n) = (3m+5n)^2$
39. Factoring the polynomial: $2x^2 + 20x + 48 = 2(x^2 + 10x + 24) = 2(x+4)(x+6)$
40. Factoring the polynomial: $a^3 - 10a^2 + 21a = a(a^2 - 10a + 21) = a(a-7)(a-3)$
41. Factoring the polynomial: $3m^3 - 18m^2 - 21m = 3m(m^2 - 6m - 7) = 3m(m-7)(m+1)$
42. Factoring the polynomial: $5y^4 + 10y^3 - 40y^2 = 5y^2(y^2 + 2y - 8) = 5y^2(y+4)(y-2)$
43. Factoring the polynomial: $8x^2 + 16x + 6 = 2(4x^2 + 8x + 3) = 2(2x+1)(2x+3)$
44. Factoring the polynomial: $3a^3 - 14a^2 - 5a = a(3a^2 - 14a - 5) = a(3a+1)(a-5)$
45. Factoring the polynomial: $20m^3 - 34m^2 + 6m = 2m(10m^2 - 17m + 3) = 2m(5m-1)(2m-3)$
46. Factoring the polynomial: $30x^2y - 55xy^2 + 15y^3 = 5y(6x^2 - 11xy + 3y^2) = 5y(3x-y)(2x-3y)$

47. Factoring the polynomial: $4x^2 + 40x + 100 = 4(x^2 + 10x + 25) = 4(x+5)(x+5) = 4(x+5)^2$
48. Factoring the polynomial: $4x^3 + 12x^2 + 9x = x(4x^2 + 12x + 9) = x(2x+3)(2x+3) = x(2x+3)^2$
49. Factoring the polynomial: $5x^2 - 45 = 5(x^2 - 9) = 5(x+3)(x-3)$
50. Factoring the polynomial: $12x^3 - 27xy^2 = 3x(4x^2 - 9y^2) = 3x(2x+3y)(2x-3y)$
51. Factoring the polynomial: $8a^3 - b^3 = (2a-b)(4a^2 + 2ab + b^2)$
52. Factoring the polynomial: $27x^3 + 8y^3 = (3x+2y)(9x^2 - 6xy + 4y^2)$
53. Factoring the polynomial: $125x^3 - 64y^3 = (5x-4y)(25x^2 + 20xy + 16y^2)$
54. Factoring the polynomial: $6a^3b + 33a^2b^2 + 15ab^3 = 3ab(2a^2 + 11ab + 5b^2) = 3ab(2a+b)(a+5b)$
55. Factoring the polynomial: $x^5 - x^3 = x^3(x^2 - 1) = x^3(x+1)(x-1)$
56. Factoring the polynomial: $4y^6 + 9y^4 = y^4(4y^2 + 9)$
57. Factoring the polynomial: $12x^5 + 20x^4y - 8x^3y^2 = 4x^3(3x^2 + 5xy - 2y^2) = 4x^3(3x-y)(x+2y)$
58. Factoring the polynomial: $30a^4b + 35a^3b^2 - 15a^2b^3 = 5a^2b(6a^2 + 7ab - 3b^2) = 5a^2b(3a-b)(2a+3b)$
59. Factoring the polynomial: $18a^3b^2 + 3a^2b^3 - 6ab^4 = 3ab^2(6a^2 + ab - 2b^2) = 3ab^2(3a+2b)(2a-b)$
60. Setting each factor equal to 0:
$x - 5 = 0 \qquad\qquad x + 2 = 0$
$x = 5 \qquad\qquad\qquad x = -2$
The solutions are $-2, 5$.
61. Setting each factor equal to 0:
$2y + 5 = 0 \qquad\qquad 2y - 5 = 0$
$2y = -5 \qquad\qquad\qquad 2y = 5$
$y = -\frac{5}{2} \qquad\qquad\qquad y = \frac{5}{2}$
The solutions are $-\frac{5}{2}, \frac{5}{2}$.
62. Solving the equation by factoring:
$m^2 + 3m = 10$
$m^2 + 3m - 10 = 0$
$(m+5)(m-2) = 0$
$m = -5, 2$
63. Solving the equation by factoring:
$a^2 - 49 = 0$
$(a+7)(a-7) = 0$
$a = -7, 7$
64. Solving the equation by factoring:
$m^2 - 9m = 0$
$m(m-9) = 0$
$m = 0, 9$
65. Solving the equation by factoring:
$6y^2 = -13y - 6$
$6y^2 + 13y + 6 = 0$
$(3y+2)(2y+3) = 0$
$y = -\frac{2}{3}, -\frac{3}{2}$
66. Solving the equation by factoring:
$9x^4 + 9x^3 = 10x^2$
$9x^4 + 9x^3 - 10x^2 = 0$
$x^2(9x^2 + 9x - 10) = 0$
$x^2(3x-2)(3x+5) = 0$
$x = 0, \frac{2}{3}, -\frac{5}{3}$

67. Let x and $x+2$ represent the two integers. The equation is:
$$x(x+2) = 120$$
$$x^2 + 2x = 120$$
$$x^2 + 2x - 120 = 0$$
$$(x+12)(x-10) = 0$$
$$x = -12, 10$$
$$x+2 = -10, 12$$
The two integers are either -12 and -10, or 10 and 12.

68. Let x and $x+1$ represent the two integers. The equation is:
$$x(x+1) = 110$$
$$x^2 + x = 110$$
$$x^2 + x - 110 = 0$$
$$(x+11)(x-10) = 0$$
$$x = -11, 10$$
$$x+1 = -10, 11$$
The two integers are either -11 and -10, or 10 and 11.

69. Let x and $x+2$ represent the two integers. The equation is:
$$x(x+2) = 3(x + x + 2) - 1$$
$$x^2 + 2x = 3(2x+2) - 1$$
$$x^2 + 2x = 6x + 6 - 1$$
$$x^2 + 2x = 6x + 5$$
$$x^2 - 4x - 5 = 0$$
$$(x-5)(x+1) = 0$$
$$x = 5, -1$$
$$x+2 = 7, 1$$
The two integers are either 5 and 7, or -1 and 1.

70. Let x and $20 - x$ represent the two numbers. The equation is:
$$x(20-x) = 75$$
$$20x - x^2 = 75$$
$$0 = x^2 - 20x + 75$$
$$0 = (x-15)(x-5)$$
$$x = 15, 5$$
$$20 - x = 5, 15$$
The two numbers are 5 and 15.

71. Let x and $2x - 1$ represent the two numbers. The equation is:
$$x(2x-1) = 66$$
$$2x^2 - x = 66$$
$$2x^2 - x - 66 = 0$$
$$(2x+11)(x-6) = 0$$
$$x = -\tfrac{11}{2}, 6$$
$$2x - 1 = -12, 11$$
The two numbers are either $-\tfrac{11}{2}$ and -12, or 6 and 11.

72. Let b represent the base, and $8b$ represent the height. The equation is:
$$\frac{1}{2}(b)(8b) = 16$$
$$4b^2 = 16$$
$$4b^2 - 16 = 0$$
$$4(b^2 - 4) = 0$$
$$4(b+2)(b-2) = 0$$
$$b = 2 \quad (b = -2 \text{ is impossible})$$

The base is 2 inches.

Chapter 6
Rational Expressions

6.1 Reducing Rational Expressions to Lowest Terms

1. a. Simplifying: $\dfrac{5+1}{25-1} = \dfrac{6}{24} = \dfrac{1}{4}$

    b. Simplifying: $\dfrac{x+1}{x^2-1} = \dfrac{1(x+1)}{(x+1)(x-1)} = \dfrac{1}{x-1}$. The variable restriction is $x \ne -1, 1$.

    c. Simplifying: $\dfrac{x^2-x}{x^2-1} = \dfrac{x(x-1)}{(x+1)(x-1)} = \dfrac{x}{x+1}$. The variable restriction is $x \ne -1, 1$.

    d. Simplifying: $\dfrac{x^3-1}{x^2-1} = \dfrac{(x-1)(x^2+x+1)}{(x+1)(x-1)} = \dfrac{x^2+x+1}{x+1}$. The variable restriction is $x \ne -1, 1$.

    e. Simplifying: $\dfrac{x^3-1}{x^3-x^2} = \dfrac{(x-1)(x^2+x+1)}{x^2(x-1)} = \dfrac{x^2+x+1}{x^2}$. The variable restriction is $x \ne 0, 1$.

3. Reducing the rational expression: $\dfrac{a-3}{a^2-9} = \dfrac{1(a-3)}{(a+3)(a-3)} = \dfrac{1}{a+3}$. The variable restriction is $a \ne -3, 3$.

5. Reducing the rational expression: $\dfrac{x+5}{x^2-25} = \dfrac{1(x+5)}{(x+5)(x-5)} = \dfrac{1}{x-5}$. The variable restriction is $x \ne -5, 5$.

7. Reducing the rational expression: $\dfrac{2x^2-8}{4} = \dfrac{2(x^2-4)}{4} = \dfrac{2(x+2)(x-2)}{4} = \dfrac{(x+2)(x-2)}{2}$
   There are no variable restrictions.

9. Reducing the rational expression: $\dfrac{2x-10}{3x-6} = \dfrac{2(x-5)}{3(x-2)}$. The variable restriction is $x \ne 2$.

11. Reducing the rational expression: $\dfrac{10a+20}{5a+10} = \dfrac{10(a+2)}{5(a+2)} = \dfrac{2}{1} = 2$

13. Reducing the rational expression: $\dfrac{5x^2-5}{4x+4} = \dfrac{5(x^2-1)}{4(x+1)} = \dfrac{5(x+1)(x-1)}{4(x+1)} = \dfrac{5(x-1)}{4}$

Chapter 6 Rational Expressions

15. Reducing the rational expression: $\dfrac{x-3}{x^2-6x+9} = \dfrac{1(x-3)}{(x-3)^2} = \dfrac{1}{x-3}$

17. Reducing the rational expression: $\dfrac{3x+15}{3x^2+24x+45} = \dfrac{3(x+5)}{3(x^2+8x+15)} = \dfrac{3(x+5)}{3(x+5)(x+3)} = \dfrac{1}{x+3}$

19. Reducing the rational expression: $\dfrac{a^2-3a}{a^3-8a^2+15a} = \dfrac{a(a-3)}{a(a^2-8a+15)} = \dfrac{a(a-3)}{a(a-3)(a-5)} = \dfrac{1}{a-5}$

21. Reducing the rational expression: $\dfrac{3x-2}{9x^2-4} = \dfrac{1(3x-2)}{(3x+2)(3x-2)} = \dfrac{1}{3x+2}$

23. Reducing the rational expression: $\dfrac{x^2+8x+15}{x^2+5x+6} = \dfrac{(x+5)(x+3)}{(x+2)(x+3)} = \dfrac{x+5}{x+2}$

25. Reducing the rational expression: $\dfrac{2m^3-2m^2-12m}{m^2-5m+6} = \dfrac{2m(m^2-m-6)}{(m-2)(m-3)} = \dfrac{2m(m-3)(m+2)}{(m-2)(m-3)} = \dfrac{2m(m+2)}{m-2}$

27. Reducing the rational expression: $\dfrac{x^3+3x^2-4x}{x^3-16x} = \dfrac{x(x^2+3x-4)}{x(x^2-16)} = \dfrac{x(x+4)(x-1)}{x(x+4)(x-4)} = \dfrac{x-1}{x-4}$

29. Reducing the rational expression: $\dfrac{4x^3-10x^2+6x}{2x^3+x^2-3x} = \dfrac{2x(2x^2-5x+3)}{x(2x^2+x-3)} = \dfrac{2x(2x-3)(x-1)}{x(2x+3)(x-1)} = \dfrac{2(2x-3)}{2x+3}$

31. Reducing the rational expression: $\dfrac{4x^2-12x+9}{4x^2-9} = \dfrac{(2x-3)^2}{(2x+3)(2x-3)} = \dfrac{2x-3}{2x+3}$

33. Reducing the rational expression: $\dfrac{x+3}{x^4-81} = \dfrac{x+3}{(x^2+9)(x^2-9)} = \dfrac{x+3}{(x^2+9)(x+3)(x-3)} = \dfrac{1}{(x^2+9)(x-3)}$

35. Reducing the rational expression: $\dfrac{3x^2+x-10}{x^4-16} = \dfrac{(3x-5)(x+2)}{(x^2+4)(x^2-4)} = \dfrac{(3x-5)(x+2)}{(x^2+4)(x+2)(x-2)} = \dfrac{3x-5}{(x^2+4)(x-2)}$

37. Reducing the rational expression: $\dfrac{42x^3-20x^2-48x}{6x^2-5x-4} = \dfrac{2x(21x^2-10x-24)}{(3x-4)(2x+1)} = \dfrac{2x(7x+6)(3x-4)}{(3x-4)(2x+1)} = \dfrac{2x(7x+6)}{2x+1}$

39. Reducing the rational expression: $\dfrac{x^3-y^3}{x^2-y^2} = \dfrac{(x-y)(x^2+xy+y^2)}{(x+y)(x-y)} = \dfrac{x^2+xy+y^2}{x+y}$

41. Reducing the rational expression: $\dfrac{x^3+8}{x^2-4} = \dfrac{(x+2)(x^2-2x+4)}{(x+2)(x-2)} = \dfrac{x^2-2x+4}{x-2}$

43. Reducing the rational expression: $\dfrac{x^3+8}{x^2+x-2} = \dfrac{(x+2)(x^2-2x+4)}{(x+2)(x-1)} = \dfrac{x^2-2x+4}{x-1}$

45. Reducing the rational expression: $\dfrac{xy+3x+2y+6}{xy+3x+5y+15} = \dfrac{x(y+3)+2(y+3)}{x(y+3)+5(y+3)} = \dfrac{(y+3)(x+2)}{(y+3)(x+5)} = \dfrac{x+2}{x+5}$

47. Reducing the rational expression: $\dfrac{x^2-3x+ax-3a}{x^2-3x+bx-3b} = \dfrac{x(x-3)+a(x-3)}{x(x-3)+b(x-3)} = \dfrac{(x-3)(x+a)}{(x-3)(x+b)} = \dfrac{x+a}{x+b}$

49. **a.** Adding: $(x^2 - 4x) + (4x - 16) = x^2 - 4x + 4x - 16 = x^2 - 16$

 b. Subtracting: $(x^2 - 4x) - (4x - 16) = x^2 - 4x - 4x + 16 = x^2 - 8x + 16$

 c. Multiplying: $(x^2 - 4x)(4x - 16) = 4x^3 - 16x^2 - 16x^2 + 64x = 4x^3 - 32x^2 + 64x$

 d. Reducing: $\dfrac{x^2 - 4x}{4x - 16} = \dfrac{x(x-4)}{4(x-4)} = \dfrac{x}{4}$

51. Writing as a ratio: $\dfrac{8}{6} = \dfrac{4}{3}$

53. Writing as a ratio: $\dfrac{200}{250} = \dfrac{4}{5}$

55. Writing as a ratio: $\dfrac{32}{4} = \dfrac{8}{1}$

57. Completing the table:

Checks Written x	Total Cost $2.00 + 0.15x$	Cost per Check $\dfrac{2.00 + 0.15x}{x}$
0	$2.00	undefined
5	$2.75	$0.55
10	$3.50	$0.35
15	$4.25	$0.28
20	$5.00	$0.25

59. The average speed is: $\dfrac{122 \text{ miles}}{3 \text{ hours}} \approx 40.7$ miles/hour

61. The average speed is: $\dfrac{785 \text{ feet}}{20 \text{ minutes}} = 39.25$ feet/minute

63. Computing the strikeouts per inning:

 Randy Johnson: $\dfrac{11.04 \text{ strikeouts}}{9 \text{ innings}} \approx 1.23$ strikeouts/inning

 Kerry Wood: $\dfrac{10.45 \text{ strikeouts}}{9 \text{ innings}} \approx 1.16$ strikeouts/inning

 Pedro Martinez: $\dfrac{10.40 \text{ strikeouts}}{9 \text{ innings}} \approx 1.16$ strikeouts/inning

 Hideo Nomo: $\dfrac{8.81 \text{ strikeouts}}{9 \text{ innings}} \approx 0.98$ strikeouts/inning

65. Her average speed on level ground is: $\dfrac{20 \text{ minutes}}{2 \text{ miles}} = 10$ minutes/mile, or $\dfrac{2 \text{ miles}}{20 \text{ minutes}} = 0.1$ miles/minute

 Her average speed downhill is: $\dfrac{40 \text{ minutes}}{6 \text{ miles}} = \dfrac{20}{3}$ minutes/mile, or $\dfrac{6 \text{ miles}}{40 \text{ minutes}} = \dfrac{3}{20}$ miles/minute

67. The average fuel consumption is: $\dfrac{168 \text{ miles}}{3.5 \text{ gallons}} = 48$ miles/gallon

69. Substituting $x = 5$ and $y = 4$: $\dfrac{x^2 - y^2}{x - y} = \dfrac{5^2 - 4^2}{5 - 4} = \dfrac{25 - 16}{5 - 4} = \dfrac{9}{1} = 9$. The result is equal to $5 + 4 = 9$.

71. Completing the table:

x	$\dfrac{x-3}{3-x}$
−2	−1
−1	−1
0	−1
1	−1
2	−1

Simplifying: $\dfrac{x-3}{3-x} = \dfrac{-(3-x)}{3-x} = -1$

73. Completing the table:

x	$\dfrac{x-5}{x^2-25}$	$\dfrac{1}{x+5}$
0	$\dfrac{1}{5}$	$\dfrac{1}{5}$
2	$\dfrac{1}{7}$	$\dfrac{1}{7}$
−2	$\dfrac{1}{3}$	$\dfrac{1}{3}$
5	undefined	$\dfrac{1}{10}$
−5	undefined	undefined

75. Completing the table:

Stock	Price	Earnings per Share	P/E
Yahoo	37.80	1.07	35.33
Google	381.24	4.51	84.53
Disney	24.96	1.34	18.63
Nike	85.46	4.88	17.51
Ebay	40.96	0.73	56.11

The stocks Disney and Nike appear to be undervalued.

77. Simplifying the expression: $\dfrac{27x^5}{9x^2} - \dfrac{45x^8}{15x^5} = 3x^3 - 3x^3 = 0$

79. Simplifying the expression: $\dfrac{72a^3b^7}{9ab^5} + \dfrac{64a^5b^3}{8a^3b} = 8a^2b^2 + 8a^2b^2 = 16a^2b^2$

81. Dividing by the monomial: $\dfrac{38x^7 + 42x^5 - 84x^3}{2x^3} = \dfrac{38x^7}{2x^3} + \dfrac{42x^5}{2x^3} - \dfrac{84x^3}{2x^3} = 19x^4 + 21x^2 - 42$

83. Dividing by the monomial: $\dfrac{28a^5b^5 + 36ab^4 - 44a^4b}{4ab} = \dfrac{28a^5b^5}{4ab} + \dfrac{36ab^4}{4ab} - \dfrac{44a^4b}{4ab} = 7a^4b^4 + 9b^3 - 11a^3$

85. Simplifying: $\dfrac{3}{4} \cdot \dfrac{10}{21} = \dfrac{3 \cdot 2 \cdot 5}{2 \cdot 2 \cdot 3 \cdot 7} = \dfrac{5}{14}$

87. Simplifying: $\dfrac{4}{5} \div \dfrac{8}{9} = \dfrac{4}{5} \cdot \dfrac{9}{8} = \dfrac{2 \cdot 2 \cdot 9}{5 \cdot 2 \cdot 2 \cdot 2} = \dfrac{9}{10}$

89. Factoring: $x^2 - 9 = (x+3)(x-3)$

91. Factoring: $3x - 9 = 3(x-3)$

93. Factoring: $x^2 - x - 20 = (x-5)(x+4)$

95. Factoring: $a^2 + 5a = a(a+5)$

97. Simplifying: $\dfrac{a(a+5)(a-5)(a+4)}{a^2+5a} = \dfrac{a(a+5)(a-5)(a+4)}{a(a+5)} = (a-5)(a+4)$

99. Multiplying: $\dfrac{5{,}603}{11} \cdot \dfrac{1}{5{,}280} \cdot \dfrac{60}{1} \approx 5.8$

6.2 Multiplication and Division of Rational Expressions

1. Simplifying the expression: $\dfrac{x+y}{3} \cdot \dfrac{6}{x+y} = \dfrac{6(x+y)}{3(x+y)} = 2$

3. Simplifying the expression: $\dfrac{2x+10}{x^2} \cdot \dfrac{x^3}{4x+20} = \dfrac{2(x+5)}{x^2} \cdot \dfrac{x^3}{4(x+5)} = \dfrac{2x^3(x+5)}{4x^2(x+5)} = \dfrac{x}{2}$

5. Simplifying the expression: $\dfrac{9}{2a-8} \div \dfrac{3}{a-4} = \dfrac{9}{2a-8} \cdot \dfrac{a-4}{3} = \dfrac{9}{2(a-4)} \cdot \dfrac{a-4}{3} = \dfrac{9(a-4)}{6(a-4)} = \dfrac{3}{2}$

7. Simplifying the expression:

$\dfrac{x+1}{x^2-9} \div \dfrac{2x+2}{x+3} = \dfrac{x+1}{x^2-9} \cdot \dfrac{x+3}{2x+2} = \dfrac{x+1}{(x+3)(x-3)} \cdot \dfrac{x+3}{2(x+1)} = \dfrac{(x+1)(x+3)}{2(x+3)(x-3)(x+1)} = \dfrac{1}{2(x-3)}$

9. Simplifying the expression: $\dfrac{a^2+5a}{7a} \cdot \dfrac{4a^2}{a^2+4a} = \dfrac{a(a+5)}{7a} \cdot \dfrac{4a^2}{a(a+4)} = \dfrac{4a^3(a+5)}{7a^2(a+4)} = \dfrac{4a(a+5)}{7(a+4)}$

11. Simplifying the expression:

$\dfrac{y^2-5y+6}{2y+4} \div \dfrac{2y-6}{y+2} = \dfrac{y^2-5y+6}{2y+4} \cdot \dfrac{y+2}{2y-6} = \dfrac{(y-2)(y-3)}{2(y+2)} \cdot \dfrac{y+2}{2(y-3)} = \dfrac{(y-2)(y-3)(y+2)}{4(y+2)(y-3)} = \dfrac{y-2}{4}$

13. Simplifying the expression:

$\dfrac{2x-8}{x^2-4} \cdot \dfrac{x^2+6x+8}{x-4} = \dfrac{2(x-4)}{(x+2)(x-2)} \cdot \dfrac{(x+4)(x+2)}{x-4} = \dfrac{2(x-4)(x+4)(x+2)}{(x+2)(x-2)(x-4)} = \dfrac{2(x+4)}{x-2}$

15. Simplifying the expression:

$\dfrac{x-1}{x^2-x-6} \cdot \dfrac{x^2+5x+6}{x^2-1} = \dfrac{x-1}{(x-3)(x+2)} \cdot \dfrac{(x+2)(x+3)}{(x+1)(x-1)} = \dfrac{(x-1)(x+2)(x+3)}{(x-3)(x+2)(x+1)(x-1)} = \dfrac{x+3}{(x-3)(x+1)}$

17. Simplifying the expression:

$\dfrac{a^2+10a+25}{a+5} \div \dfrac{a^2-25}{a-5} = \dfrac{a^2+10a+25}{a+5} \cdot \dfrac{a-5}{a^2-25} = \dfrac{(a+5)^2}{a+5} \cdot \dfrac{a-5}{(a+5)(a-5)} = \dfrac{(a+5)^2(a-5)}{(a+5)^2(a-5)} = 1$

19. Simplifying the expression:

$\dfrac{y^3-5y^2}{y^4+3y^3+2y^2} \div \dfrac{y^2-5y+6}{y^2-2y-3} = \dfrac{y^3-5y^2}{y^4+3y^3+2y^2} \cdot \dfrac{y^2-2y-3}{y^2-5y+6}$

$= \dfrac{y^2(y-5)}{y^2(y+2)(y+1)} \cdot \dfrac{(y-3)(y+1)}{(y-2)(y-3)}$

$= \dfrac{y^2(y-5)(y-3)(y+1)}{y^2(y+2)(y+1)(y-2)(y-3)}$

$= \dfrac{y-5}{(y+2)(y-2)}$

21. Simplifying the expression:

$\dfrac{2x^2+17x+21}{x^2+2x-35} \cdot \dfrac{x^3-125}{2x^2-7x-15} = \dfrac{(2x+3)(x+7)}{(x+7)(x-5)} \cdot \dfrac{(x-5)(x^2+5x+25)}{(2x+3)(x-5)}$

$= \dfrac{(2x+3)(x+7)(x-5)(x^2+5x+25)}{(x+7)(x-5)^2(2x+3)}$

$= \dfrac{x^2+5x+25}{x-5}$

158 Chapter 6 Rational Expressions

23. Simplifying the expression:

$$\frac{2x^2+10x+12}{4x^2+24x+32} \cdot \frac{2x^2+18x+40}{x^2+8x+15} = \frac{2(x^2+5x+6)}{4(x^2+6x+8)} \cdot \frac{2(x^2+9x+20)}{x^2+8x+15}$$

$$= \frac{2(x+2)(x+3)}{4(x+4)(x+2)} \cdot \frac{2(x+5)(x+4)}{(x+5)(x+3)}$$

$$= \frac{4(x+2)(x+3)(x+4)(x+5)}{4(x+2)(x+3)(x+4)(x+5)}$$

$$= 1$$

25. Simplifying the expression:

$$\frac{2a^2+7a+3}{a^2-16} \div \frac{4a^2+8a+3}{2a^2-5a-12} = \frac{2a^2+7a+3}{a^2-16} \cdot \frac{2a^2-5a-12}{4a^2+8a+3}$$

$$= \frac{(2a+1)(a+3)}{(a+4)(a-4)} \cdot \frac{(2a+3)(a-4)}{(2a+1)(2a+3)}$$

$$= \frac{(2a+1)(a+3)(2a+3)(a-4)}{(a+4)(a-4)(2a+1)(2a+3)}$$

$$= \frac{a+3}{a+4}$$

27. Simplifying the expression:

$$\frac{4y^2-12y+9}{y^2-36} \div \frac{2y^2-5y+3}{y^2+5y-6} = \frac{4y^2-12y+9}{y^2-36} \cdot \frac{y^2+5y-6}{2y^2-5y+3}$$

$$= \frac{(2y-3)^2}{(y+6)(y-6)} \cdot \frac{(y+6)(y-1)}{(2y-3)(y-1)}$$

$$= \frac{(2y-3)^2(y+6)(y-1)}{(y+6)(y-6)(2y-3)(y-1)}$$

$$= \frac{2y-3}{y-6}$$

29. Simplifying the expression:

$$\frac{x^2-1}{6x^2+42x+60} \cdot \frac{7x^2+17x+6}{x^3+1} \cdot \frac{6x+30}{7x^2-11x-6} = \frac{(x+1)(x-1)}{6(x+5)(x+2)} \cdot \frac{(7x+3)(x+2)}{(x+1)(x^2-x+1)} \cdot \frac{6(x+5)}{(7x+3)(x-2)}$$

$$= \frac{6(x+1)(x-1)(7x+3)(x+2)(x+5)}{6(x+5)(x+2)(x+1)(7x+3)(x-2)(x^2-x+1)}$$

$$= \frac{x-1}{(x-2)(x^2-x+1)}$$

31. Simplifying the expression:

$$\frac{18x^3+21x^2-60x}{21x^2-25x-4} \cdot \frac{28x^2-17x-3}{16x^3+28x^2-30x} = \frac{3x(6x^2+7x-20)}{21x^2-25x-4} \cdot \frac{28x^2-17x-3}{2x(8x^2+14x-15)}$$

$$= \frac{3x(3x-4)(2x+5)}{(7x+1)(3x-4)} \cdot \frac{(7x+1)(4x-3)}{2x(4x-3)(2x+5)}$$

$$= \frac{3x(3x-4)(2x+5)(7x+1)(4x-3)}{2x(7x+1)(3x-4)(4x-3)(2x+5)}$$

$$= \frac{3}{2}$$

33.

a. Simplifying: $\dfrac{9-1}{27-1} = \dfrac{8}{26} = \dfrac{4}{13}$

b. Reducing: $\dfrac{x^2-1}{x^3-1} = \dfrac{(x+1)(x-1)}{(x-1)(x^2+x+1)} = \dfrac{x+1}{x^2+x+1}$

c. Multiplying: $\dfrac{x^2-1}{x^3-1} \cdot \dfrac{x-1}{x+1} = \dfrac{(x+1)(x-1)}{(x-1)(x^2+x+1)} \cdot \dfrac{x-1}{x+1} = \dfrac{x-1}{x^2+x+1}$

d. Dividing: $\dfrac{x^2-1}{x^3-1} \div \dfrac{x-1}{x^2+x+1} = \dfrac{(x+1)(x-1)}{(x-1)(x^2+x+1)} \cdot \dfrac{x^2+x+1}{x-1} = \dfrac{x+1}{x-1}$

35. Simplifying the expression: $(x^2-9)\left(\dfrac{2}{x+3}\right) = \dfrac{(x+3)(x-3)}{1} \cdot \dfrac{2}{x+3} = \dfrac{2(x+3)(x-3)}{x+3} = 2(x-3)$

37. Simplifying the expression: $(x^2-x-6)\left(\dfrac{x+1}{x-3}\right) = \dfrac{(x-3)(x+2)}{1} \cdot \dfrac{x+1}{x-3} = \dfrac{(x-3)(x+2)(x+1)}{x-3} = (x+2)(x+1)$

39. Simplifying the expression: $(x^2-4x-5)\left(\dfrac{-2x}{x+1}\right) = \dfrac{(x-5)(x+1)}{1} \cdot \dfrac{-2x}{x+1} = \dfrac{-2x(x-5)(x+1)}{x+1} = -2x(x-5)$

41. Simplifying the expression:

$\dfrac{x^2-9}{x^2-3x} \cdot \dfrac{2x+10}{xy+5x+3y+15} = \dfrac{(x+3)(x-3)}{x(x-3)} \cdot \dfrac{2(x+5)}{x(y+5)+3(y+5)} = \dfrac{2(x+3)(x-3)(x+5)}{x(x-3)(y+5)(x+3)} = \dfrac{2(x+5)}{x(y+5)}$

43. Simplifying the expression:

$\dfrac{2x^2+4x}{x^2-y^2} \cdot \dfrac{x^2+3x+xy+3y}{x^2+5x+6} = \dfrac{2x(x+2)}{(x+y)(x-y)} \cdot \dfrac{x(x+3)+y(x+3)}{(x+2)(x+3)} = \dfrac{2x(x+2)(x+3)(x+y)}{(x+y)(x-y)(x+2)(x+3)} = \dfrac{2x}{x-y}$

45. Simplifying the expression:

$\dfrac{x^3-3x^2+4x-12}{x^4-16} \cdot \dfrac{3x^2+5x-2}{3x^2-10x+3} = \dfrac{x^2(x-3)+4(x-3)}{(x^2+4)(x^2-4)} \cdot \dfrac{(3x-1)(x+2)}{(3x-1)(x-3)}$

$= \dfrac{(x-3)(x^2+4)}{(x^2+4)(x+2)(x-2)} \cdot \dfrac{(3x-1)(x+2)}{(3x-1)(x-3)}$

$= \dfrac{(x-3)(x^2+4)(3x-1)(x+2)}{(x^2+4)(x+2)(x-2)(3x-1)(x-3)}$

$= \dfrac{1}{x-2}$

47. Simplifying the expression: $\left(1-\tfrac{1}{2}\right)\left(1-\tfrac{1}{3}\right)\left(1-\tfrac{1}{4}\right)\left(1-\tfrac{1}{5}\right) = \left(\tfrac{2}{2}-\tfrac{1}{2}\right)\left(\tfrac{3}{3}-\tfrac{1}{3}\right)\left(\tfrac{4}{4}-\tfrac{1}{4}\right)\left(\tfrac{5}{5}-\tfrac{1}{5}\right) = \tfrac{1}{2} \cdot \tfrac{2}{3} \cdot \tfrac{3}{4} \cdot \tfrac{4}{5} = \tfrac{1}{5}$

49. Simplifying the expression: $\left(1-\tfrac{1}{2}\right)\left(1-\tfrac{1}{3}\right)\left(1-\tfrac{1}{4}\right) \cdots \left(1-\tfrac{1}{99}\right)\left(1-\tfrac{1}{100}\right) = \tfrac{1}{2} \cdot \tfrac{2}{3} \cdot \tfrac{3}{4} \cdots \tfrac{98}{99} \cdot \tfrac{99}{100} = \tfrac{1}{100}$

51. Since 5,280 feet = 1 mile, the height is: $\dfrac{14{,}494 \text{ feet}}{5{,}280 \text{ feet/mile}} \approx 2.7$ miles

53. Converting to miles per hour: $\dfrac{1088 \text{ feet}}{1 \text{ second}} \cdot \dfrac{1 \text{ mile}}{5280 \text{ feet}} \cdot \dfrac{60 \text{ seconds}}{1 \text{ minute}} \cdot \dfrac{60 \text{ minutes}}{1 \text{ hour}} \approx 742$ miles/hour

55. Converting to miles per hour: $\dfrac{785 \text{ feet}}{20 \text{ minutes}} \cdot \dfrac{60 \text{ minutes}}{1 \text{ hour}} \cdot \dfrac{1 \text{ mile}}{5280 \text{ feet}} \approx 0.45$ miles/hour

57. Converting to miles per hour: $\dfrac{518 \text{ feet}}{40 \text{ seconds}} \cdot \dfrac{60 \text{ seconds}}{1 \text{ minute}} \cdot \dfrac{60 \text{ minutes}}{1 \text{ hour}} \cdot \dfrac{1 \text{ mile}}{5280 \text{ feet}} \approx 8.8$ miles/hour

160 Chapter 6 Rational Expressions

59. **a.** Finding his average speed: $\dfrac{10 \text{ kilometers}}{28\frac{1}{6} \text{ minutes}} \cdot \dfrac{60 \text{ minutes}}{1 \text{ hour}} \approx 21.3$ kilometers/hour

b. Converting to miles per hour: $\dfrac{21.3 \text{ kilometers}}{1 \text{ hour}} \cdot \dfrac{6.2 \text{ miles}}{10 \text{ kilometers}} \approx 13.2$ miles/hour

61. Adding the fractions: $\frac{1}{2} + \frac{5}{2} = \frac{6}{2} = 3$

63. Adding the fractions: $2 + \frac{3}{4} = \frac{2 \cdot 4}{1 \cdot 4} + \frac{3}{4} = \frac{8}{4} + \frac{3}{4} = \frac{11}{4}$

65. Simplifying the expression: $\dfrac{10x^4}{2x^2} + \dfrac{12x^6}{3x^4} = 5x^2 + 4x^2 = 9x^2$

67. Simplifying the expression: $\dfrac{12a^2b^5}{3ab^3} + \dfrac{14a^4b^7}{7a^3b^5} = 4ab^2 + 2ab^2 = 6ab^2$

69. Adding the fractions: $\frac{1}{5} + \frac{3}{5} = \frac{4}{5}$

71. Adding the fractions: $\frac{1}{10} + \frac{3}{14} = \frac{1}{10} \cdot \frac{7}{7} + \frac{3}{14} \cdot \frac{5}{5} = \frac{7}{70} + \frac{15}{70} = \frac{22}{70} = \frac{11}{35}$

73. Subtracting the fractions: $\frac{1}{10} - \frac{3}{14} = \frac{1}{10} \cdot \frac{7}{7} - \frac{3}{14} \cdot \frac{5}{5} = \frac{7}{70} - \frac{15}{70} = -\frac{8}{70} = -\frac{4}{35}$

75. Multiplying: $2(x-3) = 2x - 6$

77. Multiplying: $(x+4)(x-5) = x^2 - 5x + 4x - 20 = x^2 - x - 20$

79. Reducing the fraction: $\dfrac{x+3}{x^2-9} = \dfrac{x+3}{(x+3)(x-3)} = \dfrac{1}{x-3}$

81. Reducing the fraction: $\dfrac{x^2-x-30}{2(x+5)(x-5)} = \dfrac{(x+5)(x-6)}{2(x+5)(x-5)} = \dfrac{x-6}{2(x-5)}$

83. Simplifying: $(x+4)(x-5) - 10 = x^2 - 5x + 4x - 20 - 10 = x^2 - x - 30$

6.3 Addition and Subtraction of Rational Expressions

1. Combining the fractions: $\dfrac{3}{x} + \dfrac{4}{x} = \dfrac{7}{x}$

3. Combining the fractions: $\dfrac{9}{a} - \dfrac{5}{a} = \dfrac{4}{a}$

5. Combining the fractions: $\dfrac{1}{x+1} + \dfrac{x}{x+1} = \dfrac{1+x}{x+1} = 1$

7. Combining the fractions: $\dfrac{y^2}{y-1} - \dfrac{1}{y-1} = \dfrac{y^2-1}{y-1} = \dfrac{(y+1)(y-1)}{y-1} = y+1$

9. Combining the fractions: $\dfrac{x^2}{x+2} + \dfrac{4x+4}{x+2} = \dfrac{x^2+4x+4}{x+2} = \dfrac{(x+2)^2}{x+2} = x+2$

11. Combining the fractions: $\dfrac{x^2}{x-2} - \dfrac{4x-4}{x-2} = \dfrac{x^2-4x+4}{x-2} = \dfrac{(x-2)^2}{x-2} = x-2$

13. Combining the fractions: $\dfrac{x+2}{x+6} - \dfrac{x-4}{x+6} = \dfrac{x+2-x+4}{x+6} = \dfrac{6}{x+6}$

15. Combining the fractions: $\dfrac{y}{2} - \dfrac{2}{y} = \dfrac{y \cdot y}{2 \cdot y} - \dfrac{2 \cdot 2}{y \cdot 2} = \dfrac{y^2}{2y} - \dfrac{4}{2y} = \dfrac{y^2-4}{2y} = \dfrac{(y+2)(y-2)}{2y}$

17. Combining the fractions: $\dfrac{1}{2} + \dfrac{a}{3} = \dfrac{1 \cdot 3}{2 \cdot 3} + \dfrac{a \cdot 2}{3 \cdot 2} = \dfrac{3}{6} + \dfrac{2a}{6} = \dfrac{2a+3}{6}$

19. Combining the fractions: $\dfrac{x}{x+1} + \dfrac{3}{4} = \dfrac{x \cdot 4}{(x+1) \cdot 4} + \dfrac{3 \cdot (x+1)}{4 \cdot (x+1)} = \dfrac{4x}{4(x+1)} + \dfrac{3x+3}{4(x+1)} = \dfrac{4x+3x+3}{4(x+1)} = \dfrac{7x+3}{4(x+1)}$

21. Combining the fractions:
$$\frac{x+1}{x-2} - \frac{4x+7}{5x-10} = \frac{(x+1) \cdot 5}{(x-2) \cdot 5} - \frac{4x+7}{5(x-2)} = \frac{5x+5}{5(x-2)} - \frac{4x+7}{5(x-2)} = \frac{5x+5-4x-7}{5(x-2)} = \frac{x-2}{5(x-2)} = \frac{1}{5}$$

23. Combining the fractions:
$$\frac{4x-2}{3x+12} - \frac{x-2}{x+4} = \frac{4x-2}{3(x+4)} - \frac{(x-2) \cdot 3}{(x+4) \cdot 3} = \frac{4x-2}{3(x+4)} - \frac{3x-6}{3(x+4)} = \frac{4x-2-3x+6}{3(x+4)} = \frac{x+4}{3(x+4)} = \frac{1}{3}$$

25. Combining the fractions:
$$\frac{6}{x(x-2)} + \frac{3}{x} = \frac{6}{x(x-2)} + \frac{3 \cdot (x-2)}{x \cdot (x-2)} = \frac{6}{x(x-2)} + \frac{3x-6}{x(x-2)} = \frac{6+3x-6}{x(x-2)} = \frac{3x}{x(x-2)} = \frac{3}{x-2}$$

27. Combining the fractions:
$$\frac{4}{a} - \frac{12}{a^2+3a} = \frac{4 \cdot (a+3)}{a \cdot (a+3)} - \frac{12}{a(a+3)} = \frac{4a+12}{a(a+3)} - \frac{12}{a(a+3)} = \frac{4a+12-12}{a(a+3)} = \frac{4a}{a(a+3)} = \frac{4}{a+3}$$

29. Combining the fractions:

$$\frac{2}{x+5} - \frac{10}{x^2-25} = \frac{2 \cdot (x-5)}{(x+5) \cdot (x-5)} - \frac{10}{(x+5)(x-5)}$$
$$= \frac{2x-10}{(x+5)(x-5)} - \frac{10}{(x+5)(x-5)}$$
$$= \frac{2x-10-10}{(x+5)(x-5)}$$
$$= \frac{2x-20}{(x+5)(x-5)}$$
$$= \frac{2(x-10)}{(x+5)(x-5)}$$

31. Combining the fractions:

$$\frac{x-4}{x-3} + \frac{6}{x^2-9} = \frac{(x-4) \cdot (x+3)}{(x-3) \cdot (x+3)} + \frac{6}{(x+3)(x-3)}$$
$$= \frac{x^2-x-12}{(x+3)(x-3)} + \frac{6}{(x+3)(x-3)}$$
$$= \frac{x^2-x-12+6}{(x+3)(x-3)}$$
$$= \frac{x^2-x-6}{(x+3)(x-3)}$$
$$= \frac{(x-3)(x+2)}{(x+3)(x-3)}$$
$$= \frac{x+2}{x+3}$$

33. Combining the fractions:

$$\frac{a-4}{a-3} + \frac{5}{a^2-a-6} = \frac{(a-4) \cdot (a+2)}{(a-3) \cdot (a+2)} + \frac{5}{(a-3)(a+2)}$$
$$= \frac{a^2-2a-8}{(a-3)(a+2)} + \frac{5}{(a-3)(a+2)}$$
$$= \frac{a^2-2a-8+5}{(a-3)(a+2)}$$
$$= \frac{a^2-2a-3}{(a-3)(a+2)}$$
$$= \frac{(a-3)(a+1)}{(a-3)(a+2)}$$
$$= \frac{a+1}{a+2}$$

35. Combining the fractions:
$$\frac{8}{x^2-16} - \frac{7}{x^2-x-12} = \frac{8}{(x+4)(x-4)} - \frac{7}{(x-4)(x+3)}$$
$$= \frac{8(x+3)}{(x+4)(x-4)(x+3)} - \frac{7(x+4)}{(x+4)(x-4)(x+3)}$$
$$= \frac{8x+24}{(x+4)(x-4)(x+3)} - \frac{7x+28}{(x+4)(x-4)(x+3)}$$
$$= \frac{8x+24-7x-28}{(x+4)(x-4)(x+3)}$$
$$= \frac{x-4}{(x+4)(x-4)(x+3)}$$
$$= \frac{1}{(x+4)(x+3)}$$

37. Combining the fractions:
$$\frac{4y}{y^2+6y+5} - \frac{3y}{y^2+5y+4} = \frac{4y}{(y+5)(y+1)} - \frac{3y}{(y+4)(y+1)}$$
$$= \frac{4y(y+4)}{(y+5)(y+1)(y+4)} - \frac{3y(y+5)}{(y+5)(y+1)(y+4)}$$
$$= \frac{4y^2+16y}{(y+5)(y+1)(y+4)} - \frac{3y^2+15y}{(y+5)(y+1)(y+4)}$$
$$= \frac{4y^2+16y-3y^2-15y}{(y+5)(y+1)(y+4)}$$
$$= \frac{y^2+y}{(y+5)(y+1)(y+4)}$$
$$= \frac{y(y+1)}{(y+5)(y+1)(y+4)}$$
$$= \frac{y}{(y+5)(y+4)}$$

39. Combining the fractions:
$$\frac{4x+1}{x^2+5x+4} - \frac{x+3}{x^2+4x+3} = \frac{4x+1}{(x+4)(x+1)} - \frac{x+3}{(x+3)(x+1)}$$
$$= \frac{(4x+1)(x+3)}{(x+4)(x+1)(x+3)} - \frac{(x+3)(x+4)}{(x+4)(x+1)(x+3)}$$
$$= \frac{4x^2+13x+3}{(x+4)(x+1)(x+3)} - \frac{x^2+7x+12}{(x+4)(x+1)(x+3)}$$
$$= \frac{4x^2+13x+3-x^2-7x-12}{(x+4)(x+1)(x+3)}$$
$$= \frac{3x^2+6x-9}{(x+4)(x+1)(x+3)}$$
$$= \frac{3(x+3)(x-1)}{(x+4)(x+1)(x+3)}$$
$$= \frac{3(x-1)}{(x+4)(x+1)}$$

41. Combining the fractions:
$$\frac{1}{x}+\frac{x}{3x+9}-\frac{3}{x^2+3x}=\frac{1}{x}+\frac{x}{3(x+3)}-\frac{3}{x(x+3)}$$
$$=\frac{1\cdot 3(x+3)}{x\cdot 3(x+3)}+\frac{x\cdot x}{3(x+3)\cdot x}-\frac{3\cdot 3}{x(x+3)\cdot 3}$$
$$=\frac{3x+9}{3x(x+3)}+\frac{x^2}{3x(x+3)}-\frac{9}{3x(x+3)}$$
$$=\frac{3x+9+x^2-9}{3x(x+3)}$$
$$=\frac{x^2+3x}{3x(x+3)}$$
$$=\frac{x(x+3)}{3x(x+3)}$$
$$=\frac{1}{3}$$

43. a. Multiplying: $\dfrac{4}{9}\cdot\dfrac{1}{6}=\dfrac{2\cdot 2}{3\cdot 3}\cdot\dfrac{1}{2\cdot 3}=\dfrac{2}{3\cdot 3\cdot 3}=\dfrac{2}{27}$

b. Dividing: $\dfrac{4}{9}\div\dfrac{1}{6}=\dfrac{4}{9}\cdot\dfrac{6}{1}=\dfrac{2\cdot 2}{3\cdot 3}\cdot\dfrac{2\cdot 3}{1}=\dfrac{2\cdot 2\cdot 2}{3}=\dfrac{8}{3}$

c. Adding: $\dfrac{4}{9}+\dfrac{1}{6}=\dfrac{4}{9}\cdot\dfrac{2}{2}+\dfrac{1}{6}\cdot\dfrac{3}{3}=\dfrac{8}{18}+\dfrac{3}{18}=\dfrac{11}{18}$

d. Multiplying: $\dfrac{x+2}{x-2}\cdot\dfrac{3x+10}{x^2-4}=\dfrac{x+2}{x-2}\cdot\dfrac{3x+10}{(x+2)(x-2)}=\dfrac{3x+10}{(x-2)^2}$

e. Dividing: $\dfrac{x+2}{x-2}\div\dfrac{3x+10}{x^2-4}=\dfrac{x+2}{x-2}\cdot\dfrac{(x+2)(x-2)}{3x+10}=\dfrac{(x+2)^2}{3x+10}$

f. Subtracting:
$$\frac{x+2}{x-2}-\frac{3x+10}{x^2-4}=\frac{x+2}{x-2}\cdot\frac{x+2}{x+2}-\frac{3x+10}{(x+2)(x-2)}$$
$$=\frac{x^2+4x+4}{(x+2)(x-2)}-\frac{3x+10}{(x+2)(x-2)}$$
$$=\frac{x^2+x-6}{(x+2)(x-2)}$$
$$=\frac{(x+3)(x-2)}{(x+2)(x-2)}$$
$$=\frac{x+3}{x+2}$$

45. Completing the table:

Number x	Reciprocal $\dfrac{1}{x}$	Sum $1+\dfrac{1}{x}$	Sum $\dfrac{x+1}{x}$
1	1	2	2
2	$\dfrac{1}{2}$	$\dfrac{3}{2}$	$\dfrac{3}{2}$
3	$\dfrac{1}{3}$	$\dfrac{4}{3}$	$\dfrac{4}{3}$
4	$\dfrac{1}{4}$	$\dfrac{5}{4}$	$\dfrac{5}{4}$

47. Completing the table:

x	$x+\dfrac{4}{x}$	$\dfrac{x^2+4}{x}$	$x+4$
1	5	5	5
2	4	4	6
3	$\dfrac{13}{3}$	$\dfrac{13}{3}$	7
4	5	5	8

164 Chapter 6 Rational Expressions

49. Combining the fractions: $1+\dfrac{1}{x+2}=\dfrac{1\bullet(x+2)}{1\bullet(x+2)}+\dfrac{1}{x+2}=\dfrac{x+2}{x+2}+\dfrac{1}{x+2}=\dfrac{x+2+1}{x+2}=\dfrac{x+3}{x+2}$

51. Combining the fractions: $1-\dfrac{1}{x+3}=\dfrac{1\bullet(x+3)}{1\bullet(x+3)}-\dfrac{1}{x+3}=\dfrac{x+3}{x+3}-\dfrac{1}{x+3}=\dfrac{x+3-1}{x+3}=\dfrac{x+2}{x+3}$

53. Solving the equation:
$$2x+3(x-3)=6$$
$$2x+3x-9=6$$
$$5x-9=6$$
$$5x=15$$
$$x=3$$

55. Solving the equation:
$$x-3(x+3)=x-3$$
$$x-3x-9=x-3$$
$$-2x-9=x-3$$
$$-3x-9=-3$$
$$-3x=6$$
$$x=-2$$

57. Solving the equation:
$$7-2(3x+1)=4x+3$$
$$7-6x-2=4x+3$$
$$-6x+5=4x+3$$
$$-10x+5=3$$
$$-10x=-2$$
$$x=\tfrac{1}{5}$$

59. Solving the quadratic equation:
$$x^2+5x+6=0$$
$$(x+2)(x+3)=0$$
$$x=-2,-3$$

61. Solving the quadratic equation:
$$x^2-x=6$$
$$x^2-x-6=0$$
$$(x-3)(x+2)=0$$
$$x=-2,3$$

63. Solving the quadratic equation:
$$x^2-5x=0$$
$$x(x-5)=0$$
$$x=0,5$$

65. Simplifying: $6\left(\tfrac{1}{2}\right)=3$

67. Simplifying: $\dfrac{0}{5}=0$

69. Simplifying: $\dfrac{5}{0}$ is undefined

71. Simplifying: $1-\tfrac{5}{2}=\tfrac{2}{2}-\tfrac{5}{2}=-\tfrac{3}{2}$

73. Simplifying: $6\left(\dfrac{x}{3}+\dfrac{5}{2}\right)=6\bullet\dfrac{x}{3}+6\bullet\dfrac{5}{2}=2x+15$

75. Simplifying: $x^2\left(1-\dfrac{5}{x}\right)=x^2\bullet 1-x^2\bullet\dfrac{5}{x}=x^2-5x$

77. Solving the equation:
$$2x+15=3$$
$$2x=-12$$
$$x=-6$$

79. Solving the equation:
$$-2x-9=x-3$$
$$-3x-9=-3$$
$$-3x=6$$
$$x=-2$$

6.4 Equations Involving Rational Expressions

1. Multiplying both sides of the equation by 6:
$$6\left(\frac{x}{3}+\frac{1}{2}\right) = 6\left(-\frac{1}{2}\right)$$
$$2x+3 = -3$$
$$2x = -6$$
$$x = -3$$
Since $x=-3$ checks in the original equation, the solution is $x=-3$.

3. Multiplying both sides of the equation by $5a$:
$$5a\left(\frac{4}{a}\right) = 5a\left(\frac{1}{5}\right)$$
$$20 = a$$
Since $a=20$ checks in the original equation, the solution is $a=20$.

5. Multiplying both sides of the equation by x:
$$x\left(\frac{3}{x}+1\right) = x\left(\frac{2}{x}\right)$$
$$3+x = 2$$
$$x = -1$$
Since $x=-1$ checks in the original equation, the solution is $x=-1$.

7. Multiplying both sides of the equation by $5a$:
$$5a\left(\frac{3}{a}-\frac{2}{a}\right) = 5a\left(\frac{1}{5}\right)$$
$$15 - 10 = a$$
$$a = 5$$
Since $a=5$ checks in the original equation, the solution is $a=5$.

9. Multiplying both sides of the equation by $2x$:
$$2x\left(\frac{3}{x}+2\right) = 2x\left(\frac{1}{2}\right)$$
$$6+4x = x$$
$$6 = -3x$$
$$x = -2$$
Since $x=-2$ checks in the original equation, the solution is $x=-2$.

11. Multiplying both sides of the equation by $4y$:
$$4y\left(\frac{1}{y}-\frac{1}{2}\right) = 4y\left(-\frac{1}{4}\right)$$
$$4 - 2y = -y$$
$$4 = y$$
Since $y=4$ checks in the original equation, the solution is $y=4$.

13. Multiplying both sides of the equation by x^2:
$$x^2\left(1-\frac{8}{x}\right) = x^2\left(-\frac{15}{x^2}\right)$$
$$x^2 - 8x = -15$$
$$x^2 - 8x + 15 = 0$$
$$(x-3)(x-5) = 0$$
$$x = 3, 5$$
Both $x=3$ and $x=5$ check in the original equation.

15. Multiplying both sides of the equation by $2x$:
$$2x\left(\frac{x}{2}-\frac{4}{x}\right) = 2x\left(-\frac{7}{2}\right)$$
$$x^2 - 8 = -7x$$
$$x^2 + 7x - 8 = 0$$
$$(x+8)(x-1) = 0$$
$$x = -8, 1$$
Both $x=-8$ and $x=1$ check in the original equation.

17. Multiplying both sides of the equation by 6:
$$6\left(\frac{x-3}{2}+\frac{2x}{3}\right)=6\left(\frac{5}{6}\right)$$
$$3(x-3)+2(2x)=5$$
$$3x-9+4x=5$$
$$7x-9=5$$
$$7x=14$$
$$x=2$$
Since $x=2$ checks in the original equation, the solution is $x=2$.

19. Multiplying both sides of the equation by 12:
$$12\left(\frac{x+1}{3}+\frac{x-3}{4}\right)=12\left(\frac{1}{6}\right)$$
$$4(x+1)+3(x-3)=2$$
$$4x+4+3x-9=2$$
$$7x-5=2$$
$$7x=7$$
$$x=1$$
Since $x=1$ checks in the original equation, the solution is $x=1$.

21. Multiplying both sides of the equation by $5(x+2)$:
$$5(x+2)\cdot\frac{6}{x+2}=5(x+2)\cdot\frac{3}{5}$$
$$30=3x+6$$
$$24=3x$$
$$x=8$$
Since $x=8$ checks in the original equation, the solution is $x=8$.

23. Multiplying both sides of the equation by $(y-2)(y-3)$:
$$(y-2)(y-3)\cdot\frac{3}{y-2}=(y-2)(y-3)\cdot\frac{2}{y-3}$$
$$3(y-3)=2(y-2)$$
$$3y-9=2y-4$$
$$y=5$$
Since $y=5$ checks in the original equation, the solution is $y=5$.

25. Multiplying both sides of the equation by $3(x-2)$:
$$3(x-2)\left(\frac{x}{x-2}+\frac{2}{3}\right)=3(x-2)\left(\frac{2}{x-2}\right)$$
$$3x+2(x-2)=6$$
$$3x+2x-4=6$$
$$5x-4=6$$
$$5x=10$$
$$x=2$$
Since $x=2$ does not check in the original equation, there is no solution $(\emptyset)$.

27. Multiplying both sides of the equation by $2(x-2)$:
$$2(x-2)\left(\frac{x}{x-2}+\frac{3}{2}\right)=2(x-2)\cdot\frac{9}{2(x-2)}$$
$$2x+3(x-2)=9$$
$$2x+3x-6=9$$
$$5x-6=9$$
$$5x=15$$
$$x=3$$
Since $x=3$ checks in the original equation, the solution is $x=3$.

29. Multiplying both sides of the equation by $x^2 + 5x + 6 = (x+2)(x+3)$:

$$(x+2)(x+3)\left(\frac{5}{x+2} + \frac{1}{x+3}\right) = (x+2)(x+3) \cdot \frac{-1}{(x+2)(x+3)}$$
$$5(x+3) + 1(x+2) = -1$$
$$5x + 15 + x + 2 = -1$$
$$6x + 17 = -1$$
$$6x = -18$$
$$x = -3$$

Since $x = -3$ does not check in the original equation, there is no solution ($\varnothing$).

31. Multiplying both sides of the equation by $x^2 - 4 = (x+2)(x-2)$:

$$(x+2)(x-2)\left(\frac{8}{x^2-4} + \frac{3}{x+2}\right) = (x+2)(x-2) \cdot \frac{1}{x-2}$$
$$8 + 3(x-2) = 1(x+2)$$
$$8 + 3x - 6 = x + 2$$
$$3x + 2 = x + 2$$
$$2x = 0$$
$$x = 0$$

Since $x = 0$ checks in the original equation, the solution is $x = 0$.

33. Multiplying both sides of the equation by $2(a-3)$:

$$2(a-3)\left(\frac{a}{2} + \frac{3}{a-3}\right) = 2(a-3) \cdot \frac{a}{a-3}$$
$$a(a-3) + 6 = 2a$$
$$a^2 - 3a + 6 = 2a$$
$$a^2 - 5a + 6 = 0$$
$$(a-2)(a-3) = 0$$
$$a = 2, 3$$

Since $a = 3$ does not check in the original equation, the solution is $a = 2$.

35. Since $y^2 - 4 = (y+2)(y-2)$ and $y^2 + 2y = y(y+2)$, the LCD is $y(y+2)(y-2)$. Multiplying by the LCD:

$$y(y+2)(y-2) \cdot \frac{6}{(y+2)(y-2)} = y(y+2)(y-2) \cdot \frac{4}{y(y+2)}$$
$$6y = 4(y-2)$$
$$6y = 4y - 8$$
$$2y = -8$$
$$y = -4$$

Since $y = -4$ checks in the original equation, the solution is $y = -4$.

37. Since $a^2 - 9 = (a+3)(a-3)$ and $a^2 + a - 12 = (a+4)(a-3)$, the LCD is $(a+3)(a-3)(a+4)$. Multiplying by the LCD:

$$(a+3)(a-3)(a+4) \cdot \frac{2}{(a+3)(a-3)} = (a+3)(a-3)(a+4) \cdot \frac{3}{(a+4)(a-3)}$$
$$2(a+4) = 3(a+3)$$
$$2a + 8 = 3a + 9$$
$$-a + 8 = 9$$
$$-a = 1$$
$$a = -1$$

Since $a = -1$ checks in the original equation, the solution is $a = -1$.

39. Multiplying both sides of the equation by $x^2 - 4x - 5 = (x-5)(x+1)$:

$$(x-5)(x+1)\left(\frac{3x}{x-5} - \frac{2x}{x+1}\right) = (x-5)(x+1) \cdot \frac{-42}{(x-5)(x+1)}$$
$$3x(x+1) - 2x(x-5) = -42$$
$$3x^2 + 3x - 2x^2 + 10x = -42$$
$$x^2 + 13x + 42 = 0$$
$$(x+7)(x+6) = 0$$
$$x = -7, -6$$

Both $x = -7$ and $x = -6$ check in the original equation.

41. Multiplying both sides of the equation by $x^2 + 5x + 6 = (x+2)(x+3)$:

$$(x+2)(x+3)\left(\frac{2x}{x+2}\right) = (x+2)(x+3)\left(\frac{x}{x+3} - \frac{3}{x^2+5x+6}\right)$$
$$2x(x+3) = x(x+2) - 3$$
$$2x^2 + 6x = x^2 + 2x - 3$$
$$x^2 + 4x + 3 = 0$$
$$(x+3)(x+1) = 0$$
$$x = -3, -1$$

Since $x = -3$ does not check in the original equation, the solution is $x = -1$.

43. **a.** Solving the equation:

$$5x - 1 = 0$$
$$5x = 1$$
$$x = \tfrac{1}{5}$$

b. Solving the equation:

$$\frac{5}{x} - 1 = 0$$
$$x\left(\frac{5}{x} - 1\right) = x(0)$$
$$5 - x = 0$$
$$x = 5$$

c. Solving the equation:

$$\frac{x}{5} - 1 = \frac{2}{3}$$
$$15\left(\frac{x}{5} - 1\right) = 15\left(\frac{2}{3}\right)$$
$$3x - 15 = 10$$
$$3x = 25$$
$$x = \tfrac{25}{3}$$

d. Solving the equation:

$$\frac{5}{x} - 1 = \frac{2}{3}$$
$$3x\left(\frac{5}{x} - 1\right) = 3x\left(\frac{2}{3}\right)$$
$$15 - 3x = 2x$$
$$15 = 5x$$
$$x = 3$$

e. Solving the equation:

$$\frac{5}{x^2} + 5 = \frac{26}{x}$$
$$x^2\left(\frac{5}{x^2} + 5\right) = x^2\left(\frac{26}{x}\right)$$
$$5 + 5x^2 = 26x$$
$$5x^2 - 26x + 5 = 0$$
$$(5x - 1)(x - 5) = 0$$
$$x = \tfrac{1}{5}, 5$$

45. a. Dividing: $\dfrac{7}{a^2-5a-6} \div \dfrac{a+2}{a+1} = \dfrac{7}{(a-6)(a+1)} \cdot \dfrac{a+1}{a+2} = \dfrac{7}{(a-6)(a+2)}$

b. Adding:
$$\dfrac{7}{a^2-5a-6} + \dfrac{a+2}{a+1} = \dfrac{7}{(a-6)(a+1)} + \dfrac{a+2}{a+1} \cdot \dfrac{a-6}{a-6}$$
$$= \dfrac{7}{(a-6)(a+1)} + \dfrac{a^2-4a-12}{(a-6)(a+1)}$$
$$= \dfrac{a^2-4a-5}{(a-6)(a+1)}$$
$$= \dfrac{(a-5)(a+1)}{(a-6)(a+1)}$$
$$= \dfrac{a-5}{a-6}$$

c. Solving the equation:
$$\dfrac{7}{a^2-5a-6} + \dfrac{a+2}{a+1} = 2$$
$$\dfrac{7}{(a-6)(a+1)} + \dfrac{a+2}{a+1} = 2$$
$$(a-6)(a+1)\left(\dfrac{7}{(a-6)(a+1)} + \dfrac{a+2}{a+1}\right) = 2(a-6)(a+1)$$
$$7 + (a-6)(a+2) = 2(a-6)(a+1)$$
$$7 + a^2 - 4a - 12 = 2a^2 - 10a - 12$$
$$0 = a^2 - 6a - 7$$
$$0 = (a-7)(a+1)$$
$$a = -1, 7$$

Upon checking, only $a = 7$ checks in the original equation.

47. Let x represent the number. The equation is:
$$2(x-3) - 5 = 3$$
$$2x - 6 - 5 = 3$$
$$2x - 11 = 3$$
$$2x = 14$$
$$x = 7$$
The number is 7.

49. Let w represent the width, and $2w + 5$ represent the length. Using the perimeter formula:
$$2w + 2(2w+5) = 34$$
$$2w + 4w + 10 = 34$$
$$6w + 10 = 34$$
$$6w = 24$$
$$w = 4$$
$$2w + 5 = 13$$
The length is 13 inches and the width is 4 inches.

51. Let x and $x + 2$ represent the two integers. The equation is:
$$x(x+2) = 48$$
$$x^2 + 2x = 48$$
$$x^2 + 2x - 48 = 0$$
$$(x+8)(x-6) = 0$$
$$x = -8, 6$$
$$x + 2 = -6, 8$$
The two integers are either –8 and –6, or 6 and 8.

53. Let x and $x+2$ represent the two legs. The equation is:
$$x^2 + (x+2)^2 = 10^2$$
$$x^2 + x^2 + 4x + 4 = 100$$
$$2x^2 + 4x - 96 = 0$$
$$x^2 + 2x - 48 = 0$$
$$(x+8)(x-6) = 0$$
$$x = 6 \quad (x = -8 \text{ is impossible})$$
$$x + 2 = 8$$
The legs are 6 inches and 8 inches.

55. Multiplying both sides of the equation by $2x$:
$$2x\left(\frac{1}{x} + \frac{1}{2x}\right) = 2x\left(\frac{9}{2}\right)$$
$$2 + 1 = 9x$$
$$9x = 3$$
$$x = \tfrac{1}{3}$$

57. Multiplying both sides of the equation by $30x$:
$$30x\left(\frac{1}{10} - \frac{1}{15}\right) = 30x\left(\frac{1}{x}\right)$$
$$3x - 2x = 30$$
$$x = 30$$

59. Evaluating when $x = -6$: $y = -\dfrac{6}{-6} = 1$

61. Evaluating when $x = 2$: $y = -\dfrac{6}{2} = -3$

6.5 Applications

1. Let x and $3x$ represent the two numbers. The equation is:
$$\frac{1}{x} + \frac{1}{3x} = \frac{16}{3}$$
$$3x\left(\frac{1}{x} + \frac{1}{3x}\right) = 3x\left(\frac{16}{3}\right)$$
$$3 + 1 = 16x$$
$$16x = 4$$
$$x = \tfrac{1}{4}$$
$$3x = \tfrac{3}{4}$$
The numbers are $\tfrac{1}{4}$ and $\tfrac{3}{4}$.

3. Let x represent the number. The equation is:
$$x + \frac{1}{x} = \frac{13}{6}$$
$$6x\left(x + \frac{1}{x}\right) = 6x\left(\frac{13}{6}\right)$$
$$6x^2 + 6 = 13x$$
$$6x^2 - 13x + 6 = 0$$
$$(3x - 2)(2x - 3) = 0$$
$$x = \tfrac{2}{3}, \tfrac{3}{2}$$
The number is either $\tfrac{2}{3}$ or $\tfrac{3}{2}$.

5. Let x represent the number. The equation is:
$$\frac{7+x}{9+x} = \frac{5}{7}$$
$$7(9+x) \cdot \frac{7+x}{9+x} = 7(9+x) \cdot \frac{5}{7}$$
$$7(7+x) = 5(9+x)$$
$$49 + 7x = 45 + 5x$$
$$49 + 2x = 45$$
$$2x = -4$$
$$x = -2$$
The number is -2.

7. Let x and $x+2$ represent the two integers. The equation is:
$$\frac{1}{x} + \frac{1}{x+2} = \frac{5}{12}$$
$$12x(x+2)\left(\frac{1}{x} + \frac{1}{x+2}\right) = 12x(x+2)\left(\frac{5}{12}\right)$$
$$12(x+2) + 12x = 5x(x+2)$$
$$12x + 24 + 12x = 5x^2 + 10x$$
$$0 = 5x^2 - 14x - 24$$
$$(5x+6)(x-4) = 0$$
$$x = 4 \quad \left(x = -\tfrac{6}{5} \text{ is impossible}\right)$$
$$x+2 = 6$$
The integers are 4 and 6.

9. Let x represent the rate of the boat in still water:

	d	r	t
Upstream	26	$x-3$	$\frac{26}{x-3}$
Downstream	38	$x+3$	$\frac{38}{x+3}$

The equation is:
$$\frac{26}{x-3} = \frac{38}{x+3}$$
$$(x+3)(x-3) \cdot \frac{26}{x-3} = (x+3)(x-3) \cdot \frac{38}{x+3}$$
$$26(x+3) = 38(x-3)$$
$$26x + 78 = 38x - 114$$
$$-12x + 78 = -114$$
$$-12x = -192$$
$$x = 16$$
The speed of the boat in still water is 16 mph.

11. Let x represent the plane speed in still air:

	d	r	t
Against Wind	140	$x-20$	$\frac{140}{x-20}$
With Wind	160	$x+20$	$\frac{160}{x+20}$

The equation is:
$$\frac{140}{x-20} = \frac{160}{x+20}$$
$$(x+20)(x-20) \cdot \frac{140}{x-20} = (x+20)(x-20) \cdot \frac{160}{x+20}$$
$$140(x+20) = 160(x-20)$$
$$140x + 2800 = 160x - 3200$$
$$-20x + 2800 = -3200$$
$$-20x = -6000$$
$$x = 300$$
The plane speed in still air is 300 mph.

13. Let x and $x+20$ represent the rates of each plane:

	d	r	t
Plane 1	285	$x+20$	$\frac{285}{x+20}$
Plane 2	255	x	$\frac{255}{x}$

The equation is:
$$\frac{285}{x+20} = \frac{255}{x}$$
$$x(x+20) \cdot \frac{285}{x+20} = x(x+20) \cdot \frac{255}{x}$$
$$285x = 255(x+20)$$
$$285x = 255x + 5100$$
$$30x = 5100$$
$$x = 170$$
$$x + 20 = 190$$
The plane speeds are 170 mph and 190 mph.

15. Let x represent her rate downhill:

	d	r	t
Level Ground	2	$x-3$	$\dfrac{2}{x-3}$
Downhill	6	x	$\dfrac{6}{x}$

The equation is:
$$\frac{2}{x-3}+\frac{6}{x}=1$$
$$x(x-3)\left(\frac{2}{x-3}+\frac{6}{x}\right)=x(x-3)\cdot 1$$
$$2x+6(x-3)=x(x-3)$$
$$2x+6x-18=x^2-3x$$
$$8x-18=x^2-3x$$
$$0=x^2-11x+18$$
$$0=(x-2)(x-9)$$
$$x=9 \quad (x=2 \text{ is impossible})$$

Tina runs 9 mph on the downhill part of the course.

17. Let x represent her rate on level ground:

	d	r	t
Level Ground	4	x	$\dfrac{4}{x}$
Downhill	5	$x+2$	$\dfrac{5}{x+2}$

The equation is:
$$\frac{4}{x}+\frac{5}{x+2}=1$$
$$x(x+2)\left(\frac{4}{x}+\frac{5}{x+2}\right)=x(x+2)\cdot 1$$
$$4(x+2)+5x=x(x+2)$$
$$4x+8+5x=x^2+2x$$
$$9x+8=x^2+2x$$
$$0=x^2-7x-8$$
$$0=(x-8)(x+1)$$
$$x=8 \quad (x=-1 \text{ is impossible})$$

Jerri jogs 8 mph on level ground.

19. Let t represent the time to fill the pool with both pipes left open. The equation is:
$$\frac{1}{12}-\frac{1}{15}=\frac{1}{t}$$
$$60t\left(\frac{1}{12}-\frac{1}{15}\right)=60t\cdot\frac{1}{t}$$
$$5t-4t=60$$
$$t=60$$

It will take 60 hours to fill the pool with both pipes left open.

21. Let t represent the time to fill the bathtub with both faucets open. The equation is:
$$\frac{1}{10}+\frac{1}{12}=\frac{1}{t}$$
$$60t\left(\frac{1}{10}+\frac{1}{12}\right)=60t\cdot\frac{1}{t}$$
$$6t+5t=60$$
$$11t=60$$
$$t=\frac{60}{11}=5\frac{5}{11}$$
It will take $5\frac{5}{11}$ minutes to fill the tub with both faucets open.

23. Let t represent the time to fill the sink with both the faucet and the drain left open. The equation is:
$$\frac{1}{3}-\frac{1}{4}=\frac{1}{t}$$
$$12t\left(\frac{1}{3}-\frac{1}{4}\right)=12t\cdot\frac{1}{t}$$
$$4t-3t=12$$
$$t=12$$
It will take 12 minutes for the sink to overflow with both the faucet and drain left open.

25. Sketching the graph:

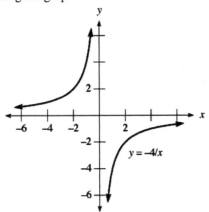

27. Sketching the graph:

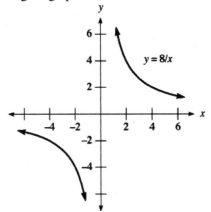

29. Sketching the graph:

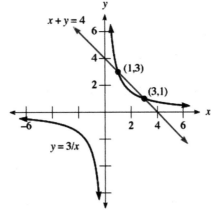

The intersection points are (1,3) and (3,1).

31. Factoring the polynomial: $15a^3b^3-20a^2b-35ab^2=5ab\left(3a^2b^2-4a-7b\right)$

33. Factoring the polynomial: $x^2-4x-12=(x-6)(x+2)$

35. Factoring the polynomial: $x^4-16=\left(x^2+4\right)\left(x^2-4\right)=\left(x^2+4\right)(x+2)(x-2)$

174   Chapter 6 Rational Expressions

37. Factoring the polynomial: $5x^3 - 25x^2 - 30x = 5x(x^2 - 5x - 6) = 5x(x-6)(x+1)$

39. Solving the equation by factoring:
$$x^2 - 6x = 0$$
$$x(x-6) = 0$$
$$x = 0, 6$$

41. Solving the equation by factoring:
$$x(x+2) = 80$$
$$x^2 + 2x = 80$$
$$x^2 + 2x - 80 = 0$$
$$(x+10)(x-8) = 0$$
$$x = -10, 8$$

43. Let x and $x+3$ represent the two integers. The equation is:
$$x^2 + (x+3)^2 = 15^2$$
$$x^2 + x^2 + 6x + 9 = 225$$
$$2x^2 + 6x - 216 = 0$$
$$x^2 + 3x - 108 = 0$$
$$(x+12)(x-9) = 0$$
$$x = 9 \quad (x = -12 \text{ is impossible})$$
$$x + 3 = 12$$
The two legs are 9 inches and 12 inches.

45. Simplifying: $\frac{1}{3} \div \frac{3}{4} = \frac{1}{3} \cdot \frac{4}{3} = \frac{4}{9}$

47. Simplifying: $1 + \frac{2}{3} = \frac{3}{3} + \frac{2}{3} = \frac{5}{3}$

49. Simplifying: $y^7 \cdot \frac{3x^5}{y^4} = \frac{3x^5 y^7}{y^4} = 3x^5 y^3$

51. Simplifying: $\frac{3x^5}{y^4} \cdot \frac{y^7}{6x^2} = \frac{3x^5 y^7}{6x^2 y^4} = \frac{x^3 y^3}{2}$

53. Factoring; $xy^2 + y = y(xy + 1)$

55. Reducing the fraction: $\frac{3x^5 y^3}{6x^2} = \frac{x^3 y^3}{2}$

57. Reducing the fraction: $\frac{x^2 - 9}{x^2 - 5x + 6} = \frac{(x+3)(x-3)}{(x-2)(x-3)} = \frac{x+3}{x-2}$

6.6 Complex Fractions

1. Simplifying the complex fraction: $\dfrac{\frac{3}{4}}{\frac{7}{8}} = \dfrac{\frac{3}{4} \cdot 8}{\frac{7}{8} \cdot 8} = \dfrac{6}{1} = 6$

3. Simplifying the complex fraction: $\dfrac{\frac{2}{3}}{4} = \dfrac{\frac{2}{3} \cdot 3}{4 \cdot 3} = \dfrac{2}{12} = \dfrac{1}{6}$

5. Simplifying the complex fraction: $\dfrac{\frac{x^2}{y}}{\frac{x}{y^3}} = \dfrac{\frac{x^2}{y} \cdot y^3}{\frac{x}{y^3} \cdot y^3} = \dfrac{x^2 y^2}{x} = xy^2$

7. Simplifying the complex fraction: $\dfrac{\frac{4x^3}{y^6}}{\frac{8x^2}{y^7}} = \dfrac{\frac{4x^3}{y^6} \cdot y^7}{\frac{8x^2}{y^7} \cdot y^7} = \dfrac{4x^3 y}{8x^2} = \dfrac{xy}{2}$

9. Simplifying the complex fraction: $\dfrac{y + \frac{1}{x}}{x + \frac{1}{y}} = \dfrac{\left(y + \frac{1}{x}\right) \cdot xy}{\left(x + \frac{1}{y}\right) \cdot xy} = \dfrac{xy^2 + y}{x^2 y + x} = \dfrac{y(xy+1)}{x(xy+1)} = \dfrac{y}{x}$

11. Simplifying the complex fraction: $\dfrac{1+\dfrac{1}{a}}{1-\dfrac{1}{a}} = \dfrac{\left(1+\dfrac{1}{a}\right)\cdot a}{\left(1-\dfrac{1}{a}\right)\cdot a} = \dfrac{a+1}{a-1}$

13. Simplifying the complex fraction: $\dfrac{\dfrac{x+1}{x^2-9}}{\dfrac{2}{x+3}} = \dfrac{\dfrac{x+1}{(x+3)(x-3)}\cdot(x+3)(x-3)}{\dfrac{2}{x+3}\cdot(x+3)(x-3)} = \dfrac{x+1}{2(x-3)}$

15. Simplifying the complex fraction: $\dfrac{\dfrac{1}{a+2}}{\dfrac{1}{a^2-a-6}} = \dfrac{\dfrac{1}{a+2}\cdot(a-3)(a+2)}{\dfrac{1}{(a-3)(a+2)}\cdot(a-3)(a+2)} = \dfrac{a-3}{1} = a-3$

17. Simplifying the complex fraction: $\dfrac{1-\dfrac{9}{y^2}}{1-\dfrac{1}{y}-\dfrac{6}{y^2}} = \dfrac{\left(1-\dfrac{9}{y^2}\right)\cdot y^2}{\left(1-\dfrac{1}{y}-\dfrac{6}{y^2}\right)\cdot y^2} = \dfrac{y^2-9}{y^2-y-6} = \dfrac{(y+3)(y-3)}{(y+2)(y-3)} = \dfrac{y+3}{y+2}$

19. Simplifying the complex fraction: $\dfrac{\dfrac{1}{y}+\dfrac{1}{x}}{\dfrac{1}{xy}} = \dfrac{\left(\dfrac{1}{y}+\dfrac{1}{x}\right)\cdot xy}{\left(\dfrac{1}{xy}\right)\cdot xy} = \dfrac{x+y}{1} = x+y$

21. Simplifying the complex fraction: $\dfrac{1-\dfrac{1}{a^2}}{1-\dfrac{1}{a}} = \dfrac{\left(1-\dfrac{1}{a^2}\right)\cdot a^2}{\left(1-\dfrac{1}{a}\right)\cdot a^2} = \dfrac{a^2-1}{a^2-a} = \dfrac{(a+1)(a-1)}{a(a-1)} = \dfrac{a+1}{a}$

23. Simplifying the complex fraction: $\dfrac{\dfrac{1}{10x}-\dfrac{y}{10x^2}}{\dfrac{1}{10}-\dfrac{y}{10x}} = \dfrac{\left(\dfrac{1}{10x}-\dfrac{y}{10x^2}\right)\cdot 10x^2}{\left(\dfrac{1}{10}-\dfrac{y}{10x}\right)\cdot 10x^2} = \dfrac{x-y}{x^2-xy} = \dfrac{1(x-y)}{x(x-y)} = \dfrac{1}{x}$

25. Simplifying the complex fraction: $\dfrac{\dfrac{1}{a+1}+2}{\dfrac{1}{a+1}+3} = \dfrac{\left(\dfrac{1}{a+1}+2\right)\cdot(a+1)}{\left(\dfrac{1}{a+1}+3\right)\cdot(a+1)} = \dfrac{1+2(a+1)}{1+3(a+1)} = \dfrac{1+2a+2}{1+3a+3} = \dfrac{2a+3}{3a+4}$

27. Simplifying each parenthesis first:
$1-\dfrac{1}{x} = \dfrac{x}{x}-\dfrac{1}{x} = \dfrac{x-1}{x}$

$1-\dfrac{1}{x+1} = \dfrac{x+1}{x+1}-\dfrac{1}{x+1} = \dfrac{x}{x+1}$

$1-\dfrac{1}{x+2} = \dfrac{x+2}{x+2}-\dfrac{1}{x+2} = \dfrac{x+1}{x+2}$

Now performing the multiplication: $\left(1-\dfrac{1}{x}\right)\left(1-\dfrac{1}{x+1}\right)\left(1-\dfrac{1}{x+2}\right) = \dfrac{x-1}{x}\cdot\dfrac{x}{x+1}\cdot\dfrac{x+1}{x+2} = \dfrac{x-1}{x+2}$

29. Simplifying each parenthesis first:
$$1+\frac{1}{x+3}=\frac{x+3}{x+3}+\frac{1}{x+3}=\frac{x+4}{x+3}$$
$$1+\frac{1}{x+2}=\frac{x+2}{x+2}+\frac{1}{x+2}=\frac{x+3}{x+2}$$
$$1+\frac{1}{x+1}=\frac{x+1}{x+1}+\frac{1}{x+1}=\frac{x+2}{x+1}$$

Now performing the multiplication: $\left(1+\frac{1}{x+3}\right)\left(1+\frac{1}{x+2}\right)\left(1+\frac{1}{x+1}\right)=\frac{x+4}{x+3}\cdot\frac{x+3}{x+2}\cdot\frac{x+2}{x+1}=\frac{x+4}{x+1}$

31. Simplifying each term in the sequence:
$$2+\frac{1}{2+1}=2+\frac{1}{3}=\frac{6}{3}+\frac{1}{3}=\frac{7}{3}$$
$$2+\cfrac{1}{2+\cfrac{1}{2+1}}=2+\cfrac{1}{\tfrac{7}{3}}=2+\frac{3}{7}=\frac{14}{7}+\frac{3}{7}=\frac{17}{7}$$
$$2+\cfrac{2}{2+\cfrac{1}{2+\cfrac{1}{2+1}}}=2+\cfrac{1}{\tfrac{17}{7}}=2+\frac{7}{17}=\frac{34}{17}+\frac{7}{17}=\frac{41}{17}$$

33. Completing the table:

Number x	Reciprocal $\frac{1}{x}$	Quotient $\frac{x}{1/x}$	Square x^2
1	1	1	1
2	$\frac{1}{2}$	4	4
3	$\frac{1}{3}$	9	9
4	$\frac{1}{4}$	16	16

35. Completing the table:

Number x	Reciprocal $\frac{1}{x}$	Sum $1+\frac{1}{x}$	Quotient $\frac{1+\frac{1}{x}}{\frac{1}{x}}$
1	1	2	2
2	$\frac{1}{2}$	$\frac{3}{2}$	3
3	$\frac{1}{3}$	$\frac{4}{3}$	4
4	$\frac{1}{4}$	$\frac{5}{4}$	5

37. Solving the inequality:
$$2x+3<5$$
$$2x+3-3<5-3$$
$$2x<2$$
$$\tfrac{1}{2}(2x)<\tfrac{1}{2}(2)$$
$$x<1$$

39. Solving the inequality:
$$-3x\leq 21$$
$$-\tfrac{1}{3}(-3x)\geq -\tfrac{1}{3}(21)$$
$$x\geq -7$$

41. Solving the inequality:
$$-2x+8>-4$$
$$-2x+8-8>-4-8$$
$$-2x>-12$$
$$-\tfrac{1}{2}(-2x)<-\tfrac{1}{2}(-12)$$
$$x<6$$

43. Solving the inequality:
$$4-2(x+1)\geq -2$$
$$4-2x-2\geq -2$$
$$-2x+2\geq -2$$
$$-2x+2-2\geq -2-2$$
$$-2x\geq -4$$
$$-\tfrac{1}{2}(-2x)\leq -\tfrac{1}{2}(-4)$$
$$x\leq 2$$

45. Solving the equation:
$$21=6x$$
$$x=\tfrac{21}{6}=\tfrac{7}{2}$$

47. Solving the equation:
$$x^2+x=6$$
$$x^2+x-6=0$$
$$(x+3)(x-2)=0$$
$$x=-3,2$$

6.7 Proportions

1. Solving the proportion:
$$\frac{x}{2} = \frac{6}{12}$$
$$12x = 12$$
$$x = 1$$

3. Solving the proportion:
$$\frac{2}{5} = \frac{4}{x}$$
$$2x = 20$$
$$x = 10$$

5. Solving the proportion:
$$\frac{10}{20} = \frac{20}{x}$$
$$10x = 400$$
$$x = 40$$

7. Solving the proportion:
$$\frac{a}{3} = \frac{5}{12}$$
$$12a = 15$$
$$a = \frac{15}{12} = \frac{5}{4}$$

9. Solving the proportion:
$$\frac{2}{x} = \frac{6}{7}$$
$$6x = 14$$
$$x = \frac{14}{6} = \frac{7}{3}$$

11. Solving the proportion:
$$\frac{x+1}{3} = \frac{4}{x}$$
$$x^2 + x = 12$$
$$x^2 + x - 12 = 0$$
$$(x+4)(x-3) = 0$$
$$x = -4, 3$$

13. Solving the proportion:
$$\frac{x}{2} = \frac{8}{x}$$
$$x^2 = 16$$
$$x^2 - 16 = 0$$
$$(x+4)(x-4) = 0$$
$$x = -4, 4$$

15. Solving the proportion:
$$\frac{4}{a+2} = \frac{a}{2}$$
$$a^2 + 2a = 8$$
$$a^2 + 2a - 8 = 0$$
$$(a+4)(a-2) = 0$$
$$a = -4, 2$$

17. Solving the proportion:
$$\frac{1}{x} = \frac{x-5}{6}$$
$$x^2 - 5x = 6$$
$$x^2 - 5x - 6 = 0$$
$$(x-6)(x+1) = 0$$
$$x = -1, 6$$

19. Comparing hits to games, the proportion is:
$$\frac{6}{18} = \frac{x}{45}$$
$$18x = 270$$
$$x = 15$$
He will get 15 hits in 45 games.

21. Comparing ml alcohol to ml water, the proportion is:
$$\frac{12}{16} = \frac{x}{28}$$
$$16x = 336$$
$$x = 21$$
The solution will have 21 ml of alcohol.

23. Comparing grams of fat to total grams, the proportion is:
$$\frac{13}{100} = \frac{x}{350}$$
$$100x = 4550$$
$$x = 45.5$$
There are 45.5 grams of fat in 350 grams of ice cream.

25. Comparing inches on the map to actual miles, the proportion is:
$$\frac{3.5}{100} = \frac{x}{420}$$
$$100x = 1470$$
$$x = 14.7$$
They are 14.7 inches apart on the map.

178 Chapter 6 Rational Expressions

27. Comparing inches on the map to actual miles, the proportion is:
$$\frac{0.5}{5} = \frac{1.25}{x}$$
$$0.5x = 6.25$$
$$x = 12.5$$
The actual distance is 12.5 miles.

29. Comparing driving time to distance:
$$\frac{46}{34} = \frac{x}{12.5}$$
$$34x = 575$$
$$x \approx 17$$
The driving time is approximately 17 minutes.

31. Comparing miles to hours, the proportion is:
$$\frac{245}{5} = \frac{x}{7}$$
$$5x = 1715$$
$$x = 343$$
He will travel 343 miles.

33. Answers will vary.

35. Reducing the fraction: $\dfrac{xy+5x+3y+15}{x^2+ax+3x+3a} = \dfrac{x(y+5)+3(y+5)}{x(x+a)+3(x+a)} = \dfrac{(y+5)(x+3)}{(x+a)(x+3)} = \dfrac{y+5}{x+a}$

37. Dividing the fractions:
$$\frac{3x+6}{x^2+4x+3} \div \frac{x^2+x-2}{x^2+2x-3} = \frac{3x+6}{x^2+4x+3} \cdot \frac{x^2+2x-3}{x^2+x-2}$$
$$= \frac{3(x+2)}{(x+3)(x+1)} \cdot \frac{(x+3)(x-1)}{(x+2)(x-1)}$$
$$= \frac{3(x+2)(x+3)(x-1)}{(x+3)(x+1)(x+2)(x-1)}$$
$$= \frac{3}{x+1}$$

39. Subtracting the fractions:
$$\frac{2}{x^2-1} - \frac{5}{x^2+3x-4} = \frac{2}{(x+1)(x-1)} - \frac{5}{(x+4)(x-1)}$$
$$= \frac{2 \cdot (x+4)}{(x+1)(x-1)(x+4)} - \frac{5 \cdot (x+1)}{(x+4)(x-1)(x+1)}$$
$$= \frac{2x+8}{(x+1)(x-1)(x+4)} - \frac{5x+5}{(x+1)(x-1)(x+4)}$$
$$= \frac{2x+8-5x-5}{(x+1)(x-1)(x+4)}$$
$$= \frac{-3x+3}{(x+1)(x-1)(x+4)}$$
$$= \frac{-3(x-1)}{(x+1)(x-1)(x+4)}$$
$$= \frac{-3}{(x+1)(x+4)}$$

41. Evaluating when $x = 3$: $y = 5(3) = 15$

43. Evaluating when $x = 5$: $y = \dfrac{20}{5} = 4$

45. Substituting $y = 72$:
$$72 = 2x^2$$
$$2x^2 - 72 = 0$$
$$2(x^2 - 36) = 0$$
$$2(x+6)(x-6) = 0$$
$$x = -6, 6$$

47. Substituting $y = 72$ and $x = 4$:

$$72 = K \cdot 4$$
$$K = 18$$

49. Substituting $y = 45$ and $x = 3$:

$$45 = K(3)^2$$
$$45 = 9K$$
$$K = 5$$

Chapter 6 Review/Test

1. Reducing the rational expression: $\dfrac{7}{14x - 28} = \dfrac{7}{14(x-2)} = \dfrac{1}{2(x-2)}$. The variable restriction is $x \neq 2$.

2. Reducing the rational expression: $\dfrac{a+6}{a^2 - 36} = \dfrac{1(a+6)}{(a+6)(a-6)} = \dfrac{1}{a-6}$. The variable restriction is $a \neq -6, 6$.

3. Reducing the rational expression: $\dfrac{8x-4}{4x+12} = \dfrac{4(2x-1)}{4(x+3)} = \dfrac{2x-1}{x+3}$. The variable restriction is $x \neq -3$.

4. Reducing the rational expression: $\dfrac{x+4}{x^2 + 8x + 16} = \dfrac{1(x+4)}{(x+4)^2} = \dfrac{1}{x+4}$. The variable restriction is $x \neq -4$.

5. Reducing the rational expression: $\dfrac{3x^3 + 16x^2 - 12x}{2x^3 + 9x^2 - 18x} = \dfrac{x(3x^2 + 16x - 12)}{x(2x^2 + 9x - 18)} = \dfrac{x(3x-2)(x+6)}{x(2x-3)(x+6)} = \dfrac{3x-2}{2x-3}$

The variable restriction is $x \neq -6, 0, \tfrac{3}{2}$.

6. Reducing the rational expression: $\dfrac{x+2}{x^4 - 16} = \dfrac{x+2}{(x^2+4)(x^2-4)} = \dfrac{x+2}{(x^2+4)(x+2)(x-2)} = \dfrac{1}{(x^2+4)(x-2)}$

The variable restriction is $x \neq -2, 2$.

7. Reducing the rational expression: $\dfrac{x^2 + 5x - 14}{x+7} = \dfrac{(x+7)(x-2)}{x+7} = x - 2$. The variable restriction is $x \neq -7$.

8. Reducing the rational expression: $\dfrac{a^2 + 16a + 64}{a+8} = \dfrac{(a+8)^2}{a+8} = a + 8$. The variable restriction is $a \neq -8$.

9. Reducing the rational expression: $\dfrac{xy + bx + ay + ab}{xy + 5x + ay + 5a} = \dfrac{x(y+b) + a(y+b)}{x(y+5) + a(y+5)} = \dfrac{(y+b)(x+a)}{(y+5)(x+a)} = \dfrac{y+b}{y+5}$

The variable restriction is $y \neq -5, x \neq -a$.

10. Performing the operations: $\dfrac{3x+9}{x^2} \cdot \dfrac{x^3}{6x+18} = \dfrac{3(x+3)}{x^2} \cdot \dfrac{x^3}{6(x+3)} = \dfrac{3x^3(x+3)}{6x^2(x+3)} = \dfrac{x}{2}$

11. Performing the operations:

$$\dfrac{x^2 + 8x + 16}{x^2 + x - 12} \div \dfrac{x^2 - 16}{x^2 - x - 6} = \dfrac{x^2 + 8x + 16}{x^2 + x - 12} \cdot \dfrac{x^2 - x - 6}{x^2 - 16}$$
$$= \dfrac{(x+4)^2}{(x+4)(x-3)} \cdot \dfrac{(x+2)(x-3)}{(x+4)(x-4)}$$
$$= \dfrac{(x+4)^2 (x+2)(x-3)}{(x+4)^2 (x-3)(x-4)}$$
$$= \dfrac{x+2}{x-4}$$

12. Performing the operations: $\left(a^2 - 4a - 12\right)\left(\dfrac{a-6}{a+2}\right) = \dfrac{(a-6)(a+2)}{1} \cdot \dfrac{a-6}{a+2} = \dfrac{(a-6)^2 (a+2)}{a+2} = (a-6)^2$

13. Performing the operations:
$$\frac{3x^2-2x-1}{x^2+6x+8} \div \frac{3x^2+13x+4}{x^2+8x+16} = \frac{3x^2-2x-1}{x^2+6x+8} \cdot \frac{x^2+8x+16}{3x^2+13x+4}$$
$$= \frac{(3x+1)(x-1)}{(x+4)(x+2)} \cdot \frac{(x+4)^2}{(3x+1)(x+4)}$$
$$= \frac{(x+4)^2(3x+1)(x-1)}{(x+4)^2(x+2)(3x+1)}$$
$$= \frac{x-1}{x+2}$$

14. Performing the operations: $\frac{2x}{2x+3} + \frac{3}{2x+3} = \frac{2x+3}{2x+3} = 1$

15. Performing the operations: $\frac{x^2}{x-9} - \frac{18x-81}{x-9} = \frac{x^2-18x+81}{x-9} = \frac{(x-9)^2}{x-9} = x-9$

16. Performing the operations: $\frac{a+4}{a+8} - \frac{a-9}{a+8} = \frac{a+4-a+9}{a+8} = \frac{13}{a+8}$

17. Performing the operations: $\frac{x}{x+9} + \frac{5}{x} = \frac{x \cdot x}{(x+9) \cdot x} + \frac{5 \cdot (x+9)}{x \cdot (x+9)} = \frac{x^2}{x(x+9)} + \frac{5x+45}{x(x+9)} = \frac{x^2+5x+45}{x(x+9)}$

18. Performing the operations: $\frac{5}{4x+20} + \frac{x}{x+5} = \frac{5}{4(x+5)} + \frac{x \cdot 4}{(x+5) \cdot 4} = \frac{5}{4(x+5)} + \frac{4x}{4(x+5)} = \frac{4x+5}{4(x+5)}$

19. Performing the operations:
$$\frac{3}{x^2-36} - \frac{2}{x^2-4x-12} = \frac{3}{(x+6)(x-6)} - \frac{2}{(x-6)(x+2)}$$
$$= \frac{3(x+2)}{(x+6)(x-6)(x+2)} - \frac{2(x+6)}{(x+6)(x-6)(x+2)}$$
$$= \frac{3x+6}{(x+6)(x-6)(x+2)} - \frac{2x+12}{(x+6)(x-6)(x+2)}$$
$$= \frac{3x+6-2x-12}{(x+6)(x-6)(x+2)}$$
$$= \frac{x-6}{(x+6)(x-6)(x+2)}$$
$$= \frac{1}{(x+6)(x+2)}$$

20. Performing the operations:
$$\frac{3a}{a^2+8a+15} - \frac{2}{a+5} = \frac{3a}{(a+5)(a+3)} - \frac{2(a+3)}{(a+5)(a+3)}$$
$$= \frac{3a}{(a+5)(a+3)} - \frac{2a+6}{(a+5)(a+3)}$$
$$= \frac{3a-2a-6}{(a+5)(a+3)}$$
$$= \frac{a-6}{(a+5)(a+3)}$$

21. Multiplying both sides of the equation by $2x$:
$$2x\left(\frac{3}{x}+\frac{1}{2}\right) = 2x\left(\frac{5}{x}\right)$$
$$6+x=10$$
$$x=4$$
Since $x=4$ checks in the original equation, the solution is $x=4$.

22. Multiplying both side of the equation by $2(a-3)$:
$$2(a-3) \cdot \frac{a}{a-3} = 2(a-3) \cdot \frac{3}{2}$$
$$2a = 3(a-3)$$
$$2a = 3a - 9$$
$$-a = -9$$
$$a = 9$$
Since $a = 9$ checks in the original equation, the solution is $a = 9$.

23. Multiplying both sides of the equation by x^2:
$$x^2\left(1 - \frac{7}{x}\right) = x^2\left(\frac{-6}{x^2}\right)$$
$$x^2 - 7x = -6$$
$$x^2 - 7x + 6 = 0$$
$$(x-6)(x-1) = 0$$
$$x = 1, 6$$
Both $x = 1$ and $x = 6$ check in the original equation.

24. Multiplying both side of the equation by $x^2 + 4x - 12 = (x+6)(x-2)$:
$$(x+6)(x-2)\left(\frac{3}{x+6} - \frac{1}{x-2}\right) = (x+6)(x-2) \cdot \frac{-8}{(x+6)(x-2)}$$
$$3(x-2) - 1(x+6) = -8$$
$$3x - 6 - x - 6 = -8$$
$$2x - 12 = -8$$
$$2x = 4$$
$$x = 2$$
Since $x = 2$ does not check in the original equation, there is no solution.

25. Since $y^2 - 16 = (y+4)(y-4)$ and $y^2 + 4y = y(y+4)$, multiply each side of the equation by $y(y+4)(y-4)$:
$$y(y+4)(y-4) \cdot \frac{2}{(y+4)(y-4)} = y(y+4)(y-4) \cdot \frac{10}{y(y+4)}$$
$$2y = 10(y-4)$$
$$2y = 10y - 40$$
$$-8y = -40$$
$$y = 5$$
Since $y = 5$ checks in the original equation, the solution is $y = 5$.

26. Let x represent the number. The equation is:
$$x + 7\left(\frac{1}{x}\right) = \frac{16}{3}$$
$$3x\left(x + \frac{7}{x}\right) = 3x \cdot \frac{16}{3}$$
$$3x^2 + 21 = 16x$$
$$3x^2 - 16x + 21 = 0$$
$$(3x-7)(x-3) = 0$$
$$x = 3, \frac{7}{3}$$
The number is either 3 or $\frac{7}{3}$.

27. Let x represent the speed of the boat in still water. Completing the table:

	d	r	t
Upstream	48	$x-3$	$\dfrac{48}{x-3}$
Downstream	72	$x+3$	$\dfrac{72}{x+3}$

The equation is:
$$\frac{48}{x-3} = \frac{72}{x+3}$$
$$(x+3)(x-3) \cdot \frac{48}{x-3} = (x+3)(x-3) \cdot \frac{72}{x+3}$$
$$48(x+3) = 72(x-3)$$
$$48x + 144 = 72x - 216$$
$$-24x + 144 = -216$$
$$-24x = -360$$
$$x = 15$$

The speed of the boat in still water is 15 mph.

28. Let t represent the time to fill the pool with both pipes left open. The equation is:
$$\frac{1}{21} - \frac{1}{28} = \frac{1}{t}$$
$$84t \cdot \left(\frac{1}{21} - \frac{1}{28}\right) = 84t \cdot \frac{1}{t}$$
$$4t - 3t = 84$$
$$t = 84$$

It will take 84 hours to fill the pool with both pipes left open.

29. Simplifying the complex fraction: $\dfrac{\frac{x+4}{x^2-16}}{\frac{2}{x-4}} = \dfrac{\frac{x+4}{(x+4)(x-4)}}{\frac{2}{x-4}} = \dfrac{\frac{1}{x-4} \cdot (x-4)}{\frac{2}{x-4} \cdot (x-4)} = \dfrac{1}{2}$

30. Simplifying the complex fraction: $\dfrac{1-\frac{9}{y^2}}{1+\frac{4}{y}-\frac{21}{y^2}} = \dfrac{\left(1-\frac{9}{y^2}\right) \cdot y^2}{\left(1+\frac{4}{y}-\frac{21}{y^2}\right) \cdot y^2} = \dfrac{y^2-9}{y^2+4y-21} = \dfrac{(y+3)(y-3)}{(y+7)(y-3)} = \dfrac{y+3}{y+7}$

31. Simplifying the complex fraction: $\dfrac{\frac{1}{a-2}+4}{\frac{1}{a-2}+1} = \dfrac{\left(\frac{1}{a-2}+4\right)(a-2)}{\left(\frac{1}{a-2}+1\right)(a-2)} = \dfrac{1+4(a-2)}{1+1(a-2)} = \dfrac{1+4a-8}{1+a-2} = \dfrac{4a-7}{a-1}$

32. Writing as a fraction: $\dfrac{40}{100} = \dfrac{2}{5}$

33. Writing as a fraction: $\dfrac{40 \text{ seconds}}{3 \text{ minutes}} = \dfrac{40 \text{ seconds}}{180 \text{ seconds}} = \dfrac{2}{9}$

34. Solving the proportion:
$$\frac{x}{9} = \frac{4}{3}$$
$$3x = 36$$
$$x = 12$$

35. Solving the proportion:
$$\frac{a}{3} = \frac{12}{a}$$
$$a^2 = 36$$
$$a^2 - 36 = 0$$
$$(a+6)(a-6) = 0$$
$$a = -6, 6$$

36. Solving the proportion:
$$\frac{8}{x-2} = \frac{x}{6}$$
$$x^2 - 2x = 48$$
$$x^2 - 2x - 48 = 0$$
$$(x+6)(x-8) = 0$$
$$x = -6, 8$$

Chapter 7
Transitions

7.1 Review of Solving Equations

1. Solving the equation:
$$2x - 4 = 6$$
$$2x = 10$$
$$x = \tfrac{10}{2} = 5$$

3. Solving the equation:
$$-300y + 100 = 500$$
$$-300y = 400$$
$$y = -\tfrac{4}{3}$$

5. Solving the equation:
$$-x = 2$$
$$x = -1 \cdot 2 = -2$$

7. Solving the equation:
$$-a = -\tfrac{3}{4}$$
$$a = -1 \cdot \left(-\tfrac{3}{4}\right) = \tfrac{3}{4}$$

9. Solving the equation:
$$-\tfrac{3}{5}a + 2 = 8$$
$$-\tfrac{3}{5}a = 6$$
$$a = -\tfrac{5}{3} \cdot 6 = -10$$

11. Solving the equation:
$$2x - 5 = 3x + 2$$
$$-x - 5 = 2$$
$$-x = 7$$
$$x = -7$$

13. Solving the equation:
$$5 - 2x = 3x + 1$$
$$5 - 5x = 1$$
$$-5x = -4$$
$$x = \tfrac{4}{5}$$

15. Solving the equation:
$$5(y+2) - 4(y+1) = 3$$
$$5y + 10 - 4y - 4 = 3$$
$$y + 6 = 3$$
$$y = -3$$

17. Solving the equation:
$$6 - 7(m - 3) = -1$$
$$6 - 7m + 21 = -1$$
$$-7m + 27 = -1$$
$$-7m = -28$$
$$m = 4$$

19. Solving the equation:
$$7 + 3(x+2) = 4(x-1)$$
$$7 + 3x + 6 = 4x - 4$$
$$3x + 13 = 4x - 4$$
$$-x + 13 = -4$$
$$-x = -17$$
$$x = 17$$

186 Chapter 7 Transititions

21. Solving the equation:
$$\tfrac{1}{2}x+\tfrac{1}{4}=\tfrac{1}{3}x+\tfrac{5}{4}$$
$$12\left(\tfrac{1}{2}x+\tfrac{1}{4}\right)=12\left(\tfrac{1}{3}x+\tfrac{5}{4}\right)$$
$$6x+3=4x+15$$
$$2x+3=15$$
$$2x=12$$
$$x=6$$

23. Solving the equation:
$$\tfrac{1}{2}x+\tfrac{1}{3}x+\tfrac{1}{4}x=13$$
$$12\left(\tfrac{1}{2}x+\tfrac{1}{3}x+\tfrac{1}{4}x\right)=12(13)$$
$$6x+4x+3x=156$$
$$13x=156$$
$$x=12$$

25. Solving the equation:
$$0.08x+0.09(9{,}000-x)=750$$
$$0.08x+810-0.09x=750$$
$$-0.01x+810=750$$
$$-0.01x=-60$$
$$x=6{,}000$$

27. Solving the equation:
$$0.35x-0.2=0.15x+0.1$$
$$0.2x-0.2=0.1$$
$$0.2x=0.3$$
$$2x=3$$
$$x=\tfrac{3}{2}$$

29. Solving the equation:
$$x^2-5x-6=0$$
$$(x+1)(x-6)=0$$
$$x=-1,6$$

31. Solving the equation:
$$x^3-5x^2+6x=0$$
$$x\left(x^2-5x+6\right)=0$$
$$x(x-2)(x-3)=0$$
$$x=0,2,3$$

33. Solving the equation:
$$3y^2+11y-4=0$$
$$(3y-1)(y+4)=0$$
$$y=-4,\tfrac{1}{3}$$

35. Solving the equation:
$$\tfrac{1}{10}t^2-\tfrac{5}{2}=0$$
$$10\left(\tfrac{1}{10}t^2-\tfrac{5}{2}\right)=10(0)$$
$$t^2-25=0$$
$$(t+5)(t-5)=0$$
$$t=-5,5$$

37. Solving the equation:
$$\tfrac{1}{5}y^2-2=-\tfrac{3}{10}y$$
$$10\left(\tfrac{1}{5}y^2-2\right)=10\left(-\tfrac{3}{10}y\right)$$
$$2y^2-20=-3y$$
$$2y^2+3y-20=0$$
$$(y+4)(2y-5)=0$$
$$y=-4,\tfrac{5}{2}$$

39. Solving the equation:
$$9x^2-12x=0$$
$$3x(3x-4)=0$$
$$x=0,\tfrac{4}{3}$$

41. Solving the equation:
$$0.02r+0.01=0.15r^2$$
$$2r+1=15r^2$$
$$15r^2-2r-1=0$$
$$(5r+1)(3r-1)=0$$
$$r=-\tfrac{1}{5},\tfrac{1}{3}$$

43. Solving the equation:
$$9a^3=16a$$
$$9a^3-16a=0$$
$$a\left(9a^2-16\right)=0$$
$$a(3a+4)(3a-4)=0$$
$$a=-\tfrac{4}{3},0,\tfrac{4}{3}$$

45. Solving the equation:
$$-100x = 10x^2$$
$$0 = 10x^2 + 100x$$
$$0 = 10x(x+10)$$
$$x = -10, 0$$

47. Solving the equation:
$$(x+6)(x-2) = -7$$
$$x^2 + 4x - 12 = -7$$
$$x^2 + 4x - 5 = 0$$
$$(x+5)(x-1) = 0$$
$$x = -5, 1$$

49. Solving the equation:
$$(x+1)^2 = 3x + 7$$
$$x^2 + 2x + 1 = 3x + 7$$
$$x^2 - x - 6 = 0$$
$$(x+2)(x-3) = 0$$
$$x = -2, 3$$

51. Solving the equation:
$$x^3 + 3x^2 - 4x - 12 = 0$$
$$x^2(x+3) - 4(x+3) = 0$$
$$(x+3)(x^2 - 4) = 0$$
$$(x+3)(x+2)(x-2) = 0$$
$$x = -3, -2, 2$$

53. Solving the equation:
$$2x^3 + 3x^2 - 8x - 12 = 0$$
$$x^2(2x+3) - 4(2x+3) = 0$$
$$(2x+3)(x^2 - 4) = 0$$
$$(2x+3)(x+2)(x-2) = 0$$
$$x = -2, -\tfrac{3}{2}, 2$$

55. Solving the equation:
$$3x - 6 = 3(x+4)$$
$$3x - 6 = 3x + 12$$
$$-6 = 12$$
Since this statement is false, there is no solution ($\emptyset$).

57. Solving the equation:
$$4y + 2 - 3y + 5 = 3 + y + 4$$
$$y + 7 = y + 7$$
$$7 = 7$$
Since this statement is true, the solution is all real numbers.

59. Solving the equation:
$$2(4t - 1) + 3 = 5t + 4 + 3t$$
$$8t - 2 + 3 = 8t + 4$$
$$8t + 1 = 8t + 4$$
$$1 = 4$$
Since this statement is false, there is no solution ($\emptyset$).

61. Solving the equation:
$$0 = 6{,}400a + 70$$
$$-70 = 6{,}400a$$
$$a = -\tfrac{70}{6{,}400} = -\tfrac{7}{640}$$

63. Solving the equation:
$$x + 2 = 2x$$
$$2 = 2x - x$$
$$x = 2$$

65. Solving the equation:
$$0.07x = 1.4$$
$$x = \tfrac{1.4}{0.07} = 20$$

67. Solving the equation:
$$5(2x + 1) = 12$$
$$10x + 5 = 12$$
$$10x = 7$$
$$x = \tfrac{7}{10} = 0.7$$

69. Solving the equation:
$$50 = \frac{K}{48}$$
$$50 \cdot 48 = \frac{K}{48} \cdot 48$$
$$K = 2,400$$

71. Solving the equation:
$$100P = 2,400$$
$$P = \frac{2,400}{100} = 24$$

73. Solving the equation:
$$x + (3x+2) = 26$$
$$4x + 2 = 26$$
$$4x = 24$$
$$x = 6$$

75. Solving the equation:
$$2x - 3(3x-5) = -6$$
$$2x - 9x + 15 = -6$$
$$-7x + 15 = -6$$
$$-7x = -21$$
$$x = 3$$

77. Solving the equation:
$$5\left(-\tfrac{19}{15}\right) + 5y = 9$$
$$-\tfrac{19}{3} + 5y = 9$$
$$5y = \tfrac{46}{3}$$
$$y = \tfrac{46}{15}$$

79. Solving the equation:
$$2\left(-\tfrac{29}{22}\right) - 3y = 4$$
$$-\tfrac{29}{11} - 3y = 4$$
$$-3y = \tfrac{73}{11}$$
$$y = -\tfrac{73}{33}$$

81. Solving the equation:
$$3x^2 + x = 10$$
$$3x^2 + x - 10 = 0$$
$$(x+2)(3x-5) = 0$$
$$x = -2, \tfrac{5}{3}$$

83. Solving the equation:
$$12(x+3) + 12(x-3) = 3(x^2 - 9)$$
$$12x + 36 + 12x - 36 = 3x^2 - 27$$
$$24x = 3x^2 - 27$$
$$0 = 3x^2 - 24x - 27$$
$$x^2 - 8x - 9 = 0$$
$$(x+1)(x-9) = 0$$
$$x = -1, 9$$

85. Let w represent the width and $2w$ represent the length. Using the perimeter formula:
$$2w + 2(2w) = 60$$
$$2w + 4w = 60$$
$$6w = 60$$
$$w = 10$$
The dimensions are 10 feet by 20 feet.

87. Let w represent the width and $2w$ represent the length. Using the perimeter formula:
$$2w + 2(2w) = 48$$
$$2w + 4w = 48$$
$$6w = 48$$
$$w = 8$$
The width is 8 feet and the length is 16 feet. Finding the cost: $C = 1.75(32) + 2.25(16) = 56 + 36 = 92$
The cost to build the pen is $92.00.

89. The total money collected is: $1204 - $250 = $954
Let x represent the amount of her sales (not including tax). Since this amount includes the tax collected, the equation is:
$$x + 0.06x = 954$$
$$1.06x = 954$$
$$x = 900$$
Her sales were $900, so the sales tax is: $0.06(900) = \$54$

91. Completing the table:

Age (years)	Maximum Heart Rate (beats per minute)
18	202
19	201
20	200
21	199
22	198
23	197

93. Completing the table:

Resting Heart Rate (beats per minute)	Training Heart Rate (beats per minute)
60	144
62	145
64	146
68	147
70	148
72	149

95. Let x and $x + 2$ represent the two odd integers. The equation is:
$$x^2 + (x+2)^2 = 34$$
$$x^2 + x^2 + 4x + 4 = 34$$
$$2x^2 + 4x - 30 = 0$$
$$x^2 + 2x - 15 = 0$$
$$(x+5)(x-3) = 0$$
$$x = -5, 3$$
$$x + 2 = -3, 5$$

The integers are either –5 and –3, or 3 and 5.

97. Simplifying: $|-3| = 3$

99. Simplifying: $-|-3| = -3$

101. $|x|$ represents the distance from x to 0 on the number line.

103. Solving the equation:
$$2a - 1 = -7$$
$$2a = -6$$
$$a = -3$$

105. Solving the equation:
$$\tfrac{2}{3}x - 3 = 7$$
$$\tfrac{2}{3}x = 10$$
$$x = 15$$

107. Solving the equation:
$$x - 5 = x - 7$$
$$-5 = -7$$

The equation has no solution $(\emptyset)$.

109. Solving the equation:
$$x - 5 = -x - 7$$
$$2x - 5 = -7$$
$$2x = -2$$
$$x = -1$$

7.2 Equations with Absolute Value

1. Solving the equation:
$$|x| = 4$$
$$x = -4, 4$$

3. Solving the equation:
$$2 = |a|$$
$$a = -2, 2$$

5. The equation $|x| = -3$ has no solution, or $\emptyset$.

7. Solving the equation:
$$|a| + 2 = 3$$
$$|a| = 1$$
$$a = -1, 1$$

9. Solving the equation:
$$|y| + 4 = 3$$
$$|y| = -1$$

The equation $|y| = -1$ has no solution, or $\emptyset$.

190 Chapter 7 Transititions

11. Solving the equation:
$$4 = |x| - 2$$
$$|x| = 6$$
$$x = -6, 6$$

13. Solving the equation:
$$|x - 2| = 5$$
$$x - 2 = -5, 5$$
$$x = -3, 7$$

15. Solving the equation:
$$|a - 4| = \tfrac{5}{3}$$
$$a - 4 = -\tfrac{5}{3}, \tfrac{5}{3}$$
$$a = \tfrac{7}{3}, \tfrac{17}{3}$$

17. Solving the equation:
$$1 = |3 - x|$$
$$3 - x = -1, 1$$
$$-x = -4, -2$$
$$x = 2, 4$$

19. Solving the equation:
$$\left|\tfrac{3}{5}a + \tfrac{1}{2}\right| = 1$$
$$\tfrac{3}{5}a + \tfrac{1}{2} = -1, 1$$
$$\tfrac{3}{5}a = -\tfrac{3}{2}, \tfrac{1}{2}$$
$$a = -\tfrac{5}{2}, \tfrac{5}{6}$$

21. Solving the equation:
$$60 = |20x - 40|$$
$$20x - 40 = -60, 60$$
$$20x = -20, 100$$
$$x = -1, 5$$

23. Since $|2x + 1| = -3$ is impossible, there is no solution, or $\emptyset$.

25. Solving the equation:
$$\left|\tfrac{3}{4}x - 6\right| = 9$$
$$\tfrac{3}{4}x - 6 = -9, 9$$
$$\tfrac{3}{4}x = -3, 15$$
$$3x = -12, 60$$
$$x = -4, 20$$

27. Solving the equation:
$$\left|1 - \tfrac{1}{2}a\right| = 3$$
$$1 - \tfrac{1}{2}a = -3, 3$$
$$-\tfrac{1}{2}a = -4, 2$$
$$a = -4, 8$$

29. Solving the equation:
$$|3x + 4| + 1 = 7$$
$$|3x + 4| = 6$$
$$3x + 4 = -6, 6$$
$$3x = -10, 2$$
$$x = -\tfrac{10}{3}, \tfrac{2}{3}$$

31. Solving the equation:
$$|3 - 2y| + 4 = 3$$
$$|3 - 2y| = -1$$

Since this equation is impossible, there is no solution, or $\emptyset$.

33. Solving the equation:
$$3 + |4t - 1| = 8$$
$$|4t - 1| = 5$$
$$4t - 1 = -5, 5$$
$$4t = -4, 6$$
$$t = -1, \tfrac{3}{2}$$

35. Solving the equation:
$$\left|9 - \tfrac{3}{5}x\right| + 6 = 12$$
$$\left|9 - \tfrac{3}{5}x\right| = 6$$
$$9 - \tfrac{3}{5}x = -6, 6$$
$$-\tfrac{3}{5}x = -15, -3$$
$$-3x = -75, -15$$
$$x = 5, 25$$

37. Solving the equation:
$$5 = \left|\frac{2x}{7} + \frac{4}{7}\right| - 3$$
$$\left|\frac{2x}{7} + \frac{4}{7}\right| = 8$$
$$\frac{2x}{7} + \frac{4}{7} = -8, 8$$
$$2x + 4 = -56, 56$$
$$2x = -60, 52$$
$$x = -30, 26$$

39. Solving the equation:
$$2 = -8 + \left|4 - \tfrac{1}{2}y\right|$$
$$\left|4 - \tfrac{1}{2}y\right| = 10$$
$$4 - \tfrac{1}{2}y = -10, 10$$
$$-\tfrac{1}{2}y = -14, 6$$
$$y = -12, 28$$

41. Solving the equation:
$$|3a + 1| = |2a - 4|$$

$3a + 1 = 2a - 4$	$3a + 1 = -2a + 4$
$a + 1 = -4$ or	$5a = 3$
$a = -5$	$a = \tfrac{3}{5}$

43. Solving the equation:
$$\left|x - \tfrac{1}{3}\right| = \left|\tfrac{1}{2}x + \tfrac{1}{6}\right|$$

$x - \tfrac{1}{3} = \tfrac{1}{2}x + \tfrac{1}{6}$	$x - \tfrac{1}{3} = -\tfrac{1}{2}x - \tfrac{1}{6}$
$6x - 2 = 3x + 1$ or	$6x - 2 = -3x - 1$
$3x - 2 = 1$	$9x - 2 = -1$
$3x = 3$	$9x = 1$
$x = 1$	$x = \tfrac{1}{9}$

45. Solving the equation:
$$|y - 2| = |y + 3|$$

$y - 2 = y + 3$	$y - 2 = -y - 3$
$-2 = -3$ or	$2y = -1$
y=impossible	$y = -\tfrac{1}{2}$

47. Solving the equation:
$$|3x - 1| = |3x + 1|$$

$3x - 1 = 3x + 1$	$3x - 1 = -3x - 1$
$-1 = 1$ or	$6x = 0$
x=impossible	$x = 0$

49. Solving the equation:
$$|3 - m| = |m + 4|$$

$3 - m = m + 4$	$3 - m = -m - 4$
$-2m = 1$ or	$3 = -4$
$m = -\tfrac{1}{2}$	$m =$ impossible

51. Solving the equation:
$$|0.03 - 0.01x| = |0.04 + 0.05x|$$

$0.03 - 0.01x = 0.04 + 0.05x$	$0.03 - 0.01x = -0.04 - 0.05x$
$-0.06x = 0.01$ or	$0.04x = -0.07$
$x = -\tfrac{1}{6}$	$x = -\tfrac{7}{4}$

53. Since $|x - 2| = |2 - x|$ is always true, the solution set is all real numbers.

55. Since $\left|\tfrac{x}{5} - 1\right| = \left|1 - \tfrac{x}{5}\right|$ is always true, the solution set is all real numbers.

57. **a.** Solving the equation:

$$4x - 5 = 0$$
$$4x = 5$$
$$x = \tfrac{5}{4} = 1.25$$

b. Solving the equation:

$$|4x - 5| = 0$$
$$4x - 5 = 0$$
$$4x = 5$$
$$x = \tfrac{5}{4} = 1.25$$

c. Solving the equation:

$$4x - 5 = 3$$
$$4x = 8$$
$$x = 2$$

d. Solving the equation:

$$|4x - 5| = 3$$
$$4x - 5 = -3, 3$$
$$4x = 2, 8$$
$$x = \tfrac{1}{2}, 2$$

e. Solving the equation:

$$|4x - 5| = |2x + 3|$$

$$4x - 5 = 2x + 3 \quad \text{or} \quad 4x - 5 = -2x - 3$$
$$2x - 5 = 3 \qquad\qquad 6x - 5 = -3$$
$$2x = 8 \qquad\qquad 6x = 2$$
$$x = 4 \qquad\qquad x = \tfrac{1}{3}$$

59. Setting $R = 722$:

$$-60|x - 11| + 962 = 722$$
$$-60|x - 11| = -240$$
$$|x - 11| = 4$$
$$x - 11 = -4, 4$$
$$x = 7, 15$$

The revenue was 722 million dollars in the years 1987 and 1995.

61. Simplifying: $\dfrac{38}{30} = \dfrac{2 \cdot 19}{2 \cdot 15} = \dfrac{19}{15}$

63. Simplifying: $\dfrac{240}{6} = \dfrac{6 \cdot 40}{6 \cdot 1} = 40$

65. Simplifying: $\dfrac{0 + 6}{0 - 3} = \dfrac{6}{-3} = -2$

67. Simplifying: $\dfrac{4 - 4}{4 - 2} = \dfrac{0}{2} = 0$

69. Solving the equation:

$$-2x - 3 = 7$$
$$-2x = 10$$
$$x = -5$$

71. Solving the equation:

$$3(2x - 4) - 7x = -3x$$
$$6x - 12 - 7x = -3x$$
$$-x - 12 = -3x$$
$$-12 = -2x$$
$$x = 6$$

73. Solving the equation:

$$|x - a| = b$$
$$x - a = -b, b$$
$$x = a - b \text{ or } x = a + b$$

75. Solving the equation:

$$|ax + b| = c$$
$$ax + b = -c, c$$
$$ax = -b - c, -b + c$$
$$x = \dfrac{-b - c}{a} \text{ or } x = \dfrac{-b + c}{a}$$

77. Solving the equation:

$$\left|\dfrac{x}{a} + \dfrac{y}{b}\right| = 1$$
$$\dfrac{x}{a} + \dfrac{y}{b} = -1, 1$$
$$\dfrac{x}{a} = -\dfrac{y}{b} - 1, -\dfrac{y}{b} + 1$$
$$x = -\dfrac{a}{b}y - a \text{ or } x = -\dfrac{a}{b}y + a$$

7.3 Compound Inequalities and Interval Notation

1. Solving the inequality:
   $2x \le 3$
   $x \le \frac{3}{2}$
   Graphing the solution set:
   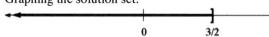

3. Solving the inequality:
   $\frac{1}{2}x > 2$
   $x > 4$
   Graphing the solution set:
   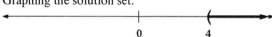

5. Solving the inequality:
   $-5x \le 25$
   $x \ge -5$
   Graphing the solution set:
   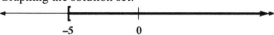

7. Solving the inequality:
   $-\frac{3}{2}x > -6$
   $-3x > -12$
   $x < 4$
   Graphing the solution set:

9. Solving the inequality:
   $-12 \le 2x$
   $x \ge -6$
   Graphing the solution set:
   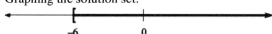

11. Solving the inequality:
    $-1 \ge -\frac{1}{4}x$
    $x \ge 4$
    Graphing the solution set:
    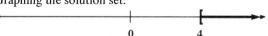

13. Solving the inequality:

    $-3x + 1 > 10$
    $-3x > 9$
    $x < -3$

    Graphing the solution set:
    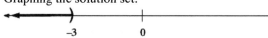

15. Solving the inequality:
    $\frac{1}{2} - \frac{m}{12} \le \frac{7}{12}$
    $12\left(\frac{1}{2} - \frac{m}{12}\right) \le 12\left(\frac{7}{12}\right)$
    $6 - m \le 7$
    $-m \le 1$
    $m \ge -1$
    Graphing the solution set:
    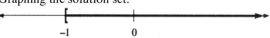

17. Solving the inequality:
    $\frac{1}{2} \ge -\frac{1}{6} - \frac{2}{9}x$
    $18\left(\frac{1}{2}\right) \ge 18\left(-\frac{1}{6} - \frac{2}{9}x\right)$
    $9 \ge -3 - 4x$
    $12 \ge -4x$
    $x \ge -3$
    Graphing the solution set:
    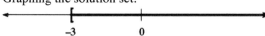

19. Solving the inequality:

    $-40 \le 30 - 20y$
    $-70 \le -20y$
    $y \le \frac{7}{2}$

    Graphing the solution set:
    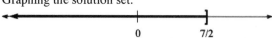

21. Solving the inequality:
    $\frac{2}{3}x - 3 < 1$
    $\frac{2}{3}x < 4$
    $2x < 12$
    $x < 6$
    Graphing the solution set:
    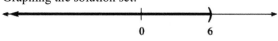

23. Solving the inequality:
    $10 - \frac{1}{2}y \le 36$
    $-\frac{1}{2}y \le 26$
    $y \ge -52$
    Graphing the solution set:
    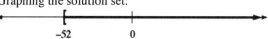

25. Solving the inequality:
$$2(3y+1) \le -10$$
$$6y+2 \le -10$$
$$6y \le -12$$
$$y \le -2$$
The solution set is $(-\infty, -2]$.

27. Solving the inequality:
$$-(a+1) - 4a \le 2a - 8$$
$$-a - 1 - 4a \le 2a - 8$$
$$-5a - 1 \le 2a - 8$$
$$-7a \le -7$$
$$a \ge 1$$
The solution set is $[1, \infty)$.

29. Solving the inequality:
$$\tfrac{1}{3}t - \tfrac{1}{2}(5-t) < 0$$
$$6\left(\tfrac{1}{3}t - \tfrac{1}{2}(5-t)\right) < 6(0)$$
$$2t - 3(5-t) < 0$$
$$2t - 15 + 3t < 0$$
$$5t - 15 < 0$$
$$5t < 15$$
$$t < 3$$
The solution set is $(-\infty, 3)$.

31. Solving the inequality:
$$-2 \le 5 - 7(2a+3)$$
$$-2 \le 5 - 14a - 21$$
$$-2 \le -16 - 14a$$
$$14 \le -14a$$
$$a \le -1$$
The solution set is $(-\infty, -1]$.

33. Solving the inequality:
$$-\tfrac{1}{3}(x+5) \le -\tfrac{2}{9}(x-1)$$
$$9\left[-\tfrac{1}{3}(x+5)\right] \le 9\left[-\tfrac{2}{9}(x-1)\right]$$
$$-3(x+5) \le -2(x-1)$$
$$-3x - 15 \le -2x + 2$$
$$-x - 15 \le 2$$
$$-x \le 17$$
$$x \ge -17$$
The solution set is $[-17, \infty)$.

35. Solving the inequality:
$$20x + 9{,}300 > 18{,}000$$
$$20x > 8{,}700$$
$$x > 435$$

37. Solving the inequality:
$$-2 \le m - 5 \le 2$$
$$3 \le m \le 7$$
The solution set is $[3,7]$. Graphing the solution set:

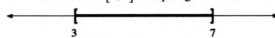

39. Solving the inequality:
$$-60 < 20a + 20 < 60$$
$$-80 < 20a < 40$$
$$-4 < a < 2$$
The solution set is $(-4, 2)$. Graphing the solution set:

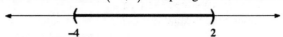

41. Solving the inequality:
$$0.5 \le 0.3a - 0.7 \le 1.1$$
$$1.2 \le 0.3a \le 1.8$$
$$4 \le a \le 6$$
The solution set is $[4,6]$. Graphing the solution set:

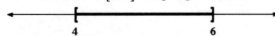

43. Solving the inequality:
$$3 < \tfrac{1}{2}x + 5 < 6$$
$$-2 < \tfrac{1}{2}x < 1$$
$$-4 < x < 2$$
The solution set is $(-4, 2)$. Graphing the solution set:

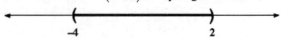

45. Solving the inequality:
$$4 < 6 + \tfrac{2}{3}x < 8$$
$$-2 < \tfrac{2}{3}x < 2$$
$$-6 < 2x < 6$$
$$-3 < x < 3$$
The solution set is $(-3, 3)$. Graphing the solution set:

47. Solving the inequality:
$$x+5 \le -2 \quad \text{or} \quad x+5 \ge 2$$
$$x \le -7 \quad \text{or} \quad x \ge -3$$
The solution set is $(-\infty, -7] \cup [-3, \infty)$. Graphing the solution set:

49. Solving the inequality:
$$5y+1 \le -4 \quad \text{or} \quad 5y+1 \ge 4$$
$$5y \le -5 \quad \text{or} \quad 5y \ge 3$$
$$y \le -1 \quad \text{or} \quad y \ge \tfrac{3}{5}$$
The solution set is $(-\infty, -1] \cup [\tfrac{3}{5}, \infty)$. Graphing the solution set:

51. Solving the inequality:
$$2x+5 < 3x-1 \quad \text{or} \quad x-4 > 2x+6$$
$$-x+5 < -1 \quad \text{or} \quad -x-4 > 6$$
$$-x < -6 \quad \text{or} \quad -x > 10$$
$$x > 6 \quad \text{or} \quad x < -10$$
The solution set is $(-\infty, -10) \cup (6, \infty)$. Graphing the solution set:

53. Writing as an inequality: $-2 < x \le 4$

55. Writing as an inequality: $x < -4$ or $x \ge 1$

57. a. Writing as an inequality: $x > 0$
    b. Writing as an inequality: $x \ge 0$
    c. Writing as an inequality: $x \ge 0$

59. There is no solution, since $x^2 \ge 0$ for all values of x.

61. This inequality is true for all values of x, since $x^2 \ge 0$ is always true.

63. There is no solution, since $\dfrac{1}{x^2} > 0$ for all values of x.

65. a. Evaluating when $x = 0$: $-\tfrac{1}{2}x + 1 = -\tfrac{1}{2}(0) + 1 = 1$
    b. Solving the equation:
$$-\tfrac{1}{2}x + 1 = -7$$
$$-\tfrac{1}{2}x = -8$$
$$x = 16$$
    c. Substituting $x = 0$: $-\tfrac{1}{2}x + 1 = -\tfrac{1}{2}(0) + 1 = 1$
    No, 0 is not a solution to the inequality.
    d. Solving the inequality:
$$-\tfrac{1}{2}x + 1 < -7$$
$$-\tfrac{1}{2}x < -8$$
$$x > 16$$

67. Solving the inequality:
$$900 - 300p \ge 300$$
$$-300p \ge -600$$
$$p \le 2$$
They should charge $2.00 per pad or less.

69. Solving the inequality:
$$900 - 300p < 525$$
$$-300p < -375$$
$$p > 1.25$$
They should charge more than $1.25 per pad.

196 Chapter 7 Transititions

71. Solving the inequality:
$$22149 - 399x > 20500$$
$$-399x > -1649$$
$$x < 4.13$$
In the years 1990 through 1994 Amtrak had more than 20,500 million passengers.

73. For adults, the inequality is $0.72 - 0.11 \le r \le 0.72 + 0.11$, or $0.61 \le r \le 0.83$. The survival rate for adults is between 61% and 83%. For juveniles, the inequality is $0.13 - 0.07 \le r \le 0.13 + 0.07$, or $0.06 \le r \le 0.20$. The survival rate for juveniles is between 6% and 20%.

75.
a. Solving the inequality:
$$95 \le \tfrac{9}{5}C + 32 \le 113$$
$$63 \le \tfrac{9}{5}C \le 81$$
$$315 \le 9C \le 405$$
$$35° \le C \le 45°$$

b. Solving the inequality:
$$68 \le \tfrac{9}{5}C + 32 \le 86$$
$$36 \le \tfrac{9}{5}C \le 54$$
$$180 \le 9C \le 270$$
$$20° \le C \le 30°$$

c. Solving the inequality:
$$-13 \le \tfrac{9}{5}C + 32 \le 14$$
$$-45 \le \tfrac{9}{5}C \le -18$$
$$-225 \le 9C \le -90$$
$$-25° \le C \le -10°$$

d. Solving the inequality:
$$-4 \le \tfrac{9}{5}C + 32 \le 23$$
$$-36 \le \tfrac{9}{5}C \le -9$$
$$-180 \le 9C \le -45$$
$$-20° \le C \le -5°$$

77. Reducing the rational expression: $\dfrac{x^2 - 16}{x+4} = \dfrac{(x+4)(x-4)}{x+4} = x - 4$

79. Reducing the rational expression: $\dfrac{10a + 20}{5a^2 - 20} = \dfrac{10(a+2)}{5(a^2 - 4)} = \dfrac{10(a+2)}{5(a+2)(a-2)} = \dfrac{2}{a-2}$

81. Reducing the rational expression: $\dfrac{2x^2 - 5x - 3}{x^2 - 3x} = \dfrac{(2x+1)(x-3)}{x(x-3)} = \dfrac{2x+1}{x}$

83. Reducing the rational expression: $\dfrac{xy + 3x + 2y + 6}{xy + 3x + ay + 3a} = \dfrac{x(y+3) + 2(y+3)}{x(y+3) + a(y+3)} = \dfrac{(y+3)(x+2)}{(y+3)(x+a)} = \dfrac{x+2}{x+a}$

85. Solving the inequality:
$$2x - 5 < 3$$
$$2x < 8$$
$$x < 4$$

87. Solving the inequality:
$$-4 \le 3a + 7$$
$$-11 \le 3a$$
$$a \ge -\tfrac{11}{3}$$

89. Solving the inequality:
$$4t - 3 \le -9$$
$$4t \le -6$$
$$t \le -\tfrac{3}{2}$$

7.4 Inequalities Involving Absolute Value

1. Solving the inequality:
   $|x| < 3$
   $-3 < x < 3$
   Graphing the solution set:
   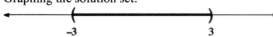

3. Solving the inequality:
   $|x| \geq 2$
   $x \leq -2$ or $x \geq 2$
   Graphing the solution set:
   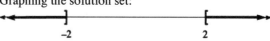

5. Solving the inequality:
   $|x| + 2 < 5$
   $|x| < 3$
   $-3 < x < 3$
   Graphing the solution set:
   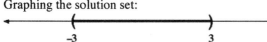

7. Solving the inequality:
   $|t| - 3 > 4$
   $|t| > 7$
   $t < -7$ or $t > 7$
   Graphing the solution set:
   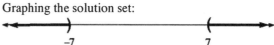

9. Since the inequality $|y| < -5$ is never true, there is no solution, or $\emptyset$. Graphing the solution set:
   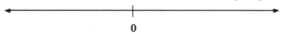

11. Since the inequality $|x| \geq -2$ is always true, the solution set is all real numbers.
    Graphing the solution set:

13. Solving the inequality:
    $|x - 3| < 7$
    $-7 < x - 3 < 7$
    $-4 < x < 10$
    Graphing the solution set:
    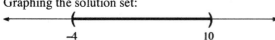

15. Solving the inequality:
    $|a + 5| \geq 4$
    $a + 5 \leq -4$ or $a + 5 \geq 4$
    $a \leq -9$ or $\quad a \geq -1$
    Graphing the solution set:
    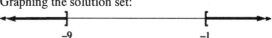

17. Since the inequality $|a - 1| < -3$ is never true, there is no solution, or $\emptyset$. Graphing the solution set:
    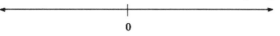

19. Solving the inequality:
    $|2x - 4| < 6$
    $-6 < 2x - 4 < 6$
    $-2 < 2x < 10$
    $-1 < x < 5$
    Graphing the solution set:
    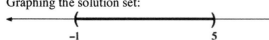

21. Solving the inequality:
    $|3y + 9| \geq 6$
    $3y + 9 \leq -6$ or $3y + 9 \geq 6$
    $3y \leq -15 \qquad 3y \geq -3$
    $y \leq -5 \qquad y \geq -1$
    Graphing the solution set:
    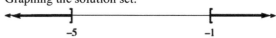

23. Solving the inequality:
    $|2k + 3| \geq 7$
    $2k + 3 \leq -7$ or $2k + 3 \geq 7$
    $2k \leq -10 \qquad 2k \geq 4$
    $k \leq -5 \qquad k \geq 2$
    Graphing the solution set:
    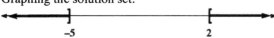

25. Solving the inequality:
    $|x - 3| + 2 < 6$
    $|x - 3| < 4$
    $-4 < x - 3 < 4$
    $-1 < x < 7$
    Graphing the solution set:
    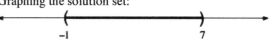

198 Chapter 7 Transititions

27. Solving the inequality:
$$|2a+1|+4 \geq 7$$
$$|2a+1| \geq 3$$
$$2a+1 \leq -3 \quad \text{or} \quad 2a+1 \geq 3$$
$$2a \leq -4 \qquad\qquad 2a \geq 2$$
$$a \leq -2 \qquad\qquad a \geq 1$$

Graphing the solution set:

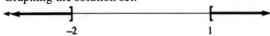

29. Solving the inequality:
$$|3x+5|-8<5$$
$$|3x+5|<13$$
$$-13<3x+5<13$$
$$-18<3x<8$$
$$-6<x<\tfrac{8}{3}$$

Graphing the solution set:

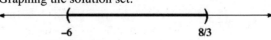

31. Solving the inequality:
$$|5-x|>3$$
$$5-x<-3 \quad \text{or} \quad 5-x>3$$
$$-x<-8 \qquad\qquad -x>-2$$
$$x>8 \qquad\qquad x<2$$

Graphing the solution set:

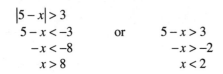

33. Solving the inequality:
$$\left|3-\tfrac{2}{3}x\right| \geq 5$$
$$3-\tfrac{2}{3}x \leq -5 \quad \text{or} \quad 3-\tfrac{2}{3}x \geq 5$$
$$-\tfrac{2}{3}x \leq -8 \qquad\qquad -\tfrac{2}{3}x \geq 2$$
$$-2x \leq -24 \qquad\qquad -2x \geq 6$$
$$x \geq 12 \qquad\qquad x \leq -3$$

Graphing the solution set:

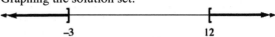

35. Solving the inequality:
$$\left|2-\tfrac{1}{2}x\right|>1$$
$$2-\tfrac{1}{2}x<-1 \quad \text{or} \quad 2-\tfrac{1}{2}x>1$$
$$-\tfrac{1}{2}x<-3 \qquad\qquad -\tfrac{1}{2}x>-1$$
$$x>6 \qquad\qquad x<2$$

Graphing the solution set:

37. Solving the inequality:
$$|x-1|<0.01$$
$$-0.01<x-1<0.01$$
$$0.99<x<1.01$$

39. Solving the inequality:
$$|2x+1| \geq \tfrac{1}{5}$$
$$2x+1 \leq -\tfrac{1}{5} \quad \text{or} \quad 2x+1 \geq \tfrac{1}{5}$$
$$2x \leq -\tfrac{6}{5} \qquad\qquad 2x \geq -\tfrac{4}{5}$$
$$x \leq -\tfrac{3}{5} \qquad\qquad x \geq -\tfrac{2}{5}$$

41. Solving the inequality:
$$\left|\tfrac{3x-2}{5}\right| \leq \tfrac{1}{2}$$
$$-\tfrac{1}{2} \leq \tfrac{3x-2}{5} \leq \tfrac{1}{2}$$
$$-\tfrac{5}{2} \leq 3x-2 \leq \tfrac{5}{2}$$
$$-\tfrac{1}{2} \leq 3x \leq \tfrac{9}{2}$$
$$-\tfrac{1}{6} \leq x \leq \tfrac{3}{2}$$

43. Solving the inequality:
$$\left|2x-\tfrac{1}{5}\right|<0.3$$
$$-0.3<2x-0.2<0.3$$
$$-0.1<2x<0.5$$
$$-0.05<x<0.25$$

45. Writing as an absolute value inequality: $|x| \leq 4$

47. Writing as an absolute value inequality: $|x-5| \leq 1$

49. **a.** Evaluating when $x = 0$: $|5x + 3| = |5(0) + 3| = |3| = 3$

b. Solving the equation:
$$|5x + 3| = 7$$
$$5x + 3 = -7, 7$$
$$5x = -10, 4$$
$$x = -2, \tfrac{4}{5}$$

c. Substituting $x = 0$: $|5x + 3| = |5(0) + 3| = |3| = 3$

No, 0 is not a solution to the inequality.

d. Solving the inequality:
$$|5x + 3| > 7$$

$5x + 3 < -7$	or	$5x + 3 > 7$
$5x < -10$		$5x > 4$
$x < -2$		$x > \tfrac{4}{5}$

51. The absolute value inequality is: $|x - 65| \le 10$

53. Performing the operations: $\dfrac{8x}{x^2 - 5x} \cdot \dfrac{x^2 - 25}{4x^2 + 4x} = \dfrac{8x}{x(x-5)} \cdot \dfrac{(x+5)(x-5)}{4x(x+1)} = \dfrac{8x(x+5)(x-5)}{4x^2(x-5)(x+1)} = \dfrac{2(x+5)}{x(x+1)}$

55. Performing the operations:
$$\dfrac{x^2 + 3x - 4}{3x^2 + 7x - 20} \div \dfrac{x^2 - 2x + 1}{3x^2 - 2x - 5} = \dfrac{x^2 + 3x - 4}{3x^2 + 7x - 20} \cdot \dfrac{3x^2 - 2x - 5}{x^2 - 2x + 1}$$
$$= \dfrac{(x+4)(x-1)}{(3x-5)(x+4)} \cdot \dfrac{(3x-5)(x+1)}{(x-1)^2}$$
$$= \dfrac{(x+4)(x-1)(3x-5)(x+1)}{(3x-5)(x+4)(x-1)^2}$$
$$= \dfrac{x+1}{x-1}$$

57. Performing the operations: $(x^2 - 36)\left(\dfrac{x+3}{x-6}\right) = \dfrac{(x+6)(x-6)}{1} \cdot \dfrac{x+3}{x-6} = \dfrac{(x+6)(x-6)(x+3)}{x-6} = (x+6)(x+3)$

59. Graphing the two equations:

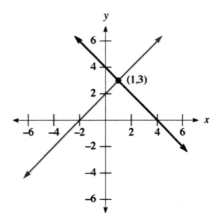

The intersection point is (1,3).

61. Adding the two equations yields:
$2x = 2$
$x = 1$
Substituting into the first equation:
$1 + y = 4$
$y = 3$
The solution is (1,3).

63. Substituting into the first equation:
$x + 2x - 1 = 2$
$3x - 1 = 2$
$3x = 3$
$x = 1$
Substituting into the first equation:
$1 + y = 2$
$y = 1$
The solution is (1,1).

7.5 Review of Systems of Linear Equations in Two Variables

1. The intersection point is (4,3).

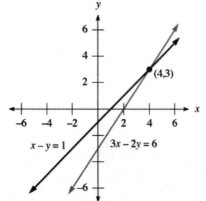

3. The intersection point is (−5,−6).

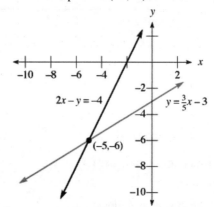

5. The intersection point is (4,2).

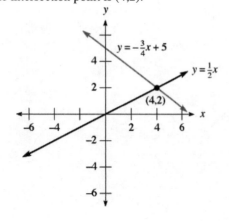

7. The lines are parallel. There is no solution to the system.

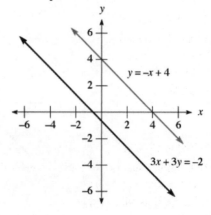

9. The lines coincide. Any solution to one of the equations is a solution to the other.

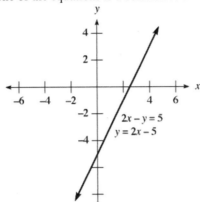

11. Solving the two equations:
$$x + y = 5$$
$$3x - y = 3$$
Adding yields:
$$4x = 8$$
$$x = 2$$
The solution is (2,3).

13. Multiply the first equation by –1:
$$-3x - y = 4$$
$$4x + y = 5$$
Adding yields: $x = 1$. The solution is (1,1).

15. Multiply the first equation by –2:
$$-6x + 4y = -12$$
$$6x - 4y = 12$$
Adding yields $0 = 0$, so the lines coincide. The solution is $\{(x, y) \mid 3x - 2y = 6\}$.

17. Multiply the first equation by 3:
$$3x + 6y = 0$$
$$2x - 6y = 5$$
Adding yields:
$$5x = 5$$
$$x = 1$$
The solution is $\left(1, -\frac{1}{2}\right)$.

19. Multiply the first equation by –2:
$$-4x + 10y = -32$$
$$4x - 3y = 11$$
Adding yields:
$$7y = -21$$
$$y = -3$$
The solution is $\left(\frac{1}{2}, -3\right)$.

21. Multiply the first equation by 3 and the second equation by –2:
$$18x + 9y = -3$$
$$-18x - 10y = -2$$
Adding yields:
$$-y = -5$$
$$y = 5$$
The solution is $\left(-\frac{8}{3}, 5\right)$.

23. Multiply the first equation by 2 and the second equation by 3:
$$8x + 6y = 28$$
$$27x - 6y = 42$$
Adding yields:
$$35x = 70$$
$$x = 2$$
The solution is (2,2).

25. Multiply the first equation by 2:
$$4x - 10y = 6$$
$$-4x + 10y = 3$$
Adding yields $0 = 9$, which is false (parallel lines). There is no solution ($\varnothing$).

27. To clear each equation of fractions, multiply the first equation by 12 and the second equation by 30:
$$3x - 2y = -24$$
$$-5x + 6y = 120$$
Multiply the first equation by 3:
$$9x - 6y = -72$$
$$-5x + 6y = 120$$
Adding yields $4x = 48$, so $x = 12$. Substituting into the first equation:
$$36 - 2y = -24$$
$$-2y = -60$$
$$y = 30$$
The solution is (12,30).

29. To clear each equation of fractions, multiply the first equation by 6 and the second equation by 20:
$$3x + 2y = 78$$
$$8x + 5y = 200$$
Multiply the first equation by 5 and the second equation by -2:
$$15x + 10y = 390$$
$$-16x - 10y = -400$$
Adding yields:
$$-x = -10$$
$$x = 10$$
The solution is (10,24).

31. Substituting into the first equation:
$$7(2y + 9) - y = 24$$
$$14y + 63 - y = 24$$
$$13y = -39$$
$$y = -3$$

The solution is (3,−3).

33. Substituting into the first equation:
$$6x - \left(-\tfrac{3}{4}x - 1\right) = 10$$
$$6x + \tfrac{3}{4}x + 1 = 10$$
$$\tfrac{27}{4}x = 9$$
$$27x = 36$$
$$x = \tfrac{4}{3}$$

The solution is $\left(\tfrac{4}{3}, -2\right)$.

35. Substituting $z = -3y + 17$ into the second equation:
$$5y + 20(-3y + 17) = 65$$
$$5y - 60y + 340 = 65$$
$$-55y + 340 = 65$$
$$-55y = -275$$
$$y = 5$$
The solution is (5,2).

37. Substituting into the first equation:
$$4x - 4 = 3x - 2$$
$$x - 4 = -2$$
$$x = 2$$

The solution is (2,4).

Problem Set 7.5

39. Solving the first equation for y yields $y = 2x - 5$. Substituting into the second equation:
$$4x - 2(2x - 5) = 10$$
$$4x - 4x + 10 = 10$$
$$10 = 10$$
Since this statement is true, the two lines coincide. The solution is $\{(x, y) \mid 2x - y = 5\}$.

41. Substituting into the first equation:
$$\tfrac{1}{3}\left(\tfrac{3}{2}y\right) - \tfrac{1}{2}y = 0$$
$$\tfrac{1}{2}y - \tfrac{1}{2}y = 0$$
$$0 = 0$$
Since this statement is true, the two lines coincide. The solution is $\{(x, y) \mid x = \tfrac{3}{2}y\}$.

43. Multiply the first equation by 2 and the second equation by 7:
$$8x - 14y = 6$$
$$35x + 14y = -21$$
Adding yields:
$$43x = -15$$
$$x = -\tfrac{15}{43}$$
Substituting into the original second equation:
$$5\left(-\tfrac{15}{43}\right) + 2y = -3$$
$$-\tfrac{75}{43} + 2y = -3$$
$$2y = -\tfrac{54}{43}$$
$$y = -\tfrac{27}{43}$$
The solution is $\left(-\tfrac{15}{43}, -\tfrac{27}{43}\right)$.

45. Multiply the first equation by 3 and the second equation by 8:
$$27x - 24y = 12$$
$$16x + 24y = 48$$
Adding yields:
$$43x = 60$$
$$x = \tfrac{60}{43}$$
Substituting into the original second equation:
$$2\left(\tfrac{60}{43}\right) + 3y = 6$$
$$\tfrac{120}{43} + 3y = 6$$
$$3y = \tfrac{138}{43}$$
$$y = \tfrac{46}{43}$$
The solution is $\left(\tfrac{60}{43}, \tfrac{46}{43}\right)$.

47. Multiply the first equation by 2 and the second equation by 5:
$$6x - 10y = 4$$
$$35x + 10y = 5$$
Adding yields:
$$41x = 9$$
$$x = \tfrac{9}{41}$$
Substituting into the original second equation:
$$7\left(\tfrac{9}{41}\right) + 2y = 1$$
$$\tfrac{63}{41} + 2y = 1$$
$$2y = -\tfrac{22}{41}$$
$$y = -\tfrac{11}{41}$$
The solution is $\left(\tfrac{9}{41}, -\tfrac{11}{41}\right)$.

49. Multiply the second equation by 3:
$$x - 3y = 7$$
$$6x + 3y = -18$$
Adding yields:
$$7x = -11$$
$$x = -\tfrac{11}{7}$$
Substituting into the original second equation:
$$2\left(-\tfrac{11}{7}\right) + y = -6$$
$$-\tfrac{22}{7} + y = -6$$
$$y = -\tfrac{20}{7}$$
The solution is $\left(-\tfrac{11}{7}, -\tfrac{20}{7}\right)$.

51. Substituting into the first equation:
$$-\tfrac{1}{3}x + 2 = \tfrac{1}{2}x + \tfrac{1}{3}$$
$$6\left(-\tfrac{1}{3}x + 2\right) = 6\left(\tfrac{1}{2}x + \tfrac{1}{3}\right)$$
$$-2x + 12 = 3x + 2$$
$$-5x = -10$$
$$x = 2$$
Substituting into the first equation: $y = \tfrac{1}{2}(2) + \tfrac{1}{3} = 1 + \tfrac{1}{3} = \tfrac{4}{3}$. The solution is $\left(2, \tfrac{4}{3}\right)$.

53. Substituting into the first equation:
$$3\left(\tfrac{2}{3}y - 4\right) - 4y = 12$$
$$2y - 12 - 4y = 12$$
$$-2y - 12 = 12$$
$$-2y = 24$$
$$y = -12$$
Substituting into the second equation: $x = \tfrac{2}{3}(-12) - 4 = -8 - 4 = -12$. The solution is $(-12, -12)$.

55. Multiply the first equation by 2:
$$8x - 6y = -14$$
$$-8x + 6y = -11$$
$$0 = -25$$
Since this statement is false, there is no solution ($\emptyset$).

57. Multiply the first equation by –20:
$$-60y - 20z = -340$$
$$5y + 20z = 65$$
Adding yields:
$$-55y = -275$$
$$y = 5$$
Substituting into the first equation:
$$3(5) + z = 17$$
$$15 + z = 17$$
$$z = 2$$
The solution is $y = 5, z = 2$.

59. Substitute into the first equation:
$$\tfrac{3}{4}x - \tfrac{1}{3}\left(\tfrac{1}{4}x\right) = 1$$
$$\tfrac{3}{4}x - \tfrac{1}{12}x = 1$$
$$\tfrac{2}{3}x = 1$$
$$x = \tfrac{3}{2}$$
Substituting into the second equation: $y = \tfrac{1}{4}\left(\tfrac{3}{2}\right) = \tfrac{3}{8}$. The solution is $\left(\tfrac{3}{2}, \tfrac{3}{8}\right)$.

61. To clear each equation of fractions, multiply the first equation by 12 and the second equation by 12:
$$3x - 6y = 4$$
$$4x - 3y = -8$$
Multiply the second equation by –2:
$$3x - 6y = 4$$
$$-8x + 6y = 16$$
Adding yields:
$$-5x = 20$$
$$x = -4$$
Substituting into the first equation:
$$3(-4) - 6y = 4$$
$$-12 - 6y = 4$$
$$-6y = 16$$
$$y = -\tfrac{8}{3}$$
The solution is $\left(-4, -\tfrac{8}{3}\right)$.

63. **a.** Simplifying: $(3x - 4y) - 3(x - y) = 3x - 4y - 3x + 3y = -y$

  b. Substituting $x = 0$:
$$3(0) - 4y = 8$$
$$-4y = 8$$
$$y = -2$$

  c. From part **b**, the y-intercept is $(0, -2)$.

d. Graphing the line:

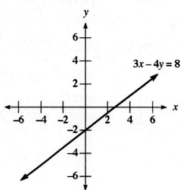

e. Multiply the second equation by –3:
$3x - 4y = 8$
$-3x + 3y = -6$
Adding yields:
$-y = 2$
$y = -2$
Substituting into the first equation:
$3x - 4(-2) = 8$
$3x + 8 = 8$
$3x = 0$
$x = 0$
The lines intersect at the point (0,–2).

65. Let x and y represent the two numbers. The system of equations is:
$y = 2x + 3$
$x + y = 18$
Substituting into the second equation:
$x + 2x + 3 = 18$
$3x = 15$
$x = 5$
$y = 2(5) + 3 = 13$
The two numbers are 5 and 13.

67. Let a represent the number of adult tickets and c represent the number of children's tickets. The system of equations is:
$a + c = 925$
$2a + c = 1150$
Multiply the first equation by –1:
$-a - c = -925$
$2a + c = 1150$
Adding yields:
$a = 225$
$c = 700$
There were 225 adult tickets and 700 children's tickets sold.

69. Let x represent the amount of 20% disinfectant and y represent the amount of 14% disinfectant.
The system of equations is:
$$x+y=15$$
$$0.20x+0.14y=0.16(15)$$
Multiplying the first equation by -0.14:
$$-0.14x-0.14y=-2.1$$
$$0.20x+0.14y=2.4$$
Adding yields:
$$0.06x=0.3$$
$$x=5$$
$$y=10$$
The mixture contains 5 gallons of 20% disinfectant and 10 gallons of 14% disinfectant.

71. Let b represent the rate of the boat and c represent the rate of the current. The system of equations is:
$$2(b+c)=24$$
$$3(b-c)=18$$
The system of equations simplifies to:
$$b+c=12$$
$$b-c=6$$
Adding yields:
$$2b=18$$
$$b=9$$
$$c=3$$
The rate of the boat is 9 mph and the rate of the current is 3 mph.

73. Let a represent the rate of the airplane and w represent the rate of the wind. The system of equations is:
$$2(a+w)=600$$
$$\tfrac{5}{2}(a-w)=600$$
The system of equations simplifies to:
$$a+w=300$$
$$a-w=240$$
Adding yields:
$$2a=540$$
$$a=270$$
$$w=30$$
The rate of the airplane is 270 mph and the rate of the wind is 30 mph.

75. Combining the rational expressions: $\dfrac{x^2}{x+5}+\dfrac{10x+25}{x+5}=\dfrac{x^2+10x+25}{x+5}=\dfrac{(x+5)^2}{x+5}=x+5$

77. Combining the rational expressions: $\dfrac{a}{3}+\dfrac{2}{5}=\dfrac{a\cdot 5}{3\cdot 5}+\dfrac{2\cdot 3}{5\cdot 3}=\dfrac{5a}{15}+\dfrac{6}{15}=\dfrac{5a+6}{15}$

79. Combining the rational expressions:
$$\frac{6}{a^2-9} - \frac{5}{a^2-a-6} = \frac{6}{(a+3)(a-3)} - \frac{5}{(a-3)(a+2)}$$
$$= \frac{6 \cdot (a+2)}{(a+3)(a-3)(a+2)} - \frac{5 \cdot (a+3)}{(a+3)(a-3)(a+2)}$$
$$= \frac{6a+12}{(a+3)(a-3)(a+2)} - \frac{5a+15}{(a+3)(a-3)(a+2)}$$
$$= \frac{6a+12-5a-15}{(a+3)(a-3)(a+2)}$$
$$= \frac{a-3}{(a+3)(a-3)(a+2)}$$
$$= \frac{1}{(a+3)(a+2)}$$

81. Simplifying: $(x+3y) - 1(x-2z) = x + 3y - x + 2z = 3y + 2z$

83. Solving the equation:
$$-9y = -9$$
$$y = 1$$

85. Solving the equation:
$$3(1) + 2z = 9$$
$$3 + 2z = 9$$
$$2z = 6$$
$$z = 3$$

87. Applying the distributive property: $2(5x - z) = 10x - 2z$

89. Applying the distributive property: $3(3x + y - 2z) = 9x + 3y - 6z$

7.6 Systems of Linear Equations in Three Variables

1. Adding the first two equations and the first and third equations results in the system:
$$2x + 3z = 5$$
$$2x - 2z = 0$$
Solving the second equation yields $x = z$, now substituting:
$$2z + 3z = 5$$
$$5z = 5$$
$$z = 1$$
So $x = 1$, now substituting into the original first equation:
$$1 + y + 1 = 4$$
$$y + 2 = 4$$
$$y = 2$$
The solution is $(1, 2, 1)$.

3. Adding the first two equations and the first and third equations results in the system:
$$2x + 3z = 13$$
$$3x - 3z = -3$$
Adding yields:
$$5x = 10$$
$$x = 2$$
Substituting to find z:
$$2(2) + 3z = 13$$
$$4 + 3z = 13$$
$$3z = 9$$
$$z = 3$$

Substituting into the original first equation:
$$2+y+3=6$$
$$y+5=6$$
$$y=1$$
The solution is (2,1,3).

5. Adding the second and third equations:
$$5x+z=11$$
Multiplying the second equation by 2:
$$x+2y+z=3$$
$$4x-2y+4z=12$$
Adding yields:
$$5x+5z=15$$
$$x+z=3$$
So the system becomes:
$$5x+z=11$$
$$x+z=3$$
Multiply the second equation by -1:
$$5x+z=11$$
$$-x-z=-3$$
Adding yields:
$$4x=8$$
$$x=2$$
Substituting to find z:
$$5(2)+z=11$$
$$z+10=11$$
$$z=1$$
Substituting into the original first equation:
$$2+2y+1=3$$
$$2y+3=3$$
$$2y=0$$
$$y=0$$
The solution is (2,0,1).

7. Multiply the second equation by -1 and add it to the first equation:
$$2x+3y-2z=4$$
$$-x-3y+3z=-4$$
Adding results in the equation $x+z=0$. Multiply the second equation by 2 and add it to the third equation:
$$2x+6y-6z=8$$
$$3x-6y+z=-3$$
Adding results in the equation:
$$5x-5z=5$$
$$x-z=1$$
So the system becomes:
$$x-z=1$$
$$x+z=0$$
Adding yields:
$$2x=1$$
$$x=\tfrac{1}{2}$$
Substituting to find z:
$$\tfrac{1}{2}+z=0$$
$$z=-\tfrac{1}{2}$$

Substituting into the original first equation:
$$2\left(\tfrac{1}{2}\right)+3y-2\left(-\tfrac{1}{2}\right)=4$$
$$1+3y+1=4$$
$$3y+2=4$$
$$3y=2$$
$$y=\tfrac{2}{3}$$
The solution is $\left(\tfrac{1}{2},\tfrac{2}{3},-\tfrac{1}{2}\right)$.

9. Multiply the first equation by 2 and add it to the second equation:
$$-2x+8y-6z=4$$
$$2x-8y+6z=1$$
Adding yields $0=5$, which is false. There is no solution (inconsistent system).

11. To clear the system of fractions, multiply the first equation by 2 and the second equation by 3:
$$x-2y+2z=0$$
$$6x+y+3z=6$$
$$x+y+z=-4$$
Multiply the third equation by 2 and add it to the first equation:
$$x-2y+2z=0$$
$$2x+2y+2z=-8$$
Adding yields the equation $3x+4z=-8$. Multiply the third equation by -1 and add it to the second equation:
$$6x+y+3z=6$$
$$-x-y-z=4$$
Adding yields the equation $5x+2z=10$. So the system becomes:
$$3x+4z=-8$$
$$5x+2z=10$$
Multiply the second equation by -2:
$$3x+4z=-8$$
$$-10x-4z=-20$$
Adding yields:
$$-7x=-28$$
$$x=4$$
Substituting to find z:
$$3(4)+4z=-8$$
$$12+4z=-8$$
$$4z=-20$$
$$z=-5$$
Substituting into the original third equation:
$$4+y-5=-4$$
$$y-1=-4$$
$$y=-3$$
The solution is $(4,-3,-5)$.

13. Multiply the first equation by -2 and add it to the third equation:
$$-4x+2y+6z=-2$$
$$4x-2y-6z=2$$
Adding yields $0=0$, which is true. Since there are now less equations than unknowns, there is no unique solution (dependent system).

15. Multiply the second equation by 3 and add it to the first equation:
$$2x - y + 3z = 4$$
$$3x + 6y - 3z = -9$$
Adding yields the equation $5x + 5y = -5$, or $x + y = -1$.

Multiply the second equation by 2 and add it to the third equation:
$$2x + 4y - 2z = -6$$
$$4x + 3y + 2z = -5$$
Adding yields the equation $6x + 7y = -11$. So the system becomes:
$$6x + 7y = -11$$
$$x + y = -1$$
Multiply the second equation by –6:
$$6x + 7y = -11$$
$$-6x - 6y = 6$$
Adding yields $y = -5$. Substituting to find x:
$$6x + 7(-5) = -11$$
$$6x - 35 = -11$$
$$6x = 24$$
$$x = 4$$
Substituting into the original first equation:
$$2(4) - (-5) + 3z = 4$$
$$13 + 3z = 4$$
$$3z = -9$$
$$z = -3$$
The solution is (4,–5,–3).

17. Adding the second and third equations results in the equation $x + y = 9$. Since this is the same as the first equation, there are less equations than unknowns. There is no unique solution (dependent system).

19. Adding the second and third equations results in the equation $4x + y = 3$. So the system becomes:
$$4x + y = 3$$
$$2x + y = 2$$
Multiplying the second equation by –1:
$$4x + y = 3$$
$$-2x - y = -2$$
Adding yields:
$$2x = 1$$
$$x = \tfrac{1}{2}$$
Substituting to find y:
$$2\left(\tfrac{1}{2}\right) + y = 2$$
$$1 + y = 2$$
$$y = 1$$
Substituting into the original second equation:
$$1 + z = 3$$
$$z = 2$$
The solution is $\left(\tfrac{1}{2}, 1, 2\right)$.

21. Multiply the third equation by 2 and adding it to the second equation:
$$6y - 4z = 1$$
$$2x + 4z = 2$$
Adding yields the equation $2x + 6y = 3$. So the system becomes:
$$2x - 3y = 0$$
$$2x + 6y = 3$$
Multiply the first equation by 2:
$$4x - 6y = 0$$
$$2x + 6y = 3$$
Adding yields:
$$6x = 3$$
$$x = \tfrac{1}{2}$$
Substituting to find y:
$$2\left(\tfrac{1}{2}\right) + 6y = 3$$
$$1 + 6y = 3$$
$$6y = 2$$
$$y = \tfrac{1}{3}$$
Substituting into the original third equation to find z:
$$\tfrac{1}{2} + 2z = 1$$
$$2z = \tfrac{1}{2}$$
$$z = \tfrac{1}{4}$$
The solution is $\left(\tfrac{1}{2}, \tfrac{1}{3}, \tfrac{1}{4}\right)$.

23. Multiply the first equation by –2 and add it to the second equation:
$$-2x - 2y + 2z = -4$$
$$2x + y + 3z = 4$$
Adding yields $-y + 5z = 0$. Multiply the first equation by –1 and add it to the third equation:
$$-x - y + z = -2$$
$$x - 2y + 2z = 6$$
Adding yields $-3y + 3z = 4$. So the system becomes:
$$-y + 5z = 0$$
$$-3y + 3z = 4$$
Multiply the first equation by –3:
$$3y - 15z = 0$$
$$-3y + 3z = 4$$
Adding yields:
$$-12z = 4$$
$$z = -\tfrac{1}{3}$$
Substituting to find y:
$$-3y + 3\left(-\tfrac{1}{3}\right) = 4$$
$$-3y - 1 = 4$$
$$-3y = 5$$
$$y = -\tfrac{5}{3}$$

Substituting into the original first equation:
$$x - \tfrac{5}{3} + \tfrac{1}{3} = 2$$
$$x - \tfrac{4}{3} = 2$$
$$x = \tfrac{10}{3}$$
The solution is $\left(\tfrac{10}{3}, -\tfrac{5}{3}, -\tfrac{1}{3}\right)$.

25. Multiply the third equation by −1 and add it to the first equation:
$$2x + 3y = -\tfrac{1}{2}$$
$$-3y - 2z = \tfrac{3}{4}$$
Adding yields the equation $2x - 2z = \tfrac{1}{4}$. So the system becomes:
$$2x - 2z = \tfrac{1}{4}$$
$$4x + 8z = 2$$
Multiply the first equation by 4:
$$8x - 8z = 1$$
$$4x + 8z = 2$$
Adding yields:
$$12x = 3$$
$$x = \tfrac{1}{4}$$
Substituting to find z:
$$4\left(\tfrac{1}{4}\right) + 8z = 2$$
$$1 + 8z = 2$$
$$8z = 1$$
$$z = \tfrac{1}{8}$$
Substituting to find y:
$$2\left(\tfrac{1}{4}\right) + 3y = -\tfrac{1}{2}$$
$$\tfrac{1}{2} + 3y = -\tfrac{1}{2}$$
$$3y = -1$$
$$y = -\tfrac{1}{3}$$
The solution is $\left(\tfrac{1}{4}, -\tfrac{1}{3}, \tfrac{1}{8}\right)$.

27. To clear each equation of fractions, multiply the first equation by 6, the second equation by 4, and the third equation by 12:
$$2x + 3y - z = 24$$
$$x - 3y + 2z = 6$$
$$6x - 8y - 3z = -64$$
Multiply the first equation by 2 and add it to the second equation:
$$4x + 6y - 2z = 48$$
$$x - 3y + 2z = 6$$
Adding yields the equation $5x + 3y = 54$. Multiply the first equation by −3 and add it to the third equation:
$$-6x - 9y + 3z = -72$$
$$6x - 8y - 3z = -64$$
Adding yields:
$$-17y = -136$$
$$y = 8$$

Substituting to find x:
$$5x + 3(8) = 54$$
$$5x + 24 = 54$$
$$5x = 30$$
$$x = 6$$
Substituting to find z:
$$6 - 3(8) + 2z = 6$$
$$-18 + 2z = 6$$
$$2z = 24$$
$$z = 12$$
The solution is $(6,8,12)$.

29. To clear each equation of fractions, multiply the first equation by 6, the second equation by 6, and the third equation by 12:
$$6x - 3y - 2z = -8$$
$$2x + 6y - 3z = 30$$
$$-3x + 8y - 12z = -9$$
Multiply the first equation by 2 and add it to the second equation:
$$12x - 6y - 4z = -16$$
$$2x + 6y - 3z = 30$$
Adding yields the equation:
$$14x - 7z = 14$$
$$2x - z = 2$$
Multiply the first equation by 8 and the third equation by 3:
$$48x - 24y - 16z = -64$$
$$-9x + 24y - 36z = -27$$
Adding yields the equation $39x - 52z = -91$. So the system becomes:
$$2x - z = 2$$
$$39x - 52z = -91$$
Multiply the first equation by -52:
$$-104x + 52z = -104$$
$$39x - 52z = -91$$
Adding yields:
$$-65x = -195$$
$$x = 3$$
Substituting to find z:
$$6 - z = 2$$
$$z = 4$$
Substituting to find y:
$$6 + 6y - 12 = 30$$
$$6y - 6 = 30$$
$$6y = 36$$
$$y = 6$$
The solution is $(3,6,4)$.

31. To clear each equation of fractions, multiply the first equation by 6, the second equation by 10, and the third equation by 12:
$$3x + 4y = 15$$
$$2x - 5z = -3$$
$$4y - 3z = 9$$
Multiply the third equation by –1 and add it to the first equation:
$$3x + 4y = 15$$
$$-4y + 3z = -9$$
Adding yields:
$$3x + 3z = 6$$
$$x + z = 2$$
So the system becomes:
$$x + z = 2$$
$$2x - 5z = -3$$
Multiply the first equation by 5:
$$5x + 5z = 10$$
$$2x - 5z = -3$$
Adding yields:
$$7x = 7$$
$$x = 1$$
Substituting yields $z = 1$. Substituting to find y:
$$3 + 4y = 15$$
$$4y = 12$$
$$y = 3$$
The solution is $(1,3,1)$.

33. To clear each equation of fractions, multiply the first equation by 4, the second equation by 12, and the third equation by 6:
$$2x - y + 2z = -8$$
$$3x - y - 4z = 3$$
$$x + 2y - 3z = 9$$
Multiply the first equation by –1 and add it to the first equation:
$$-2x + y - 2z = 8$$
$$3x - y - 4z = 3$$
Adding yields the equation $x - 6z = 11$. Multiply the first equation by 2 and add it to the third equation:
$$4x - 2y + 4z = -16$$
$$x + 2y - 3z = 9$$
Adding yields the equation $5x + z = -7$. So the system becomes:
$$5x + z = -7$$
$$x - 6z = 11$$
Multiply the first equation by 6:
$$30x + 6z = -42$$
$$x - 6z = 11$$
Adding yields:
$$31x = -31$$
$$x = -1$$
Substituting to find z:
$$-5 + z = -7$$
$$z = -2$$

Substituting to find y:
$$-1+2y+6=9$$
$$2y+5=9$$
$$2y=4$$
$$y=2$$
The solution is $(-1,2,-2)$.

35. Divide the second equation by 5 and the third equation by 10 to produce the system:
$$x-y-z=0$$
$$x+4y=16$$
$$2y-z=5$$
Multiply the third equation by -1 and add it to the first equation:
$$x-y-z=0$$
$$-2y+z=-5$$
Adding yields the equation $x-3y=-5$. So the system becomes:
$$x+4y=16$$
$$x-3y=-5$$
Multiply the second equation by -1:
$$x+4y=16$$
$$-x+3y=5$$
Adding yields:
$$7y=21$$
$$y=3$$
Substituting to find x:
$$x+12=16$$
$$x=4$$
Substituting to find z:
$$6-z=5$$
$$z=1$$
The currents are 4 amps, 3 amps, and 1 amp.

37. Let x, y and z represent the amounts invested in the three accounts. The system of equations is:
$$x+y+z=2200$$
$$z=3x$$
$$0.06x+0.08y+0.09z=178$$
Substituting into the first equation:
$$x+y+3x=2200$$
$$4x+y=2200$$
Substituting into the third equation:
$$0.06x+0.08y+0.09(3x)=178$$
$$0.33x+0.08y=178$$
The system of equations becomes:
$$4x+y=2200$$
$$0.33x+0.08y=178$$
Multiply the first equation by -0.08:
$$-0.32x-0.08y=-176$$
$$0.33x+0.08y=178$$

Adding yields:
$0.01x = 2$
$x = 200$
$z = 3(200) = 600$
$y = 2200 - 4(200) = 1400$

He invested $200 at 6%, $1,400 at 8%, and $600 at 9%.

39. Let n, d, and q represent the number of nickels, dimes, and quarters. The system of equations is:
$n + d + q = 9$
$0.05n + 0.10d + 0.25q = 1.20$
$d = n$

Substituting into the first equation:
$n + n + q = 9$
$2n + q = 9$

Substituting into the second equation:
$0.05n + 0.10n + 0.25q = 1.20$
$0.15n + 0.25q = 1.20$

The system of equations becomes:
$2n + q = 9$
$0.15n + 0.25q = 1.20$

Multiplying the first equation by –0.25:
$-0.50n - 0.25q = -2.25$
$0.15n + 0.25q = 1.20$

Adding yields:
$-0.35n = -1.05$
$n = 3$
$d = 3$
$q = 9 - 2(3) = 3$

The collection contains 3 nickels, 3 dimes, and 3 quarters.

41. Simplifying the expression: $-|-5| = -(5) = -5$
43. Simplifying the expression: $-3 - 4(-2) = -3 + 8 = 5$
45. Simplifying the expression: $5|3-8| - 6|2-5| = 5|-5| - 6|-3| = 5(5) - 6(3) = 25 - 18 = 7$
47. Simplifying the expression: $5 - 2[-3(5-7) - 8] = 5 - 2[-3(-2) - 8] = 5 - 2(6-8) = 5 - 2(-2) = 5 + 4 = 9$
49. The expression is: $-3 - (-9) = -3 + 9 = 6$
51. Since $0 + 0 \le 4$ and $4 + 0 \le 4$, but $2 + 3 > 4$, the points (0,0) and (4,0) are solutions.
53. Since $0 \le \frac{1}{2}(0)$ and $0 \le \frac{1}{2}(2)$, but $0 > \frac{1}{2}(-2)$, the points (0,0) and (2,0) are solutions.

7.7 Linear Inequalities in Two Variables

1. Graphing the solution set:

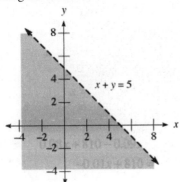

3. Graphing the solution set:

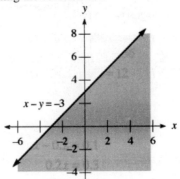

5. Graphing the solution set:

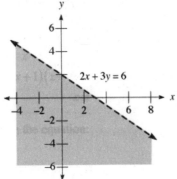

7. Graphing the solution set:

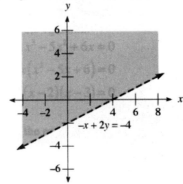

9. Graphing the solution set:

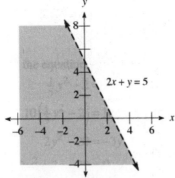

11. Graphing the solution set:

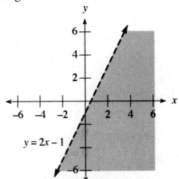

13. Graphing the solution set:

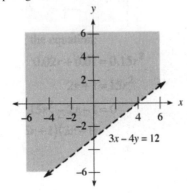

15. Graphing the solution set:

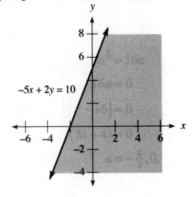

17. Graphing the solution set:

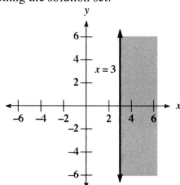

19. Graphing the solution set:

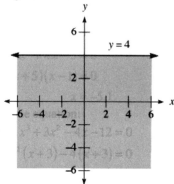

21. Graphing the solution set:

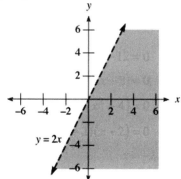

23. Graphing the solution set:

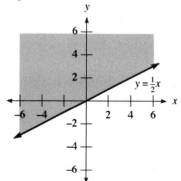

25. Graphing the solution set:

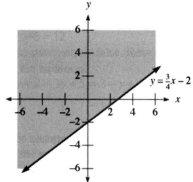

27. Graphing the solution set:

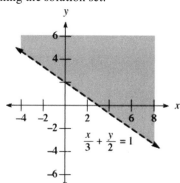

29. Graphing the solution set:

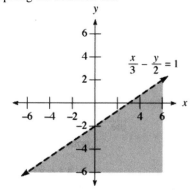

31. Graphing the solution set:

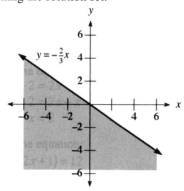

33. Graphing the solution set:

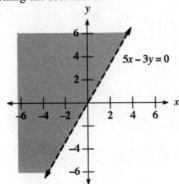

35. Graphing the solution set:

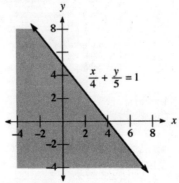

37. The inequality is $x + y > 4$.

39. The inequality is $-x + 2y \le 4$ or $y \le \frac{1}{2}x + 2$.

41. a. Solving the inequality:
$$\frac{1}{3} + \frac{y}{2} < 1$$
$$\frac{y}{2} < \frac{2}{3}$$
$$y < \frac{4}{3}$$

b. Solving the inequality:
$$\frac{1}{3} - \frac{y}{2} < 1$$
$$-\frac{y}{2} < \frac{2}{3}$$
$$y > -\frac{4}{3}$$

c. Solving for y:
$$\frac{x}{3} + \frac{y}{2} = 1$$
$$6\left(\frac{x}{3} + \frac{y}{2}\right) = 6(1)$$
$$2x + 3y = 6$$
$$3y = -2x + 6$$
$$y = -\frac{2}{3}x + 2$$

d. Graphing the inequality:

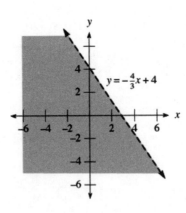

43. Graphing the region:

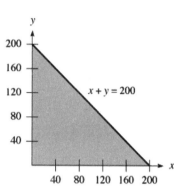

45. Graphing the region:

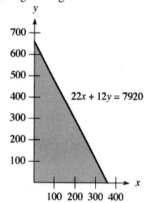

47. Solving the inequality:
$$\tfrac{1}{3} + \tfrac{y}{5} \le \tfrac{26}{15}$$
$$15\left(\tfrac{1}{3} + \tfrac{y}{5}\right) \le 15\left(\tfrac{26}{15}\right)$$
$$5 + 3y \le 26$$
$$3y \le 21$$
$$y \le 7$$

49. Solving the inequality:
$$5t - 4 > 3t - 8$$
$$2t - 4 > -8$$
$$2t > -4$$
$$t > -2$$

51. Solving the inequality:
$$-9 < -4 + 5t < 6$$
$$-5 < 5t < 10$$
$$-1 < t < 2$$

53. No, the graph does not include the boundary line.

55. Substituting $x = 4 - y$ into the second equation:
$$(4 - y) - 2y = 4$$
$$4 - 3y = 4$$
$$-3y = 0$$
$$y = 0$$
The solution is $(4, 0)$.

57. Solving the inequality:
$$20x + 9{,}300 > 18{,}000$$
$$20x > 8{,}700$$
$$x > 435$$

7.8 Systems of Linear Inequalities

1. Graphing the solution set:

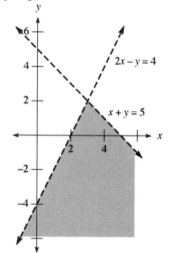

3. Graphing the solution set:

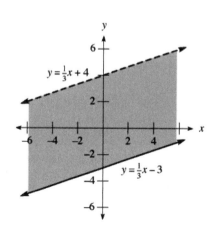

5. Graphing the solution set:

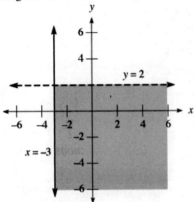

7. Graphing the solution set:

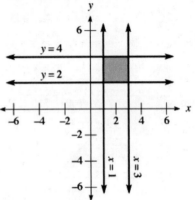

9. Graphing the solution set:

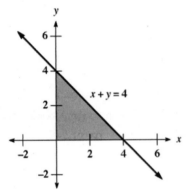

11. Graphing the solution set:

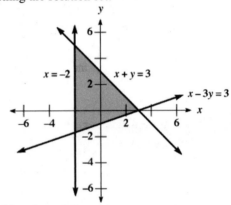

13. Graphing the solution set:

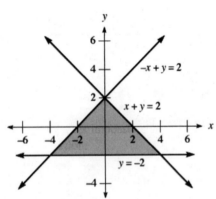

15. Graphing the solution set:

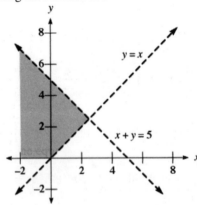

17. Graphing the solution set:

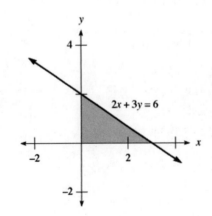

19. The system of inequalities is:
$$x + y \leq 4$$
$$-x + y < 4$$

21. The system of inequalities is:
$$x + y \geq 4$$
$$-x + y < 4$$

23. **a.** The system of inequalities is:
$$0.55x + 0.65y \leq 40$$
$$x \geq 2y$$
$$x > 15$$
$$y \geq 0$$

Graphing the solution set:

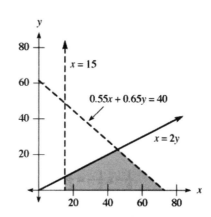

b. Substitute $x = 20$:
$$2y \leq 20$$
$$y \leq 10$$

The most he can purchase is 10 65-cent stamps.

25. Reducing the rational expression: $\dfrac{x^2 - x - 6}{x^2 - 9} = \dfrac{(x-3)(x+2)}{(x+3)(x-3)} = \dfrac{x+2}{x+3}$

27. Performing the operations:
$$\dfrac{x^2 - 25}{x + 4} \cdot \dfrac{2x + 8}{x^2 - 9x + 20} = \dfrac{(x+5)(x-5)}{x+4} \cdot \dfrac{2(x+4)}{(x-4)(x-5)} = \dfrac{2(x+5)(x-5)(x+4)}{(x+4)(x-4)(x-5)} = \dfrac{2(x+5)}{x-4}$$

29. Performing the operations: $\dfrac{x}{x^2 - 16} + \dfrac{4}{x^2 - 16} = \dfrac{x+4}{x^2 - 16} = \dfrac{x+4}{(x+4)(x-4)} = \dfrac{1}{x-4}$

31. Simplifying the complex fraction: $\dfrac{1 - \dfrac{25}{x^2}}{1 - \dfrac{8}{x} + \dfrac{15}{x^2}} \cdot \dfrac{x^2}{x^2} = \dfrac{x^2 - 25}{x^2 - 8x + 15} = \dfrac{(x+5)(x-5)}{(x-5)(x-3)} = \dfrac{x+5}{x-3}$

33. Multiplying each side of the equation by $x^2 - 9 = (x+3)(x-3)$:
$$(x+3)(x-3)\left(\dfrac{x}{x^2 - 9} - \dfrac{3}{x-3}\right) = (x+3)(x-3) \cdot \dfrac{1}{x+3}$$
$$x - 3(x+3) = x - 3$$
$$x - 3x - 9 = x - 3$$
$$-2x - 9 = x - 3$$
$$-3x = 6$$
$$x = -2$$

Since $x = -2$ checks in the original equation, the solution is $x = -2$.

Chapter 7 Review/Test

1. Solving the equation:
$$4x - 2 = 7x + 7$$
$$-3x - 2 = 7$$
$$-3x = 9$$
$$x = -3$$

2. Solving the equation:
$$\frac{3y}{4} - \frac{1}{2} + \frac{3y}{2} = 2 - y$$
$$8\left(\frac{3y}{4} - \frac{1}{2} + \frac{3y}{2}\right) = 8(2 - y)$$
$$6y - 4 + 12y = 16 - 8y$$
$$18y - 4 = -8y + 16$$
$$26y - 4 = 16$$
$$26y = 20$$
$$y = \frac{10}{13}$$

3. Solving the equation:
$$8 - 3(2t + 1) = 5(t + 2)$$
$$8 - 6t - 3 = 5t + 10$$
$$-6t + 5 = 5t + 10$$
$$-11t + 5 = 10$$
$$-11t = 5$$
$$t = -\frac{5}{11}$$

4. Solving the equation:
$$6 + 4(1 - 3t) = -3(t - 4) + 2$$
$$6 + 4 - 12t = -3t + 12 + 2$$
$$-12t + 10 = -3t + 14$$
$$-9t + 10 = 14$$
$$-9t = 4$$
$$t = -\frac{4}{9}$$

5. Solving the equation:
$$2x^2 - 5x = 12$$
$$2x^2 - 5x - 12 = 0$$
$$(2x + 3)(x - 4) = 0$$
$$x = -\frac{3}{2}, 4$$

6. Solving the equation:
$$81a^2 = 1$$
$$81a^2 - 1 = 0$$
$$(9a + 1)(9a - 1) = 0$$
$$a = -\frac{1}{9}, \frac{1}{9}$$

7. Solving the equation:
$$(x - 2)(x - 3) = 2$$
$$x^2 - 5x + 6 = 2$$
$$x^2 - 5x + 4 = 0$$
$$(x - 1)(x - 4) = 0$$
$$x = 1, 4$$

8. Solving the equation:
$$9x^3 + 18x^2 - 4x - 8 = 0$$
$$9x^2(x + 2) - 4(x + 2) = 0$$
$$(x + 2)(9x^2 - 4) = 0$$
$$(x + 2)(3x + 2)(3x - 2) = 0$$
$$x = -2, -\frac{2}{3}, \frac{2}{3}$$

9. Let w represent the width and $3w$ represent the length. Using the perimeter formula:
$$2w + 2(3w) = 32$$
$$2w + 6w = 32$$
$$8w = 32$$
$$w = 4$$
The dimensions are 4 feet by 12 feet.

10. Let x, $x + 1$, and $x + 2$ represent the three sides. Using the perimeter formula:
$$x + x + 1 + x + 2 = 12$$
$$3x + 3 = 12$$
$$3x = 9$$
$$x = 3$$
The sides are 3 meters, 4 meters, and 5 meters.

11. Solving the equation:
$$|x-3|=1$$
$$x-3=-1,1$$
$$x=2,4$$

12. Solving the equation:
$$|x-2|=3$$
$$x-2=-3,3$$
$$x=-1,5$$

13. Solving the equation:
$$|2y-3|=5$$
$$2y-3=-5,5$$
$$2y=-2,8$$
$$y=-1,4$$

14. Solving the equation:
$$|3y-2|=7$$
$$3y-2=-7,7$$
$$3y=-5,9$$
$$y=-\tfrac{5}{3},3$$

15. Solving the equation:
$$|4x-3|+2=11$$
$$|4x-3|=9$$
$$4x-3=-9,9$$
$$4x=-6,12$$
$$x=-\tfrac{3}{2},3$$

16. Solving the equation:
$$|6x-2|+4=16$$
$$|6x-2|=12$$
$$6x-2=-12,12$$
$$6x=-10,14$$
$$x=-\tfrac{5}{3},\tfrac{7}{3}$$

17. Solving the inequality:
$$\tfrac{3}{4}x+1\le 10$$
$$\tfrac{3}{4}x\le 9$$
$$3x\le 36$$
$$x\le 12$$
The solution set is $(-\infty,12]$.

18. Solving the inequality:
$$600-300x<900$$
$$-300x<300$$
$$x>-1$$
The solution set is $(-1,\infty)$.

19. Solving the inequality:
$$\tfrac{1}{3}\le \tfrac{1}{6}x\le 1$$
$$2\le x\le 6$$
The solution set is $[2,6]$.

20. Solving the inequality:
$$-\tfrac{1}{2}\le \tfrac{1}{6}x\le \tfrac{1}{3}$$
$$-3\le x\le 2$$
The solution set is $[-3,2]$.

21. Solving the inequality:
$$5t+1\le 3t-2 \quad \text{or} \quad -7t\le -21$$
$$2t\le -3 \qquad\qquad\qquad t\ge 3$$
$$t\le -\tfrac{3}{2} \qquad\qquad\qquad t\ge 3$$
The solution set is $\left(-\infty,-\tfrac{3}{2}\right]\cup[3,\infty)$.

22. Solving the inequality:
$$6t-3\le t+1 \quad \text{or} \quad -8t\le -16$$
$$5t\le 4 \qquad\qquad\qquad t\ge 2$$
$$t\le \tfrac{4}{5} \qquad\qquad\qquad t\ge 2$$
The solution set is $\left(-\infty,\tfrac{4}{5}\right]\cup[2,\infty)$.

23. Solving the inequality:
$$|y-2|<3$$
$$-3<y-2<3$$
$$-1<y<5$$

Graphing the solution set:

24. Solving the inequality:
$$|5x-1|>3$$
$$5x-1<-3 \quad \text{or} \quad 5x-1>3$$
$$5x<-2 \qquad\qquad 5x>4$$
$$x<-\tfrac{2}{5} \qquad\qquad x>\tfrac{4}{5}$$

Graphing the solution set:

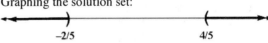

25. Solving the inequality:
$$|2t+1| - 3 < 2$$
$$|2t+1| < 5$$
$$-5 < 2t+1 < 5$$
$$-6 < 2t < 4$$
$$-3 < t < 2$$
Graphing the solution set:

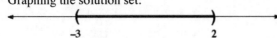

26. Solving the inequality:
$$|2t+1| - 1 < 5$$
$$|2t+1| < 6$$
$$-6 < 2t+1 < 6$$
$$-7 < 2t < 5$$
$$-\tfrac{7}{2} < t < \tfrac{5}{2}$$
Graphing the solution set:

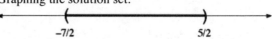

27. Solving the inequality:
$$|5+8t| + 4 \le 1$$
$$|5+8t| \le -3$$
Since this statement is never true, there is no solution, or $\varnothing$. Graphing the solution set:

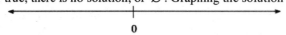

28. Since $|5x-8| \ge -2$ is true for all real numbers x, the solution is all real numbers, or $(-\infty, \infty)$.
Graphing the solution set:

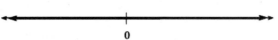

29. Multiply the second equation by –2:
$$6x - 5y = -5$$
$$-6x - 2y = -2$$
Adding yields:
$$-7y = -7$$
$$y = 1$$
Substituting to find x:
$$3x + 1 = 1$$
$$3x = 0$$
$$x = 0$$
The solution is (0,1).

30. Divide the first equation by 2 and the second equation by 3:
$$3x + 2y = 4$$
$$3x + 2y = 4$$
Since these two lines are identical, there is no unique solution (lines coincide).

31. Multiply the first equation by 3 and the second equation by 4:
$$-21x + 12y = -3$$
$$20x - 12y = 0$$
Adding yields:
$$-x = -3$$
$$x = 3$$
Substitute to find y:
$$-21 + 4y = -1$$
$$4y = 20$$
$$y = 5$$
The solution is (3,5).

32. To clear each equation of fractions, multiply the first equation by 4 and the second equation by 4:
$2x - 3y = -16$
$x + 6y = 52$
Multiply the first equation by 2:
$4x - 6y = -32$
$x + 6y = 52$
Adding yields:
$5x = 20$
$x = 4$
Substitute to find y:
$4 + 6y = 52$
$6y = 48$
$y = 8$
The solution is (4,8).

33. Substitute into the first equation:
$x + x - 1 = 2$
$2x - 1 = 2$
$2x = 3$
$x = \frac{3}{2}$
$y = \frac{3}{2} - 1 = \frac{1}{2}$
The solution is $\left(\frac{3}{2}, \frac{1}{2}\right)$.

34. Substitute into the first equation:
$2x - 3(2x - 7) = 5$
$2x - 6x + 21 = 5$
$-4x = -16$
$x = 4$
$y = 2(4) - 7 = 1$
The solution is (4,1).

35. Substitute into the first equation:
$3(-3y + 4) + 7y = 6$
$-9y + 12 + 7y = 6$
$-2y = -6$
$y = 3$
$x = -9 + 4 = -5$
The solution is (−5,3).

36. Substitute into the first equation:
$5x - (5x - 3) = 4$
$5x - 5x + 3 = 4$
$3 = 4$
Since this statement is false, there is no solution (parallel lines).

37. Adding the first and second equations yields:
$2x - 2z = -2$
$x - z = -1$
Adding the second and third equations yields $2x - 5z = -14$. So the system becomes:
$x - z = -1$
$2x - 5z = -14$
Multiply the first equation by −2:
$-2x + 2z = 2$
$2x - 5z = -14$
Adding yields:
$-3z = -12$
$z = 4$

Substitute to find x:
$$x - 4 = -1$$
$$x = 3$$
Substitute to find y:
$$3 + y + 4 = 6$$
$$y = -1$$
The solution is $(3, -1, 4)$.

38. Multiply the first equation by -1 and add it to the second equation:
$$-3x - 2y - z = -4$$
$$2x - 4y + z = -1$$
Adding yields:
$$-x - 6y = -5$$
$$x + 6y = 5$$
Multiply the first equation by -3 and add it to the third equation:
$$-9x - 6y - 3z = -12$$
$$x + 6y + 3z = -4$$
Adding yields:
$$-8x = -16$$
$$x = 2$$
Substituting to find y:
$$2 + 6y = 5$$
$$6y = 3$$
$$y = \tfrac{1}{2}$$
Substituting to find z:
$$2 + 6\left(\tfrac{1}{2}\right) + 3z = -4$$
$$3z + 5 = -4$$
$$3z = -9$$
$$z = -3$$
The solution is $\left(2, \tfrac{1}{2}, -3\right)$.

39. Multiply the second equation by 2 and add it to the third equation:
$$-6x + 8y - 2z = 4$$
$$6x - 8y + 2z = -4$$
Adding yields $0 = 0$. Since this is a true statement, there is no unique solution (dependent system).

40. Multiply the second equation by 4 and add it to the first equation:
$$4x - 6y + 8z = 4$$
$$20x + 4y - 8z = 16$$
Adding yields:
$$24x - 2y = 20$$
$$12x - y = 10$$
Multiply the second equation by 6 and add it to the third equation:
$$30x + 6y - 12z = 24$$
$$6x - 9y + 12z = 6$$
Adding yields:
$$36x - 3y = 30$$
$$12x - y = 10$$
Since these two equations are identical, there is no unique solution (dependent system).

41. Multiply the third equation by 2 and add it to the second equation:
$3x - 2z = -2$
$10y + 2z = -2$
Adding yields the equation $3x + 10y = -4$. So the system becomes:
$2x - y = 5$
$3x + 10y = -4$
Multiplying the first equation by 10:
$20x - 10y = 50$
$3x + 10y = -4$
Adding yields:
$23x = 46$
$x = 2$
Substituting to find y:
$4 - y = 5$
$y = -1$
Substituting to find z:
$-5 + z = -1$
$z = 4$
The solution is $(2, -1, 4)$.

42. Adding the first and second equations yields $x - z = -1$. Since this is the same as the third equation, there is no unique solution (dependent system).

43. Graphing the inequality:

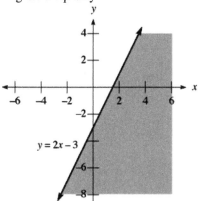

44. Graphing the inequality:

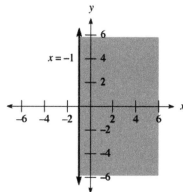

45. Graphing the solution set:

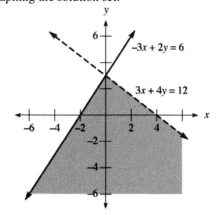

46. Graphing the solution set:

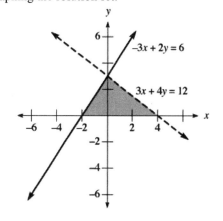

47. Graphing the solution set:

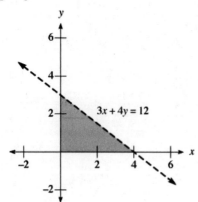

48. Graphing the solution set:

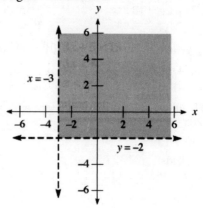

Chapter 8
Equations and Functions

8.1 The Slope of a Line

1. The slope is $\frac{3}{2}$.

3. There is no slope (undefined).

5. The slope is $\frac{2}{3}$.

7. Finding the slope: $m = \dfrac{4-1}{4-2} = \dfrac{3}{2}$
Sketching the graph:

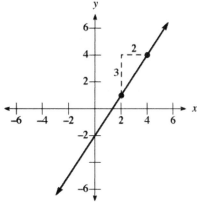

9. Finding the slope: $m = \dfrac{2-4}{5-1} = \dfrac{-2}{4} = -\dfrac{1}{2}$
Sketching the graph:

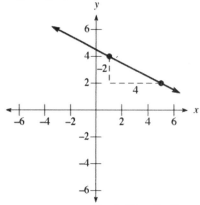

11. Finding the slope: $m = \dfrac{2-(-3)}{4-1} = \dfrac{2+3}{3} = \dfrac{5}{3}$
Sketching the graph:

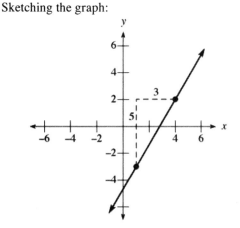

13. Finding the slope: $m = \dfrac{3-(-2)}{1-(-3)} = \dfrac{3+2}{1+3} = \dfrac{5}{4}$
Sketching the graph:

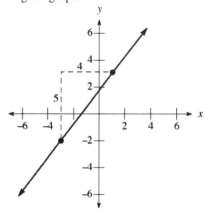

231

232 Chapter 8 Equations and Functions

15. Finding the slope: $m = \dfrac{-2-2}{3-(-3)} = \dfrac{-4}{3+3} = -\dfrac{2}{3}$

    Sketching the graph:

    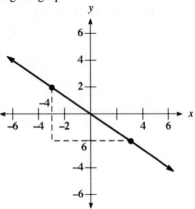

17. Finding the slope: $m = \dfrac{-2-(-5)}{3-2} = \dfrac{-2+5}{1} = 3$

    Sketching the graph:

    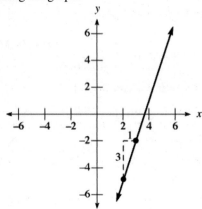

19. Using the slope formula:
    $\dfrac{-2-2}{x-5} = 2$
    $-4 = 2(x-5)$
    $-4 = 2x - 10$
    $6 = 2x$
    $x = 3$

21. Using the slope formula:
    $\dfrac{3-6}{a-2} = -1$
    $-3 = -1(a-2)$
    $-3 = -a + 2$
    $-5 = -a$
    $a = 5$

23. Using the slope formula:
    $\dfrac{4b-b}{-1-2} = -2$
    $3b = -2(-3)$
    $3b = 6$
    $b = 2$

25. Completing the table:

x	y
0	2
3	0

    Finding the slope: $m = \dfrac{2-0}{0-3} = -\dfrac{2}{3}$

27. Completing the table:

x	y
0	-5
3	-3

    Finding the slope: $m = \dfrac{-5-(-3)}{0-3} = \dfrac{-5+3}{-3} = \dfrac{2}{3}$

29. Graphing the line:

    The slope is $m = -\dfrac{1}{2}$.

31. Finding the slope of this line: $m = \dfrac{1-3}{-8-2} = \dfrac{-2}{-10} = \dfrac{1}{5}$. Since the parallel slope is the same, its slope is $\dfrac{1}{5}$.

33. Finding the slope of this line: $m = \dfrac{2-(-6)}{5-5} = \dfrac{8}{0}$, which is undefined

Since the perpendicular slope is a horizontal line, its slope is 0.

35. Finding the slope of this line: $m = \dfrac{-5-1}{4-(-2)} = \dfrac{-6}{6} = -1$. Since the parallel slope is the same, its slope is -1.

37. Finding the slope of this line: $m = \dfrac{-3-(-5)}{1-(-2)} = \dfrac{2}{3}$

Since the perpendicular slope is the negative reciprocal, its slope is $-\dfrac{3}{2}$.

39. **a.** Since the slopes between each successive pairs of points is 2, this could represent ordered pairs from a line.
  b. Since the slopes between each successive pairs of points is not the same, this could not represent ordered pairs from a line.

41. **a.** Finding the slope: $m = \dfrac{105-0}{6-0} = 17.5$ miles/hour

  b. Finding the slope: $m = \dfrac{80-0}{2-0} = 40$ kilometers/hour

  c. Finding the slope: $m = \dfrac{3600-0}{30-0} = 120$ feet/second

  d. Finding the slope: $m = \dfrac{420-0}{15-0} = 28$ meters/minute

43. For line segment A, the slope is: $m = \dfrac{250-300}{2007-2006} = -50$ million cameras/year

For line segment B, the slope is: $m = \dfrac{175-250}{2008-2007} = -75$ million cameras/year

For line segment C, the slope is: $m = \dfrac{125-150}{2010-2009} = -25$ million cameras/year

45. **a.** Let d represent the distance. Using the slope formula:
$$\dfrac{0-1106}{d-0} = -\dfrac{7}{100}$$
$$-7d = -110600$$
$$d = 15800$$
The distance to point A is 15,800 feet.

  b. The slope is $-\dfrac{7}{100} = -0.07$.

47. Computing the slope: $m = \dfrac{74\% - 9\%}{2005-1995} = \dfrac{65\%}{10} = 6.5\%$ per year

Over the 10 years from 1995 to 2005, the percent of adults that access the internet has increased at an average rate of 6.5% per year.

49. **a.** Computing the slope: $m = \dfrac{10,000-57,686}{1979-1969} = \dfrac{-47,686}{10} = -4,768.6$ cases per year

New cases of rubella decreased at an average rate of 4,768.6 cases per year from 1969 to 1979.

  b. Computing the slope: $m = \dfrac{9-10,000}{1989-1979} = \dfrac{-9991}{10} = -999.1$ cases per year

New cases of rubella decreased at an average rate of 999.1 cases per year from 1979 to 1989.

  c. Computing the slope: $m = \dfrac{9-9}{2004-1989} = \dfrac{0}{15} = 0$ cases per year

New cases of rubella did not change from 1989 to 2004, staying at a constant 9 cases per year.

51. Substituting $x = 4$:
$$3(4) + 2y = 12$$
$$12 + 2y = 12$$
$$2y = 0$$
$$y = 0$$

53. Solving for y:
$$3x + 2y = 12$$
$$2y = -3x + 12$$
$$y = -\dfrac{3}{2}x + 6$$

55. Solving for t:
$$A = P + Prt$$
$$A - P = Prt$$
$$t = \frac{A-P}{Pr}$$

57. Simplifying: $2\left(-\frac{1}{2}\right) = -1$

59. Simplifying: $\dfrac{5-(-3)}{2-6} = \dfrac{8}{-4} = -2$

61. Solving for y:
$$\frac{y-b}{x-0} = m$$
$$y - b = mx$$
$$y = mx + b$$

63. Solving for y:
$$y - 3 = -2(x + 4)$$
$$y - 3 = -2x - 8$$
$$y = -2x - 5$$

65. Solving for y: $y = -\frac{4}{3}(0) + 5 = 0 + 5 = 5$

67. Setting the slope equal to -1:
$$\frac{3x - 2 - 2}{x - 1} = -1$$
$$\frac{3x - 4}{x - 1} = -1$$
$$3x - 4 = -x + 1$$
$$4x = 5$$
$$x = \tfrac{5}{4}$$
$$y = 3\left(\tfrac{5}{4}\right) - 2 = \tfrac{15}{4} - \tfrac{8}{4} = \tfrac{7}{4}$$

The point is $\left(\tfrac{5}{4}, \tfrac{7}{4}\right)$.

69. Graphing the curves:

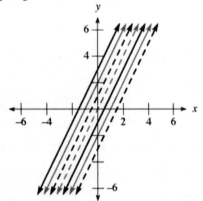

71. Graphing the curves:

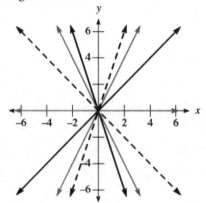

73. Graphing the curves:

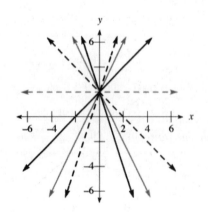

8.2 The Equation of a Line

1. Using the slope-intercept formula: $y = 2x + 3$
3. Using the slope-intercept formula: $y = x - 5$
5. Using the slope-intercept formula: $y = \frac{1}{2}x + \frac{3}{2}$
7. Using the slope-intercept formula: $y = 4$
9.   **a.** The parallel slope is 3.   **b.** The perpendicular slope is $-\frac{1}{3}$.
11. Solving for y:

    $3x + y = -2$
    $y = -3x - 2$

    a. The parallel slope is -3.
    b. The perpendicular slope is $\frac{1}{3}$.

13. Solving for y:

    $2x + 5y = -11$
    $5y = -2x - 11$
    $y = -\frac{2}{5}x - \frac{11}{5}$

    a. The parallel slope is $-\frac{2}{5}$.
    b. The perpendicular slope is $\frac{5}{2}$.

15. The slope is 3, the y-intercept is -2, and the perpendicular slope is $-\frac{1}{3}$.

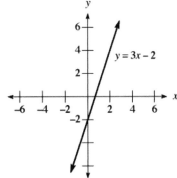

17. The slope is $\frac{2}{3}$, the y-intercept is -4, and the perpendicular slope is $-\frac{3}{2}$.

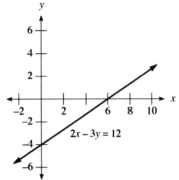

19. The slope is $-\frac{4}{5}$, the y-intercept is 4, and the perpendicular slope is $\frac{5}{4}$.

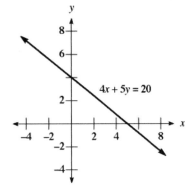

21. The slope is $\frac{1}{2}$ and the y-intercept is –4. Using the slope-intercept form, the equation is $y = \frac{1}{2}x - 4$.

23. The slope is $-\frac{2}{3}$ and the y-intercept is 3. Using the slope-intercept form, the equation is $y = -\frac{2}{3}x + 3$.

25. Using the point-slope formula:
$$y-(-5) = 2(x-(-2))$$
$$y+5 = 2(x+2)$$
$$y+5 = 2x+4$$
$$y = 2x-1$$

27. Using the point-slope formula:
$$y-1 = -\frac{1}{2}(x-(-4))$$
$$y-1 = -\frac{1}{2}(x+4)$$
$$y-1 = -\frac{1}{2}x-2$$
$$y = -\frac{1}{2}x-1$$

29. Using the point-slope formula:
$$y-2 = -3\left(x-\left(-\frac{1}{3}\right)\right)$$
$$y-2 = -3\left(x+\frac{1}{3}\right)$$
$$y-2 = -3x-1$$
$$y = -3x+1$$

31. Using the point-slope formula:
$$y-(-2) = \frac{2}{3}(x-(-4))$$
$$y+2 = \frac{2}{3}(x+4)$$
$$y+2 = \frac{2}{3}x+\frac{8}{3}$$
$$y = \frac{2}{3}x+\frac{2}{3}$$

33. Using the point-slope formula:
$$y-2 = -\frac{1}{4}(x-(-5))$$
$$y-2 = -\frac{1}{4}(x+5)$$
$$y-2 = -\frac{1}{4}x-\frac{5}{4}$$
$$y = -\frac{1}{4}x+\frac{3}{4}$$

35. First find the slope: $m = \dfrac{-1-(-4)}{1-(-2)} = \dfrac{-1+4}{1+2} = \dfrac{3}{3} = 1$. Using the point-slope formula:
$$y-(-1) = 1(x-1)$$
$$y+1 = x-1$$
$$-x+y = -2$$
$$x-y = 2$$

37. First find the slope: $m = \dfrac{1-(-5)}{2-(-1)} = \dfrac{1+5}{2+1} = \dfrac{6}{3} = 2$. Using the point-slope formula:
$$y-1 = 2(x-2)$$
$$y-1 = 2x-4$$
$$-2x+y = -3$$
$$2x-y = 3$$

39. First find the slope: $m = \dfrac{-1-\left(-\frac{1}{5}\right)}{-\frac{1}{3}-\frac{1}{3}} = \dfrac{-1+\frac{1}{5}}{-\frac{2}{3}} = \dfrac{-\frac{4}{5}}{-\frac{2}{3}} = \dfrac{4}{5} \cdot \dfrac{3}{2} = \dfrac{6}{5}$. Using the point-slope formula:
$$y-(-1) = \frac{6}{5}\left(x-\left(-\frac{1}{3}\right)\right)$$
$$y+1 = \frac{6}{5}\left(x+\frac{1}{3}\right)$$
$$5(y+1) = 6\left(x+\frac{1}{3}\right)$$
$$5y+5 = 6x+2$$
$$6x-5y = 3$$

41. **a.** For the x-intercept, substitute $y = 0$:
$$3x - 2(0) = 10$$
$$3x = 10$$
$$x = \tfrac{10}{3}$$

For the y-intercept, substitute $x = 0$:
$$3(0) - 2y = 10$$
$$-2y = 10$$
$$y = -5$$

b. Substituting $y = 1$:
$$3x - 2(1) = 10$$
$$3x - 2 = 10$$
$$3x = 12$$
$$x = 4$$
Another solution is $(4,1)$. Other answers are possible.

c. Solving for y:
$$3x - 2y = 10$$
$$-2y = -3x + 10$$
$$y = \tfrac{3}{2}x - 5$$

d. Substituting $x = 2$: $y = \tfrac{3}{2}(2) - 5 = 3 - 5 = -2$. No, the point $(2,2)$ is not a solution to the equation.

43. **a.** For the x-intercept, substitute $y = 0$:
$$\tfrac{3x}{4} - \tfrac{0}{2} = 1$$
$$\tfrac{3x}{4} = 1$$
$$3x = 4$$
$$x = \tfrac{4}{3}$$

For the y-intercept, substitute $x = 0$:
$$\tfrac{3(0)}{4} - \tfrac{y}{2} = 1$$
$$-\tfrac{y}{2} = 1$$
$$-y = 2$$
$$y = -2$$

b. Substituting $x = 2$:
$$\tfrac{3(2)}{4} - \tfrac{y}{2} = 1$$
$$\tfrac{3}{2} - \tfrac{y}{2} = 1$$
$$-\tfrac{y}{2} = -\tfrac{1}{2}$$
$$y = 1$$
Another solution is $(2,1)$. Other answers are possible.

c. Solving for y:
$$\tfrac{3x}{4} - \tfrac{y}{2} = 1$$
$$4\left(\tfrac{3x}{4} - \tfrac{y}{2}\right) = 4(1)$$
$$3x - 2y = 4$$
$$-2y = -3x + 4$$
$$y = \tfrac{3}{2}x - 2$$

d. Substituting $x = 1$: $y = \tfrac{3}{2}(1) - 2 = \tfrac{3}{2} - 2 = -\tfrac{1}{2}$. No, the point $(1,2)$ is not a solution to the equation.

45.
a. Solving for x:
$$-2x+1=-3$$
$$-2x=-4$$
$$x=2$$

b. Solving for y:
$$-2x+y=-3$$
$$y=2x-3$$

c. The slope-intercept form is $y=2x-3$, so the y-intercept is -3.

d. The slope-intercept form is $y=2x-3$, so the slope is 2.

e. Sketching the graph:

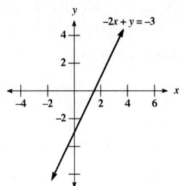

47. Two points on the line are $(0,-4)$ and $(2,0)$. Finding the slope: $m = \dfrac{0-(-4)}{2-0} = \dfrac{4}{2} = 2$

Using the slope-intercept form, the equation is $y=2x-4$.

49. Two points on the line are $(0,4)$ and $(-2,0)$. Finding the slope: $m = \dfrac{0-4}{-2-0} = \dfrac{-4}{-2} = 2$

Using the slope-intercept form, the equation is $y=2x+4$.

51. The slope is 0 and the y-intercept is -2.

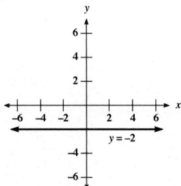

53. First find the slope:
$$3x-y=5$$
$$-y=-3x+5$$
$$y=3x-5$$

So the slope is 3. Using $(-1,4)$ in the point-slope formula:
$$y-4=3(x-(-1))$$
$$y-4=3(x+1)$$
$$y-4=3x+3$$
$$y=3x+7$$

55. First find the slope:
$$2x - 5y = 10$$
$$-5y = -2x + 10$$
$$y = \tfrac{2}{5}x - 2$$

So the perpendicular slope is $-\tfrac{5}{2}$. Using $(-4,-3)$ in the point-slope formula:
$$y - (-3) = -\tfrac{5}{2}(x - (-4))$$
$$y + 3 = -\tfrac{5}{2}(x + 4)$$
$$y + 3 = -\tfrac{5}{2}x - 10$$
$$y = -\tfrac{5}{2}x - 13$$

57. The perpendicular slope is $\tfrac{1}{4}$. Using $(-1,0)$ in the point-slope formula:
$$y - 0 = \tfrac{1}{4}(x - (-1))$$
$$y = \tfrac{1}{4}(x + 1)$$
$$y = \tfrac{1}{4}x + \tfrac{1}{4}$$

59. Using the points $(3,0)$ and $(0,2)$, first find the slope: $m = \dfrac{2-0}{0-3} = -\tfrac{2}{3}$

Using the slope-intercept formula, the equation is: $y = -\tfrac{2}{3}x + 2$

61. a. Using the points $(0,32)$ and $(25,77)$, first find the slope: $m = \dfrac{77-32}{25-0} = \dfrac{45}{25} = \tfrac{9}{5}$

Using the slope-intercept formula, the equation is: $F = \tfrac{9}{5}C + 32$

b. Substituting $C = 30$: $F = \tfrac{9}{5}(30) + 32 = 54 + 32 = 86°$

63. a. Substituting $n = 10{,}000$: $C = 125{,}000 + 6.5(10{,}000) = \$190{,}000$

b. Finding the average cost: $\dfrac{\$190{,}000}{10{,}000} = \$19$ per textbook

c. Since each textbook costs $6.50 in materials, this is the cost to produce the next textbook.

65. a. Using $(2002,75)$ and $(2004,144)$: $m = \dfrac{144 - 75}{2004 - 2002} = \dfrac{69}{2}$

Using the point-slope formula:
$$y - 75 = \tfrac{69}{2}(x - 2002)$$
$$y - 75 = \tfrac{69}{2}x - 69{,}069$$
$$y = \tfrac{69}{2}x - 68{,}994$$

b. Substituting $x = 2006$: $y = \tfrac{69}{2}(2006) - 68{,}994 = 213$ books

67. a. Using $(1995, 9\%)$ and $(2005, 74\%)$, find the slope: $m = \dfrac{74\% - 9\%}{2005 - 1995} = \dfrac{65}{10} = 6.5$

Using the point-slope formula:
$$y - 9 = 6.5(x - 1995)$$
$$y - 9 = 6.5x - 12{,}967.5$$
$$y = 6.5x - 12{,}958.5$$

b. Substituting $x = 2006$: $y = 6.5(2006) - 12{,}958.5 = 80.5\%$

In 2006, 80.5% of adults will be online.

c. Substituting $x = 2010$: $y = 6.5(2010) - 12{,}958.5 = 106.5\%$

This answer is larger than 100%, which is impossible.

69. Let w represent the width and $4w + 3$ represent the length. Using the perimeter formula:
$$2w + 2(4w + 3) = 56$$
$$2w + 8w + 6 = 56$$
$$10w = 50$$
$$w = 5$$
The width is 5 inches and the length is 23 inches.

71. The total amount collected is: $732.50 – $66 = $666.50
Let x represent the sales. Since this amount includes the sales tax:
$$x + 0.075x = 666.50$$
$$1.075x = 666.50$$
$$x = 620$$
The amount which is sales tax is therefore: $666.50 – $620 = $46.50

73. Completing the table:

x	y
0	0
10	75
20	150

75. Completing the table:

x	y
0	0
1	1
1	–1

77. First find the midpoint: $\left(\frac{1+7}{2}, \frac{4+8}{2}\right) = \left(\frac{8}{2}, \frac{12}{2}\right) = (4,6)$. Now find the slope: $m = \frac{8-4}{7-1} = \frac{4}{6} = \frac{2}{3}$

So the perpendicular slope is $-\frac{3}{2}$. Using (4,6) in the point-slope formula:
$$y - 6 = -\frac{3}{2}(x - 4)$$
$$y - 6 = -\frac{3}{2}x + 6$$
$$y = -\frac{3}{2}x + 12$$

79. First find the midpoint: $\left(\frac{-5-1}{2}, \frac{1+4}{2}\right) = \left(\frac{-6}{2}, \frac{5}{2}\right) = \left(-3, \frac{5}{2}\right)$. Now find the slope: $m = \frac{4-1}{-1+5} = \frac{3}{4}$

So the perpendicular slope is $-\frac{4}{3}$. Using $\left(-3, \frac{5}{2}\right)$ in the point-slope formula:
$$y - \frac{5}{2} = -\frac{4}{3}(x + 3)$$
$$y - \frac{5}{2} = -\frac{4}{3}x - 4$$
$$y = -\frac{4}{3}x - \frac{3}{2}$$

8.3 Introduction to Functions

1. The domain is {1,2,4} and the range is {1,3,5}. This is a function.
3. The domain is {–1,1,2} and the range is {–5,3}. This is a function.
5. The domain is {3,7} and the range is {–1,4}. This is not a function.
7. The domain is {a,b,c,d} and the range is {3,4,5}. This is a function.
9. The domain is {a} and the range is {1,2,3,4}. This is not a function.
11. Yes, since it passes the vertical line test.
13. No, since it fails the vertical line test.
15. No, since it fails the vertical line test.
17. Yes, since it passes the vertical line test.
19. Yes, since it passes the vertical line test.
21. The domain is $\{x \mid -5 \leq x \leq 5\}$ and the range is $\{y \mid 0 \leq y \leq 5\}$.
23. The domain is $\{x \mid -5 \leq x \leq 3\}$ and the range is $\{y \mid y = 3\}$.

25. The domain is all real numbers and the range is $\{y \mid y \geq -1\}$. This is a function.

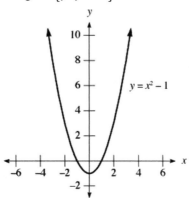

27. The domain is all real numbers and the range is $\{y \mid y \geq 4\}$. This is a function.

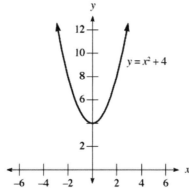

29. The domain is $\{x \mid x \geq -1\}$ and the range is all real numbers. This is not a function.

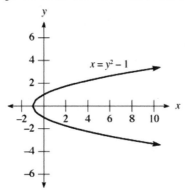

31. The domain is $\{x \mid x \geq 4\}$ and the range is all real numbers. This is not a function.

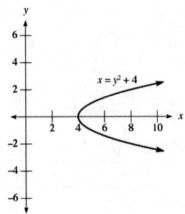

33. The domain is all real numbers and the range is $\{y \mid y \geq 0\}$. This is a function.

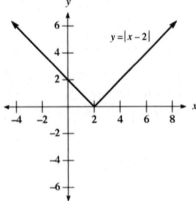

35. The domain is all real numbers and the range is $\{y \mid y \geq -2\}$. This is a function.

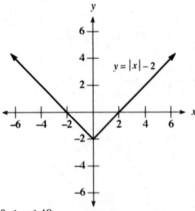

37. a. The equation is $y = 8.5x$ for $10 \leq x \leq 40$.

b. Completing the table:

Hours Worked	Function Rule	Gross Pay ($)
x	$y = 8.5x$	y
10	$y = 8.5(10) = 85$	85
20	$y = 8.5(20) = 170$	170
30	$y = 8.5(30) = 255$	255
40	$y = 8.5(40) = 340$	340

c. Constructing a line graph:

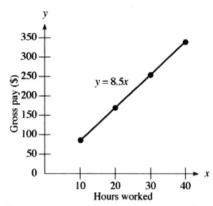

d. The domain is $\{x \mid 10 \le x \le 40\}$ and the range is $\{y \mid 85 \le y \le 340\}$.
e. The minimum is $85 and the maximum is $340.

39. a. Completing the table:

Time (sec) t	Function Rule $h = 16t - 16t^2$	Distance (ft) h
0	$h = 16(0) - 16(0)^2$	0
0.1	$h = 16(0.1) - 16(0.1)^2$	1.44
0.2	$h = 16(0.2) - 16(0.2)^2$	2.56
0.3	$h = 16(0.3) - 16(0.3)^2$	3.36
0.4	$h = 16(0.4) - 16(0.4)^2$	3.84
0.5	$h = 16(0.5) - 16(0.5)^2$	4
0.6	$h = 16(0.6) - 16(0.6)^2$	3.84
0.7	$h = 16(0.7) - 16(0.7)^2$	3.36
0.8	$h = 16(0.8) - 16(0.8)^2$	2.56
0.9	$h = 16(0.9) - 16(0.9)^2$	1.44
1	$h = 16(1) - 16(1)^2$	0

b. The domain is $\{t \mid 0 \le t \le 1\}$ and the range is $\{h \mid 0 \le h \le 4\}$.
c. Graphing the function:

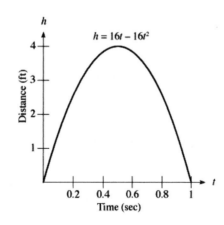

244 Chapter 8 Equations and Functions

41. a. Graphing the function:

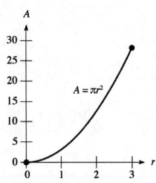

b. The domain is $\{r \mid 0 \leq r \leq 3\}$ and the range is $\{A \mid 0 \leq A \leq 9\pi\}$.

43. a. Yes, since it passes the vertical line test.
b. The domain is $\{t \mid 0 \leq t \leq 6\}$ and the range is $\{h \mid 0 \leq h \leq 60\}$.
c. At time $t = 3$ the ball reaches its maximum height.
d. The maximum height is $h = 60$.
e. At time $t = 6$ the ball hits the ground.

45. a. Figure III   b. Figure I
c. Figure II   d. Figure IV

47. Substituting $x = 4$: $y = 3(4) - 2 = 12 - 2 = 10$
49. Substituting $x = -4$: $y = 3(-4) - 2 = -12 - 2 = -14$

51. Substituting $x = 2$: $y = (2)^2 - 3 = 4 - 3 = 1$
53. Substituting $x = 0$: $y = (0)^2 - 3 = 0 - 3 = -3$

55. Solving for y:
$$\tfrac{8}{5} - 2y = 4$$
$$-2y = \tfrac{12}{5}$$
$$y = -\tfrac{6}{5}$$

57. Substituting $x = 0$ and $y = 0$:
$$0 = a(0 - 80)^2 + 70$$
$$0 = 6400a + 70$$
$$6400a = -70$$
$$a = -\tfrac{7}{640}$$

59. Simplifying: $7.5(20) = 150$
61. Simplifying: $4(3.14)(9) \approx 113$

63. Simplifying: $4(-2) - 1 = -8 - 1 = -9$

65. a. Substituting $t = 10$: $s = \dfrac{60}{10} = 6$
b. Substituting $t = 8$: $s = \dfrac{60}{8} = 7.5$

67. a. Substituting $x = 5$: $(5)^2 + 2 = 25 + 2 = 27$
b. Substituting $x = -2$: $(-2)^2 + 2 = 4 + 2 = 6$

69. The domain is all real numbers and the range is $\{y \mid y \leq 5\}$. This is a function.

Problem Set 8.4 245

71. The domain is $\{x \mid x \geq 3\}$ and the range is all real numbers. This is not a function.

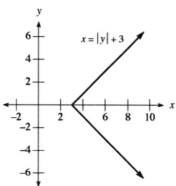

73. The domain is $\{x \mid -4 \leq x \leq 4\}$ and the range is $\{y \mid -4 \leq y \leq 4\}$. This is not a function.

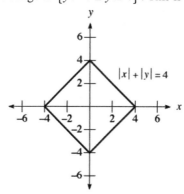

8.4 Function Notation

1. Evaluating the function: $f(2) = 2(2) - 5 = 4 - 5 = -1$
3. Evaluating the function: $f(-3) = 2(-3) - 5 = -6 - 5 = -11$
5. Evaluating the function: $g(-1) = (-1)^2 + 3(-1) + 4 = 1 - 3 + 4 = 2$
7. Evaluating the function: $g(-3) = (-3)^2 + 3(-3) + 4 = 9 - 9 + 4 = 4$
9. First evaluate each function:
$$g(4) = (4)^2 + 3(4) + 4 = 16 + 12 + 4 = 32 \qquad f(4) = 2(4) - 5 = 8 - 5 = 3$$
Now evaluating: $g(4) + f(4) = 32 + 3 = 35$
11. First evaluate each function:
$$f(3) = 2(3) - 5 = 6 - 5 = 1 \qquad g(2) = (2)^2 + 3(2) + 4 = 4 + 6 + 4 = 14$$
Now evaluating: $f(3) - g(2) = 1 - 14 = -13$
13. Evaluating the function: $f(0) = 3(0)^2 - 4(0) + 1 = 0 - 0 + 1 = 1$
15. Evaluating the function: $g(-4) = 2(-4) - 1 = -8 - 1 = -9$
17. Evaluating the function: $f(-1) = 3(-1)^2 - 4(-1) + 1 = 3 + 4 + 1 = 8$
19. Evaluating the function: $g(10) = 2(10) - 1 = 20 - 1 = 19$
21. Evaluating the function: $f(3) = 3(3)^2 - 4(3) + 1 = 27 - 12 + 1 = 16$
23. Evaluating the function: $g\left(\tfrac{1}{2}\right) = 2\left(\tfrac{1}{2}\right) - 1 = 1 - 1 = 0$ 25. Evaluating the function: $f(a) = 3a^2 - 4a + 1$

246 Chapter 8 Equations and Functions

27. $f(1) = 4$

29. $g\left(\frac{1}{2}\right) = 0$

31. $g(-2) = 2$

33. Evaluating the function: $f(0) = 2(0)^2 - 8 = 0 - 8 = -8$

35. Evaluating the function: $g(-4) = \frac{1}{2}(-4) + 1 = -2 + 1 = -1$

37. Evaluating the function: $f(a) = 2a^2 - 8$

39. Evaluating the function: $f(b) = 2b^2 - 8$

41. Evaluating the function: $f[g(2)] = f\left[\frac{1}{2}(2) + 1\right] = f(2) = 2(2)^2 - 8 = 8 - 8 = 0$

43. Evaluating the function: $g[f(-1)] = g\left[2(-1)^2 - 8\right] = g(-6) = \frac{1}{2}(-6) + 1 = -3 + 1 = -2$

45. Evaluating the function: $g[f(0)] = g\left[2(0)^2 - 8\right] = g(-8) = \frac{1}{2}(-8) + 1 = -4 + 1 = -3$

47. Graphing the function:

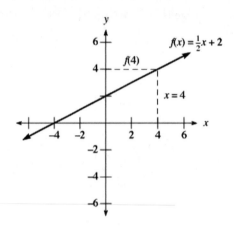

49. Finding where $f(x) = x$:
$$\frac{1}{2}x + 2 = x$$
$$2 = \frac{1}{2}x$$
$$x = 4$$

51. Graphing the function:

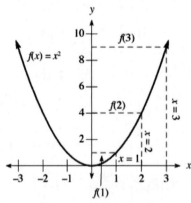

53. Evaluating: $V(3) = 150 \cdot 2^{3/3} = 150 \cdot 2 = 300$; The painting is worth $300 in 3 years.

Evaluating: $V(6) = 150 \cdot 2^{6/3} = 150 \cdot 4 = 600$; The painting is worth $600 in 6 years.

55. a. True b. True
    c. True d. False
    e. True

57.
a. Evaluating: $V(3.75) = -3300(3.75) + 18000 = \$5,625$
b. Evaluating: $V(5) = -3300(5) + 18000 = \$1,500$
c. The domain of this function is $\{t \mid 0 \le t \le 5\}$.
d. Sketching the graph:

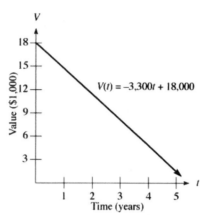

e. The range of this function is $\{V(t) \mid 1,500 \le V(t) \le 18,000\}$.
f. Solving $V(t) = 10000$:
$$-3300t + 18000 = 10000$$
$$-3300t = -8000$$
$$t \approx 2.42$$
The copier will be worth $10,000 after approximately 2.42 years.

59. Solving the equation:
$$|3x - 5| = 7$$
$$3x - 5 = -7, 7$$
$$3x = -2, 12$$
$$x = -\tfrac{2}{3}, 4$$

61. Solving the equation:
$$|4y + 2| - 8 = -2$$
$$|4y + 2| = 6$$
$$4y + 2 = -6, 6$$
$$4y = -8, 4$$
$$y = -2, 1$$

63. Solving the equation:
$$5 + |6t + 2| = 3$$
$$|6t + 2| = -2$$
Since this last equation is impossible, there is no solution, or $\varnothing$.

65. Simplifying: $(35x - 0.1x^2) - (8x + 500) = 35x - 0.1x^2 - 8x - 500 = -0.1x^2 + 27x - 500$

67. Simplifying: $(4x^2 + 3x + 2) + (2x^2 - 5x - 6) = 4x^2 + 3x + 2 + 2x^2 - 5x - 6 = 6x^2 - 2x - 4$

69. Simplifying: $(4x^2 + 3x + 2) - (2x^2 - 5x - 6) = 4x^2 + 3x + 2 - 2x^2 + 5x + 6 = 2x^2 + 8x + 8$

71. Simplifying: $0.6(m - 70) = 0.6m - 42$

73. Simplifying: $(4x - 3)(x - 1) = 4x^2 - 4x - 3x + 3 = 4x^2 - 7x + 3$

75.
a. $f(2) = 2$
b. $f(-4) = 0$

77.
a. Completing the table:

Weight (ounces)	0.6	1.0	1.1	2.5	3.0	4.8	5.0	5.3
Cost (cents)	39	39	63	87	87	135	135	159

b. The letter weighs over 2 ounces, but not over 3 ounces. As an inequality, this can be written as $2 < x \le 3$.
c. The domain is $\{x \mid 0 < x \le 6\}$.
d. The range is $\{39, 63, 87, 111, 135, 159\}$.

8.5 Algebra and Composition with Functions

1. Writing the formula: $f + g = f(x) + g(x) = (4x - 3) + (2x + 5) = 6x + 2$
3. Writing the formula: $g - f = g(x) - f(x) = (2x + 5) - (4x - 3) = -2x + 8$
5. Writing the formula: $g + f = g(x) + f(x) = (x - 2) + (3x - 5) = 4x - 7$
7. Writing the formula: $g + h = g(x) + h(x) = (x - 2) + (3x^2) = 3x^2 + x - 2$
9. Writing the formula: $g - f = g(x) - f(x) = (x - 2) - (3x - 5) = -2x + 3$
11. Writing the formula: $fh = f(x) \cdot h(x) = (3x - 5)(3x^2) = 9x^3 - 15x^2$
13. Writing the formula: $h/f = \dfrac{h(x)}{f(x)} = \dfrac{3x^2}{3x - 5}$
15. Writing the formula: $f/h = \dfrac{f(x)}{h(x)} = \dfrac{3x - 5}{3x^2}$
17. Writing the formula: $f + g + h = f(x) + g(x) + h(x) = (3x - 5) + (x - 2) + (3x^2) = 3x^2 + 4x - 7$
19. Evaluating: $(f + g)(2) = f(2) + g(2) = (2 \cdot 2 + 1) + (4 \cdot 2 + 2) = 5 + 10 = 15$
21. Evaluating: $(fg)(3) = f(3) \cdot g(3) = (2 \cdot 3 + 1)(4 \cdot 3 + 2) = 7 \cdot 14 = 98$
23. Evaluating: $(h/g)(1) = \dfrac{h(1)}{g(1)} = \dfrac{4(1)^2 + 4(1) + 1}{4(1) + 2} = \dfrac{9}{6} = \dfrac{3}{2}$
25. Evaluating: $(fh)(0) = f(0) \cdot h(0) = (2(0) + 1)(4(0)^2 + 4(0) + 1) = (1)(1) = 1$
27. Evaluating: $(f + g + h)(2) = f(2) + g(2) + h(2) = (2(2) + 1) + (4(2) + 2) + (4(2)^2 + 4(2) + 1) = 5 + 10 + 25 = 40$
29. Evaluating:

    $(h + fg)(3) = h(3) + f(3) \cdot g(3) = (4(3)^2 + 4(3) + 1) + (2(3) + 1) \cdot (4(3) + 2) = 49 + 7 \cdot 14 = 49 + 98 = 147$

31. a. Evaluating: $(f \circ g)(5) = f(g(5)) = f(5 + 4) = f(9) = 9^2 = 81$
    b. Evaluating: $(g \circ f)(5) = g(f(5)) = g(5^2) = g(25) = 25 + 4 = 29$
    c. Evaluating: $(f \circ g)(x) = f(g(x)) = f(x + 4) = (x + 4)^2$
    d. Evaluating: $(g \circ f)(x) = g(f(x)) = g(x^2) = x^2 + 4$

33. a. Evaluating: $(f \circ g)(0) = f(g(0)) = f(4 \cdot 0 - 1) = f(-1) = (-1)^2 + 3(-1) = 1 - 3 = -2$
    b. Evaluating: $(g \circ f)(0) = g(f(0)) = g(0^2 + 3 \cdot 0) = g(0) = 4(0) - 1 = -1$
    c. Evaluating: $(g \circ f)(x) = g(f(x)) = g(x^2 + 3x) = 4(x^2 + 3x) - 1 = 4x^2 + 12x - 1$

35. Evaluating each composition:

    $(f \circ g)(x) = f(g(x)) = f\left(\dfrac{x + 4}{5}\right) = 5\left(\dfrac{x + 4}{5}\right) - 4 = x + 4 - 4 = x$

    $(g \circ f)(x) = g(f(x)) = g(5x - 4) = \dfrac{5x - 4 + 4}{5} = \dfrac{5x}{5} = x$

    Thus $(f \circ g)(x) = (g \circ f)(x) = x$.

37. Using the graph: $f(2) + 5 = 1 + 5 = 6$
39. Using the graph: $f(-3) + g(-3) = 3 + (-1) = 2$
41. Using the graph: $(f \circ g)(0) = f(g(0)) = f(-3) = 3$
43. Using the graph: $f(x) = -3$ when $x = -8$
45. Using the graph: $f(-3) + 2 = 4 + 2 = 6$
47. Using the graph: $f(2) + g(2) = 3 + 2 = 5$
49. Using the graph: $(f \circ g)(0) = f(g(0)) = f(2) = 3$
51. Using the graph: $f(x) = 1$ when $x = -6$

53.
a. Finding the revenue: $R(x) = x(11.5 - 0.05x) = 11.5x - 0.05x^2$
b. Finding the cost: $C(x) = 2x + 200$
c. Finding the profit: $P(x) = R(x) - C(x) = (11.5x - 0.05x^2) - (2x + 200) = -0.05x^2 + 9.5x - 200$
d. Finding the average cost: $\bar{C}(x) = \dfrac{C(x)}{x} = \dfrac{2x + 200}{x} = 2 + \dfrac{200}{x}$

55.
a. The function is $M(x) = 220 - x$.
b. Evaluating: $M(24) = 220 - 24 = 196$ beats per minute
c. The training heart rate function is: $T(M) = 62 + 0.6(M - 62) = 0.6M + 24.8$
Finding the composition: $T(M(x)) = T(220 - x) = 0.6(220 - x) + 24.8 = 156.8 - 0.6x$
Evaluating: $T(M(24)) = 156.8 - 0.6(24) \approx 142$ beats per minute
d. Evaluating: $T(M(36)) = 156.8 - 0.6(36) \approx 135$ beats per minute
e. Evaluating: $T(M(48)) = 156.8 - 0.6(48) \approx 128$ beats per minute

57. Solving the inequality:
$|x - 3| < 1$
$-1 < x - 3 < 1$
$2 < x < 4$

59. Solving the inequality:
$|6 - x| > 2$
$6 - x < -2$ or $6 - x > 2$
$-x < -8$ $-x > -4$
$x > 8$ $x < 4$

61. Solving the inequality:
$|7x - 1| \le 6$
$-6 \le 7x - 1 \le 6$
$-5 \le 7x \le 7$
$-\dfrac{5}{7} \le x \le 1$

63. Simplifying: $16(3.5)^2 = 16(12.25) = 196$

65. Simplifying: $\dfrac{180}{45} = 4$

67. Simplifying: $\dfrac{0.0005(200)}{(0.25)^2} = \dfrac{0.1}{0.0625} = 1.6$

69. Solving for K:
$15 = K(5)$
$K = 3$

71. Solving for K:
$50 = \dfrac{K}{48}$
$K = 50 \cdot 48 = 2{,}400$

8.6 Variation

1. Direct variation
3. Direct variation
5. Direct variation
7. Direct variation
9. Direct variation
11. Inverse variation
13. The variation equation is $y = Kx$. Substituting $x = 2$ and $y = 10$:
$10 = K \cdot 2$
$K = 5$
So $y = 5x$. Substituting $x = 6$: $y = 5 \cdot 6 = 30$

15. The variation equation is $y = Kx$. Substituting $x = 4$ and $y = -32$:
$-32 = K \cdot 4$
$K = -8$
So $y = -8x$. Substituting $y = -40$:
$-40 = -8x$
$x = 5$

250 Chapter 8 Equations and Functions

17. The variation equation is $r = \dfrac{K}{s}$. Substituting $s = 4$ and $r = -3$:

    $-3 = \dfrac{K}{4}$
    $K = -12$

    So $r = \dfrac{-12}{s}$. Substituting $s = 2$: $r = \dfrac{-12}{2} = -6$

19. The variation equation is $r = \dfrac{K}{s}$. Substituting $s = 3$ and $r = 8$:

    $8 = \dfrac{K}{3}$
    $K = 24$

    So $r = \dfrac{24}{s}$. Substituting $r = 48$:

    $48 = \dfrac{24}{s}$
    $48s = 24$
    $s = \dfrac{1}{2}$

21. The variation equation is $d = Kr^2$. Substituting $r = 5$ and $d = 10$:

    $10 = K \cdot 5^2$
    $10 = 25K$
    $K = \dfrac{2}{5}$

    So $d = \dfrac{2}{5}r^2$. Substituting $r = 10$: $d = \dfrac{2}{5}(10)^2 = \dfrac{2}{5} \cdot 100 = 40$

23. The variation equation is $d = Kr^2$. Substituting $r = 2$ and $d = 100$:

    $100 = K \cdot 2^2$
    $100 = 4K$
    $K = 25$

    So $d = 25r^2$. Substituting $r = 3$: $d = 25(3)^2 = 25 \cdot 9 = 225$

25. The variation equation is $y = \dfrac{K}{|x|}$. Substituting $x = 3$ and $y = 6$:

    $6 = \dfrac{K}{|3|}$
    $6 = \dfrac{K}{3}$
    $K = 18$

    So $y = \dfrac{18}{|x|}$. Substituting $x = 9$: $y = \dfrac{18}{|9|} = \dfrac{18}{9} = 2$

27. The variation equation is $y = \dfrac{K}{|x|}$. Substituting $x = -5$ and $y = 20$:

    $20 = \dfrac{K}{|-5|}$
    $20 = \dfrac{K}{5}$
    $K = 100$

    So $y = \dfrac{100}{|x|}$. Substituting $x = 10$: $y = \dfrac{100}{|10|} = \dfrac{100}{10} = 10$

29. The variation equation is $y = \dfrac{K}{x^2}$. Substituting $x = 3$ and $y = 45$:

$$45 = \dfrac{K}{3^2}$$
$$45 = \dfrac{K}{9}$$
$$K = 405$$

So $y = \dfrac{405}{x^2}$. Substituting $x = 5$: $y = \dfrac{405}{5^2} = \dfrac{405}{25} = \dfrac{81}{5}$

31. The variation equation is $y = \dfrac{K}{x^2}$. Substituting $x = 3$ and $y = 18$:

$$18 = \dfrac{K}{3^2}$$
$$18 = \dfrac{K}{9}$$
$$K = 162$$

So $y = \dfrac{162}{x^2}$. Substituting $x = 2$: $y = \dfrac{162}{2^2} = \dfrac{162}{4} = 40.5$

33. The variation equation is $z = Kxy^2$. Substituting $x = 3$, $y = 3$, and $z = 54$:

$$54 = K(3)(3)^2$$
$$54 = 27K$$
$$K = 2$$

So $z = 2xy^2$. Substituting $x = 2$ and $y = 4$: $z = 2(2)(4)^2 = 64$

35. The variation equation is $z = Kxy^2$. Substituting $x = 1$, $y = 4$, and $z = 64$:

$$64 = K(1)(4)^2$$
$$64 = 16K$$
$$K = 4$$

So $z = 4xy^2$. Substituting $z = 32$ and $y = 1$:

$$32 = 4x(1)^2$$
$$32 = 4x$$
$$x = 8$$

37. Let l represent the length and f represent the force. The variation equation is $l = Kf$. Substituting $f = 5$ and $l = 3$:

$$3 = K \cdot 5$$
$$K = \dfrac{3}{5}$$

So $l = \dfrac{3}{5}f$. Substituting $l = 10$:

$$10 = \dfrac{3}{5}f$$
$$50 = 3f$$
$$f = \dfrac{50}{3}$$

The force required is $\dfrac{50}{3}$ pounds.

39. a. The variation equation is $T = 4P$.
    b. Graphing the equation:

    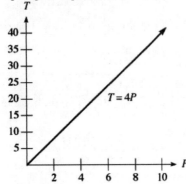

    c. Substituting $T = 280$:
    $$280 = 4P$$
    $$P = 70$$
    The pressure is 70 pounds per square inch.

41. Let v represent the volume and p represent the pressure. The variation equation is $v = \dfrac{K}{p}$.

    Substituting $p = 36$ and $v = 25$:
    $$25 = \dfrac{K}{36}$$
    $$K = 900$$

    The equation is $v = \dfrac{900}{p}$. Substituting $v = 75$:
    $$75 = \dfrac{900}{p}$$
    $$75p = 900$$
    $$p = 12$$
    The pressure is 12 pounds per square inch.

43. a. The variation equation is $f = \dfrac{80}{d}$.
    b. Graphing the equation:

    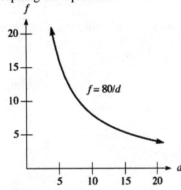

    c. Substituting $d = 10$:
    $$f = \dfrac{80}{10}$$
    $$f = 8$$
    The f-stop is 8.

45. Let A represent the surface area, h represent the height, and r represent the radius. The variation equation is $A = Khr$.
    Substituting $A = 94$, $r = 3$, and $h = 5$:
    $$94 = K(3)(5)$$
    $$94 = 15K$$
    $$K = \tfrac{94}{15}$$
    The equation is $A = \tfrac{94}{15}hr$. Substituting $r = 2$ and $h = 8$: $A = \tfrac{94}{15}(8)(2) = \tfrac{1504}{15}$. The surface area is $\tfrac{1504}{15}$ square inches

47. Let R represent the resistance, l represent the length, and d represent the diameter. The variation equation is $R = \dfrac{Kl}{d^2}$.
    Substituting $R = 10$, $l = 100$, and $d = 0.01$:
    $$10 = \frac{K(100)}{(0.01)^2}$$
    $$0.001 = 100K$$
    $$K = 0.00001$$
    The equation is $R = \dfrac{0.000001l}{d^2}$. Substituting $l = 60$ and $d = 0.02$: $R = \dfrac{0.00001(60)}{(0.02)^2} = 1.5$. The resistance is 1.5 ohms.

49. Let p represent the pitch and w represent the wavelength. The variation equation is $p = \dfrac{K}{w}$.
    Substituting $p = 420$ and $w = 2.2$:
    $$420 = \frac{K}{2.2}$$
    $$K = 924$$
    The equation is $p = \dfrac{924}{w}$. Substituting $p = 720$:
    $$720 = \frac{924}{w}$$
    $$720w = 924$$
    $$w \approx 1.28$$
    The wavelength is approximately 1.28 meters.

51. The variation equation is $F = \dfrac{Gm_1 m_2}{d^2}$.

53. Solving the equation:
    $$x - 5 = 7$$
    $$x = 12$$

55. Solving the equation:
    $$5 - \tfrac{4}{7}a = -11$$
    $$7\left(5 - \tfrac{4}{7}a\right) = 7(-11)$$
    $$35 - 4a = -77$$
    $$-4a = -112$$
    $$a = 28$$

57. Solving the equation:
    $$5(x-1) - 2(2x+3) = 5x - 4$$
    $$5x - 5 - 4x - 6 = 5x - 4$$
    $$x - 11 = 5x - 4$$
    $$-4x = 7$$
    $$x = -\tfrac{7}{4}$$

59. Solving for w:
    $$P = 2l + 2w$$
    $$P - 2l = 2w$$
    $$w = \frac{P - 2l}{2}$$

254 Chapter 8 Equations and Functions

61. Solving the inequality:

$$-5t \le 30$$
$$t \ge -6$$

The solution set is $[-6, \infty)$. Graphing the solution set:

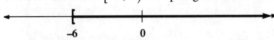

63. Solving the inequality:
$$1.6x - 2 < 0.8x + 2.8$$
$$0.8x - 2 < 2.8$$
$$0.8x < 4.8$$
$$x < 6$$

The solution set is $(-\infty, 6)$. Graphing the solution set:

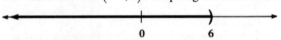

65. Solving the equation:
$$\left|\tfrac{1}{4}x - 1\right| = \tfrac{1}{2}$$
$$\tfrac{1}{4}x - 1 = -\tfrac{1}{2}, \tfrac{1}{2}$$
$$\tfrac{1}{4}x = \tfrac{1}{2}, \tfrac{3}{2}$$
$$x = 2, 6$$

67. Solving the equation:

$$|3 - 2x| + 5 = 2$$
$$|3 - 2x| = -3$$

Since this statement is false, there is no solution, or $\emptyset$.

69. Solving the inequality:

$$\left|\tfrac{x}{5} + 1\right| \ge \tfrac{4}{5}$$

$$\tfrac{x}{5} + 1 \le -\tfrac{4}{5} \qquad \text{or} \qquad \tfrac{x}{5} + 1 \ge \tfrac{4}{5}$$
$$x + 5 \le -4 \qquad\qquad\qquad x + 5 \ge 4$$
$$x \le -9 \qquad\qquad\qquad\quad x \ge -1$$

Graphing the solution set:

71. Since $|3 - 4t| > -5$ is always true, the solution set is all real numbers. Graphing the solution set:

73. **a.** The distance appears to vary directly with the square of the speed.

  b. Let $d = Ks^2$. Substituting $d = 108$ and $s = 40$:
  $$108 = K(40)^2$$
  $$1600K = 108$$
  $$K = 0.0675$$
  So the equation is $d = 0.0675s^2$.

  c. Substituting $s = 55$: $d = 0.0675(55)^2 \approx 204.2$ feet away

  d. Substituting $s = 56$: $d = 0.0675(56)^2 \approx 211.7$ feet. This is approximately 7.5 feet further away.

Chapter 8 Review/Test

1. Graphing the line:

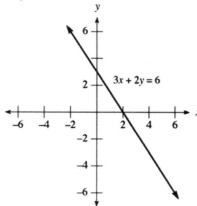

2. Graphing the line:

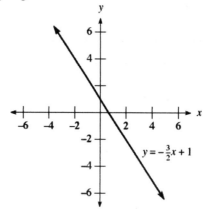

3. Graphing the line:

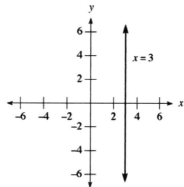

4. Finding the slope: $m = \dfrac{6-2}{3-5} = \dfrac{4}{-2} = -2$

5. Finding the slope: $m = \dfrac{2-2}{3-(-4)} = \dfrac{0}{7} = 0$

6. Solving for x:
$$\dfrac{-3-x}{1-4} = 2$$
$$\dfrac{-3-x}{-3} = 2$$
$$-3-x = -6$$
$$-x = -3$$
$$x = 3$$

7. Solving for x:
$$\dfrac{x-7}{2+4} = -\dfrac{1}{3}$$
$$\dfrac{x-7}{6} = -\dfrac{1}{3}$$
$$x-7 = -2$$
$$x = 5$$

8. Finding the slope: $m = \dfrac{-2-8}{5-3} = \dfrac{-10}{2} = -5$

9. Solving for y:
$$\dfrac{y-3y}{2-5} = 4$$
$$\dfrac{-2y}{-3} = 4$$
$$-2y = -12$$
$$y = 6$$

10. Using the slope-intercept formula, the slope is $y = 3x + 5$.
11. Using the slope-intercept formula, the slope is $y = -2x$.

256 Chapter 8 Equations and Functions

12. Solving for y:
$$3x - y = 6$$
$$-y = -3x + 6$$
$$y = 3x - 6$$

The slope is $m = 3$ and the y-intercept is $b = -6$.

13. Solving for y:
$$2x - 3y = 9$$
$$-3y = -2x + 9$$
$$y = \tfrac{2}{3}x - 3$$

The slope is $m = \tfrac{2}{3}$ and the y-intercept is $b = -3$.

14. Using the point-slope formula:
$$y - 4 = 2(x - 2)$$
$$y - 4 = 2x - 4$$
$$y = 2x$$

15. Using the point-slope formula:
$$y - 1 = -\tfrac{1}{3}(x + 3)$$
$$y - 1 = -\tfrac{1}{3}x - 1$$
$$y = -\tfrac{1}{3}x$$

16. First find the slope: $m = \dfrac{-5 - 5}{-3 - 2} = \dfrac{-10}{-5} = 2$. Now using the point-slope formula:
$$y - 5 = 2(x - 2)$$
$$y - 5 = 2x - 4$$
$$y = 2x + 1$$

17. First find the slope: $m = \dfrac{7 - 7}{4 - (-3)} = \dfrac{0}{7} = 0$. Since the line is horizontal, its equation is $y = 7$.

18. First find the slope: $m = \dfrac{-4 - (-1)}{-3 - (-5)} = \dfrac{-4 + 1}{-3 + 5} = -\tfrac{3}{2}$. Now using the point-slope formula:
$$y + 1 = -\tfrac{3}{2}(x + 5)$$
$$y + 1 = -\tfrac{3}{2}x - \tfrac{15}{2}$$
$$y = -\tfrac{3}{2}x - \tfrac{17}{2}$$

19. First find the slope by solving for y:
$$2x - y = 4$$
$$-y = -2x + 4$$
$$y = 2x - 4$$

The parallel slope is also $m = 2$. Now using the point-slope formula:
$$y + 3 = 2(x - 2)$$
$$y + 3 = 2x - 4$$
$$y = 2x - 7$$

20. The perpendicular slope is $m = \tfrac{1}{3}$. Using the point-slope formula:
$$y - 0 = \tfrac{1}{3}(x - 2)$$
$$y = \tfrac{1}{3}x - \tfrac{2}{3}$$

21. The domain is $\{2, 3, 4\}$ and the range is $\{2, 3, 4\}$. This is a function.
22. The domain is $\{-4, -2, 6\}$ and the range is $\{0, 3\}$. This is a function.
23. $f(\pi) = 2$
24. $f(-3) + g(-1) = 0 + 4 = 4$
25. $f(-3) = 0$
26. $f(2) + g(2) = -1 + 2 = 1$
27. Evaluating the function: $f(0) = 2(0)^2 - 4(0) + 1 = 0 - 0 + 1 = 1$
28. Evaluating the function: $g(a) = 3a + 2$
29. Evaluating the function: $f[g(0)] = f[3(0) + 2] = f(2) = 2(2)^2 - 4(2) + 1 = 8 - 8 + 1 = 1$
30. Evaluating the function: $f[g(1)] = f[3(1) + 2] = f(5) = 2(5)^2 - 4(5) + 1 = 50 - 20 + 1 = 31$

31. The variation equation is $y = Kx$. Substituting $x = 2$ and $y = 6$:
    $6 = K \cdot 2$
    $K = 3$
    The equation is $y = 3x$. Substituting $x = 8$: $y = 3 \cdot 8 = 24$

32. The variation equation is $y = Kx$. Substituting $x = 5$ and $y = -3$:
    $-3 = K \cdot 5$
    $K = -\frac{3}{5}$
    The equation is $y = -\frac{3}{5}x$. Substituting $x = -10$: $y = -\frac{3}{5}(-10) = 6$

33. The variation equation is $y = \dfrac{K}{x^2}$. Substituting $x = 2$ and $y = 9$:
    $9 = \dfrac{K}{2^2}$
    $9 = \dfrac{K}{4}$
    $K = 36$
    The equation is $y = \dfrac{36}{x^2}$. Substituting $x = 3$: $y = \dfrac{36}{3^2} = \dfrac{36}{9} = 4$

34. The variation equation is $y = \dfrac{K}{x^2}$. Substituting $x = 5$ and $y = 4$:
    $4 = \dfrac{K}{5^2}$
    $4 = \dfrac{K}{25}$
    $K = 100$
    The equation is $y = \dfrac{100}{x^2}$. Substituting $x = 2$: $y = \dfrac{100}{2^2} = \dfrac{100}{4} = 25$

35. The variation equation is $t = Kd$. Substituting $t = 42$ and $d = 2$:
    $42 = K \cdot 2$
    $K = 21$
    The equation is $t = 21d$. Substituting $d = 4$: $t = 21 \cdot 4 = 84$. The tension is 84 pounds.

36. Let I represent the intensity and d represent the distance. The variation equation is $I = \dfrac{K}{d^2}$.
    Substituting $I = 9$ and $d = 4$:
    $9 = \dfrac{K}{4^2}$
    $9 = \dfrac{K}{16}$
    $K = 144$
    The equation is $I = \dfrac{144}{d^2}$. Substituting $d = 3$: $I = \dfrac{144}{3^2} = \dfrac{144}{9} = 16$. The intensity is 16 foot-candles.

Chapter 9
Rational Exponents and Roots

9.1 Rational Exponents

1. Finding the root: $\sqrt{144} = 12$
3. Finding the root: $\sqrt{-144}$ is not a real number
5. Finding the root: $-\sqrt{49} = -7$
7. Finding the root: $\sqrt[3]{-27} = -3$
9. Finding the root: $\sqrt[4]{16} = 2$
11. Finding the root: $\sqrt[4]{-16}$ is not a real number
13. Finding the root: $\sqrt{0.04} = 0.2$
15. Finding the root: $\sqrt[3]{0.008} = 0.2$
17. Finding the root: $\sqrt[3]{125} = 5$
19. Finding the root: $-\sqrt[3]{216} = -6$
21. Finding the root: $\sqrt{\frac{1}{36}} = \frac{1}{6}$
23. Finding the root: $\sqrt[3]{\frac{8}{125}} = \frac{2}{5}$
25. Simplifying: $\sqrt{36a^8} = 6a^4$
27. Simplifying: $\sqrt[3]{27a^{12}} = 3a^4$
29. Simplifying: $\sqrt[5]{32x^{10}y^5} = 2x^2y$
31. Simplifying: $\sqrt[4]{16a^{12}b^{20}} = 2a^3b^5$
33. Writing as a root and simplifying: $36^{1/2} = \sqrt{36} = 6$
35. Writing as a root and simplifying: $-9^{1/2} = -\sqrt{9} = -3$
37. Writing as a root and simplifying: $8^{1/3} = \sqrt[3]{8} = 2$
39. Writing as a root and simplifying: $(-8)^{1/3} = \sqrt[3]{-8} = -2$
41. Writing as a root and simplifying: $32^{1/5} = \sqrt[5]{32} = 2$
43. Writing as a root and simplifying: $\left(\frac{81}{25}\right)^{1/2} = \sqrt{\frac{81}{25}} = \frac{9}{5}$
45. Simplifying: $27^{2/3} = \left(27^{1/3}\right)^2 = 3^2 = 9$
47. Simplifying: $25^{3/2} = \left(25^{1/2}\right)^3 = 5^3 = 125$
49. Simplifying: $27^{-1/3} = \left(27^{1/3}\right)^{-1} = 3^{-1} = \frac{1}{3}$
51. Simplifying: $81^{-3/4} = \left(81^{1/4}\right)^{-3} = 3^{-3} = \frac{1}{3^3} = \frac{1}{27}$
53. Simplifying: $\left(\frac{25}{36}\right)^{-1/2} = \left(\frac{36}{25}\right)^{1/2} = \frac{6}{5}$
55. Simplifying: $\left(\frac{81}{16}\right)^{-3/4} = \left(\frac{16}{81}\right)^{3/4} = \left[\left(\frac{16}{81}\right)^{1/4}\right]^3 = \left(\frac{2}{3}\right)^3 = \frac{8}{27}$
57. Simplifying: $16^{1/2} + 27^{1/3} = 4 + 3 = 7$
59. Simplifying: $8^{-2/3} + 4^{-1/2} = \left(8^{1/3}\right)^{-2} + \left(4^{1/2}\right)^{-1} = 2^{-2} + 2^{-1} = \frac{1}{4} + \frac{1}{2} = \frac{3}{4}$
61. Using properties of exponents: $x^{3/5} \cdot x^{1/5} = x^{3/5+1/5} = x^{4/5}$

63. Using properties of exponents: $\left(a^{3/4}\right)^{4/3} = a^{3/4 \cdot 4/3} = a$

65. Using properties of exponents: $\dfrac{x^{1/5}}{x^{3/5}} = x^{1/5-3/5} = x^{-2/5} = \dfrac{1}{x^{2/5}}$

67. Using properties of exponents: $\dfrac{x^{5/6}}{x^{2/3}} = x^{5/6-2/3} = x^{5/6-4/6} = x^{1/6}$

69. Using properties of exponents: $\left(x^{3/5} y^{5/6} z^{1/3}\right)^{3/5} = x^{3/5 \cdot 3/5} y^{5/6 \cdot 3/5} z^{1/3 \cdot 3/5} = x^{9/25} y^{1/2} z^{1/5}$

71. Using properties of exponents: $\dfrac{a^{3/4} b^2}{a^{7/8} b^{1/4}} = a^{3/4-7/8} b^{2-1/4} = a^{6/8-7/8} b^{8/4-1/4} = a^{-1/8} b^{7/4} = \dfrac{b^{7/4}}{a^{1/8}}$

73. Using properties of exponents: $\dfrac{\left(y^{2/3}\right)^{3/4}}{\left(y^{1/3}\right)^{3/5}} = \dfrac{y^{1/2}}{y^{1/5}} = y^{1/2-1/5} = y^{5/10-2/10} = y^{3/10}$

75. Using properties of exponents: $\left(\dfrac{a^{-1/4}}{b^{1/2}}\right)^8 = \dfrac{a^{-1/4 \cdot 8}}{b^{1/2 \cdot 8}} = \dfrac{a^{-2}}{b^4} = \dfrac{1}{a^2 b^4}$

77. Using properties of exponents: $\dfrac{\left(r^{-2} s^{1/3}\right)^6}{r^8 s^{3/2}} = \dfrac{r^{-12} s^2}{r^8 s^{3/2}} = r^{-12-8} s^{2-3/2} = r^{-20} s^{1/2} = \dfrac{s^{1/2}}{r^{20}}$

79. Using properties of exponents: $\dfrac{\left(25 a^6 b^4\right)^{1/2}}{\left(8 a^{-9} b^3\right)^{-1/3}} = \dfrac{25^{1/2} a^3 b^2}{8^{-1/3} a^3 b^{-1}} = \dfrac{5}{1/2} a^{3-3} b^{2+1} = 10 b^3$

81. Substituting $r = 250$: $v = \left(\dfrac{5 \cdot 250}{2}\right)^{1/2} = 625^{1/2} = 25$. The maximum speed is 25 mph.

83. a. The length of the side is $126 + 86 + 86 + 126 = 424$ pm
    b. Let d represent the diagonal. Using the Pythagorean theorem:
    $$d^2 = 424^2 + 424^2 = 359,552$$
    $$d = \sqrt{359,552} \approx 600 \text{ pm}$$
    c. Converting to meters: $600 \text{ pm} \cdot \dfrac{1 \text{ m}}{10^{12} \text{ pm}} = 6 \times 10^{-10}$ m

85. a. This graph is B.   b. This graph is A.
    c. This graph is C.   d. The points of intersection are (0,0) and (1,1).

87. Multiplying: $x^2 \left(x^4 - x\right) = x^2 \cdot x^4 - x^2 \cdot x = x^6 - x^3$

89. Multiplying: $(x-3)(x+5) = x^2 + 5x - 3x - 15 = x^2 + 2x - 15$

91. Multiplying: $\left(x^2 - 5\right)^2 = \left(x^2\right)^2 - 2\left(x^2\right)(5) + 5^2 = x^4 - 10x^2 + 25$

93. Multiplying: $(x-3)\left(x^2 + 3x + 9\right) = x^3 + 3x^2 + 9x - 3x^2 - 9x - 27 = x^3 - 27$

95. Simplifying: $x^2 \left(x^4 - x^3\right) = x^2 \cdot x^4 - x^2 \cdot x^3 = x^6 - x^5$

97. Simplifying: $(3a - 2b)(4a - b) = 12a^2 - 3ab - 8ab + 2b^2 = 12a^2 - 11ab + 2b^2$

99. Simplifying: $\left(x^3 - 2\right)\left(x^3 + 2\right) = \left(x^3\right)^2 - (2)^2 = x^6 - 4$

101. Simplifying: $\dfrac{15x^2 y - 20x^4 y^2}{5xy} = \dfrac{15x^2 y}{5xy} - \dfrac{20x^4 y^2}{5xy} = 3x - 4x^3 y$

103. Factoring: $x^2 - 3x - 10 = (x-5)(x+2)$

105. Factoring: $6x^2 + 11x - 10 = (3x-2)(2x+5)$

107. Simplifying: $x^{2/3} \cdot x^{4/3} = x^{2/3 + 4/3} = x^2$

109. Simplifying: $\left(t^{1/2}\right)^2 = t^{1/2 \cdot 2} = t^1 = t$

111. Simplifying: $\dfrac{x^{2/3}}{x^{1/3}} = x^{2/3 - 1/3} = x^{1/3}$

113. Simplifying each expression:
$\left(9^{1/2} + 4^{1/2}\right)^2 = (3+2)^2 = 5^2 = 25$
$9 + 4 = 13$
Note that the values are not equal.

115. Rewriting with exponents: $\sqrt{\sqrt{a}} = \sqrt{a^{1/2}} = \left(a^{1/2}\right)^{1/2} = a^{1/4} = \sqrt[4]{a}$

9.2 More Expressions Involving Rational Exponents

1. Multiplying: $x^{2/3}\left(x^{1/3} + x^{4/3}\right) = x^{2/3} \cdot x^{1/3} + x^{2/3} \cdot x^{4/3} = x + x^2$

3. Multiplying: $a^{1/2}\left(a^{3/2} - a^{1/2}\right) = a^{1/2} \cdot a^{3/2} - a^{1/2} \cdot a^{1/2} = a^2 - a$

5. Multiplying: $2x^{1/3}\left(3x^{8/3} - 4x^{5/3} + 5x^{2/3}\right) = 2x^{1/3} \cdot 3x^{8/3} - 2x^{1/3} \cdot 4x^{5/3} + 2x^{1/3} \cdot 5x^{2/3} = 6x^3 - 8x^2 + 10x$

7. Multiplying:
$4x^{1/2}y^{3/5}\left(3x^{3/2}y^{-3/5} - 9x^{-1/2}y^{7/5}\right) = 4x^{1/2}y^{3/5} \cdot 3x^{3/2}y^{-3/5} - 4x^{1/2}y^{3/5} \cdot 9x^{-1/2}y^{7/5} = 12x^2 - 36y^2$

9. Multiplying: $\left(x^{2/3} - 4\right)\left(x^{2/3} + 2\right) = x^{2/3} \cdot x^{2/3} + 2x^{2/3} - 4x^{2/3} - 8 = x^{4/3} - 2x^{2/3} - 8$

11. Multiplying: $\left(a^{1/2} - 3\right)\left(a^{1/2} - 7\right) = a^{1/2} \cdot a^{1/2} - 7a^{1/2} - 3a^{1/2} + 21 = a - 10a^{1/2} + 21$

13. Multiplying: $\left(4y^{1/3} - 3\right)\left(5y^{1/3} + 2\right) = 20y^{2/3} + 8y^{1/3} - 15y^{1/3} - 6 = 20y^{2/3} - 7y^{1/3} - 6$

15. Multiplying: $\left(5x^{2/3} + 3y^{1/2}\right)\left(2x^{2/3} + 3y^{1/2}\right) = 10x^{4/3} + 15x^{2/3}y^{1/2} + 6x^{2/3}y^{1/2} + 9y = 10x^{4/3} + 21x^{2/3}y^{1/2} + 9y$

17. Multiplying: $\left(t^{1/2} + 5\right)^2 = \left(t^{1/2} + 5\right)\left(t^{1/2} + 5\right) = t + 5t^{1/2} + 5t^{1/2} + 25 = t + 10t^{1/2} + 25$

19. Multiplying: $\left(x^{3/2} + 4\right)^2 = \left(x^{3/2} + 4\right)\left(x^{3/2} + 4\right) = x^3 + 4x^{3/2} + 4x^{3/2} + 16 = x^3 + 8x^{3/2} + 16$

21. Multiplying: $\left(a^{1/2} - b^{1/2}\right)^2 = \left(a^{1/2} - b^{1/2}\right)\left(a^{1/2} - b^{1/2}\right) = a - a^{1/2}b^{1/2} - a^{1/2}b^{1/2} + b = a - 2a^{1/2}b^{1/2} + b$

23. Multiplying:
$\left(2x^{1/2} - 3y^{1/2}\right)^2 = \left(2x^{1/2} - 3y^{1/2}\right)\left(2x^{1/2} - 3y^{1/2}\right)$
$= 4x - 6x^{1/2}y^{1/2} - 6x^{1/2}y^{1/2} + 9y$
$= 4x - 12x^{1/2}y^{1/2} + 9y$

25. Multiplying: $\left(a^{1/2} - 3^{1/2}\right)\left(a^{1/2} + 3^{1/2}\right) = \left(a^{1/2}\right)^2 - \left(3^{1/2}\right)^2 = a - 3$

27. Multiplying: $\left(x^{3/2} + y^{3/2}\right)\left(x^{3/2} - y^{3/2}\right) = \left(x^{3/2}\right)^2 - \left(y^{3/2}\right)^2 = x^3 - y^3$

29. Multiplying: $\left(t^{1/2} - 2^{3/2}\right)\left(t^{1/2} + 2^{3/2}\right) = \left(t^{1/2}\right)^2 - \left(2^{3/2}\right)^2 = t - 2^3 = t - 8$

31. Multiplying: $\left(2x^{3/2} + 3^{1/2}\right)\left(2x^{3/2} - 3^{1/2}\right) = \left(2x^{3/2}\right)^2 - \left(3^{1/2}\right)^2 = 4x^3 - 3$

33. Multiplying: $\left(x^{1/3} + y^{1/3}\right)\left(x^{2/3} - x^{1/3}y^{1/3} + y^{2/3}\right) = \left(x^{1/3}\right)^3 + \left(y^{1/3}\right)^3 = x + y$

35. Multiplying: $(a^{1/3} - 2)(a^{2/3} + 2a^{1/3} + 4) = (a^{1/3})^3 - (2)^3 = a - 8$

37. Multiplying: $(2x^{1/3} + 1)(4x^{2/3} - 2x^{1/3} + 1) = (2x^{1/3})^3 + (1)^3 = 8x + 1$

39. Multiplying: $(t^{1/4} - 1)(t^{1/4} + 1)(t^{1/2} + 1) = (t^{1/2} - 1)(t^{1/2} + 1) = t - 1$

41. Dividing: $\dfrac{18x^{3/4} + 27x^{1/4}}{9x^{1/4}} = \dfrac{18x^{3/4}}{9x^{1/4}} + \dfrac{27x^{1/4}}{9x^{1/4}} = 2x^{1/2} + 3$

43. Dividing: $\dfrac{12x^{2/3}y^{1/3} - 16x^{1/3}y^{2/3}}{4x^{1/3}y^{1/3}} = \dfrac{12x^{2/3}y^{1/3}}{4x^{1/3}y^{1/3}} - \dfrac{16x^{1/3}y^{2/3}}{4x^{1/3}y^{1/3}} = 3x^{1/3} - 4y^{1/3}$

45. Dividing: $\dfrac{21a^{7/5}b^{3/5} - 14a^{2/5}b^{8/5}}{7a^{2/5}b^{3/5}} = \dfrac{21a^{7/5}b^{3/5}}{7a^{2/5}b^{3/5}} - \dfrac{14a^{2/5}b^{8/5}}{7a^{2/5}b^{3/5}} = 3a - 2b$

47. Factoring: $12(x-2)^{3/2} - 9(x-2)^{1/2} = 3(x-2)^{1/2}[4(x-2) - 3] = 3(x-2)^{1/2}(4x - 8 - 3) = 3(x-2)^{1/2}(4x - 11)$

49. Factoring: $5(x-3)^{12/5} - 15(x-3)^{7/5} = 5(x-3)^{7/5}[(x-3) - 3] = 5(x-3)^{7/5}(x - 6)$

51. Factoring: $9x(x+1)^{3/2} + 6(x+1)^{1/2} = 3(x+1)^{1/2}[3x(x+1) + 2] = 3(x+1)^{1/2}(3x^2 + 3x + 2)$

53. Factoring: $x^{2/3} - 5x^{1/3} + 6 = (x^{1/3} - 2)(x^{1/3} - 3)$

55. Factoring: $a^{2/5} - 2a^{1/5} - 8 = (a^{1/5} - 4)(a^{1/5} + 2)$

57. Factoring: $2y^{2/3} - 5y^{1/3} - 3 = (2y^{1/3} + 1)(y^{1/3} - 3)$

59. Factoring: $9t^{2/5} - 25 = (3t^{1/5} + 5)(3t^{1/5} - 5)$

61. Factoring: $4x^{2/7} + 20x^{1/7} + 25 = (2x^{1/7} + 5)^2$

63. Evaluating: $f(4) = 4 - 2\sqrt{4} - 8 = 4 - 4 - 8 = -8$

65. Evaluating: $f(25) = 2(25) + 9\sqrt{25} - 5 = 50 + 45 - 5 = 90$

67. Evaluating: $g\left(\tfrac{9}{4}\right) = 2\left(\tfrac{9}{4}\right) - \sqrt{\tfrac{9}{4}} - 6 = \tfrac{9}{2} - \tfrac{3}{2} - 6 = 3 - 6 = -3$

69. Evaluating: $f(-8) = (-8)^{2/3} - 2(-8)^{1/3} - 8 = 4 + 4 - 8 = 0$

71. Writing as a single fraction: $\dfrac{3}{x^{1/2}} + x^{1/2} = \dfrac{3}{x^{1/2}} + x^{1/2} \cdot \dfrac{x^{1/2}}{x^{1/2}} = \dfrac{3 + x}{x^{1/2}}$

73. Writing as a single fraction: $x^{2/3} + \dfrac{5}{x^{1/3}} = x^{2/3} \cdot \dfrac{x^{1/3}}{x^{1/3}} + \dfrac{5}{x^{1/3}} = \dfrac{x + 5}{x^{1/3}}$

75. Writing as a single fraction:

$$\dfrac{3x^2}{(x^3+1)^{1/2}} + (x^3+1)^{1/2} = \dfrac{3x^2}{(x^3+1)^{1/2}} + (x^3+1)^{1/2} \cdot \dfrac{(x^3+1)^{1/2}}{(x^3+1)^{1/2}} = \dfrac{3x^2 + x^3 + 1}{(x^3+1)^{1/2}} = \dfrac{x^3 + 3x^2 + 1}{(x^3+1)^{1/2}}$$

77. Writing as a single fraction:

$$\frac{x^2}{(x^2+4)^{1/2}} - (x^2+4)^{1/2} = \frac{x^2}{(x^2+4)^{1/2}} - (x^2+4)^{1/2} \cdot \frac{(x^2+4)^{1/2}}{(x^2+4)^{1/2}}$$

$$= \frac{x^2 - (x^2+4)}{(x^2+4)^{1/2}}$$

$$= \frac{x^2 - x^2 - 4}{(x^2+4)^{1/2}}$$

$$= \frac{-4}{(x^2+4)^{1/2}}$$

79. Using the formula $r = \left(\frac{A}{P}\right)^{1/t} - 1$ to find the annual rate of return: $r = \left(\frac{900}{500}\right)^{1/4} - 1 \approx 0.158$

The annual rate of return is approximately 15.8%.

81. Finding the slope: $m = \frac{5-(-1)}{-2-(-4)} = \frac{5+1}{-2+4} = \frac{6}{2} = 3$

83. Solving for y:
$2x - 3y = 6$
$-3y = -2x + 6$
$y = \frac{2}{3}x - 2$

The slope is $m = \frac{2}{3}$ and the y-intercept is $b = -2$.

85. Using the point-slope formula:
$y - 2 = \frac{2}{3}(x + 6)$
$y - 2 = \frac{2}{3}x + 4$
$y = \frac{2}{3}x + 6$

87. First find the slope: $m = \frac{-2-0}{0-3} = \frac{2}{3}$. The slope-intercept form is $y = \frac{2}{3}x - 2$.

89. Simplifying: $\sqrt{25} = 5$

91. Simplifying: $\sqrt{6^2} = 6$

93. Simplifying: $\sqrt{16x^4 y^2} = 4x^2 y$

95. Simplifying: $\sqrt{(5y)^2} = 5y$

97. Simplifying: $\sqrt[3]{27} = 3$

99. Simplifying: $\sqrt[3]{2^3} = 2$

101. Simplifying: $\sqrt[3]{8a^3 b^3} = 2ab$

103. Filling in the blank: $50 = 25 \cdot 2$

105. Filling in the blank: $48x^4 y^3 = 48x^4 y^2 \cdot y$

107. Filling in the blank: $12x^7 y^6 = 4x^6 y^6 \cdot 3x$

9.3 Simplified Form for Radicals

1. Simplifying the radical: $\sqrt{8} = \sqrt{4 \cdot 2} = 2\sqrt{2}$
3. Simplifying the radical: $\sqrt{98} = \sqrt{49 \cdot 2} = 7\sqrt{2}$
5. Simplifying the radical: $\sqrt{288} = \sqrt{144 \cdot 2} = 12\sqrt{2}$
7. Simplifying the radical: $\sqrt{80} = \sqrt{16 \cdot 5} = 4\sqrt{5}$
9. Simplifying the radical: $\sqrt{48} = \sqrt{16 \cdot 3} = 4\sqrt{3}$
11. Simplifying the radical: $\sqrt{675} = \sqrt{225 \cdot 3} = 15\sqrt{3}$
13. Simplifying the radical: $\sqrt[3]{54} = \sqrt[3]{27 \cdot 2} = 3\sqrt[3]{2}$
15. Simplifying the radical: $\sqrt[3]{128} = \sqrt[3]{64 \cdot 2} = 4\sqrt[3]{2}$
17. Simplifying the radical: $\sqrt[3]{432} = \sqrt[3]{216 \cdot 2} = 6\sqrt[3]{2}$
19. Simplifying the radical: $\sqrt[5]{64} = \sqrt[5]{32 \cdot 2} = 2\sqrt[5]{2}$
21. Simplifying the radical: $\sqrt{18x^3} = \sqrt{9x^2 \cdot 2x} = 3x\sqrt{2x}$
23. Simplifying the radical: $\sqrt[4]{32y^7} = \sqrt[4]{16y^4 \cdot 2y^3} = 2y\sqrt[4]{2y^3}$
25. Simplifying the radical: $\sqrt[3]{40x^4y^7} = \sqrt[3]{8x^3y^6 \cdot 5xy} = 2xy^2\sqrt[3]{5xy}$
27. Simplifying the radical: $\sqrt{48a^2b^3c^4} = \sqrt{16a^2b^2c^4 \cdot 3b} = 4abc^2\sqrt{3b}$
29. Simplifying the radical: $\sqrt[3]{48a^2b^3c^4} = \sqrt[3]{8b^3c^3 \cdot 6a^2c} = 2bc\sqrt[3]{6a^2c}$
31. Simplifying the radical: $\sqrt[5]{64x^8y^{12}} = \sqrt[5]{32x^5y^{10} \cdot 2x^3y^2} = 2xy^2\sqrt[5]{2x^3y^2}$
33. Simplifying the radical: $\sqrt[5]{243x^7y^{10}z^5} = \sqrt[5]{243x^5y^{10}z^5 \cdot x^2} = 3xy^2z\sqrt[5]{x^2}$
35. Substituting into the expression: $\sqrt{b^2 - 4ac} = \sqrt{(-6)^2 - 4(2)(3)} = \sqrt{36 - 24} = \sqrt{12} = 2\sqrt{3}$
37. Substituting into the expression: $\sqrt{b^2 - 4ac} = \sqrt{(2)^2 - 4(1)(6)} = \sqrt{4 - 24} = \sqrt{-20}$, which is not a real number
39. Substituting into the expression: $\sqrt{b^2 - 4ac} = \sqrt{\left(-\frac{1}{2}\right)^2 - 4\left(\frac{1}{2}\right)\left(-\frac{5}{4}\right)} = \sqrt{\frac{1}{4} + \frac{5}{2}} = \sqrt{\frac{11}{4}} = \frac{\sqrt{11}}{2}$
41. Rationalizing the denominator: $\frac{2}{\sqrt{3}} = \frac{2}{\sqrt{3}} \cdot \frac{\sqrt{3}}{\sqrt{3}} = \frac{2\sqrt{3}}{3}$
43. Rationalizing the denominator: $\frac{5}{\sqrt{6}} = \frac{5}{\sqrt{6}} \cdot \frac{\sqrt{6}}{\sqrt{6}} = \frac{5\sqrt{6}}{6}$
45. Rationalizing the denominator: $\sqrt{\frac{1}{2}} = \frac{1}{\sqrt{2}} \cdot \frac{\sqrt{2}}{\sqrt{2}} = \frac{\sqrt{2}}{2}$
47. Rationalizing the denominator: $\sqrt{\frac{1}{5}} = \frac{1}{\sqrt{5}} \cdot \frac{\sqrt{5}}{\sqrt{5}} = \frac{\sqrt{5}}{5}$
49. Rationalizing the denominator: $\frac{4}{\sqrt[3]{2}} = \frac{4}{\sqrt[3]{2}} \cdot \frac{\sqrt[3]{4}}{\sqrt[3]{4}} = \frac{4\sqrt[3]{4}}{2} = 2\sqrt[3]{4}$
51. Rationalizing the denominator: $\frac{2}{\sqrt[3]{9}} = \frac{2}{\sqrt[3]{9}} \cdot \frac{\sqrt[3]{3}}{\sqrt[3]{3}} = \frac{2\sqrt[3]{3}}{3}$
53. Rationalizing the denominator: $\sqrt[4]{\frac{3}{2x^2}} = \frac{\sqrt[4]{3}}{\sqrt[4]{2x^2}} \cdot \frac{\sqrt[4]{8x^2}}{\sqrt[4]{8x^2}} = \frac{\sqrt[4]{24x^2}}{2x}$
55. Rationalizing the denominator: $\sqrt[4]{\frac{8}{y}} = \frac{\sqrt[4]{8}}{\sqrt[4]{y}} \cdot \frac{\sqrt[4]{y^3}}{\sqrt[4]{y^3}} = \frac{\sqrt[4]{8y^3}}{y}$
57. Rationalizing the denominator: $\sqrt[3]{\frac{4x}{3y}} = \frac{\sqrt[3]{4x}}{\sqrt[3]{3y}} \cdot \frac{\sqrt[3]{9y^2}}{\sqrt[3]{9y^2}} = \frac{\sqrt[3]{36xy^2}}{3y}$
59. Rationalizing the denominator: $\sqrt[3]{\frac{2x}{9y}} = \frac{\sqrt[3]{2x}}{\sqrt[3]{9y}} \cdot \frac{\sqrt[3]{3y^2}}{\sqrt[3]{3y^2}} = \frac{\sqrt[3]{6xy^2}}{3y}$
61. Rationalizing the denominator: $\sqrt[4]{\frac{1}{8x^3}} = \frac{1}{\sqrt[4]{8x^3}} \cdot \frac{\sqrt[4]{2x}}{\sqrt[4]{2x}} = \frac{\sqrt[4]{2x}}{2x}$

63. Simplifying: $\sqrt{\dfrac{27x^3}{5y}} = \dfrac{\sqrt{27x^3}}{\sqrt{5y}} \cdot \dfrac{\sqrt{5y}}{\sqrt{5y}} = \dfrac{\sqrt{135x^3y}}{5y} = \dfrac{3x\sqrt{15xy}}{5y}$

65. Simplifying: $\sqrt{\dfrac{75x^3y^2}{2z}} = \dfrac{\sqrt{75x^3y^2}}{\sqrt{2z}} \cdot \dfrac{\sqrt{2z}}{\sqrt{2z}} = \dfrac{\sqrt{150x^3y^2z}}{2z} = \dfrac{5xy\sqrt{6xz}}{2z}$

67. Simplifying: $\sqrt[3]{\dfrac{16a^4b^3}{9c}} = \dfrac{\sqrt[3]{16a^4b^3}}{\sqrt[3]{9c}} \cdot \dfrac{\sqrt[3]{3c^2}}{\sqrt[3]{3c^2}} = \dfrac{\sqrt[3]{48a^4b^3c^2}}{3c} = \dfrac{2ab\sqrt[3]{6ac^2}}{3c}$

69. Simplifying: $\sqrt[3]{\dfrac{8x^3y^6}{9z}} = \dfrac{\sqrt[3]{8x^3y^6}}{\sqrt[3]{9z}} \cdot \dfrac{\sqrt[3]{3z^2}}{\sqrt[3]{3z^2}} = \dfrac{\sqrt[3]{24x^3y^6z^2}}{3z} = \dfrac{2xy^2\sqrt[3]{3z^2}}{3z}$

71. Simplifying: $\sqrt{\sqrt{x^2}} = \sqrt{x}$

73. Simplifying: $\sqrt[3]{\sqrt{xy}} = \left((xy)^{1/2}\right)^{1/3} = (xy)^{1/6} = \sqrt[6]{xy}$

75. Simplifying: $\sqrt[3]{\sqrt[4]{a}} = \left(a^{1/4}\right)^{1/3} = a^{1/12} = \sqrt[12]{a}$

77. Simplifying: $\sqrt[3]{\sqrt[3]{6x^{10}}} = \left(\left(6x^{10}\right)^{1/3}\right)^{1/3} = \left(6x^{10}\right)^{1/9} = x(6x)^{1/9} = x\sqrt[9]{6x}$

79. Simplifying: $\sqrt[4]{\sqrt[3]{a^{12}b^{24}c^{14}}} = \left(\left(a^{12}b^{24}c^{14}\right)^{1/3}\right)^{1/4} = \left(a^{12}b^{24}c^{14}\right)^{1/12} = ab^2c\left(c^2\right)^{1/12} = ab^2c\left(c^{1/6}\right) = ab^2c\sqrt[6]{c}$

81. Simplifying: $\sqrt[3]{\sqrt[5]{3a^{17}b^{16}c^{30}}} = \left(\left(3a^{17}b^{16}c^{30}\right)^{1/3}\right)^{1/5} = \left(3a^{17}b^{16}c^{30}\right)^{1/15} = abc^2\left(3a^2b\right)^{1/15} = abc^2\sqrt[15]{3a^2b}$

83. Simplifying: $\left(\sqrt{\sqrt[3]{8ab^6}}\right)^2 = \sqrt[3]{8ab^6} = 2b^2\sqrt[3]{a}$

85. Simplifying: $\sqrt{25x^2} = 5|x|$

87. Simplifying: $\sqrt{27x^3y^2} = \sqrt{9x^2y^2 \cdot 3x} = 3|xy|\sqrt{3x}$

89. Simplifying: $\sqrt{x^2-10x+25} = \sqrt{(x-5)^2} = |x-5|$

91. Simplifying: $\sqrt{4x^2+12x+9} = \sqrt{(2x+3)^2} = |2x+3|$

93. Simplifying: $\sqrt{4a^4+16a^3+16a^2} = \sqrt{4a^2\left(a^2+4a+4\right)} = \sqrt{4a^2(a+2)^2} = 2|a(a+2)|$

95. Simplifying: $\sqrt{4x^3-8x^2} = \sqrt{4x^2(x-2)} = 2|x|\sqrt{x-2}$

97. Substituting $a = 9$ and $b = 16$:
$\sqrt{a+b} = \sqrt{9+16} = \sqrt{25} = 5$
$\sqrt{a}+\sqrt{b} = \sqrt{9}+\sqrt{16} = 3+4 = 7$
Thus $\sqrt{a+b} \ne \sqrt{a}+\sqrt{b}$.

99. Substituting $w = 10$ and $l = 15$: $d = \sqrt{l^2+w^2} = \sqrt{15^2+10^2} = \sqrt{225+100} = \sqrt{325} = \sqrt{25 \cdot 13} = 5\sqrt{13}$ feet

101. a. Substituting $w = 3$, $l = 4$, and $h = 12$:
$d = \sqrt{l^2+w^2+h^2} = \sqrt{4^2+3^2+12^2} = \sqrt{16+9+144} = \sqrt{169} = 13$ feet
b. Substituting $w = 2$, $l = 4$, and $h = 6$:
$d = \sqrt{l^2+w^2+h^2} = \sqrt{4^2+2^2+6^2} = \sqrt{16+4+36} = \sqrt{56} = 2\sqrt{14} \approx 7.5$ feet

103. Answers will vary.

105. Performing the operations: $\dfrac{8xy^3}{9x^2y} \div \dfrac{16x^2y^2}{18xy^3} = \dfrac{8xy^3}{9x^2y} \cdot \dfrac{18xy^3}{16x^2y^2} = \dfrac{144x^2y^6}{144x^4y^3} = \dfrac{y^3}{x^2}$

107. Performing the operations: $\dfrac{12a^2-4a-5}{2a+1} \cdot \dfrac{7a+3}{42a^2-17a-15} = \dfrac{(6a-5)(2a+1)}{2a+1} \cdot \dfrac{7a+3}{(7a+3)(6a-5)} = 1$

109. Performing the operations:
$$\frac{8x^3+27}{27x^3+1} \div \frac{6x^2+7x-3}{9x^2-1} = \frac{8x^3+27}{27x^3+1} \cdot \frac{9x^2-1}{6x^2+7x-3}$$
$$= \frac{(2x+3)(4x^2-6x+9)}{(3x+1)(9x^2-3x+1)} \cdot \frac{(3x+1)(3x-1)}{(2x+3)(3x-1)}$$
$$= \frac{4x^2-6x+9}{9x^2-3x+1}$$

111. Simplifying: $5x - 4x + 6x = 7x$

113. Simplifying: $35xy^2 - 8xy^2 = 27xy^2$

115. Simplifying: $\frac{1}{2}x + \frac{1}{3}x = \frac{3}{6}x + \frac{2}{6}x = \frac{5}{6}x$

117. Simplifying: $\sqrt{18} = \sqrt{9 \cdot 2} = 3\sqrt{2}$

119. Simplifying: $\sqrt{75xy^3} = \sqrt{25y^2 \cdot 3xy} = 5y\sqrt{3xy}$

121. Simplifying: $\sqrt[3]{8a^4b^2} = \sqrt[3]{8a^3 \cdot ab^2} = 2a\sqrt[3]{ab^2}$

123. The prime factorization of 8,640 is: $8640 = 2^6 \cdot 3^3 \cdot 5$. Therefore: $\sqrt[3]{8640} = \sqrt[3]{2^6 \cdot 3^3 \cdot 5} = 2^2 \cdot 3\sqrt[3]{5} = 12\sqrt[3]{5}$

125. The prime factorization of 10,584 is: $10584 = 2^3 \cdot 3^3 \cdot 7^2$. Therefore: $\sqrt[3]{10584} = \sqrt[3]{2^3 \cdot 3^3 \cdot 7^2} = 2 \cdot 3\sqrt[3]{7^2} = 6\sqrt[3]{49}$

127. Rationalizing the denominator: $\frac{1}{\sqrt[10]{a^3}} = \frac{1}{\sqrt[10]{a^3}} \cdot \frac{\sqrt[10]{a^7}}{\sqrt[10]{a^7}} = \frac{\sqrt[10]{a^7}}{a}$

129. Rationalizing the denominator: $\frac{1}{\sqrt[20]{a^{11}}} = \frac{1}{\sqrt[20]{a^{11}}} \cdot \frac{\sqrt[20]{a^9}}{\sqrt[20]{a^9}} = \frac{\sqrt[20]{a^9}}{a}$

131. Graphing each function:

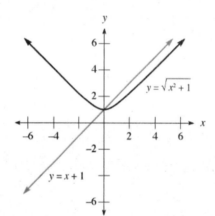

Note that the two graphs are not the same.

133. When $x = 2$, the distance apart is: $(2+1) - \sqrt{2^2+1} = 3 - \sqrt{5} \approx 0.76$

135. The two expressions are equal when $x = 0$.

9.4 Addition and Subtraction of Radical Expressions

1. Combining radicals: $3\sqrt{5} + 4\sqrt{5} = 7\sqrt{5}$
3. Combining radicals: $3x\sqrt{7} - 4x\sqrt{7} = -x\sqrt{7}$
5. Combining radicals: $5\sqrt[3]{10} - 4\sqrt[3]{10} = \sqrt[3]{10}$
7. Combining radicals: $8\sqrt[5]{6} - 2\sqrt[5]{6} + 3\sqrt[5]{6} = 9\sqrt[5]{6}$
9. Combining radicals: $3x\sqrt{2} - 4x\sqrt{2} + x\sqrt{2} = 0$
11. Combining radicals: $\sqrt{20} - \sqrt{80} + \sqrt{45} = 2\sqrt{5} - 4\sqrt{5} + 3\sqrt{5} = \sqrt{5}$
13. Combining radicals: $4\sqrt{8} - 2\sqrt{50} - 5\sqrt{72} = 8\sqrt{2} - 10\sqrt{2} - 30\sqrt{2} = -32\sqrt{2}$
15. Combining radicals: $5x\sqrt{8} + 3\sqrt{32x^2} - 5\sqrt{50x^2} = 10x\sqrt{2} + 12x\sqrt{2} - 25x\sqrt{2} = -3x\sqrt{2}$
17. Combining radicals: $5\sqrt[3]{16} - 4\sqrt[3]{54} = 10\sqrt[3]{2} - 12\sqrt[3]{2} = -2\sqrt[3]{2}$
19. Combining radicals: $\sqrt[3]{x^4 y^2} + 7x\sqrt[3]{xy^2} = x\sqrt[3]{xy^2} + 7x\sqrt[3]{xy^2} = 8x\sqrt[3]{xy^2}$
21. Combining radicals: $5a^2\sqrt{27ab^3} - 6b\sqrt{12a^5 b} = 15a^2 b\sqrt{3ab} - 12a^2 b\sqrt{3ab} = 3a^2 b\sqrt{3ab}$
23. Combining radicals: $b\sqrt[3]{24a^5 b} + 3a\sqrt[3]{81a^2 b^4} = 2ab\sqrt[3]{3a^2 b} + 9ab\sqrt[3]{3a^2 b} = 11ab\sqrt[3]{3a^2 b}$
25. Combining radicals: $5x\sqrt[4]{3y^5} + y\sqrt[4]{243x^4 y} + \sqrt[4]{48x^4 y^5} = 5xy\sqrt[4]{3y} + 3xy\sqrt[4]{3y} + 2xy\sqrt[4]{3y} = 10xy\sqrt[4]{3y}$
27. Combining radicals: $\dfrac{\sqrt{2}}{2} + \dfrac{1}{\sqrt{2}} = \dfrac{\sqrt{2}}{2} + \dfrac{1}{\sqrt{2}} \cdot \dfrac{\sqrt{2}}{\sqrt{2}} = \dfrac{\sqrt{2}}{2} + \dfrac{\sqrt{2}}{2} = \sqrt{2}$
29. Combining radicals: $\dfrac{\sqrt{5}}{3} + \dfrac{1}{\sqrt{5}} = \dfrac{\sqrt{5}}{3} + \dfrac{1}{\sqrt{5}} \cdot \dfrac{\sqrt{5}}{\sqrt{5}} = \dfrac{\sqrt{5}}{3} + \dfrac{\sqrt{5}}{5} = \dfrac{5\sqrt{5}}{15} + \dfrac{3\sqrt{5}}{15} = \dfrac{8\sqrt{5}}{15}$
31. Combining radicals: $\sqrt{x} - \dfrac{1}{\sqrt{x}} = \sqrt{x} - \dfrac{1}{\sqrt{x}} \cdot \dfrac{\sqrt{x}}{\sqrt{x}} = \sqrt{x} - \dfrac{\sqrt{x}}{x} = \dfrac{x\sqrt{x}}{x} - \dfrac{\sqrt{x}}{x} = \dfrac{(x-1)\sqrt{x}}{x}$
33. Combining radicals: $\dfrac{\sqrt{18}}{6} + \sqrt{\dfrac{1}{2}} + \dfrac{\sqrt{2}}{2} = \dfrac{3\sqrt{2}}{6} + \dfrac{1}{\sqrt{2}} \cdot \dfrac{\sqrt{2}}{\sqrt{2}} + \dfrac{\sqrt{2}}{2} = \dfrac{\sqrt{2}}{2} + \dfrac{\sqrt{2}}{2} + \dfrac{\sqrt{2}}{2} = \dfrac{3\sqrt{2}}{2}$
35. Combining radicals: $\sqrt{6} - \sqrt{\dfrac{2}{3}} + \sqrt{\dfrac{1}{6}} = \sqrt{6} - \dfrac{\sqrt{2}}{\sqrt{3}} \cdot \dfrac{\sqrt{3}}{\sqrt{3}} + \dfrac{1}{\sqrt{6}} \cdot \dfrac{\sqrt{6}}{\sqrt{6}} = \sqrt{6} - \dfrac{\sqrt{6}}{3} + \dfrac{\sqrt{6}}{6} = \dfrac{6\sqrt{6}}{6} - \dfrac{2\sqrt{6}}{6} + \dfrac{\sqrt{6}}{6} = \dfrac{5\sqrt{6}}{6}$
37. Combining radicals: $\sqrt[3]{25} + \dfrac{3}{\sqrt[3]{5}} = \sqrt[3]{25} + \dfrac{3}{\sqrt[3]{5}} \cdot \dfrac{\sqrt[3]{25}}{\sqrt[3]{25}} = \sqrt[3]{25} + \dfrac{3\sqrt[3]{25}}{5} = \dfrac{5\sqrt[3]{25}}{5} + \dfrac{3\sqrt[3]{25}}{5} = \dfrac{8\sqrt[3]{25}}{5}$
39. a. Finding the function: $f(x) + g(x) = \sqrt{8x} + \sqrt{72x} = 2\sqrt{2x} + 6\sqrt{2x} = 8\sqrt{2x}$
    b. Finding the function: $f(x) - g(x) = \sqrt{8x} - \sqrt{72x} = 2\sqrt{2x} - 6\sqrt{2x} = -4\sqrt{2x}$
41. a. Finding the function: $f(x) + g(x) = 3\sqrt{2x} + \sqrt{2x} = 4\sqrt{2x}$
    b. Finding the function: $f(x) - g(x) = 3\sqrt{2x} - \sqrt{2x} = 2\sqrt{2x}$
43. a. Finding the function: $f(x) + g(x) = x\sqrt{2} + 2x\sqrt{2} = 3x\sqrt{2}$
    b. Finding the function: $f(x) - g(x) = x\sqrt{2} - 2x\sqrt{2} = -x\sqrt{2}$
45. a. Finding the function: $f(x) + g(x) = \left(\sqrt{2x} - 2\right) + \left(2\sqrt{2x} + 5\right) = 3\sqrt{2x} + 3$
    b. Finding the function: $f(x) - g(x) = \left(\sqrt{2x} - 2\right) - \left(2\sqrt{2x} + 5\right) = -\sqrt{2x} - 7$
47. Using a calculator:
    $\sqrt{12} \approx 3.464 \quad 2\sqrt{3} \approx 3.464$
49. It is equal to the decimal approximation for $\sqrt{50}$:
    $\sqrt{8} + \sqrt{18} \approx 7.071 \approx \sqrt{50} \quad \sqrt{26} \approx 5.099$
51. Answers will vary.
53. Answers will vary.

55. First use the Pythagorean theorem to find the height h:
$$x^2 + h^2 = (2x)^2$$
$$x^2 + h^2 = 4x^2$$
$$h^2 = 3x^2$$
$$h = x\sqrt{3}$$
Therefore the ratio is: $\dfrac{x\sqrt{3}}{2x} = \dfrac{\sqrt{3}}{2}$

57. Graphing the inequality:

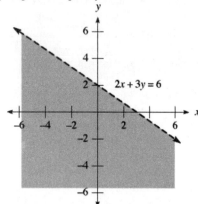

59. Graphing the inequality:

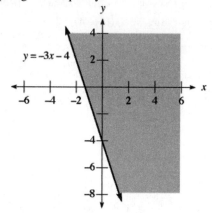

61. Graphing the inequality:

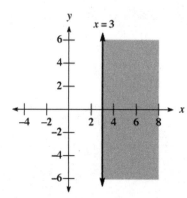

63. Solving the inequality:
$$-2 \leq 5x - 1 \leq 2$$
$$-1 \leq 5x \leq 3$$
$$-\tfrac{1}{5} \leq x \leq \tfrac{3}{5}$$

65. Simplifying: $3 \cdot 2 = 6$

67. Simplifying: $(x+y)(4x-y) = 4x^2 - xy + 4xy - y^2 = 4x^2 + 3xy - y^2$

69. Simplifying: $(x+3)^2 = x^2 + 2(3x) + 3^2 = x^2 + 6x + 9$

71. Simplifying: $(x-2)(x+2) = x^2 - 2^2 = x^2 - 4$

73. Simplifying: $2\sqrt{18} = 2\sqrt{9 \cdot 2} = 2 \cdot 3\sqrt{2} = 6\sqrt{2}$

75. Simplifying: $\left(\sqrt{6}\right)^2 = 6$

77. Simplifying: $\left(3\sqrt{x}\right)^2 = 9x$

79. Rationalizing the denominator: $\dfrac{\sqrt{3}}{\sqrt{2}} = \dfrac{\sqrt{3}}{\sqrt{2}} \cdot \dfrac{\sqrt{2}}{\sqrt{2}} = \dfrac{\sqrt{6}}{2}$

81. Simplifying: $\sqrt[5]{32x^5 y^5} - y\sqrt[3]{27x^3} = 2xy - 3xy = -xy$

83. Simplifying: $3\sqrt[9]{x^9 y^{18} z^{27}} - 4\sqrt[6]{x^6 y^{12} z^{18}} = 3xy^2 z^3 - 4xy^2 z^3 = -xy^2 z^3$

85. Simplifying:
$$3c\sqrt[8]{4a^6b^{18}} + b\sqrt[4]{32a^3b^5c^4} = 3c(2^2 a^6 b^{18})^{1/8} + b \cdot 2bc\sqrt[4]{2a^3b}$$
$$= 3c(2a^3b^9)^{1/4} + 2b^2c\sqrt[4]{2b}$$
$$= 3c(2a^3b^9)^{1/4} + 2b^2c\sqrt[4]{2a^3b}$$
$$= 3b^2c\sqrt[4]{2a^3b} + 2b^2c\sqrt[4]{2a^3b}$$
$$= 5b^2c\sqrt[4]{2a^3b}$$

87. Simplifying:
$$3\sqrt[9]{8a^{12}b^9} + b\sqrt[3]{16a^4} - 8\sqrt[6]{4a^8b^6} = 3(2^3 a^{12} b^9)^{1/9} + b \cdot 2a\sqrt[3]{2a} - 8(2^2 a^8 b^6)^{1/6}$$
$$= 3b(2a^4)^{1/3} + 2ab\sqrt[3]{2a} - 8b(2a^4)^{1/3}$$
$$= 3ab\sqrt[3]{2a} + 2ab\sqrt[3]{2a} - 8ab\sqrt[3]{2a}$$
$$= -3ab\sqrt[3]{2a}$$

9.5 Multiplication and Division of Radical Expressions

1. Multiplying: $\sqrt{6}\sqrt{3} = \sqrt{18} = 3\sqrt{2}$
3. Multiplying: $(2\sqrt{3})(5\sqrt{7}) = 10\sqrt{21}$
5. Multiplying: $(4\sqrt{6})(2\sqrt{15})(3\sqrt{10}) = 24\sqrt{900} = 24 \cdot 30 = 720$
7. Multiplying: $(3\sqrt[3]{3})(6\sqrt[3]{9}) = 18\sqrt[3]{27} = 18 \cdot 3 = 54$
9. Multiplying: $\sqrt{3}(\sqrt{2} - 3\sqrt{3}) = \sqrt{6} - 3\sqrt{9} = \sqrt{6} - 9$
11. Multiplying: $6\sqrt[3]{4}(2\sqrt[3]{2} + 1) = 12\sqrt[3]{8} + 6\sqrt[3]{4} = 24 + 6\sqrt[3]{4}$
13. Multiplying: $\sqrt[3]{4}(\sqrt[3]{2} + \sqrt[3]{6}) = \sqrt[3]{8} + \sqrt[3]{24} = 2 + 2\sqrt[3]{2}$
15. Multiplying: $\sqrt[3]{x}(\sqrt[3]{x^2 y^4} + \sqrt[3]{x^5 y}) = \sqrt[3]{x^3 y^4} + \sqrt[3]{x^6 y} = xy\sqrt[3]{y} + x^2 \sqrt[3]{y}$
17. Multiplying: $\sqrt[4]{2x^3}(\sqrt[4]{8x^6} + \sqrt[4]{16x^9}) = \sqrt[4]{16x^9} + \sqrt[4]{32x^{12}} = 2x^2\sqrt[4]{x} + 2x^3\sqrt[4]{2}$
19. Multiplying: $(\sqrt{3} + \sqrt{2})(3\sqrt{3} - \sqrt{2}) = 3\sqrt{9} - \sqrt{6} + 3\sqrt{6} - \sqrt{4} = 9 + 2\sqrt{6} - 2 = 7 + 2\sqrt{6}$
21. Multiplying: $(\sqrt{x} + 5)(\sqrt{x} - 3) = x - 3\sqrt{x} + 5\sqrt{x} - 15 = x + 2\sqrt{x} - 15$
23. Multiplying: $(3\sqrt{6} + 4\sqrt{2})(\sqrt{6} + 2\sqrt{2}) = 3\sqrt{36} + 4\sqrt{12} + 6\sqrt{12} + 8\sqrt{4} = 18 + 8\sqrt{3} + 12\sqrt{3} + 16 = 34 + 20\sqrt{3}$
25. Multiplying: $(\sqrt{3} + 4)^2 = (\sqrt{3} + 4)(\sqrt{3} + 4) = \sqrt{9} + 4\sqrt{3} + 4\sqrt{3} + 16 = 19 + 8\sqrt{3}$
27. Multiplying: $(\sqrt{x} - 3)^2 = (\sqrt{x} - 3)(\sqrt{x} - 3) = x - 3\sqrt{x} - 3\sqrt{x} + 9 = x - 6\sqrt{x} + 9$
29. Multiplying: $(2\sqrt{a} - 3\sqrt{b})^2 = (2\sqrt{a} - 3\sqrt{b})(2\sqrt{a} - 3\sqrt{b}) = 4a - 6\sqrt{ab} - 6\sqrt{ab} + 9b = 4a - 12\sqrt{ab} + 9b$
31. Multiplying: $(\sqrt{x-4} + 2)^2 = (\sqrt{x-4} + 2)(\sqrt{x-4} + 2) = x - 4 + 2\sqrt{x-4} + 2\sqrt{x-4} + 4 = x + 4\sqrt{x-4}$
33. Multiplying: $(\sqrt{x-5} - 3)^2 = (\sqrt{x-5} - 3)(\sqrt{x-5} - 3) = x - 5 - 3\sqrt{x-5} - 3\sqrt{x-5} + 9 = x + 4 - 6\sqrt{x-5}$
35. Multiplying: $(\sqrt{3} - \sqrt{2})(\sqrt{3} + \sqrt{2}) = (\sqrt{3})^2 - (\sqrt{2})^2 = 3 - 2 = 1$
37. Multiplying: $(\sqrt{a} + 7)(\sqrt{a} - 7) = (\sqrt{a})^2 - (7)^2 = a - 49$
39. Multiplying: $(5 - \sqrt{x})(5 + \sqrt{x}) = (5)^2 - (\sqrt{x})^2 = 25 - x$

41. Multiplying: $\left(\sqrt{x-4}+2\right)\left(\sqrt{x-4}-2\right) = \left(\sqrt{x-4}\right)^2 - (2)^2 = x-4-4 = x-8$

43. Multiplying: $\left(\sqrt{3}+1\right)^3 = \left(\sqrt{3}+1\right)\left(3+2\sqrt{3}+1\right) = \left(\sqrt{3}+1\right)\left(4+2\sqrt{3}\right) = 4\sqrt{3}+4+6+2\sqrt{3} = 10+6\sqrt{3}$

45. Multiplying: $\left(\sqrt[3]{3}+\sqrt[3]{2}\right)\left(\sqrt[3]{9}+\sqrt[3]{4}\right) = \sqrt[3]{27}+\sqrt[3]{12}+\sqrt[3]{18}+\sqrt[3]{8} = 3+\sqrt[3]{12}+\sqrt[3]{18}+2 = 5+\sqrt[3]{12}+\sqrt[3]{18}$

47. Multiplying: $\left(\sqrt[3]{x^5}+\sqrt[3]{y}\right)\left(\sqrt[3]{x}+\sqrt[3]{y^2}\right) = \sqrt[3]{x^6}+\sqrt[3]{x^5 y^2}+\sqrt[3]{xy}+\sqrt[3]{y^3} = x^2+x\sqrt[3]{x^2 y^2}+\sqrt[3]{xy}+y$

49. Rationalizing the denominator: $\dfrac{1}{\sqrt{2}} = \dfrac{1}{\sqrt{2}} \cdot \dfrac{\sqrt{2}}{\sqrt{2}} = \dfrac{\sqrt{2}}{2}$

51. Rationalizing the denominator: $\dfrac{1}{\sqrt{x}} = \dfrac{1}{\sqrt{x}} \cdot \dfrac{\sqrt{x}}{\sqrt{x}} = \dfrac{\sqrt{x}}{x}$

53. Rationalizing the denominator: $\dfrac{4}{\sqrt{3}} = \dfrac{4}{\sqrt{3}} \cdot \dfrac{\sqrt{3}}{\sqrt{3}} = \dfrac{4\sqrt{3}}{3}$

55. Rationalizing the denominator: $\dfrac{2x}{\sqrt{6}} = \dfrac{2x}{\sqrt{6}} \cdot \dfrac{\sqrt{6}}{\sqrt{6}} = \dfrac{2x\sqrt{6}}{6} = \dfrac{x\sqrt{6}}{3}$

57. Rationalizing the denominator: $\dfrac{4}{\sqrt{10x}} = \dfrac{4}{\sqrt{10x}} \cdot \dfrac{\sqrt{10x}}{\sqrt{10x}} = \dfrac{4\sqrt{10x}}{10x} = \dfrac{2\sqrt{10x}}{5x}$

59. Rationalizing the denominator: $\dfrac{2}{\sqrt{8}} = \dfrac{2}{\sqrt{8}} \cdot \dfrac{\sqrt{2}}{\sqrt{2}} = \dfrac{2\sqrt{2}}{4} = \dfrac{\sqrt{2}}{2}$

61. Rationalizing the denominator: $\sqrt{\dfrac{32x^3 y}{4xy^2}} = \sqrt{\dfrac{8x^2}{y}} = \dfrac{2x\sqrt{2}}{\sqrt{y}} \cdot \dfrac{\sqrt{y}}{\sqrt{y}} = \dfrac{2x\sqrt{2y}}{y}$

63. Rationalizing the denominator: $\sqrt{\dfrac{12a^3 b^3 c}{8ab^5 c^2}} = \sqrt{\dfrac{3a^2}{2b^2 c}} = \dfrac{a\sqrt{3}}{\sqrt{2b^2 c}} \cdot \dfrac{\sqrt{2c}}{\sqrt{2c}} = \dfrac{a\sqrt{6c}}{2bc}$

65. Rationalizing the denominator: $\sqrt[3]{\dfrac{16a^4 b^2 c^4}{2ab^3 c}} = \sqrt[3]{\dfrac{8a^3 c^3}{b}} = \dfrac{2ac}{\sqrt[3]{b}} \cdot \dfrac{\sqrt[3]{b^2}}{\sqrt[3]{b^2}} = \dfrac{2ac\sqrt[3]{b^2}}{b}$

67. Rationalizing the denominator: $\dfrac{\sqrt{2}}{\sqrt{6}-\sqrt{2}} = \dfrac{\sqrt{2}}{\sqrt{6}-\sqrt{2}} \cdot \dfrac{\sqrt{6}+\sqrt{2}}{\sqrt{6}+\sqrt{2}} = \dfrac{\sqrt{12}+2}{6-2} = \dfrac{2\sqrt{3}+2}{4} = \dfrac{1+\sqrt{3}}{2}$

69. Rationalizing the denominator: $\dfrac{\sqrt{5}}{\sqrt{5}+1} = \dfrac{\sqrt{5}}{\sqrt{5}+1} \cdot \dfrac{\sqrt{5}-1}{\sqrt{5}-1} = \dfrac{5-\sqrt{5}}{5-1} = \dfrac{5-\sqrt{5}}{4}$

71. Rationalizing the denominator: $\dfrac{\sqrt{x}}{\sqrt{x}-3} = \dfrac{\sqrt{x}}{\sqrt{x}-3} \cdot \dfrac{\sqrt{x}+3}{\sqrt{x}+3} = \dfrac{x+3\sqrt{x}}{x-9}$

73. Rationalizing the denominator: $\dfrac{\sqrt{5}}{2\sqrt{5}-3} = \dfrac{\sqrt{5}}{2\sqrt{5}-3} \cdot \dfrac{2\sqrt{5}+3}{2\sqrt{5}+3} = \dfrac{2\sqrt{25}+3\sqrt{5}}{20-9} = \dfrac{10+3\sqrt{5}}{11}$

75. Rationalizing the denominator: $\dfrac{3}{\sqrt{x}-\sqrt{y}} = \dfrac{3}{\sqrt{x}-\sqrt{y}} \cdot \dfrac{\sqrt{x}+\sqrt{y}}{\sqrt{x}+\sqrt{y}} = \dfrac{3\sqrt{x}+3\sqrt{y}}{x-y}$

77. Rationalizing the denominator: $\dfrac{\sqrt{6}+\sqrt{2}}{\sqrt{6}-\sqrt{2}} = \dfrac{\sqrt{6}+\sqrt{2}}{\sqrt{6}-\sqrt{2}} \cdot \dfrac{\sqrt{6}+\sqrt{2}}{\sqrt{6}+\sqrt{2}} = \dfrac{6+2\sqrt{12}+2}{6-2} = \dfrac{8+4\sqrt{3}}{4} = 2+\sqrt{3}$

79. Rationalizing the denominator: $\dfrac{\sqrt{7}-2}{\sqrt{7}+2} = \dfrac{\sqrt{7}-2}{\sqrt{7}+2} \cdot \dfrac{\sqrt{7}-2}{\sqrt{7}-2} = \dfrac{7-4\sqrt{7}+4}{7-4} = \dfrac{11-4\sqrt{7}}{3}$

81. Rationalizing the denominator: $\dfrac{\sqrt{a}+\sqrt{b}}{\sqrt{a}-\sqrt{b}} = \dfrac{\sqrt{a}+\sqrt{b}}{\sqrt{a}-\sqrt{b}} \cdot \dfrac{\sqrt{a}+\sqrt{b}}{\sqrt{a}+\sqrt{b}} = \dfrac{a+2\sqrt{ab}+b}{a-b}$

83. Rationalizing the denominator: $\dfrac{\sqrt{x}+2}{\sqrt{x}-2} = \dfrac{\sqrt{x}+2}{\sqrt{x}-2} \cdot \dfrac{\sqrt{x}+2}{\sqrt{x}+2} = \dfrac{x+4\sqrt{x}+4}{x-4}$

85. Rationalizing the denominator:
$\dfrac{2\sqrt{3}-\sqrt{7}}{3\sqrt{3}+\sqrt{7}} = \dfrac{2\sqrt{3}-\sqrt{7}}{3\sqrt{3}+\sqrt{7}} \cdot \dfrac{3\sqrt{3}-\sqrt{7}}{3\sqrt{3}-\sqrt{7}} = \dfrac{18-3\sqrt{21}-2\sqrt{21}+7}{27-7} = \dfrac{25-5\sqrt{21}}{20} = \dfrac{5-\sqrt{21}}{4}$

87. Rationalizing the denominator: $\dfrac{3\sqrt{x}+2}{1+\sqrt{x}} = \dfrac{3\sqrt{x}+2}{1+\sqrt{x}} \cdot \dfrac{1-\sqrt{x}}{1-\sqrt{x}} = \dfrac{3\sqrt{x}+2-3x-2\sqrt{x}}{1-x} = \dfrac{\sqrt{x}-3x+2}{1-x}$

89. Simplifying: $\dfrac{2}{\sqrt{3}} + \sqrt{12} = \dfrac{2}{\sqrt{3}} \cdot \dfrac{\sqrt{3}}{\sqrt{3}} + 2\sqrt{3} = \dfrac{2\sqrt{3}}{3} + \dfrac{2\sqrt{3}}{1} \cdot \dfrac{3}{3} = \dfrac{2\sqrt{3}}{3} + \dfrac{6\sqrt{3}}{3} = \dfrac{8\sqrt{3}}{3}$

91. Simplifying: $\dfrac{1}{\sqrt{5}} + \sqrt{20} = \dfrac{1}{\sqrt{5}} \cdot \dfrac{\sqrt{5}}{\sqrt{5}} + 2\sqrt{5} = \dfrac{\sqrt{5}}{5} + \dfrac{2\sqrt{5}}{1} \cdot \dfrac{5}{5} = \dfrac{\sqrt{5}}{5} + \dfrac{10\sqrt{5}}{5} = \dfrac{11\sqrt{5}}{5}$

93. Simplifying: $\dfrac{6}{\sqrt{12}} - \sqrt{75} = \dfrac{6}{2\sqrt{3}} \cdot \dfrac{\sqrt{3}}{\sqrt{3}} - 5\sqrt{3} = \dfrac{6\sqrt{3}}{6} - 5\sqrt{3} = \sqrt{3} - 5\sqrt{3} = -4\sqrt{3}$

95. Simplifying:
$\dfrac{1}{\sqrt{3}} + \sqrt{48} + \dfrac{4}{\sqrt{12}} = \dfrac{1}{\sqrt{3}} \cdot \dfrac{\sqrt{3}}{\sqrt{3}} + 4\sqrt{3} + \dfrac{4}{2\sqrt{3}} \cdot \dfrac{\sqrt{3}}{\sqrt{3}}$
$= \dfrac{\sqrt{3}}{3} + \dfrac{4\sqrt{3}}{1} \cdot \dfrac{3}{3} + \dfrac{4\sqrt{3}}{6}$
$= \dfrac{\sqrt{3}}{3} + \dfrac{12\sqrt{3}}{3} + \dfrac{2\sqrt{3}}{3}$
$= \dfrac{15\sqrt{3}}{3}$
$= 5\sqrt{3}$

97. Simplifying the product: $\left(\sqrt[3]{2}+\sqrt[3]{3}\right)\left(\sqrt[3]{4}-\sqrt[3]{6}+\sqrt[3]{9}\right) = \sqrt[3]{8} - \sqrt[3]{12} + \sqrt[3]{18} + \sqrt[3]{12} - \sqrt[3]{18} + \sqrt[3]{27} = 2+3 = 5$

99. The correct statement is: $5\left(2\sqrt{3}\right) = 10\sqrt{3}$

101. The correct statement is: $\left(\sqrt{x}+3\right)^2 = \left(\sqrt{x}+3\right)\left(\sqrt{x}+3\right) = x+6\sqrt{x}+9$

103. The correct statement is: $\left(5\sqrt{3}\right)^2 = \left(5\sqrt{3}\right)\left(5\sqrt{3}\right) = 25 \cdot 3 = 75$

105. Substituting $h = 50$: $t = \dfrac{\sqrt{100-50}}{4} = \dfrac{\sqrt{50}}{4} = \dfrac{5\sqrt{2}}{4}$ second

Substituting $h = 0$: $t = \dfrac{\sqrt{100-0}}{4} = \dfrac{\sqrt{100}}{4} = \dfrac{10}{4} = \dfrac{5}{2}$ second

107. Since the large rectangle is a golden rectangle and $AC = 6$, then $CE = 6\left(\dfrac{1+\sqrt{5}}{2}\right) = 3+3\sqrt{5}$. Since $CD = 6$, then

$DE = 3+3\sqrt{5}-6 = 3\sqrt{5}-3$. Now computing the ratio:

$\dfrac{EF}{DE} = \dfrac{6}{3\sqrt{5}-3} \cdot \dfrac{3\sqrt{5}+3}{3\sqrt{5}+3} = \dfrac{18\left(\sqrt{5}+1\right)}{45-9} = \dfrac{18\left(\sqrt{5}+1\right)}{36} = \dfrac{1+\sqrt{5}}{2}$

Therefore the smaller rectangle $BDEF$ is also a golden rectangle.

109. Since the large rectangle is a golden rectangle and $AC = 2x$, then $CE = 2x\left(\dfrac{1+\sqrt{5}}{2}\right) = x\left(1+\sqrt{5}\right)$. Since $CD = 2x$, then

$DE = x\left(1+\sqrt{5}\right) - 2x = x\left(-1+\sqrt{5}\right)$. Now computing the ratio:

$$\dfrac{EF}{DE} = \dfrac{2x}{x\left(-1+\sqrt{5}\right)} = \dfrac{2}{-1+\sqrt{5}} \cdot \dfrac{-1-\sqrt{5}}{-1-\sqrt{5}} = \dfrac{-2\left(\sqrt{5}+1\right)}{1-5} = \dfrac{-2\left(\sqrt{5}+1\right)}{-4} = \dfrac{1+\sqrt{5}}{2}$$

Therefore the smaller rectangle $BDEF$ is also a golden rectangle.

111. The variation equation is $y = Kx^2$. Substituting $x = 5$ and $y = 75$:

$75 = K \cdot 5^2$
$75 = 25K$
$K = 3$

So $y = 3x^2$. Substituting $x = 7$: $y = 3 \cdot 7^2 = 3 \cdot 49 = 147$

113. The variation equation is $y = \dfrac{K}{x}$. Substituting $x = 25$ and $y = 10$:

$10 = \dfrac{K}{25}$
$K = 250$

So $y = \dfrac{250}{x}$. Substituting $y = 5$:

$5 = \dfrac{250}{x}$
$5x = 250$
$x = 50$

115. The variation equation is $z = Kxy^2$. Substituting $z = 40$, $x = 5$, and $y = 2$:

$40 = K \cdot 5 \cdot 2^2$
$40 = 20K$
$K = 2$

So $z = 2xy^2$. Substituting $x = 2$ and $y = 5$: $z = 2 \cdot 2 \cdot 5^2 = 100$

117. Simplifying: $(t+5)^2 = t^2 + 2(5t) + 5^2 = t^2 + 10t + 25$

119. Simplifying: $\sqrt{x} \cdot \sqrt{x} = \sqrt{x^2} = x$

121. Solving the equation:
$3x + 4 = 5^2$
$3x + 4 = 25$
$3x = 21$
$x = 7$

123. Solving the equation:
$t^2 + 7t + 12 = 0$
$(t+4)(t+3) = 0$
$t = -4, -3$

125. Solving the equation:

$t^2 + 10t + 25 = t + 7$
$t^2 + 9t + 18 = 0$
$(t+6)(t+3) = 0$
$t = -6, -3$

127. Solving the equation:
$(x+4)^2 = x + 6$
$x^2 + 8x + 16 = x + 6$
$x^2 + 7x + 10 = 0$
$(x+5)(x+2) = 0$
$x = -5, -2$

129. Rationalizing the denominator: $\dfrac{x}{\sqrt{x-2}+4} = \dfrac{x}{\sqrt{x-2}+4} \cdot \dfrac{\sqrt{x-2}-4}{\sqrt{x-2}-4} = \dfrac{x\left(\sqrt{x-2}-4\right)}{x-2-16} = \dfrac{x\left(\sqrt{x-2}-4\right)}{x-18}$

131. Rationalizing the denominator: $\dfrac{x}{\sqrt{x+5}-5} = \dfrac{x}{\sqrt{x+5}-5} \cdot \dfrac{\sqrt{x+5}+5}{\sqrt{x+5}+5} = \dfrac{x\left(\sqrt{x+5}+5\right)}{x+5-25} = \dfrac{x\left(\sqrt{x+5}+5\right)}{x-20}$

133. Rationalizing the denominator: $\dfrac{3x}{\sqrt{5x}+x} = \dfrac{3x}{\sqrt{5x}+x} \cdot \dfrac{\sqrt{5x}-x}{\sqrt{5x}-x} = \dfrac{3x\left(\sqrt{5x}-x\right)}{5x-x^2} = \dfrac{3x\left(\sqrt{5x}-x\right)}{x(5-x)} = \dfrac{3\left(\sqrt{5x}-x\right)}{5-x}$

9.6 Equations with Radicals

1. Solving the equation:
$$\sqrt{2x+1} = 3$$
$$\left(\sqrt{2x+1}\right)^2 = 3^2$$
$$2x+1 = 9$$
$$2x = 8$$
$$x = 4$$

3. Solving the equation:
$$\sqrt{4x+1} = -5$$
$$\left(\sqrt{4x+1}\right)^2 = (-5)^2$$
$$4x+1 = 25$$
$$4x = 24$$
$$x = 6$$
Since this value does not check, the solution set is $\varnothing$.

5. Solving the equation:
$$\sqrt{2y-1} = 3$$
$$\left(\sqrt{2y-1}\right)^2 = 3^2$$
$$2y-1 = 9$$
$$2y = 10$$
$$y = 5$$

7. Solving the equation:
$$\sqrt{5x-7} = -1$$
$$\left(\sqrt{5x-7}\right)^2 = (-1)^2$$
$$5x-7 = 1$$
$$5x = 8$$
$$x = \tfrac{8}{5}$$
Since this value does not check, the solution set is $\varnothing$.

9. Solving the equation:
$$\sqrt{2x-3}-2 = 4$$
$$\sqrt{2x-3} = 6$$
$$\left(\sqrt{2x-3}\right)^2 = 6^2$$
$$2x-3 = 36$$
$$2x = 39$$
$$x = \tfrac{39}{2}$$

11. Solving the equation:
$$\sqrt{4a+1}+3 = 2$$
$$\sqrt{4a+1} = -1$$
$$\left(\sqrt{4a+1}\right)^2 = (-1)^2$$
$$4a+1 = 1$$
$$4a = 0$$
$$a = 0$$
Since this value does not check, the solution set is $\varnothing$.

13. Solving the equation:
$$\sqrt[4]{3x+1} = 2$$
$$\left(\sqrt[4]{3x+1}\right)^4 = 2^4$$
$$3x+1 = 16$$
$$3x = 15$$
$$x = 5$$

15. Solving the equation:
$$\sqrt[3]{2x-5} = 1$$
$$\left(\sqrt[3]{2x-5}\right)^3 = 1^3$$
$$2x-5 = 1$$
$$2x = 6$$
$$x = 3$$

17. Solving the equation:
$$\sqrt[3]{3a+5} = -3$$
$$\left(\sqrt[3]{3a+5}\right)^3 = (-3)^3$$
$$3a+5 = -27$$
$$3a = -32$$
$$a = -\tfrac{32}{3}$$

19. Solving the equation:
$$\sqrt{y-3} = y-3$$
$$\left(\sqrt{y-3}\right)^2 = (y-3)^2$$
$$y-3 = y^2-6y+9$$
$$0 = y^2-7y+12$$
$$0 = (y-3)(y-4)$$
$$y = 3, 4$$

21. Solving the equation:

$$\sqrt{a+2} = a+2$$
$$\left(\sqrt{a+2}\right)^2 = (a+2)^2$$
$$a+2 = a^2 + 4a + 4$$
$$0 = a^2 + 3a + 2$$
$$0 = (a+2)(a+1)$$
$$a = -2, -1$$

23. Solving the equation:

$$\sqrt{2x+3} = \frac{2x-7}{3}$$
$$\left(\sqrt{2x+3}\right)^2 = \left(\frac{2x-7}{3}\right)^2$$
$$2x+3 = \frac{4x^2 - 28x + 49}{9}$$
$$18x + 27 = 4x^2 - 28x + 49$$
$$0 = 4x^2 - 46x + 22$$
$$0 = 2x^2 - 23x + 11$$
$$0 = (2x-1)(x-11)$$
$$x = \tfrac{1}{2}, 11$$

The solution is 11 (1/2 does not check).

25. Solving the equation:

$$\sqrt{4x-3} = \frac{x+3}{2}$$
$$\left(\sqrt{4x-3}\right)^2 = \left(\frac{x+3}{2}\right)^2$$
$$4x - 3 = \frac{x^2 + 6x + 9}{4}$$
$$16x - 12 = x^2 + 6x + 9$$
$$0 = x^2 - 10x + 21$$
$$0 = (x-3)(x-7)$$
$$x = 3, 7$$

27. Solving the equation:

$$\sqrt{7x+2} = \frac{2x+2}{3}$$
$$\left(\sqrt{7x+2}\right)^2 = \left(\frac{2x+2}{3}\right)^2$$
$$7x + 2 = \frac{4x^2 + 8x + 4}{9}$$
$$63x + 18 = 4x^2 + 8x + 4$$
$$0 = 4x^2 - 55x - 14$$
$$0 = (4x+1)(x-14)$$
$$x = -\tfrac{1}{4}, 14$$

29. Solving the equation:

$$\sqrt{2x+4} = \sqrt{1-x}$$
$$\left(\sqrt{2x+4}\right)^2 = \left(\sqrt{1-x}\right)^2$$
$$2x + 4 = 1 - x$$
$$3x = -3$$
$$x = -1$$

31. Solving the equation:

$$\sqrt{4a+7} = -\sqrt{a+2}$$
$$\left(\sqrt{4a+7}\right)^2 = \left(-\sqrt{a+2}\right)^2$$
$$4a + 7 = a + 2$$
$$3a = -5$$
$$a = -\tfrac{5}{3}$$

Since this value does not check, the solution set is $\varnothing$.

33. Solving the equation:

$$\sqrt[4]{5x-8} = \sqrt[4]{4x-1}$$
$$\left(\sqrt[4]{5x-8}\right)^4 = \left(\sqrt[4]{4x-1}\right)^4$$
$$5x - 8 = 4x - 1$$
$$x = 7$$

35. Solving the equation:

$$x + 1 = \sqrt{5x+1}$$
$$(x+1)^2 = \left(\sqrt{5x+1}\right)^2$$
$$x^2 + 2x + 1 = 5x + 1$$
$$x^2 - 3x = 0$$
$$x(x-3) = 0$$
$$x = 0, 3$$

37. Solving the equation:

$$t + 5 = \sqrt{2t+9}$$
$$(t+5)^2 = \left(\sqrt{2t+9}\right)^2$$
$$t^2 + 10t + 25 = 2t + 9$$
$$t^2 + 8t + 16 = 0$$
$$(t+4)^2 = 0$$
$$t = -4$$

39. Solving the equation:

$$\sqrt{y-8} = \sqrt{8-y}$$
$$\left(\sqrt{y-8}\right)^2 = \left(\sqrt{8-y}\right)^2$$
$$y - 8 = 8 - y$$
$$2y = 16$$
$$y = 8$$

41. Solving the equation:
$$\sqrt[3]{3x+5} = \sqrt[3]{5-2x}$$
$$\left(\sqrt[3]{3x+5}\right)^3 = \left(\sqrt[3]{5-2x}\right)^3$$
$$3x+5 = 5-2x$$
$$5x = 0$$
$$x = 0$$

43. Solving the equation:
$$\sqrt{x-8} = \sqrt{x}-2$$
$$\left(\sqrt{x-8}\right)^2 = \left(\sqrt{x}-2\right)^2$$
$$x-8 = x-4\sqrt{x}+4$$
$$-12 = -4\sqrt{x}$$
$$\sqrt{x} = 3$$
$$x = 9$$

45. Solving the equation:
$$\sqrt{x+1} = \sqrt{x}+1$$
$$\left(\sqrt{x+1}\right)^2 = \left(\sqrt{x}+1\right)^2$$
$$x+1 = x+2\sqrt{x}+1$$
$$0 = 2\sqrt{x}$$
$$\sqrt{x} = 0$$
$$x = 0$$

47. Solving the equation:
$$\sqrt{x+8} = \sqrt{x-4}+2$$
$$\left(\sqrt{x+8}\right)^2 = \left(\sqrt{x-4}+2\right)^2$$
$$x+8 = x-4+4\sqrt{x-4}+4$$
$$8 = 4\sqrt{x-4}$$
$$\sqrt{x-4} = 2$$
$$x-4 = 4$$
$$x = 8$$

49. Solving the equation:
$$\sqrt{x-5}-3 = \sqrt{x-8}$$
$$\left(\sqrt{x-5}-3\right)^2 = \left(\sqrt{x-8}\right)^2$$
$$x-5-6\sqrt{x-5}+9 = x-8$$
$$-6\sqrt{x-5} = -12$$
$$\sqrt{x-5} = 2$$
$$x-5 = 4$$
$$x = 9$$

Since this value does not check, the solution set is $\varnothing$.

51. Solving the equation:
$$\sqrt{x+4} = 2-\sqrt{2x}$$
$$\left(\sqrt{x+4}\right)^2 = \left(2-\sqrt{2x}\right)^2$$
$$x+4 = 4-4\sqrt{2x}+2x$$
$$-x = -4\sqrt{2x}$$
$$(-x)^2 = \left(-4\sqrt{2x}\right)^2$$
$$x^2 = 32x$$
$$x^2 - 32x = 0$$
$$x(x-32) = 0$$
$$x = 0, 32$$

The solution is 0 (32 does not check).

53. Solving the equation:
$$\sqrt{2x+4} = \sqrt{x+3}+1$$
$$\left(\sqrt{2x+4}\right)^2 = \left(\sqrt{x+3}+1\right)^2$$
$$2x+4 = x+3+2\sqrt{x+3}+1$$
$$x = 2\sqrt{x+3}$$
$$x^2 = \left(2\sqrt{x+3}\right)^2$$
$$x^2 = 4x+12$$
$$x^2 - 4x - 12 = 0$$
$$(x-6)(x+2) = 0$$
$$x = -2, 6$$

The solution is 6 (–2 does not check).

55. Solving the equation:
$$f(x) = 0$$
$$\sqrt{2x-1} = 0$$
$$\left(\sqrt{2x-1}\right)^2 = (0)^2$$
$$2x-1 = 0$$
$$2x = 1$$
$$x = \tfrac{1}{2}$$

57. Solving the equation:
$$f(x) = 2x - 1$$
$$\sqrt{2x-1} = 2x - 1$$
$$\left(\sqrt{2x-1}\right)^2 = (2x-1)^2$$
$$2x - 1 = 4x^2 - 4x + 1$$
$$0 = 4x^2 - 6x + 2$$
$$0 = 2x^2 - 3x + 1$$
$$0 = (2x-1)(x-1)$$
$$x = \tfrac{1}{2}, 1$$

59. Solving the equation:
$$f(x) = \sqrt{x-4} + 2$$
$$\sqrt{2x-1} = \sqrt{x-4} + 2$$
$$\left(\sqrt{2x-1}\right)^2 = \left(\sqrt{x-4} + 2\right)^2$$
$$2x - 1 = x - 4 + 4\sqrt{x-4} + 4$$
$$x - 1 = 4\sqrt{x-4}$$
$$(x-1)^2 = \left(4\sqrt{x-4}\right)^2$$
$$x^2 - 2x + 1 = 16(x-4)$$
$$x^2 - 2x + 1 = 16x - 64$$
$$x^2 - 18x + 65 = 0$$
$$(x-5)(x-13) = 0$$
$$x = 5, 13$$

61. Solving the equation:
$$g(x) = 0$$
$$\sqrt{2x+3} = 0$$
$$\left(\sqrt{2x+3}\right)^2 = (0)^2$$
$$2x + 3 = 0$$
$$2x = -3$$
$$x = -\tfrac{3}{2}$$

63. Solving the equation:
$$g(x) = -\sqrt{5x}$$
$$\sqrt{2x+3} = -\sqrt{5x}$$
$$\left(\sqrt{2x+3}\right)^2 = \left(-\sqrt{5x}\right)^2$$
$$2x + 3 = 5x$$
$$3 = 3x$$
$$x = 1$$
Since this value does not check, the solution set is $\varnothing$.

65. Solving the equation:
$$f(x) = g(x)$$
$$\sqrt{2x} - 1 = 0$$
$$\sqrt{2x} = 1$$
$$\left(\sqrt{2x}\right)^2 = (1)^2$$
$$2x = 1$$
$$x = \tfrac{1}{2}$$

67. Solving the equation:
$$f(x) = g(x)$$
$$\sqrt{2x} - 1 = \sqrt{2x+5}$$
$$\left(\sqrt{2x} - 1\right)^2 = \left(\sqrt{2x+5}\right)^2$$
$$2x - 2\sqrt{2x} + 1 = 2x + 5$$
$$-2\sqrt{2x} = 4$$
$$\sqrt{2x} = -2$$
$$\left(\sqrt{2x}\right)^2 = (-2)^2$$
$$2x = 4$$
$$x = 2$$
Since this value does not check, the solution set is $\varnothing$.

69. Solving the equation:
$$h(x) = f(x)$$
$$\sqrt[3]{3x+5} = 2$$
$$\left(\sqrt[3]{3x+5}\right)^3 = (2)^3$$
$$3x + 5 = 8$$
$$3x = 3$$
$$x = 1$$

71. Solving the equation:
$$h(x) = f(x)$$
$$\sqrt[3]{5-2x} = 3$$
$$\left(\sqrt[3]{5-2x}\right)^3 = (3)^3$$
$$5 - 2x = 27$$
$$-2x = 22$$
$$x = -11$$

73. Graphing the equation:

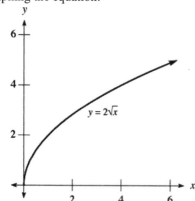

75. Graphing the equation:

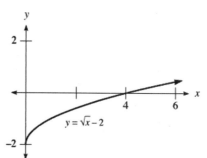

77. Graphing the equation:

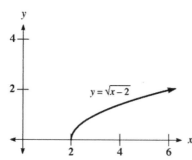

79. Graphing the equation:

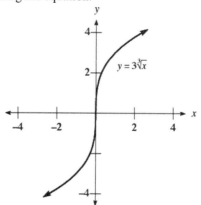

81. Graphing the equation:

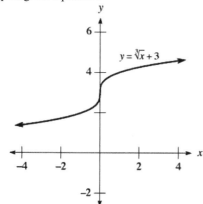

83. Graphing the equation:

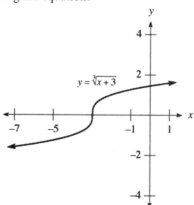

85. Solving for h:

$$t = \frac{\sqrt{100-h}}{4}$$
$$4t = \sqrt{100-h}$$
$$16t^2 = 100 - h$$
$$h = 100 - 16t^2$$

87. Solving for L:

$$2 = 2\left(\tfrac{22}{7}\right)\sqrt{\tfrac{L}{32}}$$
$$\tfrac{7}{22} = \sqrt{\tfrac{L}{32}}$$
$$\left(\tfrac{7}{22}\right)^2 = \tfrac{L}{32}$$
$$L = 32\left(\tfrac{7}{22}\right)^2 \approx 3.24 \text{ feet}$$

89. The width is $\sqrt{25} = 5$ meters.

91. Solving the equation:
$$\sqrt{x} = 50$$
$$x = 2500$$
The plume is 2,500 meters down river.

93. Multiplying: $\sqrt{2}(\sqrt{3}-\sqrt{2}) = \sqrt{6} - \sqrt{4} = \sqrt{6} - 2$

95. Multiplying: $(\sqrt{x}+5)^2 = (\sqrt{x}+5)(\sqrt{x}+5) = x+5\sqrt{x}+5\sqrt{x}+25 = x+10\sqrt{x}+25$

97. Rationalizing the denominator: $\dfrac{\sqrt{x}}{\sqrt{x}+3} = \dfrac{\sqrt{x}}{\sqrt{x}+3} \cdot \dfrac{\sqrt{x}-3}{\sqrt{x}-3} = \dfrac{x-3\sqrt{x}}{x-9}$

99. Simplifying: $\sqrt{25} = 5$

101. Simplifying: $\sqrt{12} = \sqrt{4 \bullet 3} = 2\sqrt{3}$

103. Simplifying: $(-1)^{15} = -1$

105. Simplifying: $(-1)^{50} = 1$

107. Solving the equation:
$$3x = 12$$
$$x = 4$$

109. Solving the equation:
$$4x - 3 = 5$$
$$4x = 8$$
$$x = 2$$

111. Performing the operations: $(3+4x)+(7-6x) = 10-2x$

113. Performing the operations: $(7+3x)-(5+6x) = 7+3x-5-6x = 2-3x$

115. Performing the operations: $(3-4x)(2+5x) = 6+15x-8x-20x^2 = 6+7x-20x^2$

117. Performing the operations: $2x(4-6x) = 8x-12x^2$

119. Performing the operations: $(2+3x)^2 = 2^2 + 2(2)(3x) + (3x)^2 = 4+12x+9x^2$

121. Performing the operations: $(2-3x)(2+3x) = 2^2 - (3x)^2 = 4-9x^2$

9.7 Complex Numbers

1. Writing in terms of i: $\sqrt{-36} = 6i$

3. Writing in terms of i: $-\sqrt{-25} = -5i$

5. Writing in terms of i: $\sqrt{-72} = 6i\sqrt{2}$

7. Writing in terms of i: $-\sqrt{-12} = -2i\sqrt{3}$

9. Rewriting the expression: $i^{28} = (i^4)^7 = (1)^7 = 1$

11. Rewriting the expression: $i^{26} = i^{24}i^2 = (i^4)^6 i^2 = (1)^6(-1) = -1$

13. Rewriting the expression: $i^{75} = i^{72}i^3 = (i^4)^{18} i^2 i = (1)^{18}(-1)i = -i$

15. Setting real and imaginary parts equal:
$$2x = 6 \qquad 3y = -3$$
$$x = 3 \qquad y = -1$$

17. Setting real and imaginary parts equal:
$$-x = 2 \qquad 10y = -5$$
$$x = -2 \qquad y = -\tfrac{1}{2}$$

19. Setting real and imaginary parts equal:
$$2x = -16 \qquad -2y = 10$$
$$x = -8 \qquad y = -5$$

21. Setting real and imaginary parts equal:
$$2x - 4 = 10 \qquad -6y = -3$$
$$2x = 14 \qquad y = \tfrac{1}{2}$$
$$x = 7$$

23. Setting real and imaginary parts equal:
$$7x - 1 = 2 \qquad 5y + 2 = 4$$
$$7x = 3 \qquad 5y = 2$$
$$x = \tfrac{3}{7} \qquad y = \tfrac{2}{5}$$

25. Combining the numbers: $(2+3i)+(3+6i) = 5+9i$

27. Combining the numbers: $(3-5i)+(2+4i) = 5-i$

29. Combining the numbers: $(5+2i)-(3+6i) = 5+2i-3-6i = 2-4i$

31. Combining the numbers: $(3-5i)-(2+i) = 3-5i-2-i = 1-6i$

33. Combining the numbers: $\left[(3+2i)-(6+i)\right]+(5+i) = 3+2i-6-i+5+i = 2+2i$

35. Combining the numbers: $\left[(7-i)-(2+4i)\right]-(6+2i) = 7-i-2-4i-6-2i = -1-7i$

37. Combining the numbers:
$(3+2i)-\left[(3-4i)-(6+2i)\right] = (3+2i)-(3-4i-6-2i) = (3+2i)-(-3-6i) = 3+2i+3+6i = 6+8i$

39. Combining the numbers: $(4-9i)+\left[(2-7i)-(4+8i)\right] = (4-9i)+(2-7i-4-8i) = (4-9i)+(-2-15i) = 2-24i$

41. Finding the product: $3i(4+5i) = 12i+15i^2 = -15+12i$

43. Finding the product: $6i(4-3i) = 24i-18i^2 = 18+24i$

45. Finding the product: $(3+2i)(4+i) = 12+8i+3i+2i^2 = 12+11i-2 = 10+11i$

47. Finding the product: $(4+9i)(3-i) = 12+27i-4i-9i^2 = 12+23i+9 = 21+23i$

49. Finding the product: $(1+i)^3 = (1+i)(1+i)^2 = (1+i)(1+2i-1) = (1+i)(2i) = -2+2i$

51. Finding the product: $(2-i)^3 = (2-i)(2-i)^2 = (2-i)(4-4i-1) = (2-i)(3-4i) = 6-11i-4 = 2-11i$

53. Finding the product: $(2+5i)^2 = (2+5i)(2+5i) = 4+10i+10i-25 = -21+20i$

55. Finding the product: $(1-i)^2 = (1-i)(1-i) = 1-i-i-1 = -2i$

57. Finding the product: $(3-4i)^2 = (3-4i)(3-4i) = 9-12i-12i-16 = -7-24i$

59. Finding the product: $(2+i)(2-i) = 4-i^2 = 4+1 = 5$

61. Finding the product: $(6-2i)(6+2i) = 36-4i^2 = 36+4 = 40$

63. Finding the product: $(2+3i)(2-3i) = 4-9i^2 = 4+9 = 13$

65. Finding the product: $(10+8i)(10-8i) = 100-64i^2 = 100+64 = 164$

67. Finding the quotient: $\dfrac{2-3i}{i} = \dfrac{2-3i}{i} \cdot \dfrac{i}{i} = \dfrac{2i+3}{-1} = -3-2i$

69. Finding the quotient: $\dfrac{5+2i}{-i} = \dfrac{5+2i}{-i} \cdot \dfrac{i}{i} = \dfrac{5i-2}{1} = -2+5i$

71. Finding the quotient: $\dfrac{4}{2-3i} = \dfrac{4}{2-3i} \cdot \dfrac{2+3i}{2+3i} = \dfrac{8+12i}{4+9} = \dfrac{8+12i}{13} = \dfrac{8}{13}+\dfrac{12}{13}i$

73. Finding the quotient: $\dfrac{6}{-3+2i} = \dfrac{6}{-3+2i} \cdot \dfrac{-3-2i}{-3-2i} = \dfrac{-18-12i}{9+4} = \dfrac{-18-12i}{13} = -\dfrac{18}{13}-\dfrac{12}{13}i$

75. Finding the quotient: $\dfrac{2+3i}{2-3i} = \dfrac{2+3i}{2-3i} \cdot \dfrac{2+3i}{2+3i} = \dfrac{4+12i-9}{4+9} = \dfrac{-5+12i}{13} = -\dfrac{5}{13}+\dfrac{12}{13}i$

77. Finding the quotient: $\dfrac{5+4i}{3+6i} = \dfrac{5+4i}{3+6i} \cdot \dfrac{3-6i}{3-6i} = \dfrac{15-18i+24}{9+36} = \dfrac{39-18i}{45} = \dfrac{13}{15}-\dfrac{2}{5}i$

79. Dividing to find R: $R = \dfrac{80+20i}{-6+2i} = \dfrac{80+20i}{-6+2i} \cdot \dfrac{-6-2i}{-6-2i} = \dfrac{-480-280i+40}{36+4} = \dfrac{-440-280i}{40} = (-11-7i)$ ohms

81. The domain is $\{1,3,4\}$ and the range is $\{2,4\}$. This is a function.

83. The domain is $\{1,2,3\}$ and the range is $\{1,2,3\}$. This is a function.

85. Since this passes the vertical line test, it is a function.

87. Since this fails the vertical line test, it is not a function.

Chapter 9 Review/Test

1. Simplifying: $\sqrt{49} = 7$

2. Simplifying: $(-27)^{1/3} = -3$

3. Simplifying: $16^{1/4} = 2$

4. Simplifying: $9^{3/2} = \left(9^{1/2}\right)^3 = 3^3 = 27$

5. Simplifying: $\sqrt[5]{32x^{15}y^{10}} = 2x^3 y^2$

6. Simplifying: $8^{-4/3} = \left(8^{1/3}\right)^{-4} = 2^{-4} = \dfrac{1}{2^4} = \dfrac{1}{16}$

7. Simplifying: $x^{2/3} \cdot x^{4/3} = x^{2/3+4/3} = x^2$

8. Simplifying: $\left(a^{2/3}b^{4/3}\right)^3 = a^{3 \cdot 2/3}b^{3 \cdot 4/3} = a^2 b^4$

9. Simplifying: $\dfrac{a^{3/5}}{a^{1/4}} = a^{3/5-1/4} = a^{12/20-5/20} = a^{7/20}$

10. Simplifying: $\dfrac{a^{2/3}b^3}{a^{1/4}b^{1/3}} = a^{2/3-1/4}b^{3-1/3} = a^{8/12-3/12}b^{9/3-1/3} = a^{5/12}b^{8/3}$

11. Multiplying: $\left(3x^{1/2}+5y^{1/2}\right)\left(4x^{1/2}-3y^{1/2}\right) = 12x - 9x^{1/2}y^{1/2} + 20x^{1/2}y^{1/2} - 15y = 12x + 11x^{1/2}y^{1/2} - 15y$

12. Multiplying: $\left(a^{1/3}-5\right)^2 = \left(a^{1/3}-5\right)\left(a^{1/3}-5\right) = a^{2/3} - 5a^{1/3} - 5a^{1/3} + 25 = a^{2/3} - 10a^{1/3} + 25$

13. Dividing: $\dfrac{28x^{5/6}+14x^{7/6}}{7x^{1/3}} = \dfrac{28x^{5/6}}{7x^{1/3}} + \dfrac{14x^{7/6}}{7x^{1/3}} = 4x^{5/6-1/3} + 2x^{7/6-1/3} = 4x^{1/2} + 2x^{5/6}$

14. Factoring: $8(x-3)^{5/4} - 2(x-3)^{1/4} = 2(x-3)^{1/4}\left[4(x-3)-1\right] = 2(x-3)^{1/4}(4x-12-1) = 2(x-3)^{1/4}(4x-13)$

15. Simplifying: $x^{3/4} + \dfrac{5}{x^{1/4}} = x^{3/4} \cdot \dfrac{x^{1/4}}{x^{1/4}} + \dfrac{5}{x^{1/4}} = \dfrac{x+5}{x^{1/4}}$

16. Simplifying: $\sqrt{12} = \sqrt{4 \cdot 3} = 2\sqrt{3}$

17. Simplifying: $\sqrt{50} = \sqrt{25 \cdot 2} = 5\sqrt{2}$

18. Simplifying: $\sqrt[3]{16} = \sqrt[3]{8 \cdot 2} = 2\sqrt[3]{2}$

19. Simplifying: $\sqrt{18x^2} = \sqrt{9x^2 \cdot 2} = 3x\sqrt{2}$

20. Simplifying: $\sqrt{80a^3b^4c^2} = \sqrt{16a^2b^4c^2 \cdot 5a} = 4ab^2c\sqrt{5a}$

21. Simplifying: $\sqrt[4]{32a^4b^5c^6} = \sqrt[4]{16a^4b^4c^4 \cdot 2bc^2} = 2abc\sqrt[4]{2bc^2}$

22. Rationalizing the denominator: $\dfrac{3}{\sqrt{2}} = \dfrac{3}{\sqrt{2}} \cdot \dfrac{\sqrt{2}}{\sqrt{2}} = \dfrac{3\sqrt{2}}{2}$

23. Rationalizing the denominator: $\dfrac{6}{\sqrt[3]{2}} = \dfrac{6}{\sqrt[3]{2}} \cdot \dfrac{\sqrt[3]{4}}{\sqrt[3]{4}} = \dfrac{6\sqrt[3]{4}}{2} = 3\sqrt[3]{4}$

24. Simplifying: $\sqrt{\dfrac{48x^3}{7y}} = \dfrac{4x\sqrt{3x}}{\sqrt{7y}} \cdot \dfrac{\sqrt{7y}}{\sqrt{7y}} = \dfrac{4x\sqrt{21xy}}{7y}$

25. Simplifying: $\sqrt[3]{\dfrac{40x^2y^3}{3z}} = \dfrac{2y\sqrt[3]{5x^2}}{\sqrt[3]{3z}} \cdot \dfrac{\sqrt[3]{9z^2}}{\sqrt[3]{9z^2}} = \dfrac{2y\sqrt[3]{45x^2z^2}}{3z}$

26. Combining the expressions: $5x\sqrt{6} + 2x\sqrt{6} - 9x\sqrt{6} = -2x\sqrt{6}$

27. Combining the expressions: $\sqrt{12} + \sqrt{3} = 2\sqrt{3} + \sqrt{3} = 3\sqrt{3}$

28. Combining the expressions: $\dfrac{3}{\sqrt{5}} + \sqrt{5} = \dfrac{3}{\sqrt{5}} \cdot \dfrac{\sqrt{5}}{\sqrt{5}} + \sqrt{5} = \dfrac{3\sqrt{5}}{5} + \dfrac{5\sqrt{5}}{5} = \dfrac{8\sqrt{5}}{5}$

29. Combining the expressions: $3\sqrt{8} - 4\sqrt{72} + 5\sqrt{50} = 6\sqrt{2} - 24\sqrt{2} + 25\sqrt{2} = 7\sqrt{2}$

30. Combining the expressions: $3b\sqrt{27a^5b} + 2a\sqrt{3a^3b^3} = 9a^2b\sqrt{3ab} + 2a^2b\sqrt{3ab} = 11a^2b\sqrt{3ab}$

31. Combining the expressions: $2x\sqrt[3]{xy^3z^2} - 6y\sqrt[3]{x^4z^2} = 2xy\sqrt[3]{xz^2} - 6xy\sqrt[3]{xz^2} = -4xy\sqrt[3]{xz^2}$

32. Multiplying: $\sqrt{2}\left(\sqrt{3}-2\sqrt{2}\right)=\sqrt{6}-2\sqrt{4}=\sqrt{6}-4$

33. Multiplying: $\left(\sqrt{x}-2\right)\left(\sqrt{x}-3\right)=x-2\sqrt{x}-3\sqrt{x}+6=x-5\sqrt{x}+6$

34. Rationalizing the denominator: $\dfrac{3}{\sqrt{5}-2}=\dfrac{3}{\sqrt{5}-2}\cdot\dfrac{\sqrt{5}+2}{\sqrt{5}+2}=\dfrac{3\left(\sqrt{5}+2\right)}{5-4}=3\sqrt{5}+6$

35. Rationalizing the denominator: $\dfrac{\sqrt{7}+\sqrt{5}}{\sqrt{7}-\sqrt{5}}=\dfrac{\sqrt{7}+\sqrt{5}}{\sqrt{7}-\sqrt{5}}\cdot\dfrac{\sqrt{7}+\sqrt{5}}{\sqrt{7}+\sqrt{5}}=\dfrac{7+2\sqrt{35}+5}{7-5}=\dfrac{12+2\sqrt{35}}{2}=6+\sqrt{35}$

36. Rationalizing the denominator: $\dfrac{3\sqrt{7}}{3\sqrt{7}-4}=\dfrac{3\sqrt{7}}{3\sqrt{7}-4}\cdot\dfrac{3\sqrt{7}+4}{3\sqrt{7}+4}=\dfrac{9\cdot7+12\sqrt{7}}{63-16}=\dfrac{63+12\sqrt{7}}{47}$

37. Solving the equation:
$\sqrt{4a+1}=1$
$\left(\sqrt{4a+1}\right)^2=(1)^2$
$4a+1=1$
$4a=0$
$a=0$

38. Solving the equation:
$\sqrt[3]{3x-8}=1$
$\left(\sqrt[3]{3x-8}\right)^3=(1)^3$
$3x-8=1$
$3x=9$
$x=3$

39. Solving the equation:
$\sqrt{3x+1}-3=1$
$\sqrt{3x+1}=4$
$\left(\sqrt{3x+1}\right)^2=(4)^2$
$3x+1=16$
$3x=15$
$x=5$

40. Solving the equation:
$\sqrt{x+4}=\sqrt{x}-2$
$\left(\sqrt{x+4}\right)^2=\left(\sqrt{x}-2\right)^2$
$x+4=x-2\sqrt{x}+4$
$0=-2\sqrt{x}$
$x=0$

There is no solution (0 does not check).

41. Graphing the equation:

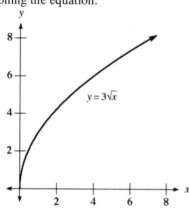

42. Graphing the equation:

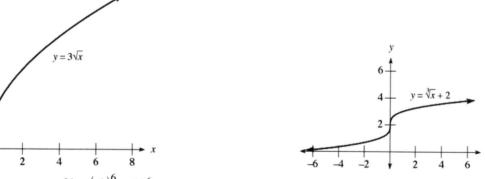

43. Writing in terms of i: $i^{24}=\left(i^4\right)^6=(1)^6=1$

44. Writing in terms of i: $i^{27}=i^{24}\cdot i^2\cdot i=\left(i^4\right)^6\cdot(-1)\cdot i=(1)^6(-i)=-i$

45. Setting real and imaginary parts equal:
$-2x=3 \qquad 8y=-4$
$x=-\dfrac{3}{2} \qquad y=-\dfrac{1}{2}$

46. Setting real and imaginary parts equal:
$3x+2=-4 \qquad 2y=-8$
$3x=-6 \qquad y=-4$
$x=-2$

47. Combining the numbers: $(3+5i)+(6-2i)=9+3i$

48. Combining the numbers: $(2+5i)-\left[(3+2i)+(6-i)\right]=(2+5i)-(9+i)=2+5i-9-i=-7+4i$

49. Multiplying: $3i(4+2i)=12i+6i^2=-6+12i$

50. Multiplying: $(2+3i)(4+i) = 8+12i+2i+3i^2 = 8+14i-3 = 5+14i$

51. Multiplying: $(4+2i)^2 = (4+2i)(4+2i) = 16+8i+8i+4i^2 = 16+16i-4 = 12+16i$

52. Multiplying: $(4+3i)(4-3i) = 16-9i^2 = 16+9 = 25$

53. Dividing: $\dfrac{3+i}{i} = \dfrac{3+i}{i} \cdot \dfrac{i}{i} = \dfrac{3i-1}{-1} = 1-3i$

54. Dividing: $\dfrac{-3}{2+i} = \dfrac{-3}{2+i} \cdot \dfrac{2-i}{2-i} = \dfrac{-6+3i}{4+1} = -\dfrac{6}{5}+\dfrac{3}{5}i$

55. Let l represent the desired length. Using the Pythagorean theorem:
$$18^2 + 13.5^2 = l^2$$
$$l^2 = 506.25$$
$$l = \sqrt{506.25} = 22.5$$
The length of one side of the roof is 22.5 feet.

56. Let d represent the distance across the pond. Using the Pythagorean theorem:
$$25^2 + 60^2 = d^2$$
$$d^2 = 4225$$
$$d = \sqrt{4225} = 65$$
The distance across the pond is 65 yards.

Chapter 10
Quadratic Functions

10.1 Completing the Square

1. Solving the equation:
$$x^2 = 25$$
$$x = \pm\sqrt{25} = \pm 5$$

3. Solving the equation:
$$y^2 = \tfrac{3}{4}$$
$$y = \pm\sqrt{\tfrac{3}{4}} = \pm\tfrac{\sqrt{3}}{2}$$

5. Solving the equation:
$$x^2 + 12 = 0$$
$$x^2 = -12$$
$$x = \pm\sqrt{-12} = \pm 2i\sqrt{3}$$

7. Solving the equation:
$$4a^2 - 45 = 0$$
$$4a^2 = 45$$
$$a^2 = \tfrac{45}{4}$$
$$a = \pm\sqrt{\tfrac{45}{4}} = \pm\tfrac{3\sqrt{5}}{2}$$

9. Solving the equation:
$$(2y-1)^2 = 25$$
$$2y - 1 = \pm\sqrt{25} = \pm 5$$
$$2y - 1 = -5, 5$$
$$2y = -4, 6$$
$$y = -2, 3$$

11. Solving the equation:
$$(2a+3)^2 = -9$$
$$2a + 3 = \pm\sqrt{-9} = \pm 3i$$
$$2a = -3 \pm 3i$$
$$a = \tfrac{-3 \pm 3i}{2}$$

13. Solving the equation:
$$x^2 + 8x + 16 = -27$$
$$(x+4)^2 = -27$$
$$x + 4 = \pm\sqrt{-27} = \pm 3i\sqrt{3}$$
$$x = -4 \pm 3i\sqrt{3}$$

15. Solving the equation:
$$4a^2 - 12a + 9 = -4$$
$$(2a-3)^2 = -4$$
$$2a - 3 = \pm\sqrt{-4} = \pm 2i$$
$$2a = 3 \pm 2i$$
$$a = \tfrac{3 \pm 2i}{2}$$

17. Completing the square: $x^2 + 12x + 36 = (x+6)^2$

19. Completing the square: $x^2 - 4x + 4 = (x-2)^2$

21. Completing the square: $a^2 - 10a + 25 = (a-5)^2$

23. Completing the square: $x^2 + 5x + \tfrac{25}{4} = \left(x + \tfrac{5}{2}\right)^2$

25. Completing the square: $y^2 - 7y + \tfrac{49}{4} = \left(y - \tfrac{7}{2}\right)^2$

27. Completing the square: $x^2 + \tfrac{1}{2}x + \tfrac{1}{16} = \left(x + \tfrac{1}{4}\right)^2$

284 Chapter 10 Quadratic Functions

29. Completing the square: $x^2 + \frac{2}{3}x + \frac{1}{9} = \left(x + \frac{1}{3}\right)^2$

31. Solving the equation:
$$x^2 + 12x = -27$$
$$x^2 + 12x + 36 = -27 + 36$$
$$(x+6)^2 = 9$$
$$x + 6 = \pm\sqrt{9} = \pm 3$$
$$x + 6 = -3, 3$$
$$x = -9, -3$$

33. Solving the equation:
$$a^2 - 2a + 5 = 0$$
$$a^2 - 2a + 1 = -5 + 1$$
$$(a-1)^2 = -4$$
$$a - 1 = \pm\sqrt{-4} = \pm 2i$$
$$a = 1 \pm 2i$$

35. Solving the equation:
$$y^2 - 8y + 1 = 0$$
$$y^2 - 8y + 16 = -1 + 16$$
$$(y-4)^2 = 15$$
$$y - 4 = \pm\sqrt{15}$$
$$y = 4 \pm \sqrt{15}$$

37. Solving the equation:
$$x^2 - 5x - 3 = 0$$
$$x^2 - 5x + \frac{25}{4} = 3 + \frac{25}{4}$$
$$\left(x - \frac{5}{2}\right)^2 = \frac{37}{4}$$
$$x - \frac{5}{2} = \pm\frac{\sqrt{37}}{2}$$
$$x = \frac{5 \pm \sqrt{37}}{2}$$

39. Solving the equation:
$$2x^2 - 4x - 8 = 0$$
$$x^2 - 2x - 4 = 0$$
$$x^2 - 2x + 1 = 4 + 1$$
$$(x-1)^2 = 5$$
$$x - 1 = \pm\sqrt{5}$$
$$x = 1 \pm \sqrt{5}$$

41. Solving the equation:
$$3t^2 - 8t + 1 = 0$$
$$t^2 - \frac{8}{3}t + \frac{1}{3} = 0$$
$$t^2 - \frac{8}{3}t + \frac{16}{9} = -\frac{1}{3} + \frac{16}{9}$$
$$\left(t - \frac{4}{3}\right)^2 = \frac{13}{9}$$
$$t - \frac{4}{3} = \pm\sqrt{\frac{13}{9}} = \pm\frac{\sqrt{13}}{3}$$
$$t = \frac{4 \pm \sqrt{13}}{3}$$

43. Solving the equation:
$$4x^2 - 3x + 5 = 0$$
$$x^2 - \frac{3}{4}x + \frac{5}{4} = 0$$
$$x^2 - \frac{3}{4}x + \frac{9}{64} = -\frac{5}{4} + \frac{9}{64}$$
$$\left(x - \frac{3}{8}\right)^2 = -\frac{71}{64}$$
$$x - \frac{3}{8} = \pm\sqrt{-\frac{71}{64}} = \pm\frac{i\sqrt{71}}{8}$$
$$x = \frac{3 \pm i\sqrt{71}}{8}$$

45. a. No, it cannot be solved by factoring.
b. Solving the equation:
$$x^2 = -9$$
$$x = \pm\sqrt{-9}$$
$$x = \pm 3i$$

47. a. Solving by factoring:
$$x^2 - 6x = 0$$
$$x(x-6) = 0$$
$$x = 0, 6$$

b. Solving by completing the square:
$$x^2 - 6x = 0$$
$$x^2 - 6x + 9 = 0 + 9$$
$$(x-3)^2 = 9$$
$$x - 3 = \pm\sqrt{9}$$
$$x - 3 = -3, 3$$
$$x = 0, 6$$

49. a. Solving by factoring:
$$x^2 + 2x = 35$$
$$x^2 + 2x - 35 = 0$$
$$(x+7)(x-5) = 0$$
$$x = -7, 5$$

b. Solving by completing the square:
$$x^2 + 2x = 35$$
$$x^2 + 2x + 1 = 35 + 1$$
$$(x+1)^2 = 36$$
$$x + 1 = \pm\sqrt{36}$$
$$x + 1 = -6, 6$$
$$x = -7, 5$$

51. Substituting: $x^2 - 6x - 7 = (-3+\sqrt{2})^2 - 6(-3+\sqrt{2}) - 7 = 9 - 6\sqrt{2} + 2 + 18 - 6\sqrt{2} - 7 = 22 - 12\sqrt{2}$
No, $x = -3 + \sqrt{2}$ is not a solution to the equation.

53. a. Solving the equation:
$$5x - 7 = 0$$
$$5x = 7$$
$$x = \tfrac{7}{5}$$

b. Solving the equation:
$$5x - 7 = 8$$
$$5x = 15$$
$$x = 3$$

c. Solving the equation:
$$(5x-7)^2 = 8$$
$$5x - 7 = \pm\sqrt{8}$$
$$5x - 7 = \pm 2\sqrt{2}$$
$$5x = 7 \pm 2\sqrt{2}$$
$$x = \frac{7 \pm 2\sqrt{2}}{5}$$

d. Solving the equation:
$$\sqrt{5x-7} = 8$$
$$(\sqrt{5x-7})^2 = (8)^2$$
$$5x - 7 = 64$$
$$5x = 71$$
$$x = \tfrac{71}{5}$$

e. Solving the equation:
$$\frac{5}{2} - \frac{7}{2x} = \frac{4}{x}$$
$$2x\left(\frac{5}{2} - \frac{7}{2x}\right) = 2x\left(\frac{4}{x}\right)$$
$$5x - 7 = 8$$
$$5x = 15$$
$$x = 3$$

55. Solving the equation:
$$(x+5)^2 + (x-5)^2 = 52$$
$$x^2 + 10x + 25 + x^2 - 10x + 25 = 52$$
$$2x^2 + 50 = 52$$
$$2x^2 = 2$$
$$x^2 = 1$$
$$x = \pm\sqrt{1} = \pm 1$$

57. Solving the equation:
$$(2x+3)^2 + (2x-3)^2 = 26$$
$$4x^2 + 12x + 9 + 4x^2 - 12x + 9 = 26$$
$$8x^2 + 18 = 26$$
$$8x^2 = 8$$
$$x^2 = 1$$
$$x = \pm\sqrt{1} = \pm 1$$

59. Solving the equation:
$$(3x+4)(3x-4)-(x+2)(x-2)=-4$$
$$9x^2-16-x^2+4=-4$$
$$8x^2-12=-4$$
$$8x^2=8$$
$$x^2=1$$
$$x=\pm\sqrt{1}=\pm 1$$

61. Completing the table:

x	$f(x)$	$g(x)$	$h(x)$
−2	49	49	25
−1	25	25	13
0	9	9	9
1	1	1	13
2	1	1	25

63. Solving the equation:
$$f(x)=0$$
$$(x-3)^2=0$$
$$x-3=0$$
$$x=3$$

65. a. Finding the x-intercepts:
$$x^2-5x-6=0$$
$$(x+1)(x-6)=0$$
$$x=-1,6$$

b. Setting $f(x)=0$:
$$f(x)=0$$
$$x^2-5x-6=0$$
$$(x+1)(x-6)=0$$
$$x=-1,6$$

c. Finding the value: $f(0)=(0)^2-5(0)-6=-6$

d. Finding the value: $f(1)=(1)^2-5(1)-6=-10$

67. The other two sides are $\frac{\sqrt{3}}{2}$ inch, 1 inch.

69. The hypotenuse is $\sqrt{2}$ inches.

71. Let x represent the horizontal distance. Using the Pythagorean theorem:
$$x^2+120^2=790^2$$
$$x^2+14400=624100$$
$$x^2=609700$$
$$x=\sqrt{609700}\approx 781 \text{ feet}$$

73. Solving for r:
$$3456=3000(1+r)^2$$
$$(1+r)^2=1.152$$
$$1+r=\sqrt{1.152}$$
$$r=\sqrt{1.152}-1\approx 0.073$$
The annual interest rate is 7.3%.

75. Its length is $20\sqrt{2}\approx 28$ feet.

77. Simplifying: $\sqrt{45}=\sqrt{9\cdot 5}=3\sqrt{5}$

79. Simplifying: $\sqrt{27y^5}=\sqrt{9y^4\cdot 3y}=3y^2\sqrt{3y}$

81. Simplifying: $\sqrt[3]{54x^6y^5}=\sqrt[3]{27x^6y^3\cdot 2y^2}=3x^2y\sqrt[3]{2y^2}$

83. Rationalizing the denominator: $\frac{3}{\sqrt{2}}=\frac{3}{\sqrt{2}}\cdot\frac{\sqrt{2}}{\sqrt{2}}=\frac{3\sqrt{2}}{2}$

85. Rationalizing the denominator: $\frac{2}{\sqrt[3]{4}}=\frac{2}{\sqrt[3]{4}}\cdot\frac{\sqrt[3]{2}}{\sqrt[3]{2}}=\frac{2\sqrt[3]{2}}{2}=\sqrt[3]{2}$

87. Simplifying: $\sqrt{49-4(6)(-5)}=\sqrt{49+120}=\sqrt{169}=13$

89. Simplifying: $\sqrt{(-27)^2-4(0.1)(1,700)}=\sqrt{729-680}=\sqrt{49}=7$

91. Simplifying: $\dfrac{-7+\sqrt{169}}{12} = \dfrac{-7+13}{12} = \dfrac{6}{12} = \dfrac{1}{2}$

93. Substituting values: $\sqrt{b^2 - 4ac} = \sqrt{7^2 - 4(6)(-5)} = \sqrt{49+120} = \sqrt{169} = 13$

95. Factoring: $27t^3 - 8 = (3t-2)(9t^2 + 6t + 4)$

97. Solving for x:

$$(x+a)^2 + (x-a)^2 = 10a^2$$
$$x^2 + 2ax + a^2 + x^2 - 2ax + a^2 = 10a^2$$
$$2x^2 + 2a^2 = 10a^2$$
$$2x^2 = 8a^2$$
$$x^2 = 4a^2$$
$$x = \pm 2a$$

99. Solving for x:

$$x^2 + px + q = 0$$
$$x^2 + px + \dfrac{p^2}{4} = -q + \dfrac{p^2}{4}$$
$$\left(x + \dfrac{p}{2}\right)^2 = \dfrac{p^2 - 4q}{4}$$
$$x + \dfrac{p}{2} = \pm\sqrt{\dfrac{p^2 - 4q}{4}} = \pm\dfrac{\sqrt{p^2 - 4q}}{2}$$
$$x = \dfrac{-p \pm \sqrt{p^2 - 4q}}{2}$$

101. Solving for x:

$$3x^2 + px + q = 0$$
$$x^2 + \dfrac{p}{3}x + \dfrac{q}{3} = 0$$
$$x^2 + \dfrac{p}{3}x + \dfrac{p^2}{36} = -\dfrac{q}{3} + \dfrac{p^2}{36}$$
$$\left(x + \dfrac{p}{6}\right)^2 = \dfrac{p^2 - 12q}{36}$$
$$x + \dfrac{p}{6} = \pm\sqrt{\dfrac{p^2 - 12q}{36}} = \pm\dfrac{\sqrt{p^2 - 12q}}{6}$$
$$x = \dfrac{-p \pm \sqrt{p^2 - 12q}}{6}$$

103. Completing the square:

$$x^2 - 10x + y^2 - 6y = -30$$
$$(x^2 - 10x + 25) + (y^2 - 6y + 9) = -30 + 25 + 9$$
$$(x-5)^2 + (y-3)^2 = 4$$
$$(x-5)^2 + (y-3)^2 = 2^2$$

10.2 The Quadratic Formula

1.
  a. Using the quadratic formula:
  $$x = \dfrac{-4 \pm \sqrt{4^2 - 4(3)(-2)}}{2(3)} = \dfrac{-4 \pm \sqrt{16+24}}{6} = \dfrac{-4 \pm \sqrt{40}}{6} = \dfrac{-4 \pm 2\sqrt{10}}{6} = \dfrac{-2 \pm \sqrt{10}}{3}$$

  b. Using the quadratic formula: $x = \dfrac{4 \pm \sqrt{(-4)^2 - 4(3)(-2)}}{2(3)} = \dfrac{4 \pm \sqrt{16+24}}{6} = \dfrac{4 \pm \sqrt{40}}{6} = \dfrac{4 \pm 2\sqrt{10}}{6} = \dfrac{2 \pm \sqrt{10}}{3}$

  c. Using the quadratic formula: $x = \dfrac{-4 \pm \sqrt{4^2 - 4(3)(2)}}{2(3)} = \dfrac{-4 \pm \sqrt{16-24}}{6} = \dfrac{-4 \pm \sqrt{-8}}{6} = \dfrac{-4 \pm 2i\sqrt{2}}{6} = \dfrac{-2 \pm i\sqrt{2}}{3}$

  d. Using the quadratic formula:
  $$x = \dfrac{-4 \pm \sqrt{4^2 - 4(2)(-3)}}{2(2)} = \dfrac{-4 \pm \sqrt{16+24}}{4} = \dfrac{-4 \pm \sqrt{40}}{4} = \dfrac{-4 \pm 2\sqrt{10}}{4} = \dfrac{-2 \pm \sqrt{10}}{2}$$

  e. Using the quadratic formula: $x = \dfrac{4 \pm \sqrt{(-4)^2 - 4(2)(3)}}{2(2)} = \dfrac{4 \pm \sqrt{16-24}}{4} = \dfrac{4 \pm \sqrt{-8}}{4} = \dfrac{4 \pm 2i\sqrt{2}}{4} = \dfrac{2 \pm i\sqrt{2}}{2}$

288 Chapter 10 Quadratic Functions

3. a. Using the quadratic formula: $x = \dfrac{2 \pm \sqrt{(-2)^2 - 4(1)(2)}}{2(1)} = \dfrac{2 \pm \sqrt{4-8}}{2} = \dfrac{2 \pm \sqrt{-4}}{2} = \dfrac{2 \pm 2i}{2} = 1 \pm i$

   b. Using the quadratic formula: $x = \dfrac{2 \pm \sqrt{(-2)^2 - 4(1)(5)}}{2(1)} = \dfrac{2 \pm \sqrt{4-20}}{2} = \dfrac{2 \pm \sqrt{-16}}{2} = \dfrac{2 \pm 4i}{2} = 1 \pm 2i$

   c. Using the quadratic formula: $x = \dfrac{-2 \pm \sqrt{2^2 - 4(1)(2)}}{2(1)} = \dfrac{-2 \pm \sqrt{4-8}}{2} = \dfrac{-2 \pm \sqrt{-4}}{2} = \dfrac{-2 \pm 2i}{2} = -1 \pm i$

5. Solving the equation:
$$\tfrac{1}{6}x^2 - \tfrac{1}{2}x + \tfrac{1}{3} = 0$$
$$x^2 - 3x + 2 = 0$$
$$(x-1)(x-2) = 0$$
$$x = 1, 2$$

7. Solving the equation:
$$\tfrac{x^2}{2} + 1 = \tfrac{2x}{3}$$
$$3x^2 + 6 = 4x$$
$$3x^2 - 4x + 6 = 0$$
$$x = \dfrac{4 \pm \sqrt{16-72}}{6} = \dfrac{4 \pm \sqrt{-56}}{6} = \dfrac{4 \pm 2i\sqrt{14}}{6} = \dfrac{2 \pm i\sqrt{14}}{3}$$

9. Solving the equation:
$$y^2 - 5y = 0$$
$$y(y-5) = 0$$
$$y = 0, 5$$

11. Solving the equation:
$$30x^2 + 40x = 0$$
$$10x(3x+4) = 0$$
$$x = -\tfrac{4}{3}, 0$$

13. Solving the equation:
$$\tfrac{2t^2}{3} - t = -\tfrac{1}{6}$$
$$4t^2 - 6t = -1$$
$$4t^2 - 6t + 1 = 0$$
$$t = \dfrac{6 \pm \sqrt{36-16}}{8} = \dfrac{6 \pm \sqrt{20}}{8} = \dfrac{6 \pm 2\sqrt{5}}{8} = \dfrac{3 \pm \sqrt{5}}{4}$$

15. Solving the equation:
$$0.01x^2 + 0.06x - 0.08 = 0$$
$$x^2 + 6x - 8 = 0$$
$$x = \dfrac{-6 \pm \sqrt{36+32}}{2} = \dfrac{-6 \pm \sqrt{68}}{2} = \dfrac{-6 \pm 2\sqrt{17}}{2} = -3 \pm \sqrt{17}$$

17. Solving the equation:
$$2x + 3 = -2x^2$$
$$2x^2 + 2x + 3 = 0$$
$$x = \dfrac{-2 \pm \sqrt{4-24}}{4} = \dfrac{-2 \pm \sqrt{-20}}{4} = \dfrac{-2 \pm 2i\sqrt{5}}{4} = \dfrac{-1 \pm i\sqrt{5}}{2}$$

19. Solving the equation:
$$100x^2 - 200x + 100 = 0$$
$$100(x^2 - 2x + 1) = 0$$
$$100(x-1)^2 = 0$$
$$x = 1$$

21. Solving the equation:
$$\tfrac{1}{2}r^2 = \tfrac{1}{6}r - \tfrac{2}{3}$$
$$3r^2 = r - 4$$
$$3r^2 - r + 4 = 0$$
$$r = \frac{1 \pm \sqrt{1-48}}{6} = \frac{1 \pm \sqrt{-47}}{6} = \frac{1 \pm i\sqrt{47}}{6}$$

23. Solving the equation:
$$(x-3)(x-5) = 1$$
$$x^2 - 8x + 15 = 1$$
$$x^2 - 8x + 14 = 0$$
$$x = \frac{8 \pm \sqrt{64-56}}{2} = \frac{8 \pm \sqrt{8}}{2} = \frac{8 \pm 2\sqrt{2}}{2} = 4 \pm \sqrt{2}$$

25. Solving the equation:
$$(x+3)^2 + (x-8)(x-1) = 16$$
$$x^2 + 6x + 9 + x^2 - 9x + 8 = 16$$
$$2x^2 - 3x + 1 = 0$$
$$(2x-1)(x-1) = 0$$
$$x = \tfrac{1}{2}, 1$$

27. Solving the equation:
$$\frac{x^2}{3} - \frac{5x}{6} = \tfrac{1}{2}$$
$$2x^2 - 5x = 3$$
$$2x^2 - 5x - 3 = 0$$
$$(2x+1)(x-3) = 0$$
$$x = -\tfrac{1}{2}, 3$$

29. Solving the equation:
$$\frac{1}{x+1} - \frac{1}{x} = \tfrac{1}{2}$$
$$2x(x+1)\left(\frac{1}{x+1} - \frac{1}{x}\right) = 2x(x+1) \cdot \tfrac{1}{2}$$
$$2x - (2x+2) = x^2 + x$$
$$2x - 2x - 2 = x^2 + x$$
$$x^2 + x + 2 = 0$$
$$x = \frac{-1 \pm \sqrt{1-8}}{2} = \frac{-1 \pm \sqrt{-7}}{2} = \frac{-1 \pm i\sqrt{7}}{2}$$

31. Solving the equation:
$$\frac{1}{y-1} + \frac{1}{y+1} = 1$$
$$(y+1)(y-1)\left(\frac{1}{y-1} + \frac{1}{y+1}\right) = (y+1)(y-1) \cdot 1$$
$$y+1+y-1 = y^2 - 1$$
$$2y = y^2 - 1$$
$$y^2 - 2y - 1 = 0$$
$$y = \frac{2 \pm \sqrt{4+4}}{2} = \frac{2 \pm \sqrt{8}}{2} = \frac{2 \pm 2\sqrt{2}}{2} = 1 \pm \sqrt{2}$$

290 Chapter 10 Quadratic Functions

33. Solving the equation:
$$\frac{1}{x+2} + \frac{1}{x+3} = 1$$
$$(x+2)(x+3)\left(\frac{1}{x+2} + \frac{1}{x+3}\right) = (x+2)(x+3) \cdot 1$$
$$x+3+x+2 = x^2 + 5x + 6$$
$$2x+5 = x^2 + 5x + 6$$
$$x^2 + 3x + 1 = 0$$
$$x = \frac{-3 \pm \sqrt{9-4}}{2} = \frac{-3 \pm \sqrt{5}}{2}$$

35. Solving the equation:
$$\frac{6}{r^2-1} - \frac{1}{2} = \frac{1}{r+1}$$
$$2(r+1)(r-1)\left(\frac{6}{(r+1)(r-1)} - \frac{1}{2}\right) = 2(r+1)(r-1) \cdot \frac{1}{r+1}$$
$$12 - (r^2 - 1) = 2r - 2$$
$$12 - r^2 + 1 = 2r - 2$$
$$r^2 + 2r - 15 = 0$$
$$(r+5)(r-3) = 0$$
$$r = -5, 3$$

37. Solving the equation:
$$x^3 - 8 = 0$$
$$(x-2)(x^2 + 2x + 4) = 0$$
$$x = 2 \quad \text{or} \quad x = \frac{-2 \pm \sqrt{4-16}}{2} = \frac{-2 \pm \sqrt{-12}}{2} = \frac{-2 \pm 2i\sqrt{3}}{2} = -1 \pm i\sqrt{3}$$
$$x = 2, -1 \pm i\sqrt{3}$$

39. Solving the equation:
$$8a^3 + 27 = 0$$
$$(2a+3)(4a^2 - 6a + 9) = 0$$
$$a = -\frac{3}{2} \quad \text{or} \quad a = \frac{6 \pm \sqrt{36 - 144}}{8} = \frac{6 \pm \sqrt{-108}}{8} = \frac{6 \pm 6i\sqrt{3}}{8} = \frac{3 \pm 3i\sqrt{3}}{4}$$
$$a = -\frac{3}{2}, \frac{3 \pm 3i\sqrt{3}}{4}$$

41. Solving the equation:
$$125t^3 - 1 = 0$$
$$(5t-1)(25t^2 + 5t + 1) = 0$$
$$t = \frac{1}{5} \quad \text{or} \quad t = \frac{-5 \pm \sqrt{25 - 100}}{50} = \frac{-5 \pm \sqrt{-75}}{50} = \frac{-5 \pm 5i\sqrt{3}}{50} = \frac{-1 \pm i\sqrt{3}}{10}$$
$$t = \frac{1}{5}, \frac{-1 \pm i\sqrt{3}}{10}$$

43. Solving the equation:
$$2x^3 + 2x^2 + 3x = 0$$
$$x(2x^2 + 2x + 3) = 0$$

$x = 0$ or $x = \dfrac{-2 \pm \sqrt{4-24}}{4} = \dfrac{-2 \pm \sqrt{-20}}{4} = \dfrac{-2 \pm 2i\sqrt{5}}{4} = \dfrac{-1 \pm i\sqrt{5}}{2}$

$x = 0, \dfrac{-1 \pm i\sqrt{5}}{2}$

45. Solving the equation:
$$3y^4 = 6y^3 - 6y^2$$
$$3y^4 - 6y^3 + 6y^2 = 0$$
$$3y^2(y^2 - 2y + 2) = 0$$

$y = 0$ or $y = \dfrac{2 \pm \sqrt{4-8}}{2} = \dfrac{2 \pm \sqrt{-4}}{2} = \dfrac{2 \pm 2i}{2} = 1 \pm i$

$y = 0, 1 \pm i$

47. Solving the equation:
$$6t^5 + 4t^4 = -2t^3$$
$$6t^5 + 4t^4 + 2t^3 = 0$$
$$2t^3(3t^2 + 2t + 1) = 0$$

$t = 0$ or $t = \dfrac{-2 \pm \sqrt{4-12}}{6} = \dfrac{-2 \pm \sqrt{-8}}{6} = \dfrac{-2 \pm 2i\sqrt{2}}{6} = \dfrac{-1 \pm i\sqrt{2}}{3}$

$t = 0, \dfrac{-1 \pm i\sqrt{2}}{3}$

49. The expressions from **a** and **b** are equivalent, since: $\dfrac{6 + 2\sqrt{3}}{4} = \dfrac{2(3+\sqrt{3})}{4} = \dfrac{3+\sqrt{3}}{2}$

51. a. Solving by factoring:
$$3x^2 - 5x = 0$$
$$x(3x - 5) = 0$$
$$x = 0, \tfrac{5}{3}$$

b. Using the quadratic formula: $x = \dfrac{5 \pm \sqrt{(-5)^2 - 4(3)(0)}}{2(3)} = \dfrac{5 \pm \sqrt{25-0}}{6} = \dfrac{5 \pm 5}{6} = 0, \tfrac{5}{3}$

53. No, it cannot be solved by factoring. Using the quadratic formula:
$$x = \dfrac{4 \pm \sqrt{(-4)^2 - 4(1)(7)}}{2(1)} = \dfrac{4 \pm \sqrt{16-28}}{2} = \dfrac{4 \pm \sqrt{-12}}{2} = \dfrac{4 \pm 2i\sqrt{3}}{2} = 2 \pm i\sqrt{3}$$

55. Substituting: $x^2 + 2x = (-1+i)^2 + 2(-1+i) = 1 - 2i + i^2 - 2 + 2i = 1 - 2i - 1 - 2 + 2i = -2$

Yes, $x = -1+i$ is a solution to the equation.

57. a. Solving the equation:
$$f(x) = 0$$
$$x^2 - 2x - 3 = 0$$
$$(x+1)(x-3) = 0$$
$$x = -1, 3$$

b. Solving the equation:
$$f(x) = -11$$
$$x^2 - 2x - 3 = -11$$
$$x^2 - 2x = -8$$
$$x^2 - 2x + 1 = -8 + 1$$
$$(x-1)^2 = -7$$
$$x - 1 = \pm i\sqrt{7}$$
$$x = 1 \pm i\sqrt{7}$$

c. Solving the equation:
$$f(x) = -2x+1$$
$$x^2 - 2x - 3 = -2x + 1$$
$$x^2 - 4 = 0$$
$$(x+2)(x-2) = 0$$
$$x = -2, 2$$

d. Solving the equation:
$$f(x) = 2x+1$$
$$x^2 - 2x - 3 = 2x + 1$$
$$x^2 - 4x = 4$$
$$x^2 - 4x + 4 = 4 + 4$$
$$(x-2)^2 = 8$$
$$x - 2 = \pm 2\sqrt{2}$$
$$x = 2 \pm 2\sqrt{2}$$

59. a. Solving the equation:
$$f(x) = g(x)$$
$$\frac{10}{x^2} = 3 + \frac{1}{x}$$
$$x^2 \left(\frac{10}{x^2}\right) = x^2 \left(3 + \frac{1}{x}\right)$$
$$10 = 3x^2 + x$$
$$0 = 3x^2 + x - 10$$
$$0 = (3x - 5)(x + 2)$$
$$x = -2, \tfrac{5}{3}$$

b. Solving the equation:
$$f(x) = g(x)$$
$$\frac{10}{x^2} = 8x - \frac{17}{x^2}$$
$$x^2 \left(\frac{10}{x^2}\right) = x^2 \left(8x - \frac{17}{x^2}\right)$$
$$10 = 8x^3 - 17$$
$$0 = 8x^3 - 27$$
$$0 = (2x - 3)(4x^2 + 6x + 9)$$
$$x = \tfrac{3}{2}, x = \frac{-6 \pm \sqrt{6^2 - 4(4)(9)}}{2(4)} = \frac{-6 \pm \sqrt{-108}}{8} = \frac{-6 \pm 6i\sqrt{3}}{8} = \frac{-3 \pm 3i\sqrt{3}}{4}$$

c. Solving the equation:
$$f(x) = g(x)$$
$$\frac{10}{x^2} = 0$$
$$10 = 0$$
(impossible)

There is no solution ($\emptyset$).

d. Solving the equation:
$$f(x) = g(x)$$
$$\frac{10}{x^2} = 10$$
$$x^2 \left(\frac{10}{x^2}\right) = x^2 (10)$$
$$10 = 10x^2$$
$$x^2 = 1$$
$$x = \pm 1$$

61. Substituting $s = 74$:
$$5t + 16t^2 = 74$$
$$16t^2 + 5t - 74 = 0$$
$$(16t + 37)(t - 2) = 0$$
$$t = 2 \quad \left(t = -\tfrac{37}{16} \text{ is impossible}\right)$$

It will take 2 seconds for the object to fall 74 feet.

63. Since profit is revenue minus the cost, the equation is:
$$100x - 0.5x^2 - (60x + 300) = 300$$
$$100x - 0.5x^2 - 60x - 300 = 300$$
$$-0.5x^2 + 40x - 600 = 0$$
$$x^2 - 80x + 1{,}200 = 0$$
$$(x-20)(x-60) = 0$$
$$x = 20, 60$$
The weekly profit is $300 if 20 items or 60 items are sold.

65. Let x represent the width of strip being cut off. After removing the strip, the overall area is 80% of its original area. The equation is:
$$(10.5 - 2x)(8.2 - 2x) = 0.80(10.5 \times 8.2)$$
$$86.1 - 37.4x + 4x^2 = 68.88$$
$$4x^2 - 37.4x + 17.22 = 0$$
$$x = \frac{37.4 \pm \sqrt{(-37.4)^2 - 4(4)(17.22)}}{8} = \frac{37.4 \pm \sqrt{1123.24}}{8} \approx \frac{37.4 \pm 33.5}{8} \approx 0.49, 8.86$$
The width of strip is 0.49 centimeter (8.86 cm is impossible).

67. **a.** The two equations are: $l + w = 10, lw = 15$
b. Since $l = 10 - w$, the equation is:
$$w(10 - w) = 15$$
$$10w - w^2 = 15$$
$$w^2 - 10w + 15 = 0$$
$$w = \frac{10 \pm \sqrt{100 - 60}}{2} = \frac{10 \pm \sqrt{40}}{2} = \frac{10 \pm 2\sqrt{10}}{2} = 5 \pm \sqrt{10} \approx 1.84, 8.16$$
The length and width are 8.16 yards and 1.84 yards.
c. Two answers are possible because either dimension (long or short) may be considered the length.

69. Dividing using long division:

$$\begin{array}{r} 4y+1 \\ 2y-7{\overline{\smash{\big)}\,8y^2 - 26y - 9}} \\ \underline{8y^2 - 28y} \\ 2y - 9 \\ \underline{2y - 7} \\ -2 \end{array}$$

The quotient is $4y + 1 - \dfrac{2}{2y - 7}$.

71. Dividing using long division:

$$\begin{array}{r} x^2 + 7x + 12 \\ x+2{\overline{\smash{\big)}\,x^3 + 9x^2 + 26x + 24}} \\ \underline{x^3 + 2x^2} \\ 7x^2 + 26x \\ \underline{7x^2 + 14x} \\ 12x + 24 \\ \underline{12x + 24} \\ 0 \end{array}$$

The quotient is $x^2 + 7x + 12$.

73. Simplifying: $25^{1/2} = \sqrt{25} = 5$

75. Simplifying: $\left(\frac{9}{25}\right)^{3/2} = \left(\frac{3}{5}\right)^3 = \frac{27}{125}$

77. Simplifying: $8^{-2/3} = \left(8^{-1/3}\right)^2 = \left(\frac{1}{2}\right)^2 = \frac{1}{4}$

79. Simplifying: $\dfrac{\left(49x^8 y^{-4}\right)^{1/2}}{\left(27x^{-3}y^9\right)^{-1/3}} = \dfrac{7x^4 y^{-2}}{\frac{1}{3}xy^{-3}} = 21x^{4-1}y^{-2+3} = 21x^3 y$

81. Evaluating $b^2 - 4ac$: $b^2 - 4ac = (-3)^2 - 4(1)(-40) = 9 + 160 = 169$

83. Evaluating $b^2 - 4ac$: $b^2 - 4ac = 12^2 - 4(4)(9) = 144 - 144 = 0$

85. Solving the equation:
$$k^2 - 144 = 0$$
$$(k+12)(k-12) = 0$$
$$k = -12, 12$$

87. Multiplying: $(x-3)(x+2) = x^2 + 2x - 3x - 6 = x^2 - x - 6$

89. Multiplying:
$$(x-3)(x-3)(x+2) = (x^2 - 6x + 9)(x+2)$$
$$= x^3 - 6x^2 + 9x + 2x^2 - 12x + 18$$
$$= x^3 - 4x^2 - 3x + 18$$

91. Solving the equation:
$$x^2 + \sqrt{3}x - 6 = 0$$
$$(x + 2\sqrt{3})(x - \sqrt{3}) = 0$$
$$x = -2\sqrt{3}, \sqrt{3}$$

93. Solving the equation:
$$\sqrt{2}x^2 + 2x - \sqrt{2} = 0$$
$$x = \frac{-2 \pm \sqrt{4 + 4(2)}}{2\sqrt{2}} = \frac{-2 \pm \sqrt{12}}{2\sqrt{2}} = \frac{-2 \pm 2\sqrt{3}}{2\sqrt{2}} \cdot \frac{\sqrt{2}}{\sqrt{2}} = \frac{-2\sqrt{2} \pm 2\sqrt{6}}{4} = \frac{-\sqrt{2} \pm \sqrt{6}}{2}$$

95. Solving the equation:
$$x^2 + ix + 2 = 0$$
$$x = \frac{-i \pm \sqrt{-1 - 8}}{2} = \frac{-i \pm 3i}{2} = -2i, i$$

10.3 Additional Items Involving Solutions to Equations

1. Computing the discriminant: $D = (-6)^2 - 4(1)(5) = 36 - 20 = 16$. The equation will have two rational solutions.

3. First write the equation as $4x^2 - 4x + 1 = 0$. Computing the discriminant: $D = (-4)^2 - 4(4)(1) = 16 - 16 = 0$
The equation will have one rational solution.

5. Computing the discriminant: $D = 1^2 - 4(1)(-1) = 1 + 4 = 5$. The equation will have two irrational solutions.

7. First write the equation as $2y^2 - 3y - 1 = 0$. Computing the discriminant: $D = (-3)^2 - 4(2)(-1) = 9 + 8 = 17$
The equation will have two irrational solutions.

9. Computing the discriminant: $D = 0^2 - 4(1)(-9) = 36$. The equation will have two rational solutions.

11. First write the equation as $5a^2 - 4a - 5 = 0$. Computing the discriminant: $D = (-4)^2 - 4(5)(-5) = 16 + 100 = 116$
The equation will have two irrational solutions.

13. Setting the discriminant equal to 0:
$$(-k)^2 - 4(1)(25) = 0$$
$$k^2 - 100 = 0$$
$$k^2 = 100$$
$$k = \pm 10$$

15. First write the equation as $x^2 - kx + 36 = 0$. Setting the discriminant equal to 0:
$$(-k)^2 - 4(1)(36) = 0$$
$$k^2 - 144 = 0$$
$$k^2 = 144$$
$$k = \pm 12$$

17. Setting the discriminant equal to 0:
$$(-12)^2 - 4(4)(k) = 0$$
$$144 - 16k = 0$$
$$16k = 144$$
$$k = 9$$

19. First write the equation as $kx^2 - 40x - 25 = 0$. Setting the discriminant equal to 0:
$$(-40)^2 - 4(k)(-25) = 0$$
$$1600 + 100k = 0$$
$$100k = -1600$$
$$k = -16$$

21. Setting the discriminant equal to 0:
$$(-k)^2 - 4(3)(2) = 0$$
$$k^2 - 24 = 0$$
$$k^2 = 24$$
$$k = \pm\sqrt{24} = \pm 2\sqrt{6}$$

23. Writing the equation:
$$(x-5)(x-2) = 0$$
$$x^2 - 7x + 10 = 0$$

25. Writing the equation:
$$(t+3)(t-6) = 0$$
$$t^2 - 3t - 18 = 0$$

27. Writing the equation:
$$(y-2)(y+2)(y-4) = 0$$
$$(y^2 - 4)(y-4) = 0$$
$$y^3 - 4y^2 - 4y + 16 = 0$$

29. Writing the equation:
$$(2x-1)(x-3) = 0$$
$$2x^2 - 7x + 3 = 0$$

31. Writing the equation:
$$(4t+3)(t-3) = 0$$
$$4t^2 - 9t - 9 = 0$$

33. Writing the equation:
$$(x-3)(x+3)(6x-5) = 0$$
$$(x^2 - 9)(6x-5) = 0$$
$$6x^3 - 5x^2 - 54x + 45 = 0$$

35. Writing the equation:
$$(2a+1)(5a-3) = 0$$
$$10a^2 - a - 3 = 0$$

37. Writing the equation:
$$(3x+2)(3x-2)(x-1) = 0$$
$$(9x^2 - 4)(x-1) = 0$$
$$9x^3 - 9x^2 - 4x + 4 = 0$$

39. Writing the equation:
$$(x-2)(x+2)(x-3)(x+3) = 0$$
$$(x^2 - 4)(x^2 - 9) = 0$$
$$x^4 - 13x^2 + 36 = 0$$

41. Writing the equation:
$$(x-1)(x+2) = 0$$
$$x^2 + x - 2 = 0$$

So $f(x) = x^2 + x - 2$.

43. Starting with the solutions:
$$x = 3 \pm \sqrt{2}$$
$$x - 3 = \pm\sqrt{2}$$
$$(x-3)^2 = (\pm\sqrt{2})^2$$
$$x^2 - 6x + 9 = 2$$
$$x^2 - 6x + 7 = 0$$
So $f(x) = x^2 - 6x + 7$.

45. Starting with the solutions:
$$x = 2 \pm i$$
$$x - 2 = \pm i$$
$$(x-2)^2 = (\pm i)^2$$
$$x^2 - 4x + 4 = -1$$
$$x^2 - 4x + 5 = 0$$
So $f(x) = x^2 - 4x + 5$.

47. Starting with the solutions:
$$x = \frac{5 \pm \sqrt{7}}{2}$$
$$2x = 5 \pm \sqrt{7}$$
$$(2x-5)^2 = (\pm\sqrt{7})^2$$
$$4x^2 - 20x + 25 = 7$$
$$4x^2 - 20x + 18 = 0$$
$$2x^2 - 10x + 9 = 0$$
So $f(x) = 2x^2 - 10x + 9$.

49. Starting with the solutions:
$$x = \frac{3 \pm i\sqrt{5}}{2}$$
$$2x = 3 \pm i\sqrt{5}$$
$$(2x-3)^2 = (\pm i\sqrt{5})^2$$
$$4x^2 - 12x + 9 = -5$$
$$4x^2 - 12x + 14 = 0$$
$$2x^2 - 6x + 7 = 0$$
So $f(x) = 2x^2 - 6x + 7$.

51. Starting with the solutions:
$$x = 2 \pm i\sqrt{3}$$
$$(x-2)^2 = (\pm i\sqrt{3})^2$$
$$x^2 - 4x + 4 = -3$$
$$x^2 - 4x + 7 = 0$$
So $f(x) = x(x^2 - 4x + 7) = x^3 - 4x^2 + 7x$.

53.
  a. This graph has one x-intercept, which is $y = g(x)$.
  b. This graph has no x-intercepts, which is $y = h(x)$.
  c. This graph has two x-intercepts, which is $y = f(x)$.

55. Multiplying: $(a^{1/2} - 5)(a^{1/2} + 3) = a - 2a^{1/2} - 15$

57. Multiplying: $(x^{1/2} - 8)(x^{1/2} + 8) = (x^{1/2})^2 - (8)^2 = x - 64$

59. Dividing: $\dfrac{45x^{5/3}y^{7/3} - 36x^{8/3}y^{4/3}}{9x^{2/3}y^{1/3}} = \dfrac{45x^{5/3}y^{7/3}}{9x^{2/3}y^{1/3}} - \dfrac{36x^{8/3}y^{4/3}}{9x^{2/3}y^{1/3}} = 5xy^2 - 4x^2y$

61. Factoring: $8(x+1)^{4/3} - 2(x+1)^{1/3} = 2(x+1)^{1/3}[4(x+1) - 1] = 2(x+1)^{1/3}(4x+3)$

63. Factoring: $9x^{2/3} + 12x^{1/3} + 4 = (3x^{1/3} + 2)^2$

65. Simplifying: $(x-2)^2 - 3(x-2) - 10 = x^2 - 4x + 4 - 3x + 6 - 10 = x^2 - 7x$

67. Simplifying: $(3a-2)^2 + 2(3a-2) - 3 = 9a^2 - 12a + 4 + 6a - 4 - 3 = 9a^2 - 6a - 3$

69. Simplifying:
$$6(2a+4)^2 - (2a+4) - 2 = 6(4a^2 + 16a + 16) - (2a+4) - 2$$
$$= 24a^2 + 96a + 96 - 2a - 4 - 2$$
$$= 24a^2 + 94a + 90$$

71. Solving the equation:
$$x^2 = -2$$
$$x = \pm\sqrt{-2} = \pm i\sqrt{2}$$

73. Solving the equation:
$$\sqrt{x} = 2$$
$$x = 2^2 = 4$$

75. Solving the equation:
$$x + 3 = -2$$
$$x = -5$$

77. Solving the equation:
$$y^2 + y - 6 = 0$$
$$(y+3)(y-2) = 0$$
$$y = -3, 2$$

79. Solving the equation:
$$6x^2 - 13x - 5 = 0$$
$$(3x+1)(2x-5) = 0$$
$$x = -\tfrac{1}{3}, \tfrac{5}{2}$$

81. Solving the equation:
$$4x^2 = 8x + 5$$
$$4x^2 - 8x - 5 = 0$$
$$(2x-5)(2x+1) = 0$$
$$x = -\tfrac{1}{2}, \tfrac{5}{2}$$

83. Solving the equation:
$$4x^2 - 1 = 4x$$
$$4x^2 - 4x - 1 = 0$$
$$x = \frac{4 \pm \sqrt{16+16}}{8} = \frac{4 \pm \sqrt{32}}{8} = \frac{4 \pm 4\sqrt{2}}{8} = \frac{1 \pm \sqrt{2}}{2} \approx -0.21, 1.21$$

85. Solving the equation:
$$3x^3 - 2x^2 - 3x + 2 = 0$$
$$x^2(3x-2) - 1(3x-2) = 0$$
$$(3x-2)(x^2-1) = 0$$
$$(3x-2)(x+1)(x-1) = 0$$
$$x = -1, \tfrac{2}{3}, 1$$

87. Solving the equation:
$$3x^3 - 9x = 2x^2 - 6$$
$$3x^3 - 2x^2 - 9x + 6 = 0$$
$$x^2(3x-2) - 3(3x-2) = 0$$
$$(3x-2)(x^2-3) = 0$$
$$x = \tfrac{2}{3}, \pm\sqrt{3}$$
$$x \approx -1.73, 0.67, 1.73$$

10.4 Equations Quadratic in Form

1. Solving the equation:
$$(x-3)^2 + 3(x-3) + 2 = 0$$
$$(x-3+2)(x-3+1) = 0$$
$$(x-1)(x-2) = 0$$
$$x = 1, 2$$

3. Solving the equation:
$$2(x+4)^2 + 5(x+4) - 12 = 0$$
$$[2(x+4) - 3][(x+4) + 4] = 0$$
$$(2x+5)(x+8) = 0$$
$$x = -8, -\tfrac{5}{2}$$

5. Solving the equation:
$$x^4 - 10x^2 + 9 = 0$$
$$(x^2 - 1)(x^2 - 9) = 0$$
$$x^2 = 1, 9$$
$$x = \pm 1, \pm 3$$

7. Solving the equation:
$$x^4 - 7x^2 + 12 = 0$$
$$(x^2 - 4)(x^2 - 3) = 0$$
$$x^2 = 4, 3$$
$$x = \pm 2, \pm\sqrt{3}$$

9. Solving the equation:
$$x^4 - 6x^2 - 27 = 0$$
$$(x^2 - 9)(x^2 + 3) = 0$$
$$x^2 = 9, -3$$
$$x = \pm 3, \pm i\sqrt{3}$$

11. Solving the equation:
$$x^4 + 9x^2 = -20$$
$$x^4 + 9x^2 + 20 = 0$$
$$(x^2 + 4)(x^2 + 5) = 0$$
$$x^2 = -4, -5$$
$$x = \pm 2i, \pm i\sqrt{5}$$

13. Solving the equation:
$$(2a-3)^2 - 9(2a-3) = -20$$
$$(2a-3)^2 - 9(2a-3) + 20 = 0$$
$$(2a-3-4)(2a-3-5) = 0$$
$$(2a-7)(2a-8) = 0$$
$$a = \tfrac{7}{2}, 4$$

15. Solving the equation:
$$2(4a+2)^2 = 3(4a+2) + 20$$
$$2(4a+2)^2 - 3(4a+2) - 20 = 0$$
$$[2(4a+2)+5][(4a+2)-4] = 0$$
$$(8a+9)(4a-2) = 0$$
$$a = -\tfrac{9}{8}, \tfrac{1}{2}$$

17. Solving the equation:
$$6t^4 = -t^2 + 5$$
$$6t^4 + t^2 - 5 = 0$$
$$(6t^2 - 5)(t^2 + 1) = 0$$
$$t^2 = \tfrac{5}{6}, -1$$
$$t = \pm\sqrt{\tfrac{5}{6}} = \pm\tfrac{\sqrt{30}}{6}, \pm i$$

19. Solving the equation:
$$9x^4 - 49 = 0$$
$$(3x^2 - 7)(3x^2 + 7) = 0$$
$$x^2 = \tfrac{7}{3}, -\tfrac{7}{3}$$
$$x = \pm\sqrt{\tfrac{7}{3}}, \pm\sqrt{-\tfrac{7}{3}}$$
$$t = \pm\tfrac{\sqrt{21}}{3}, \pm\tfrac{i\sqrt{21}}{3}$$

21. Solving the equation:
$$x - 7\sqrt{x} + 10 = 0$$
$$(\sqrt{x} - 5)(\sqrt{x} - 2) = 0$$
$$\sqrt{x} = 2, 5$$
$$x = 4, 25$$
Both values check in the original equation.

23. Solving the equation:
$$t - 2\sqrt{t} - 15 = 0$$
$$(\sqrt{t} - 5)(\sqrt{t} + 3) = 0$$
$$\sqrt{t} = -3, 5$$
$$t = 9, 25$$
Only $t = 25$ checks in the original equation.

25. Solving the equation:
$$6x + 11\sqrt{x} = 35$$
$$6x + 11\sqrt{x} - 35 = 0$$
$$(3\sqrt{x} - 5)(2\sqrt{x} + 7) = 0$$
$$\sqrt{x} = \tfrac{5}{3}, -\tfrac{7}{2}$$
$$x = \tfrac{25}{9}, \tfrac{49}{4}$$
Only $x = \tfrac{25}{9}$ checks in the original equation.

27. Solving the equation:
$$x - 2\sqrt{x} - 8 = 0$$
$$(\sqrt{x} - 4)(\sqrt{x} + 2) = 0$$
$$\sqrt{x} = -2, 4$$
$$x = 4, 16$$
Only $x = 16$ checks in the original equation.

29. Solving the equation:
$$x + 3\sqrt{x} - 18 = 0$$
$$(\sqrt{x} - 3)(\sqrt{x} + 6) = 0$$
$$\sqrt{x} = -6, 3$$
$$x = 36, 9$$
Only $x = 9$ checks in the original equation.

31. Solving the equation:
$$2x + 9\sqrt{x} - 5 = 0$$
$$(2\sqrt{x} - 1)(\sqrt{x} + 5) = 0$$
$$\sqrt{x} = -5, \tfrac{1}{2}$$
$$x = 25, \tfrac{1}{4}$$
Only $x = \tfrac{1}{4}$ checks in the original equation.

33. Solving the equation:
$$(a-2) - 11\sqrt{a-2} + 30 = 0$$
$$(\sqrt{a-2} - 6)(\sqrt{a-2} - 5) = 0$$
$$\sqrt{a-2} = 5, 6$$
$$a - 2 = 25, 36$$
$$a = 27, 38$$

35. Solving the equation:
$$(2x+1) - 8\sqrt{2x+1} + 15 = 0$$
$$(\sqrt{2x+1} - 3)(\sqrt{2x+1} - 5) = 0$$
$$\sqrt{2x+1} = 3, 5$$
$$2x + 1 = 9, 25$$
$$2x = 8, 24$$
$$x = 4, 12$$

37. Solving the equation:
$$(x^2+1)^2 - 2(x^2+1) - 15 = 0$$
$$(x^2+1-5)(x^2+1+3) = 0$$
$$(x^2-4)(x^2+4) = 0$$
$$x^2 = 4, -4$$
$$x = \pm 2, \pm 2i$$

39. Solving the equation:
$$(x^2+5)^2 - 6(x^2+5) - 27 = 0$$
$$(x^2+5-9)(x^2+5+3) = 0$$
$$(x^2-4)(x^2+8) = 0$$
$$x^2 = 4, -8$$
$$x = \pm 2, \pm 2i\sqrt{2}$$

41. Solving the equation:
$$x^{-2} - 3x^{-1} + 2 = 0$$
$$(x^{-1} - 2)(x^{-1} - 1) = 0$$
$$x^{-1} = 2, 1$$
$$x = \tfrac{1}{2}, 1$$

43. Solving the equation:
$$y^{-4} - 5y^{-2} = 0$$
$$y^{-2}(y^{-2} - 5) = 0$$
$$y^{-2} = 0, 5$$
$$y^2 = \tfrac{1}{5}$$
$$y = \pm\sqrt{\tfrac{1}{5}} = \pm\tfrac{\sqrt{5}}{5}$$

45. Solving the equation:
$$x^{2/3} + 4x^{1/3} - 12 = 0$$
$$(x^{1/3} + 6)(x^{1/3} - 2) = 0$$
$$x^{1/3} = -6, 2$$
$$x = (-6)^3, 2^3$$
$$x = -216, 8$$

47. Solving the equation:
$$x^{4/3} + 6x^{2/3} + 9 = 0$$
$$(x^{2/3} + 3)^2 = 0$$
$$x^{2/3} = -3$$
$$x^2 = (-3)^3 = -27$$
$$x = \pm\sqrt{-27} = \pm 3i\sqrt{3}$$

49. Solving for t:
$$16t^2 - vt - h = 0$$
$$t = \frac{v \pm \sqrt{v^2 - 4(16)(-h)}}{32} = \frac{v \pm \sqrt{v^2 + 64h}}{32}$$

51. Solving for x:
$$kx^2 + 8x + 4 = 0$$
$$x = \frac{-8 \pm \sqrt{64 - 16k}}{2k} = \frac{-8 \pm 4\sqrt{4-k}}{2k} = \frac{-4 \pm 2\sqrt{4-k}}{k}$$

53. Solving for x:
$$x^2 + 2xy + y^2 = 0$$
$$x = \frac{-2y \pm \sqrt{4y^2 - 4y^2}}{2} = \frac{-2y}{2} = -y$$

55. Solving for t (note that $t > 0$):
$$16t^2 - 8t - h = 0$$
$$t = \frac{8 + \sqrt{64 + 64h}}{32} = \frac{8 + 8\sqrt{1+h}}{32} = \frac{1 + \sqrt{1+h}}{4}$$

57. a. Sketching the graph:

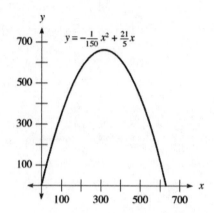

b. Finding the x-intercepts:
$$-\frac{1}{150}x^2 + \frac{21}{5}x = 0$$
$$-\frac{1}{150}x(x-630) = 0$$
$$x = 0, 630$$
The width is 630 feet.

59. Let $x = BC$. Solving the proportion:
$$\frac{AB}{BC} = \frac{BC}{AC}$$
$$\frac{4}{x} = \frac{x}{4+x}$$
$$16 + 4x = x^2$$
$$0 = x^2 - 4x - 16$$
$$x = \frac{4 \pm \sqrt{16+64}}{2} = \frac{4 \pm \sqrt{80}}{2} = \frac{4 \pm 4\sqrt{5}}{2} = 2 \pm 2\sqrt{5}$$
Thus $BC = 2 + 2\sqrt{5} = 4\left(\frac{1+\sqrt{5}}{2}\right)$.

61. Combining: $5\sqrt{7} - 2\sqrt{7} = 3\sqrt{7}$

63. Combining: $\sqrt{18} - \sqrt{8} + \sqrt{32} = 3\sqrt{2} - 2\sqrt{2} + 4\sqrt{2} = 5\sqrt{2}$

65. Combining: $9x\sqrt{20x^3y^2} + 7y\sqrt{45x^5} = 9x \cdot 2xy\sqrt{5x} + 7y \cdot 3x^2\sqrt{5x} = 18x^2y\sqrt{5x} + 21x^2y\sqrt{5x} = 39x^2y\sqrt{5x}$

67. Multiplying: $(\sqrt{5}-2)(\sqrt{5}+8) = 5 - 2\sqrt{5} + 8\sqrt{5} - 16 = -11 + 6\sqrt{5}$

69. Multiplying: $(\sqrt{x}+2)^2 = (\sqrt{x})^2 + 4\sqrt{x} + 4 = x + 4\sqrt{x} + 4$

71. Rationalizing the denominator: $\frac{\sqrt{7}}{\sqrt{7}-2} \cdot \frac{\sqrt{7}+2}{\sqrt{7}+2} = \frac{7+2\sqrt{7}}{7-4} = \frac{7+2\sqrt{7}}{3}$

73. Evaluating when $x = 1$: $y = 3(1)^2 - 6(1) + 1 = 3 - 6 + 1 = -2$

75. Evaluating: $P(135) = -0.1(135)^2 + 27(135) - 500 = -1,822.5 + 3,645 - 500 = 1,322.5$

77. Solving the equation:
$$0 = a(80)^2 + 70$$
$$0 = 6400a + 70$$
$$6400a = -70$$
$$a = -\frac{7}{640}$$

79. Solving the equation:
$$x^2 - 6x + 5 = 0$$
$$(x-1)(x-5) = 0$$
$$x = 1, 5$$

81. Solving the equation:
$$-x^2 - 2x + 3 = 0$$
$$x^2 + 2x - 3 = 0$$
$$(x+3)(x-1) = 0$$
$$x = -3, 1$$

83. Solving the equation:
$$2x^2 - 6x + 5 = 0$$
$$x = \frac{6 \pm \sqrt{(-6)^2 - 4(2)(5)}}{2(2)} = \frac{6 \pm \sqrt{36-40}}{4} = \frac{6 \pm \sqrt{-4}}{4} = \frac{6 \pm 2i}{4} = \frac{3 \pm i}{2} = \frac{3}{2} \pm \frac{1}{2}i$$

85. Completing the square: $x^2 - 6x + 9 = (x-3)^2$

87. Completing the square: $y^2 + 2y + 1 = (y+1)^2$

89. To find the x-intercepts, set $y = 0$:
$$x^3 - 4x = 0$$
$$x(x^2 - 4) = 0$$
$$x(x+2)(x-2) = 0$$
$$x = -2, 0, 2$$
To find the y-intercept, set $x = 0$: $y = 0^3 - 4(0) = 0$. The x-intercepts are $-2, 0, 2$ and the y-intercept is 0.

91. To find the x-intercepts, set $y = 0$:
$$3x^3 + x^2 - 27x - 9 = 0$$
$$x^2(3x+1) - 9(3x+1) = 0$$
$$(3x+1)(x^2 - 9) = 0$$
$$(3x+1)(x+3)(x-3) = 0$$
$$x = -3, -\tfrac{1}{3}, 3$$
To find the y-intercept, set $x = 0$: $y = 3(0)^3 + (0)^2 - 27(0) - 9 = -9$

The x-intercepts are $-3, -\tfrac{1}{3}, 3$ and the y-intercept is -9.

93. Using long division:

$$\begin{array}{r}
2x^2 + x - 1 \\
x-4 \overline{) 2x^3 - 7x^2 - 5x + 4} \\
\underline{2x^3 - 8x^2} \\
x^2 - 5x \\
\underline{x^2 - 4x} \\
-x + 4 \\
\underline{-x + 4} \\
0
\end{array}$$

Now solve the equation:
$$2x^2 + x - 1 = 0$$
$$(2x-1)(x+1) = 0$$
$$x = -1, \tfrac{1}{2}$$
It also crosses the x-axis at $\tfrac{1}{2}, -1$.

302   Chapter 10 Quadratic Functions

10.5 Graphing Parabolas

1. First complete the square: $y = x^2 + 2x - 3 = \left(x^2 + 2x + 1\right) - 1 - 3 = (x+1)^2 - 4$

   The x-intercepts are $-3, 1$ and the vertex is $(-1, -4)$. Graphing the parabola:

   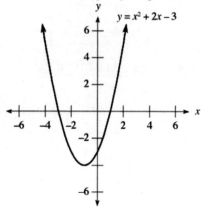

3. First complete the square: $y = -x^2 - 4x + 5 = -\left(x^2 + 4x + 4\right) + 4 + 5 = -(x+2)^2 + 9$

   The x-intercepts are $-5, 1$ and the vertex is $(-2, 9)$. Graphing the parabola:

   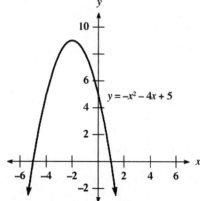

5. The x-intercepts are $-1, 1$ and the vertex is $(0, -1)$. Graphing the parabola:

   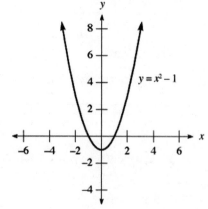

7. The x-intercepts are –3,3 and the vertex is (0,9). Graphing the parabola:

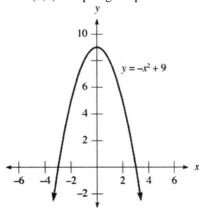

9. First complete the square: $y = 2x^2 - 4x - 6 = 2(x^2 - 2x + 1) - 2 - 6 = 2(x-1)^2 - 8$

The x-intercepts are –1,3 and the vertex is (1,–8). Graphing the parabola:

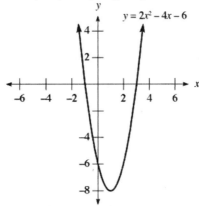

11. First complete the square: $y = x^2 - 2x - 4 = (x^2 - 2x + 1) - 1 - 4 = (x-1)^2 - 5$

The x-intercepts are $1 \pm \sqrt{5}$ and the vertex is (1,–5). Graphing the parabola:

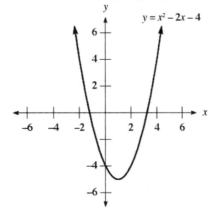

304 Chapter 10 Quadratic Functions

13. First complete the square: $y = x^2 - 4x - 4 = (x^2 - 4x + 4) - 4 - 4 = (x-2)^2 - 8$

    The vertex is (2,–8). Graphing the parabola:

    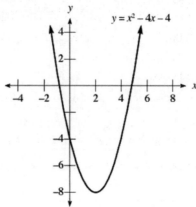

15. First complete the square: $y = -x^2 + 2x - 5 = -(x^2 - 2x + 1) + 1 - 5 = -(x-1)^2 - 4$

    The vertex is (1,–4). Graphing the parabola:

    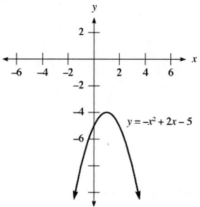

17. The vertex is (0,1). Graphing the parabola:

    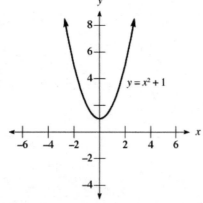

19. The vertex is (0,–3). Graphing the parabola:

    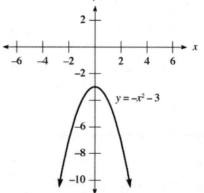

21. Completing the square: $y = x^2 - 6x + 5 = (x^2 - 6x + 9) - 9 + 5 = (x-3)^2 - 4$

    The vertex is (3,–4), which is the lowest point on the graph.

23. Completing the square: $y = -x^2 + 2x + 8 = -(x^2 - 2x + 1) + 1 + 8 = -(x-1)^2 + 9$

    The vertex is (1,9), which is the highest point on the graph.

25. Completing the square: $y = -x^2 + 4x + 12 = -(x^2 - 4x + 4) + 4 + 12 = -(x-2)^2 + 16$

    The vertex is (2,16), which is the highest point on the graph.

27. Completing the square: $y = -x^2 - 8x = -(x^2 + 8x + 16) + 16 = -(x+4)^2 + 16$

The vertex is (–4,16), which is the highest point on the graph.

29. First complete the square:
$$P(x) = -0.002x^2 + 3.5x - 800 = -0.002(x^2 - 1750x + 765{,}625) + 1{,}531.25 - 800 = -0.002(x - 875)^2 + 731.25$$

It must sell 875 patterns to obtain a maximum profit of $731.25.

31. The ball is in her hand at times 0 sec and 2 sec.

Completing the square: $h(t) = -16t^2 + 32t = -16(t^2 - 2t + 1) + 16 = -16(t-1)^2 + 16$

The maximum height of the ball is 16 feet.

33. Completing the square: $R = xp = 1200p - 100p^2 = -100(p^2 - 12p + 36) + 3600 = -100(p-6)^2 + 3600$

The price is $6.00 and the maximum revenue is $3,600. Sketching the graph:

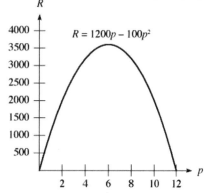

35. Completing the square: $R = xp = 1700p - 100p^2 = -100(p^2 - 17p + 72.25) + 7225 = -100(p-8.5)^2 + 7225$

The price is $8.50 and the maximum revenue is $7,225. Sketching the graph:

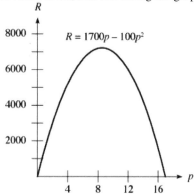

37. The equation is given on the graph:

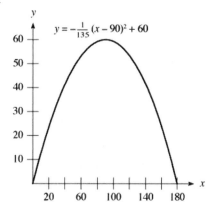

306 Chapter 10 Quadratic Functions

39. Performing the operations: $(3-5i)-(2-4i) = 3-5i-2+4i = 1-i$

41. Performing the operations: $(3+2i)(7-3i) = 21+5i-6i^2 = 27+5i$

43. Performing the operations: $\dfrac{i}{3+i} = \dfrac{i}{3+i} \cdot \dfrac{3-i}{3-i} = \dfrac{3i-i^2}{9-i^2} = \dfrac{3i+1}{9+1} = \dfrac{1}{10} + \dfrac{3}{10}i$

45. Solving the equation:
$$x^2 - 2x - 8 = 0$$
$$(x-4)(x+2) = 0$$
$$x = -2, 4$$

47. Solving the equation:
$$6x^2 - x = 2$$
$$6x^2 - x - 2 = 0$$
$$(2x+1)(3x-2) = 0$$
$$x = -\tfrac{1}{2}, \tfrac{2}{3}$$

49. Solving the equation:
$$x^2 - 6x + 9 = 0$$
$$(x-3)^2 = 0$$
$$x = 3$$

51. The equation is $y = (x-2)^2 - 4$.

10.6 Quadratic Inequalities

1. Factoring the inequality:
$$x^2 + x - 6 > 0$$
$$(x+3)(x-2) > 0$$
Forming the sign chart:

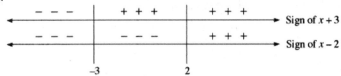

The solution set is $x < -3$ or $x > 2$. Graphing the solution set:

3. Factoring the inequality:
$$x^2 - x - 12 \le 0$$
$$(x+3)(x-4) \le 0$$
Forming the sign chart:

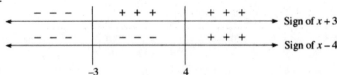

The solution set is $-3 \le x \le 4$. Graphing the solution set:

5. Factoring the inequality:
$$x^2 + 5x \geq -6$$
$$x^2 + 5x + 6 \geq 0$$
$$(x+2)(x+3) \geq 0$$
Forming the sign chart:

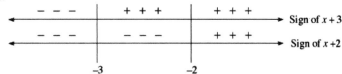

The solution set is $x \leq -3$ or $x \geq -2$. Graphing the solution set:

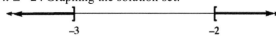

7. Factoring the inequality:
$$6x^2 < 5x - 1$$
$$6x^2 - 5x + 1 < 0$$
$$(3x-1)(2x-1) < 0$$
Forming the sign chart:

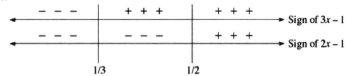

The solution set is $\frac{1}{3} < x < \frac{1}{2}$. Graphing the solution set:

9. Factoring the inequality:
$$x^2 - 9 < 0$$
$$(x+3)(x-3) < 0$$
Forming the sign chart:

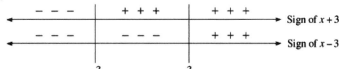

The solution set is $-3 < x < 3$. Graphing the solution set:

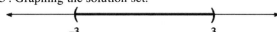

11. Factoring the inequality:
$$4x^2 - 9 \geq 0$$
$$(2x+3)(2x-3) \geq 0$$
Forming the sign chart:

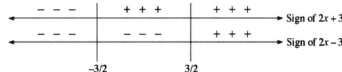

The solution set is $x \leq -\frac{3}{2}$ or $x \geq \frac{3}{2}$. Graphing the solution set:

308 Chapter 10 Quadratic Functions

13. Factoring the inequality:
$$2x^2 - x - 3 < 0$$
$$(2x-3)(x+1) < 0$$
Forming the sign chart:

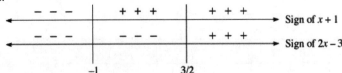

The solution set is $-1 < x < \frac{3}{2}$. Graphing the solution set:

15. Factoring the inequality:
$$x^2 - 4x + 4 \geq 0$$
$$(x-2)^2 \geq 0$$
Since this inequality is always true, the solution set is all real numbers. Graphing the solution set:

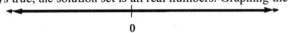

17. Factoring the inequality:
$$x^2 - 10x + 25 < 0$$
$$(x-5)^2 < 0$$
Since this inequality is never true, there is no solution.

19. Forming the sign chart:

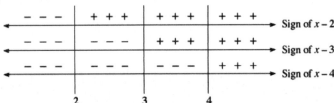

The solution set is $2 < x < 3$ or $x > 4$. Graphing the solution set:

21. Forming the sign chart:

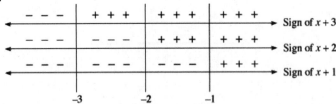

The solution set is $x \leq -3$ or $-2 \leq x \leq -1$. Graphing the solution set:

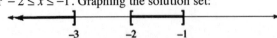

23. Forming the sign chart:

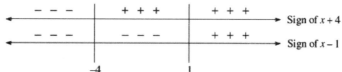

The solution set is $-4 < x \le 1$. Graphing the solution set:

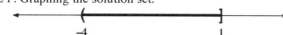

25. Forming the sign chart:

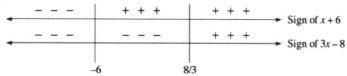

The solution set is $-6 < x < \frac{8}{3}$. Graphing the solution set:

27. Forming the sign chart:

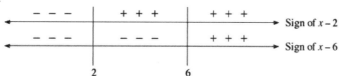

The solution set is $x < 2$ or $x > 6$. Graphing the solution set:

29. Forming the sign chart:

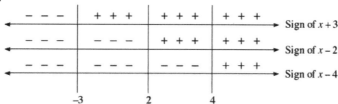

The solution set is $x < -3$ or $2 < x < 4$. Graphing the solution set:

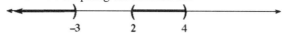

31. Forming the sign chart:

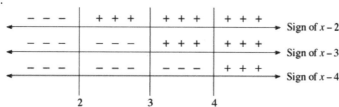

The solution set is $2 < x < 3$ or $x > 4$. Graphing the solution set:

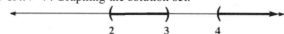

33.   **a.** Writing the statement: $x - 1 > 0$ **b.** Writing the statement: $x - 1 \ge 0$
    c. Writing the statement: $x - 1 \ge 0$
35. This statement is never true, so there is no solution ($\varnothing$).
37. This statement is only true if $x = 1$.   39. This statement is true for all real numbers.
41. This statement is true if $x \ne 1$, or $x < 1$ or $x > 1$.
43. Since $x^2 - 6x + 9 = (x - 3)^2$, this statement is never true, so there is no solution ($\varnothing$).

45. Since $x^2 - 6x + 9 = (x-3)^2$, this statement is true if $x \neq 3$, or $x < 3$ or $x > 3$.

47.
  a. The solution set is $-2 < x < 2$.
  b. The solution set is $x < -2$ or $x > 2$.
  c. The solution set is $x = -2, 2$.

49.
  a. The solution set is $-2 < x < 5$.
  b. The solution set is $x < -2$ or $x > 5$.
  c. The solution set is $x = -2, 5$.

51.
  a. The solution set is $x < -1$ or $1 < x < 3$.
  c. The solution set is $x = -1, 1, 3$.
  b. The solution set is $-1 < x < 1$ or $x > 3$.

53. Let w represent the width and $2w + 3$ represent the length. Using the area formula:
$$w(2w+3) \geq 44$$
$$2w^2 + 3w \geq 44$$
$$2w^2 + 3w - 44 \geq 0$$
$$(2w+11)(w-4) \geq 0$$

Forming the sign chart:

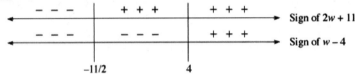

The width is at least 4 inches.

55. Solving the inequality:
$$1300p - 100p^2 \geq 4000$$
$$-100p^2 + 1300p - 4000 \geq 0$$
$$p^2 - 13p + 40 \leq 0$$
$$(p-8)(p-5) \leq 0$$

Forming the sign chart:

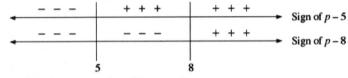

She should charge at least $5 but no more than $8 per radio.

57. Using a calculator: $\dfrac{50,000}{32,000} = 1.5625$

59. Using a calculator: $\dfrac{1}{2}\left(\dfrac{4.5926}{1.3876} - 2\right) \approx 0.6549$

61. Solving the equation:
$$\sqrt{3t-1} = 2$$
$$\left(\sqrt{3t-1}\right)^2 = (2)^2$$
$$3t - 1 = 4$$
$$3t = 5$$
$$t = \tfrac{5}{3}$$

The solution is $\tfrac{5}{3}$.

63. Solving the equation:
$$\sqrt{x+3} = x - 3$$
$$\left(\sqrt{x+3}\right)^2 = (x-3)^2$$
$$x + 3 = x^2 - 6x + 9$$
$$0 = x^2 - 7x + 6$$
$$0 = (x-6)(x-1)$$
$$x = 1, 6 \qquad (x = 1 \text{ does not check})$$

The solution is 6.

65. Graphing the equation:

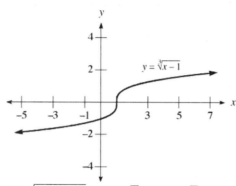

67. Using the quadratic formula: $x = \dfrac{2 \pm \sqrt{4-4(-1)}}{2(1)} = \dfrac{2 \pm \sqrt{8}}{2} = \dfrac{2 \pm 2\sqrt{2}}{2} = 1 \pm \sqrt{2}$

The inequality is satisfied when $1-\sqrt{2} < x < 1+\sqrt{2}$.

69. Using the quadratic formula: $x = \dfrac{8 \pm \sqrt{64-4(13)}}{2(1)} = \dfrac{8 \pm \sqrt{12}}{2} = \dfrac{8 \pm 2\sqrt{3}}{2} = 4 \pm \sqrt{3}$

The inequality is satisfied when $x < 4-\sqrt{3}$ or $x > 4+\sqrt{3}$.

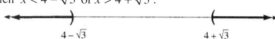

Chapter 10 Review/Test

1. Solving the equation:
$(2t-5)^2 = 25$
$2t-5 = \pm 5$
$2t-5 = -5, 5$
$2t = 0, 10$
$t = 0, 5$

2. Solving the equation:
$(3t-2)^2 = 4$
$3t-2 = \pm 2$
$3t-2 = -2, 2$
$3t = 0, 4$
$t = 0, \dfrac{4}{3}$

3. Solving the equation:
$(3y-4)^2 = -49$
$3y-4 = \pm\sqrt{-49}$
$3y-4 = \pm 7i$
$3y = 4 \pm 7i$
$y = \dfrac{4 \pm 7i}{3}$

4. Solving the equation:
$(2x+6)^2 = 12$
$2x+6 = \pm\sqrt{12}$
$2x+6 = \pm 2\sqrt{3}$
$2x = -6 \pm 2\sqrt{3}$
$x = -3 \pm \sqrt{3}$

5. Solving by completing the square:
$2x^2 + 6x - 20 = 0$
$x^2 + 3x = 10$
$x^2 + 3x + \dfrac{9}{4} = 10 + \dfrac{9}{4}$
$\left(x + \dfrac{3}{2}\right)^2 = \dfrac{49}{4}$
$x + \dfrac{3}{2} = -\dfrac{7}{2}, \dfrac{7}{2}$
$x = -5, 2$

6. Solving by completing the square:
$3x^2 + 15x = -18$
$x^2 + 5x = -6$
$x^2 + 5x + \dfrac{25}{4} = -6 + \dfrac{25}{4}$
$\left(x + \dfrac{5}{2}\right)^2 = \dfrac{1}{4}$
$x + \dfrac{5}{2} = -\dfrac{1}{2}, \dfrac{1}{2}$
$x = -3, -2$

7. Solving by completing the square:
$$a^2 + 9 = 6a$$
$$a^2 - 6a = -9$$
$$a^2 - 6a + 9 = -9 + 9$$
$$(a-3)^2 = 0$$
$$a - 3 = 0$$
$$a = 3$$

8. Solving by completing the square:
$$a^2 + 4 = 4a$$
$$a^2 - 4a = -4$$
$$a^2 - 4a + 4 = -4 + 4$$
$$(a-2)^2 = 0$$
$$a - 2 = 0$$
$$a = 2$$

9. Solving by completing the square:
$$2y^2 + 6y = -3$$
$$y^2 + 3y = -\tfrac{3}{2}$$
$$y^2 + 3y + \tfrac{9}{4} = -\tfrac{3}{2} + \tfrac{9}{4}$$
$$\left(y + \tfrac{3}{2}\right)^2 = \tfrac{3}{4}$$
$$y + \tfrac{3}{2} = \pm \tfrac{\sqrt{3}}{2}$$
$$y = \tfrac{-3 \pm \sqrt{3}}{2}$$

10. Solving by completing the square:
$$3y^2 + 3 = 9y$$
$$3y^2 - 9y = -3$$
$$y^2 - 3y = -1$$
$$y^2 - 3y + \tfrac{9}{4} = -1 + \tfrac{9}{4}$$
$$\left(y - \tfrac{3}{2}\right)^2 = \tfrac{5}{4}$$
$$y - \tfrac{3}{2} = \pm \tfrac{\sqrt{5}}{2}$$
$$y = \tfrac{3 \pm \sqrt{5}}{2}$$

11. Solving the equation:
$$\tfrac{1}{6}x^2 + \tfrac{1}{2}x - \tfrac{5}{3} = 0$$
$$x^2 + 3x - 10 = 0$$
$$(x+5)(x-2) = 0$$
$$x = -5, 2$$

12. Solving the equation:
$$8x^2 - 18x = 0$$
$$2x(4x - 9) = 0$$
$$x = 0, \tfrac{9}{4}$$

13. Solving the equation:
$$4t^2 - 8t + 19 = 0$$
$$t = \frac{8 \pm \sqrt{64 - 304}}{8} = \frac{8 \pm \sqrt{-240}}{8} = \frac{8 \pm 4i\sqrt{15}}{8} = \frac{2 \pm i\sqrt{15}}{2}$$

14. Solving the equation:
$$100x^2 - 200x = 100$$
$$x^2 - 2x - 1 = 0$$
$$x = \frac{2 \pm \sqrt{4+4}}{2} = \frac{2 \pm \sqrt{8}}{2} = \frac{2 \pm 2\sqrt{2}}{2} = 1 \pm \sqrt{2}$$

15. Solving the equation:
$$0.06a^2 + 0.05a = 0.04$$
$$0.06a^2 + 0.05a - 0.04 = 0$$
$$6a^2 + 5a - 4 = 0$$
$$(2a - 1)(3a + 4) = 0$$
$$a = -\tfrac{4}{3}, \tfrac{1}{2}$$

16. Solving the equation:
$$9 - 6x = -x^2$$
$$x^2 - 6x + 9 = 0$$
$$(x - 3)^2 = 0$$
$$x = 3$$

17. Solving the equation:
$$(2x+1)(x-5) - (x+3)(x-2) = -17$$
$$2x^2 - 9x - 5 - x^2 - x + 6 = -17$$
$$x^2 - 10x + 1 = -17$$
$$x^2 - 10x + 18 = 0$$
$$x = \frac{10 \pm \sqrt{100 - 72}}{2} = \frac{10 \pm \sqrt{28}}{2} = \frac{10 \pm 2\sqrt{7}}{2} = 5 \pm \sqrt{7}$$

18. Solving the equation:
$$2y^3 + 2y = 10y^2$$
$$2y^3 - 10y^2 + 2y = 0$$
$$2y(y^2 - 5y + 1) = 0$$
$$y = 0, \frac{5 \pm \sqrt{25-4}}{2} = 0, \frac{5 \pm \sqrt{21}}{2}$$

19. Solving the equation:
$$5x^2 = -2x + 3$$
$$5x^2 + 2x - 3 = 0$$
$$(x+1)(5x-3) = 0$$
$$x = -1, \frac{3}{5}$$

20. Solving the equation:
$$x^3 - 27 = 0$$
$$(x-3)(x^2 + 3x + 9) = 0$$
$$x = 3, \frac{-3 \pm \sqrt{9-36}}{2} = \frac{-3 \pm \sqrt{-27}}{2} = \frac{-3 \pm 3i\sqrt{3}}{2}$$

21. Solving the equation:
$$3 - \frac{2}{x} + \frac{1}{x^2} = 0$$
$$3x^2 - 2x + 1 = 0$$
$$x = \frac{2 \pm \sqrt{4-12}}{6} = \frac{2 \pm \sqrt{-8}}{6} = \frac{2 \pm 2i\sqrt{2}}{6} = \frac{1 \pm i\sqrt{2}}{3}$$

22. Solving the equation:
$$\frac{1}{x-3} + \frac{1}{x+2} = 1$$
$$x + 2 + x - 3 = (x-3)(x+2)$$
$$2x - 1 = x^2 - x - 6$$
$$0 = x^2 - 3x - 5$$
$$x = \frac{3 \pm \sqrt{9+20}}{2} = \frac{3 \pm \sqrt{29}}{2}$$

23. The profit equation is given by:
$$34x - 0.1x^2 - 7x - 400 = 1300$$
$$-0.1x^2 + 27x - 1700 = 0$$
$$x^2 - 270x + 17000 = 0$$
$$(x - 100)(x - 170) = 0$$
$$x = 100, 170$$
The company must sell either 100 or 170 items for its weekly profit to be $1,300.

24. The profit equation is given by:
$$110x - 0.5x^2 - 70x - 300 = 300$$
$$-0.5x^2 + 40x - 600 = 0$$
$$x^2 - 80x + 1200 = 0$$
$$(x - 20)(x - 60) = 0$$
$$x = 20, 60$$
The company must sell either 20 or 60 items for its weekly profit to be $300.

25. First write the equation as $2x^2 - 8x + 8 = 0$. Computing the discriminant: $D = (-8)^2 - 4(2)(8) = 64 - 64 = 0$
The equation will have one rational solution.

314 Chapter 10 Quadratic Functions

26. First write the equation as $4x^2 - 8x + 4 = 0$. Computing the discriminant: $D = (-8)^2 - 4(4)(4) = 64 - 64 = 0$
The equation will have one rational solution.

27. Computing the discriminant: $D = (1)^2 - 4(2)(-3) = 1 + 24 = 25$. The equation will have two rational solutions.

28. First write the equation as $5x^2 + 11x - 12 = 0$. Computing the discriminant:
$$D = (11)^2 - 4(5)(-12) = 121 + 240 = 361$$
The equation will have two rational solutions.

29. First write the equation as $x^2 - x - 1 = 0$. Computing the discriminant: $D = (-1)^2 - 4(1)(-1) = 1 + 4 = 5$
The equation will have two irrational solutions.

30. First write the equation as $x^2 - 5x + 5 = 0$. Computing the discriminant: $D = (-5)^2 - 4(1)(5) = 25 - 20 = 5$
The equation will have two irrational solutions.

31. First write the equation as $3x^2 + 5x + 4 = 0$. Computing the discriminant: $D = (5)^2 - 4(3)(4) = 25 - 48 = -23$
The equation will have two complex solutions.

32. First write the equation as $4x^2 - 3x + 6 = 0$. Computing the discriminant: $D = (-3)^2 - 4(4)(6) = 9 - 96 = -87$
The equation will have two complex solutions.

33. Setting the discriminant equal to 0:
$$(-k)^2 - 4(25)(4) = 0$$
$$k^2 - 400 = 0$$
$$k^2 = 400$$
$$k = \pm 20$$

34. Setting the discriminant equal to 0:
$$k^2 - 4(4)(25) = 0$$
$$k^2 - 400 = 0$$
$$k^2 = 400$$
$$k = \pm 20$$

35. Setting the discriminant equal to 0:
$$(12)^2 - 4(k)(9) = 0$$
$$144 - 36k = 0$$
$$36k = 144$$
$$k = 4$$

36. Setting the discriminant equal to 0:
$$(-16)^2 - 4(k)(16) = 0$$
$$256 - 64k = 0$$
$$64k = 256$$
$$k = 4$$

37. Setting the discriminant equal to 0:
$$30^2 - 4(9)(k) = 0$$
$$900 - 36k = 0$$
$$36k = 900$$
$$k = 25$$

38. Setting the discriminant equal to 0:
$$28^2 - 4(4)(k) = 0$$
$$784 - 16k = 0$$
$$16k = 784$$
$$k = 49$$

39. Writing the equation:
$$(x-3)(x-5) = 0$$
$$x^2 - 8x + 15 = 0$$

40. Writing the equation:
$$(x+2)(x-4) = 0$$
$$x^2 - 2x - 8 = 0$$

41. Writing the equation:
$$(2y-1)(y+4) = 0$$
$$2y^2 + 7y - 4 = 0$$

42. Writing the equation:
$$(t-3)(t+3)(t-5) = 0$$
$$(t^2 - 9)(t-5) = 0$$
$$t^3 - 5t^2 - 9t + 45 = 0$$

43. Solving the equation:
$$(x-2)^2 - 4(x-2) - 60 = 0$$
$$(x-2-10)(x-2+6) = 0$$
$$(x-12)(x+4) = 0$$
$$x = -4, 12$$

44. Solving the equation:
$$6(2y+1)^2 - (2y+1) - 2 = 0$$
$$[3(2y+1) - 2][2(2y+1) + 1] = 0$$
$$(6y+1)(4y+3) = 0$$
$$y = -\tfrac{3}{4}, -\tfrac{1}{6}$$

45. Solving the equation:
$$x^4 - x^2 = 12$$
$$x^4 - x^2 - 12 = 0$$
$$(x^2 - 4)(x^2 + 3) = 0$$
$$x^2 = 4, -3$$
$$x = \pm 2, \pm i\sqrt{3}$$

46. Solving the equation:
$$x - \sqrt{x} - 2 = 0$$
$$(\sqrt{x} - 2)(\sqrt{x} + 1) = 0$$
$$\sqrt{x} = 2, -1$$
$$x = 4, 1 \qquad (x = 1 \text{ does not check})$$

47. Solving the equation:
$$2x - 11\sqrt{x} = -12$$
$$2x - 11\sqrt{x} + 12 = 0$$
$$(2\sqrt{x} - 3)(\sqrt{x} - 4) = 0$$
$$\sqrt{x} = \tfrac{3}{2}, 4$$
$$x = \tfrac{9}{4}, 16$$

48. Solving the equation:
$$\sqrt{x+5} = \sqrt{x} + 1$$
$$(\sqrt{x+5})^2 = (\sqrt{x} + 1)^2$$
$$x + 5 = x + 2\sqrt{x} + 1$$
$$2\sqrt{x} = 4$$
$$\sqrt{x} = 2$$
$$x = 4$$

49. Solving the equation:
$$\sqrt{y+21} + \sqrt{y} = 7$$
$$(\sqrt{y+21})^2 = (7 - \sqrt{y})^2$$
$$y + 21 = 49 - 14\sqrt{y} + y$$
$$14\sqrt{y} = 28$$
$$\sqrt{y} = 2$$
$$y = 4$$

50. Solving the equation:
$$\sqrt{y+9} - \sqrt{y-6} = 3$$
$$(\sqrt{y+9})^2 = (3 + \sqrt{y-6})^2$$
$$y + 9 = 9 + 6\sqrt{y-6} + y - 6$$
$$6\sqrt{y-6} = 6$$
$$\sqrt{y-6} = 1$$
$$y - 6 = 1$$
$$y = 7$$

51. Solving for t:
$$16t^2 - 10t - h = 0$$
$$t = \frac{10 \pm \sqrt{100 + 64h}}{32} = \frac{10 \pm 2\sqrt{25 + 16h}}{32} = \frac{5 \pm \sqrt{25 + 16h}}{16}$$

52. Solving for t:
$$16t^2 - vt - 10 = 0$$
$$t = \frac{v \pm \sqrt{v^2 - 4(16)(-10)}}{32} = \frac{v \pm \sqrt{v^2 + 640}}{32}$$

53. Factoring the inequality:
$$x^2 - x - 2 < 0$$
$$(x-2)(x+1) < 0$$
Forming a sign chart:

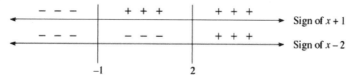

The solution set is $-1 < x < 2$. Graphing the solution set:

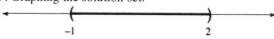

316 Chapter 10 Quadratic Functions

54. Factoring the inequality:
$$3x^2 - 14x + 8 \le 0$$
$$(3x - 2)(x - 4) \le 0$$
Forming a sign chart:

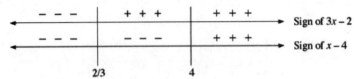

The solution set is $\frac{2}{3} \le x \le 4$. Graphing the solution set:

55. Factoring the inequality:
$$2x^2 + 5x - 12 \ge 0$$
$$(2x - 3)(x + 4) \ge 0$$
Forming a sign chart:

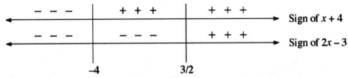

The solution set is $x \le -4$ or $x \ge \frac{3}{2}$. Graphing the solution set:

56. Forming a sign chart:

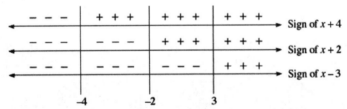

The solution set is $-4 < x < -2$ or $x > 3$. Graphing the solution set:

57. First complete the square: $y = x^2 - 6x + 8 = \left(x^2 - 6x + 9\right) + 8 - 9 = (x - 3)^2 - 1$
    The x-intercepts are 2,4, and the vertex is (3,–1). Graphing the parabola:

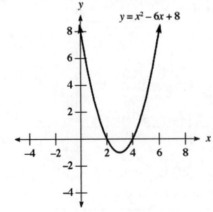

58. The x-intercepts are ± 2, and the vertex is $(0,-4)$. Graphing the parabola:

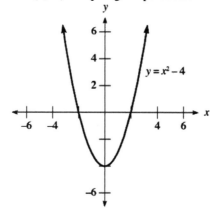

Chapter 11
Exponential and Logarithmic Functions

11.1 Exponential Functions

1. Evaluating: $g(0) = \left(\frac{1}{2}\right)^0 = 1$

3. Evaluating: $g(-1) = \left(\frac{1}{2}\right)^{-1} = 2$

5. Evaluating: $f(-3) = 3^{-3} = \frac{1}{27}$

7. Evaluating: $f(2) + g(-2) = 3^2 + \left(\frac{1}{2}\right)^{-2} = 9 + 4 = 13$

9. Graphing the function:

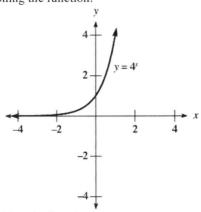

11. Graphing the function:

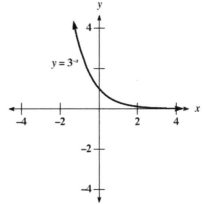

13. Graphing the function:

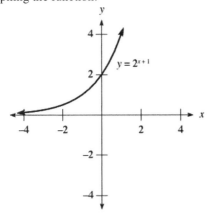

15. Graphing the function:

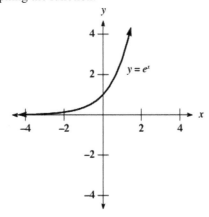

320 Chapter 11 Exponential and Logarithmic Functions

17. Graphing the functions:

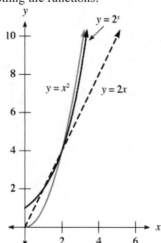

19. Graphing the functions:

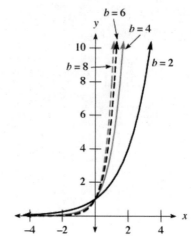

21. The equation is: $h(n) = 6\left(\frac{2}{3}\right)^n$. Substituting $n = 5$: $h(5) = 6\left(\frac{2}{3}\right)^5 \approx 0.79$ feet

23. After 8 days, there will be: $1400 \cdot 2^{-8/8} = 1400 \cdot \frac{1}{2} = 700$ micrograms

    After 11 days, there will be: $1400 \cdot 2^{-11/8} \approx 539.8$ micrograms

25. a. The equation is $A(t) = 1200\left(1 + \frac{0.06}{4}\right)^{4t}$.   b. Substitute $t = 8$: $A(8) = 1200\left(1 + \frac{0.06}{4}\right)^{32} \approx \$1,932.39$

    c. Using a graphing calculator, the time is approximately 11.64 years.

    d. Substitute $t = 8$ into the compound interest formula: $A(8) = 1200e^{0.06 \times 8} \approx \$1,939.29$

27. a. Substitute $t = 20$: $E(20) = 78.16(1.11)^{20} \approx \$630$ billion. The estimate is \$69 billion too low.

    b. For 2008, substitute $t = 38$: $E(38) = 78.16(1.11)^{38} \approx \$4,123$ billion

    For 2009, substitute $t = 39$: $E(39) = 78.16(1.11)^{39} \approx \$4,577$ billion

    For 2010, substitute $t = 40$: $E(40) = 78.16(1.11)^{40} \approx \$5,080$ billion

29. Graphing the function:

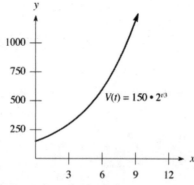

31. The painting will be worth \$600 after approximately 6 years.

33.   a.   Substitute $t = 3.5$: $V(5) = 450,000(1 - 0.30)^5 \approx \$129,138.48$
    b.   The domain is $\{t \mid 0 \le t \le 6\}$.
    c.   Sketching the graph:

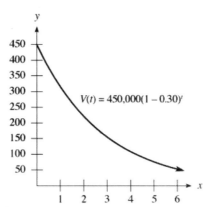

    d.   The range is $\{V(t) \mid 52,942.05 \le V(t) \le 450,000\}$.
    e.   From the graph, the crane will be worth \$85,000 after approximately 4.7 years, or 4 years 8 months.
35. The domain is $\{1, 3, 4\}$ and the range is $\{1, 2, 4\}$. This is a function.
37. Find where the quantity inside the radical is non-negative:
$$3x + 1 \ge 0$$
$$3x \ge -1$$
$$x \ge -\tfrac{1}{3}$$
The domain is $\{x \mid x \ge -\tfrac{1}{3}\}$.
39. Evaluating the function: $f(0) = 2(0)^2 - 18 = 0 - 18 = -18$
41. Simplifying the function: $\dfrac{g(x+h) - g(x)}{h} = \dfrac{2(x+h) - 6 - (2x - 6)}{h} = \dfrac{2x + 2h - 6 - 2x + 6}{h} = \dfrac{2h}{h} = 2$

43. Solving for y:
$$x = 2y - 3$$
$$2y = x + 3$$
$$y = \dfrac{x+3}{2}$$

45. Solving for y:
$$x = y^2 - 2$$
$$y^2 = x + 2$$
$$y = \pm\sqrt{x+2}$$

47. Solving for y:
$$x = \dfrac{y-4}{y-2}$$
$$x(y-2) = y - 4$$
$$xy - 2x = y - 4$$
$$xy - y = 2x - 4$$
$$y(x-1) = 2x - 4$$
$$y = \dfrac{2x-4}{x-1}$$

49. Solving for y:
$$x = \sqrt{y-3}$$
$$x^2 = y - 3$$
$$y = x^2 + 3$$

51. Graphing the function:

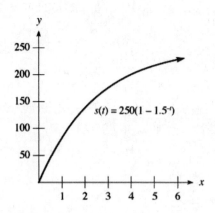

11.2 The Inverse of a Function

1. Let $y = f(x)$. Switch x and y and solve for y:
$$3y - 1 = x$$
$$3y = x + 1$$
$$y = \frac{x+1}{3}$$
The inverse is $f^{-1}(x) = \frac{x+1}{3}$.

3. Let $y = f(x)$. Switch x and y and solve for y:
$$y^3 = x$$
$$y = \sqrt[3]{x}$$
The inverse is $f^{-1}(x) = \sqrt[3]{x}$.

5. Let $y = f(x)$. Switch x and y and solve for y:
$$\frac{y-3}{y-1} = x$$
$$y - 3 = xy - x$$
$$y - xy = 3 - x$$
$$y(1 - x) = 3 - x$$
$$y = \frac{3-x}{1-x} = \frac{x-3}{x-1}$$
The inverse is $f^{-1}(x) = \frac{x-3}{x-1}$.

7. Let $y = f(x)$. Switch x and y and solve for y:
$$\frac{y-3}{4} = x$$
$$y - 3 = 4x$$
$$y = 4x + 3$$
The inverse is $f^{-1}(x) = 4x + 3$.

9. Let $y = f(x)$. Switch x and y and solve for y:
$$\tfrac{1}{2}y - 3 = x$$
$$y - 6 = 2x$$
$$y = 2x + 6$$

The inverse is $f^{-1}(x) = 2x + 6$.

11. Let $y = f(x)$. Switch x and y and solve for y:
$$\frac{2y+1}{3y+1} = x$$
$$2y + 1 = 3xy + x$$
$$2y - 3xy = x - 1$$
$$y(2 - 3x) = x - 1$$
$$y = \frac{x-1}{2-3x} = \frac{1-x}{3x-2}$$
The inverse is $f^{-1}(x) = \frac{1-x}{3x-2}$.

13. Finding the inverse:
$$2y - 1 = x$$
$$2y = x + 1$$
$$y = \frac{x+1}{2}$$

The inverse is $y^{-1} = \frac{x+1}{2}$. Graphing each curve:

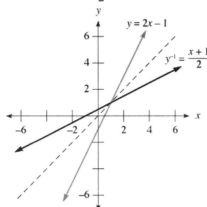

15. Finding the inverse:
$$y^2 - 3 = x$$
$$y^2 = x + 3$$
$$y = \pm\sqrt{x+3}$$

The inverse is $y^{-1} = \pm\sqrt{x+3}$. Graphing each curve:

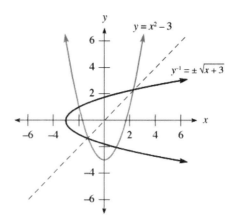

17. Finding the inverse:
$$y^2 - 2y - 3 = x$$
$$y^2 - 2y + 1 = x + 3 + 1$$
$$(y-1)^2 = x + 4$$
$$y - 1 = \pm\sqrt{x+4}$$
$$y = 1 \pm \sqrt{x+4}$$

The inverse is $y^{-1} = 1 \pm \sqrt{x+4}$. Graphing each curve:

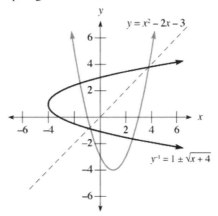

324 Chapter 11 Exponential and Logarithmic Functions

19. The inverse is $x = 3^y$. Graphing each curve:

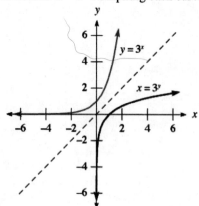

21. The inverse is $x = 4$. Graphing each curve:

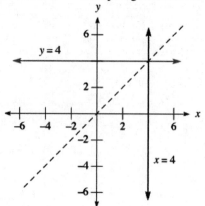

23. Finding the inverse:
$$\tfrac{1}{2} y^3 = x$$
$$y^3 = 2x$$
$$y = \sqrt[3]{2x}$$

The inverse is $y^{-1} = \sqrt[3]{2x}$. Graphing each curve:

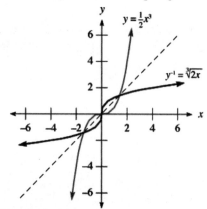

25. Finding the inverse:
$$\tfrac{1}{2} y + 2 = x$$
$$y + 4 = 2x$$
$$y = 2x - 4$$

The inverse is $y^{-1} = 2x - 4$. Graphing each curve:

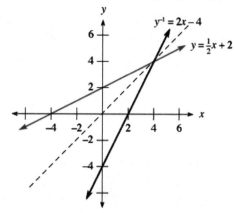

27. Finding the inverse:
$$\sqrt{y+2} = x$$
$$y + 2 = x^2$$
$$y = x^2 - 2$$

The inverse is $y^{-1} = x^2 - 2, x \geq 0$. Graphing each curve:

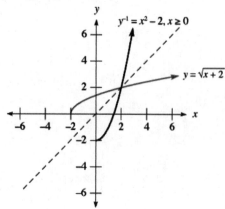

29. a. Yes, this function is one-to-one. b. No, this function is not one-to-one.
    c. Yes, this function is one-to-one.

31. a. Evaluating the function: $f(2) = 3(2) - 2 = 6 - 2 = 4$
    b. Evaluating the function: $f^{-1}(2) = \frac{2+2}{3} = \frac{4}{3}$
    c. Evaluating the function: $f\left[f^{-1}(2)\right] = f\left(\frac{4}{3}\right) = 3\left(\frac{4}{3}\right) - 2 = 4 - 2 = 2$
    d. Evaluating the function: $f^{-1}\left[f(2)\right] = f^{-1}(4) = \frac{4+2}{3} = \frac{6}{3} = 2$

33. Let $y = f(x)$. Switch x and y and solve for y:

$$\frac{1}{y} = x$$

$$y = \frac{1}{x}$$

The inverse is $f^{-1}(x) = \frac{1}{x}$.

35. a. The value is –3.   b. The value is –6.
    c. The value is 2.   d. The value is 3.
    e. The value is –2.   f. The value is 3.
    g. Each is an inverse of the other.

37. Solving the equation:

$$(2x - 1)^2 = 25$$
$$2x - 1 = \pm\sqrt{25}$$
$$2x - 1 = -5, 5$$
$$2x = -4, 6$$
$$x = -2, 3$$

39. The number is 25, since $x^2 - 10x + 25 = (x - 5)^2$.

41. Solving the equation:

$$x^2 - 10x + 8 = 0$$
$$x^2 - 10x + 25 = -8 + 25$$
$$(x - 5)^2 = 17$$
$$x - 5 = \pm\sqrt{17}$$
$$x = 5 \pm \sqrt{17}$$

43. Solving the equation:

$$3x^2 - 6x + 6 = 0$$
$$x^2 - 2x + 2 = 0$$
$$x^2 - 2x + 1 = -2 + 1$$
$$(x - 1)^2 = -1$$
$$x - 1 = \pm i$$
$$x = 1 \pm i$$

45. Simplifying: $3^{-2} = \frac{1}{3^2} = \frac{1}{9}$

47. Solving the equation:
$$2 = 3x$$
$$x = \frac{2}{3}$$

49. Solving the equation:
$$4 = x^3$$
$$x = \sqrt[3]{4}$$

51. Completing the statement: $8 = 2^3$
53. Completing the statement: $10,000 = 10^4$
55. Completing the statement: $81 = 3^4$
57. Completing the statement: $6 = 6^1$
59. Finding the inverse:

$$3y + 5 = x$$
$$3y = x - 5$$
$$y = \frac{x - 5}{3}$$

So $f^{-1}(x) = \frac{x - 5}{3}$. Now verifying the inverse: $f\left[f^{-1}(x)\right] = f\left(\frac{x - 5}{3}\right) = 3\left(\frac{x - 5}{3}\right) + 5 = x - 5 + 5 = x$

61. Finding the inverse:
$$y^3 + 1 = x$$
$$y^3 = x - 1$$
$$y = \sqrt[3]{x-1}$$

So $f^{-1}(x) = \sqrt[3]{x-1}$. Now verifying the inverse: $f\left[f^{-1}(x)\right] = f\left(\sqrt[3]{x-1}\right) = \left(\sqrt[3]{x-1}\right)^3 + 1 = x - 1 + 1 = x$

63. Finding the inverse:
$$\frac{y-4}{y-2} = x$$
$$y - 4 = xy - 2x$$
$$y - xy = 4 - 2x$$
$$y(1-x) = 4 - 2x$$
$$y = \frac{4-2x}{1-x} = \frac{2x-4}{x-1}$$

So $f^{-1}(x) = \frac{2x-4}{x-1}$. Now verifying the inverse:

$$f\left[f^{-1}(x)\right] = f\left(\frac{2x-4}{x-1}\right) = \frac{\frac{2x-4}{x-1} - 4}{\frac{2x-4}{x-1} - 2} = \frac{2x-4-4(x-1)}{2x-4-2(x-1)} = \frac{2x-4-4x+4}{2x-4-2x+2} = \frac{-2x}{-2} = x$$

65. a. From the graph: $f(0) = 1$ b. From the graph: $f(1) = 2$
   c. From the graph: $f(2) = 5$ d. From the graph: $f^{-1}(1) = 0$
   e. From the graph: $f^{-1}(2) = 1$ f. From the graph: $f^{-1}(5) = 2$
   g. From the graph: $f^{-1}[f(2)] = 2$ h. From the graph: $f\left[f^{-1}(5)\right] = 5$

11.3 Logarithms Are Exponents

1. Writing in logarithmic form: $\log_2 16 = 4$
3. Writing in logarithmic form: $\log_5 125 = 3$
5. Writing in logarithmic form: $\log_{10} 0.01 = -2$
7. Writing in logarithmic form: $\log_2 \frac{1}{32} = -5$
9. Writing in logarithmic form: $\log_{1/2} 8 = -3$
11. Writing in logarithmic form: $\log_3 27 = 3$
13. Writing in exponential form: $10^2 = 100$
15. Writing in exponential form: $2^6 = 64$
17. Writing in exponential form: $8^0 = 1$
19. Writing in exponential form: $10^{-3} = 0.001$
21. Writing in exponential form: $6^2 = 36$
23. Writing in exponential form: $5^{-2} = \frac{1}{25}$
25. Solving the equation:
$$\log_3 x = 2$$
$$x = 3^2 = 9$$
27. Solving the equation:
$$\log_5 x = -3$$
$$x = 5^{-3} = \frac{1}{125}$$
29. Solving the equation:
$$\log_2 16 = x$$
$$2^x = 16$$
$$x = 4$$
31. Solving the equation:
$$\log_8 2 = x$$
$$8^x = 2$$
$$x = \frac{1}{3}$$
33. Solving the equation:
$$\log_x 4 = 2$$
$$x^2 = 4$$
$$x = 2$$
35. Solving the equation:
$$\log_x 5 = 3$$
$$x^3 = 5$$
$$x = \sqrt[3]{5}$$

37. Sketching the graph:

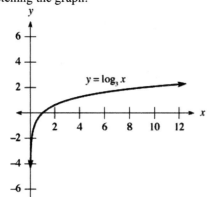

39. Sketching the graph:

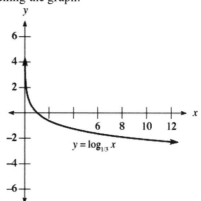

41. Sketching the graph:

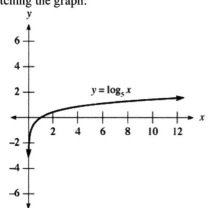

43. Sketching the graph:

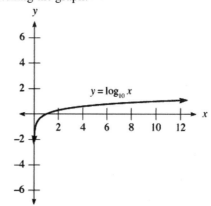

45. Simplifying the logarithm:

$$x = \log_2 16$$
$$2^x = 16$$
$$x = 4$$

47. Simplifying the logarithm:

$$x = \log_{25} 125$$
$$25^x = 125$$
$$5^{2x} = 5^3$$
$$2x = 3$$
$$x = \tfrac{3}{2}$$

49. Simplifying the logarithm:

$$x = \log_{10} 1000$$
$$10^x = 1000$$
$$x = 3$$

51. Simplifying the logarithm:

$$x = \log_3 3$$
$$3^x = 3$$
$$x = 1$$

53. Simplifying the logarithm:

$$x = \log_5 1$$
$$5^x = 1$$
$$x = 0$$

328 Chapter 11 Exponential and Logarithmic Functions

55. First find $\log_6 6$:
$$x = \log_6 6$$
$$6^x = 6$$
$$x = 1$$
Now find $\log_3 1$:
$$x = \log_3 1$$
$$3^x = 1$$
$$x = 0$$

57. First find $\log_2 16$:
$$x = \log_2 16$$
$$2^x = 16$$
$$x = 4$$
Now find $\log_2 4$:
$$x = \log_2 4$$
$$2^x = 4$$
$$x = 2$$
Now find $\log_4 2$:
$$x = \log_4 2$$
$$4^x = 2$$
$$2^{2x} = 2$$
$$2x = 1$$
$$x = \tfrac{1}{2}$$

59. The pH is given by: $\text{pH} = -\log_{10}\left(10^{-7}\right) = -(-7) = 7$

61. The $\left[H^+\right]$ is given by:
$$-\log_{10}\left[H^+\right] = 6$$
$$\log_{10}\left[H^+\right] = -6$$
$$\left[H^+\right] = 10^{-6}$$

63. Using the relationship $M = \log_{10} T$:
$$M = \log_{10} 100$$
$$10^M = 100$$
$$M = 2$$

65. It is 10^8 times as large.

67. Completing the square: $x^2 + 10x + 25 = (x+5)^2$

69. Completing the square: $y^2 - 2y + 1 = (y-1)^2$

71. Solving the equation:
$$-y^2 = 9$$
$$y^2 = -9$$
$$y = \pm\sqrt{-9} = \pm 3i$$

73. Solving the equation:
$$-x^2 - 8 = -4$$
$$-x^2 = 4$$
$$x^2 = -4$$
$$x = \pm\sqrt{-4} = \pm 2i$$

75. Solving the equation:
$$2x^2 + 4x - 3 = 0$$
$$x = \frac{-4 \pm \sqrt{16+24}}{4} = \frac{-4 \pm \sqrt{40}}{4} = \frac{-4 \pm 2\sqrt{10}}{4} = \frac{-2 \pm \sqrt{10}}{2}$$

77. Solving the equation:
$$(2y-3)(2y-1) = -4$$
$$4y^2 - 8y + 3 = -4$$
$$4y^2 - 8y + 7 = 0$$
$$y = \frac{8 \pm \sqrt{64-112}}{8} = \frac{8 \pm \sqrt{-48}}{8} = \frac{8 \pm 4i\sqrt{3}}{8} = \frac{2 \pm i\sqrt{3}}{2}$$

79. Solving the equation:
$$t^3 - 125 = 0$$
$$(t-5)\left(t^2 + 5t + 25\right) = 0$$
$$t = 5, \frac{-5 \pm \sqrt{25-100}}{2} = \frac{-5 \pm \sqrt{-75}}{2} = \frac{-5 \pm 5i\sqrt{3}}{2}$$

81. Solving the equation:
$$4x^5 - 16x^4 = 20x^3$$
$$4x^5 - 16x^4 - 20x^3 = 0$$
$$4x^3(x^2 - 4x - 5) = 0$$
$$4x^3(x-5)(x+1) = 0$$
$$x = -1, 0, 5$$

83. Solving the equation:
$$\frac{1}{x-3} + \frac{1}{x+2} = 1$$
$$x+2+x-3 = (x-3)(x+2)$$
$$2x-1 = x^2 - x - 6$$
$$x^2 - 3x - 5 = 0$$
$$x = \frac{3 \pm \sqrt{9+20}}{2} = \frac{3 \pm \sqrt{29}}{2}$$

85. Simplifying; $8^{2/3} = \left(8^{1/3}\right)^2 = \left(\sqrt[3]{8}\right)^2 = 2^2 = 4$

87. Solving the equation:
$$(x+2)(x) = 2^3$$
$$x^2 + 2x = 8$$
$$x^2 + 2x - 8 = 0$$
$$(x+4)(x-2) = 0$$
$$x = -4, 2$$

89. Solving the equation:
$$\frac{x-2}{x+1} = 9$$
$$x - 2 = 9(x+1)$$
$$x - 2 = 9x + 9$$
$$-8x = 11$$
$$x = -\frac{11}{8}$$

91. Writing in exponential form: $2^3 = (x+2)(x)$

93. Writing in exponential form: $3^4 = \frac{x-2}{x+1}$

95. a. Completing the table:

x	-1	0	1	2
$f(x)$	$\frac{1}{8}$	1	8	64

b. Completing the table:

x	$\frac{1}{8}$	1	8	64
$f^{-1}(x)$	-1	0	1	2

c. The equation is $f(x) = 8^x$.

d. The equation is $f^{-1}(x) = \log_8 x$.

11.4 Properties of Logarithms

1. Using properties of logarithms: $\log_3 4x = \log_3 4 + \log_3 x$

3. Using properties of logarithms: $\log_6 \frac{5}{x} = \log_6 5 - \log_6 x$

5. Using properties of logarithms: $\log_2 y^5 = 5 \log_2 y$

7. Using properties of logarithms: $\log_9 \sqrt[3]{z} = \log_9 z^{1/3} = \frac{1}{3} \log_9 z$

9. Using properties of logarithms: $\log_6 x^2 y^4 = \log_6 x^2 + \log_6 y^4 = 2\log_6 x + 4\log_6 y$

11. Using properties of logarithms: $\log_5 \sqrt{x} \cdot y^4 = \log_5 x^{1/2} + \log_5 y^4 = \frac{1}{2}\log_5 x + 4\log_5 y$

13. Using properties of logarithms: $\log_b \frac{xy}{z} = \log_b xy - \log_b z = \log_b x + \log_b y - \log_b z$

15. Using properties of logarithms: $\log_{10} \frac{4}{xy} = \log_{10} 4 - \log_{10} xy = \log_{10} 4 - \log_{10} x - \log_{10} y$

17. Using properties of logarithms: $\log_{10} \frac{x^2 y}{\sqrt{z}} = \log_{10} x^2 + \log_{10} y - \log_{10} z^{1/2} = 2\log_{10} x + \log_{10} y - \frac{1}{2}\log_{10} z$

330 Chapter 11 Exponential and Logarithmic Functions

19. Using properties of logarithms: $\log_{10} \dfrac{x^3 \sqrt{y}}{z^4} = \log_{10} x^3 + \log_{10} y^{1/2} - \log_{10} z^4 = 3\log_{10} x + \tfrac{1}{2} \log_{10} y - 4\log_{10} z$

21. Using properties of logarithms:

   $\log_b \sqrt[3]{\dfrac{x^2 y}{z^4}} = \log_b \dfrac{x^{2/3} y^{1/3}}{z^{4/3}} = \log_b x^{2/3} + \log_b y^{1/3} - \log_b z^{4/3} = \tfrac{2}{3} \log_b x + \tfrac{1}{3} \log_b y - \tfrac{4}{3} \log_b z$

23. Writing as a single logarithm: $\log_b x + \log_b z = \log_b xz$

25. Writing as a single logarithm: $2\log_3 x - 3\log_3 y = \log_3 x^2 - \log_3 y^3 = \log_3 \dfrac{x^2}{y^3}$

27. Writing as a single logarithm: $\tfrac{1}{2} \log_{10} x + \tfrac{1}{3} \log_{10} y = \log_{10} x^{1/2} + \log_{10} y^{1/3} = \log_{10} \sqrt{x} \sqrt[3]{y}$

29. Writing as a single logarithm: $3\log_2 x + \tfrac{1}{2} \log_2 y - \log_2 z = \log_2 x^3 + \log_2 y^{1/2} - \log_2 z = \log_2 \dfrac{x^3 \sqrt{y}}{z}$

31. Writing as a single logarithm: $\tfrac{1}{2} \log_2 x - 3\log_2 y - 4\log_2 z = \log_2 x^{1/2} - \log_2 y^3 - \log_2 z^4 = \log_2 \dfrac{\sqrt{x}}{y^3 z^4}$

33. Writing as a single logarithm:

   $\tfrac{3}{2} \log_{10} x - \tfrac{3}{4} \log_{10} y - \tfrac{4}{5} \log_{10} z = \log_{10} x^{3/2} - \log_{10} y^{3/4} - \log_{10} z^{4/5} = \log_{10} \dfrac{x^{3/2}}{y^{3/4} z^{4/5}}$

35. Solving the equation:

   $\log_2 x + \log_2 3 = 1$
   $\log_2 3x = 1$
   $3x = 2^1$
   $3x = 2$
   $x = \tfrac{2}{3}$

37. Solving the equation:

   $\log_3 x - \log_3 2 = 2$
   $\log_3 \dfrac{x}{2} = 2$
   $\dfrac{x}{2} = 3^2$
   $\dfrac{x}{2} = 9$
   $x = 18$

39. Solving the equation:

   $\log_3 x + \log_3 (x - 2) = 1$
   $\log_3 (x^2 - 2x) = 1$
   $x^2 - 2x = 3^1$
   $x^2 - 2x - 3 = 0$
   $(x - 3)(x + 1) = 0$
   $x = 3, -1$

   The solution is 3 (−1 does not check).

41. Solving the equation:

   $\log_3 (x + 3) - \log_3 (x - 1) = 1$
   $\log_3 \dfrac{x + 3}{x - 1} = 1$
   $\dfrac{x + 3}{x - 1} = 3^1$
   $x + 3 = 3x - 3$
   $-2x = -6$
   $x = 3$

43. Solving the equation:

   $\log_2 x + \log_2 (x - 2) = 3$
   $\log_2 (x^2 - 2x) = 3$
   $x^2 - 2x = 2^3$
   $x^2 - 2x - 8 = 0$
   $(x - 4)(x + 2) = 0$
   $x = 4, -2$

   The solution is 4 (−2 does not check).

45. Solving the equation:

   $\log_8 x + \log_8 (x - 3) = \tfrac{2}{3}$
   $\log_8 (x^2 - 3x) = \tfrac{2}{3}$
   $x^2 - 3x = 8^{2/3}$
   $x^2 - 3x - 4 = 0$
   $(x - 4)(x + 1) = 0$
   $x = 4, -1$

   The solution is 4 (−1 does not check).

47. Solving the equation:
$$\log_5 \sqrt{x} + \log_5 \sqrt{6x+5} = 1$$
$$\log_5 \sqrt{6x^2+5x} = 1$$
$$\tfrac{1}{2}\log_5(6x^2+5x) = 1$$
$$\log_5(6x^2+5x) = 2$$
$$6x^2+5x = 5^2$$
$$6x^2+5x-25 = 0$$
$$(3x-5)(2x+5) = 0$$
$$x = \tfrac{5}{3}, -\tfrac{5}{2}$$
The solution is $\tfrac{5}{3}$ ($-\tfrac{5}{2}$ does not check).

49. Rewriting the formula:
$$D = 10\log_{10}\left(\frac{I}{I_0}\right)$$
$$D = 10\left(\log_{10} I - \log_{10} I_0\right)$$

51. Solving for N:
$$N = \log_{10}\frac{100}{1} = \log_{10} 10^2 = 2$$
$$N = \log_{10} 100 - \log_{10} 1 = \log_{10} 10^2 = 2$$
So $N = 2$ in both cases.

53. Dividing: $\dfrac{12x^2+y^2}{36} = \dfrac{12x^2}{36} + \dfrac{y^2}{36} = \dfrac{x^2}{3} + \dfrac{y^2}{36}$

55. Dividing: $\dfrac{25x^2+4y^2}{100} = \dfrac{25x^2}{100} + \dfrac{4y^2}{100} = \dfrac{x^2}{4} + \dfrac{y^2}{25}$

57. Computing the discriminant: $D = (-5)^2 - 4(2)(4) = 25 - 32 = -7$. There are two complex solutions.

59. Writing the equation:
$$(x+3)(x-5) = 0$$
$$x^2 - 2x - 15 = 0$$

61. Writing the equation:
$$(3y-2)(y-3) = 0$$
$$3y^2 - 11y + 6 = 0$$

63. Simplifying: $5^0 = 1$

65. Simplifying: $\log_3 3 = \log_3 3^1 = 1$

67. Simplifying: $\log_b b^4 = 4$

69. Using a calculator: $10^{-5.6} \approx 2.5 \times 10^{-6}$

71. Using a calculator: $\dfrac{2.00 \times 10^8}{3.96 \times 10^6} \approx 51$

11.5 Common Logarithms and Natural Logarithms

1. Evaluating the logarithm: $\log 378 \approx 2.5775$

3. Evaluating the logarithm: $\log 37.8 \approx 1.5775$

5. Evaluating the logarithm: $\log 3{,}780 \approx 3.5775$

7. Evaluating the logarithm: $\log 0.0378 \approx -1.4225$

9. Evaluating the logarithm: $\log 37{,}800 \approx 4.5775$

11. Evaluating the logarithm: $\log 600 \approx 2.7782$

13. Evaluating the logarithm: $\log 2{,}010 \approx 3.3032$

15. Evaluating the logarithm: $\log 0.00971 \approx -2.0128$

17. Evaluating the logarithm: $\log 0.0314 \approx -1.5031$

19. Evaluating the logarithm: $\log 0.399 \approx -0.3990$

21. Solving for x:
$$\log x = 2.8802$$
$$x = 10^{2.8802} \approx 759$$

23. Solving for x:
$$\log x = -2.1198$$
$$x = 10^{-2.1198} \approx 0.00759$$

25. Solving for x:
$$\log x = 3.1553$$
$$x = 10^{3.1553} \approx 1{,}430$$

27. Solving for x:
$$\log x = -5.3497$$
$$x = 10^{-5.3497} \approx 0.00000447$$

29. Solving for x:
$$\log x = -7.0372$$
$$x = 10^{-7.0372} \approx 0.0000000918$$

31. Solving for x:
$$\log x = 10$$
$$x = 10^{10}$$

33. Solving for x:
$$\log x = -10$$
$$x = 10^{-10}$$

35. Solving for x:
$$\log x = 20$$
$$x = 10^{20}$$

37. Solving for x:
$$\log x = -2$$
$$x = 10^{-2} = \frac{1}{100}$$

39. Solving for x:
$$\log x = \log_2 8$$
$$\log x = 3$$
$$x = 10^3 = 1{,}000$$

41. Simplifying the logarithm: $\ln e = \ln e^1 = 1$

43. Simplifying the logarithm: $\ln e^5 = 5$

45. Simplifying the logarithm: $\ln e^x = x$

47. Using properties of logarithms: $\ln 10 e^{3t} = \ln 10 + \ln e^{3t} = \ln 10 + 3t$

49. Using properties of logarithms: $\ln Ae^{-2t} = \ln A + \ln e^{-2t} = \ln A - 2t$

51. Evaluating the logarithm: $\ln 15 = \ln(3 \cdot 5) = \ln 3 + \ln 5 = 1.0986 + 1.6094 = 2.7080$

53. Evaluating the logarithm: $\ln \frac{1}{3} = \ln 3^{-1} = -\ln 3 = -1.0986$

55. Evaluating the logarithm: $\ln 9 = \ln 3^2 = 2 \ln 3 = 2(1.0986) = 2.1972$

57. Evaluating the logarithm: $\ln 16 = \ln 2^4 = 4 \ln 2 = 4(0.6931) = 2.7724$

59. **a.** Using a calculator: $\frac{\log 25}{\log 15} \approx 1.1886$ **b.** Using a calculator: $\log \frac{25}{15} \approx 0.2218$

61. **a.** Using a calculator: $\frac{\log 4}{\log 8} \approx 0.6667$ **b.** Using a calculator: $\log \frac{4}{8} \approx -0.3010$

63. Computing the pH: $\text{pH} = -\log(6.50 \times 10^{-4}) \approx 3.19$

65. Finding the concentration:
$$4.75 = -\log\left[H^+\right]$$
$$-4.75 = \log\left[H^+\right]$$
$$\left[H^+\right] = 10^{-4.75} \approx 1.78 \times 10^{-5}$$

67. Finding the magnitude:
$$5.5 = \log T$$
$$T = 10^{5.5} \approx 3.16 \times 10^5$$

69. Finding the magnitude:
$$8.3 = \log T$$
$$T = 10^{8.3} \approx 2.00 \times 10^8$$

71. For the first earthquake:
$$\log T_1 = 6.5$$
$$T_1 = 10^{6.5}$$

For the second earthquake:
$$\log T_2 = 5.5$$
$$T_2 = 10^{5.5}$$

The ratio is $\dfrac{T_1}{T_2} = \dfrac{10^{6.5}}{10^{5.5}} = 10$ times stronger.

73. Completing the table:

Location	Date	Magnitude (M)	Shockwave (T)
Moresby Island	January 23	4.0	1.00×10^4
Vancouver Island	April 30	5.3	1.99×10^5
Quebec City	June 29	3.2	1.58×10^3
Mould Bay	November 13	5.2	1.58×10^5
St. Lawrence	December 14	3.7	5.01×10^3

75. Finding the rate of depreciation:
$$\log(1-r) = \tfrac{1}{5} \log \frac{4500}{9000}$$
$$\log(1-r) \approx -0.0602$$
$$1 - r \approx 10^{-0.0602}$$
$$r = 1 - 10^{-0.0602}$$
$$r \approx 0.129 = 12.9\%$$

77. Finding the rate of depreciation:
$$\log(1-r) = \tfrac{1}{5} \log \frac{5750}{7550}$$
$$\log(1-r) \approx -0.0237$$
$$1 - r \approx 10^{-0.0237}$$
$$r = 1 - 10^{-0.0237}$$
$$r \approx 0.053 = 5.3\%$$

79. It appears to approach e. Completing the table:

x	$(1+x)^{1/x}$
1	2
0.5	2.25
0.1	2.5937
0.01	2.7048
0.001	2.7169
0.0001	2.7181
0.00001	2.7183

81. Solving the equation:
$$(y+3)^2 + y^2 = 9$$
$$y^2 + 6y + 9 + y^2 = 9$$
$$2y^2 + 6y = 0$$
$$2y(y+3) = 0$$
$$y = -3, 0$$

83. Solving the equation:
$$(x+3)^2 + 1^2 = 2$$
$$x^2 + 6x + 9 + 1 = 2$$
$$x^2 + 6x + 8 = 0$$
$$(x+4)(x+2) = 0$$
$$x = -4, -2$$

85. Solving the equation:
$$x^4 - 2x^2 - 8 = 0$$
$$(x^2 - 4)(x^2 + 2) = 0$$
$$x^2 = 4, -2$$
$$x = \pm 2, \pm i\sqrt{2}$$

87. Solving the equation:
$$2x - 5\sqrt{x} + 3 = 0$$
$$(2\sqrt{x} - 3)(\sqrt{x} - 1) = 0$$
$$\sqrt{x} = 1, \tfrac{3}{2}$$
$$x = 1, \tfrac{9}{4}$$

89. Solving the equation:
$$5(2x+1) = 12$$
$$10x + 5 = 12$$
$$10x = 7$$
$$x = \tfrac{7}{10} = 0.7$$

91. Using a calculator: $\dfrac{100{,}000}{32{,}000} = 3.125$

93. Using a calculator: $\dfrac{1}{2}\left(\dfrac{-0.6931}{1.4289} + 3\right) \approx 1.2575$

95. Rewriting the logarithm: $\log 1.05^t = t \log 1.05$

97. Simplifying: $\ln e^{0.05t} = 0.05t$

11.6 Exponential Equations and Change of Base

1. Solving the equation:
$$3^x = 5$$
$$\ln 3^x = \ln 5$$
$$x \ln 3 = \ln 5$$
$$x = \dfrac{\ln 5}{\ln 3} \approx 1.4650$$

3. Solving the equation:
$$5^x = 3$$
$$\ln 5^x = \ln 3$$
$$x \ln 5 = \ln 3$$
$$x = \dfrac{\ln 3}{\ln 5} \approx 0.6826$$

5. Solving the equation:
$$5^{-x} = 12$$
$$\ln 5^{-x} = \ln 12$$
$$-x \ln 5 = \ln 12$$
$$x = -\dfrac{\ln 12}{\ln 5} \approx -1.5440$$

7. Solving the equation:
$$12^{-x} = 5$$
$$\ln 12^{-x} = \ln 5$$
$$-x \ln 12 = \ln 5$$
$$x = -\dfrac{\ln 5}{\ln 12} \approx -0.6477$$

9. Solving the equation:
$$8^{x+1} = 4$$
$$2^{3x+3} = 2^2$$
$$3x + 3 = 2$$
$$3x = -1$$
$$x = -\tfrac{1}{3} \approx -0.3333$$

11. Solving the equation:
$$4^{x-1} = 4$$
$$4^{x-1} = 4^1$$
$$x - 1 = 1$$
$$x = 2 = 2.0000$$

13. Solving the equation:
$$3^{2x+1} = 2$$
$$\ln 3^{2x+1} = \ln 2$$
$$(2x+1)\ln 3 = \ln 2$$
$$2x+1 = \frac{\ln 2}{\ln 3}$$
$$2x = \frac{\ln 2}{\ln 3} - 1$$
$$x = \tfrac{1}{2}\left(\frac{\ln 2}{\ln 3} - 1\right) \approx -0.1845$$

15. Solving the equation:
$$3^{1-2x} = 2$$
$$\ln 3^{1-2x} = \ln 2$$
$$(1-2x)\ln 3 = \ln 2$$
$$1-2x = \frac{\ln 2}{\ln 3}$$
$$-2x = \frac{\ln 2}{\ln 3} - 1$$
$$x = \tfrac{1}{2}\left(1 - \frac{\ln 2}{\ln 3}\right) \approx 0.1845$$

17. Solving the equation:
$$15^{3x-4} = 10$$
$$\ln 15^{3x-4} = \ln 10$$
$$(3x-4)\ln 15 = \ln 10$$
$$3x-4 = \frac{\ln 10}{\ln 15}$$
$$3x = \frac{\ln 10}{\ln 15} + 4$$
$$x = \tfrac{1}{3}\left(\frac{\ln 10}{\ln 15} + 4\right) \approx 1.6168$$

19. Solving the equation:
$$6^{5-2x} = 4$$
$$\ln 6^{5-2x} = \ln 4$$
$$(5-2x)\ln 6 = \ln 4$$
$$5-2x = \frac{\ln 4}{\ln 6}$$
$$-2x = \frac{\ln 4}{\ln 6} - 5$$
$$x = \tfrac{1}{2}\left(5 - \frac{\ln 4}{\ln 6}\right) \approx 2.1131$$

21. Evaluating the logarithm: $\log_8 16 = \dfrac{\log 16}{\log 8} \approx 1.3333$

23. Evaluating the logarithm: $\log_{16} 8 = \dfrac{\log 8}{\log 16} = 0.7500$

25. Evaluating the logarithm: $\log_7 15 = \dfrac{\log 15}{\log 7} \approx 1.3917$

27. Evaluating the logarithm: $\log_{15} 7 = \dfrac{\log 7}{\log 15} \approx 0.7186$

29. Evaluating the logarithm: $\log_8 240 = \dfrac{\log 240}{\log 8} \approx 2.6356$

31. Evaluating the logarithm: $\log_4 321 = \dfrac{\log 321}{\log 4} \approx 4.1632$

33. Evaluating the logarithm: $\ln 345 \approx 5.8435$

35. Evaluating the logarithm: $\ln 0.345 \approx -1.0642$

37. Evaluating the logarithm: $\ln 10 \approx 2.3026$

39. Evaluating the logarithm: $\ln 45{,}000 \approx 10.7144$

41. Using the compound interest formula:
$$500\left(1+\frac{0.06}{2}\right)^{2t} = 1000$$
$$\left(1+\frac{0.06}{2}\right)^{2t} = 2$$
$$\ln\left(1+\frac{0.06}{2}\right)^{2t} = \ln 2$$
$$2t\ln\left(1+\frac{0.06}{2}\right) = \ln 2$$
$$t = \frac{\ln 2}{2\ln\left(1+\frac{0.06}{2}\right)} \approx 11.7$$

It will take 11.7 years.

43. Using the compound interest formula:
$$1000\left(1+\frac{0.12}{6}\right)^{6t} = 3000$$
$$\left(1+\frac{0.12}{6}\right)^{6t} = 3$$
$$\ln\left(1+\frac{0.12}{6}\right)^{6t} = \ln 3$$
$$6t\ln\left(1+\frac{0.12}{6}\right) = \ln 3$$
$$t = \frac{\ln 3}{6\ln\left(1+\frac{0.12}{6}\right)} \approx 9.25$$

It will take 9.25 years.

45. Using the compound interest formula:

$$P\left(1+\frac{0.08}{4}\right)^{4t} = 2P$$

$$\left(1+\frac{0.08}{4}\right)^{4t} = 2$$

$$\ln\left(1+\frac{0.08}{4}\right)^{4t} = \ln 2$$

$$4t \ln\left(1+\frac{0.08}{4}\right) = \ln 2$$

$$t = \frac{\ln 2}{4\ln\left(1+\frac{0.08}{4}\right)} \approx 8.75$$

It will take 8.75 years.

47. Using the compound interest formula:

$$25\left(1+\frac{0.06}{2}\right)^{2t} = 75$$

$$\left(1+\frac{0.06}{2}\right)^{2t} = 3$$

$$\ln\left(1+\frac{0.06}{2}\right)^{2t} = \ln 3$$

$$2t \ln\left(1+\frac{0.06}{2}\right) = \ln 3$$

$$t = \frac{\ln 3}{2\ln\left(1+\frac{0.06}{2}\right)} \approx 18.58$$

It was invested 18.58 years ago.

49. Using the continuous interest formula:

$$500e^{0.06t} = 1000$$
$$e^{0.06t} = 2$$
$$0.06t = \ln 2$$
$$t = \frac{\ln 2}{0.06} \approx 11.55$$

It will take 11.55 years.

51. Using the continuous interest formula:

$$500e^{0.06t} = 1500$$
$$e^{0.06t} = 3$$
$$0.06t = \ln 3$$
$$t = \frac{\ln 3}{0.06} \approx 18.31$$

It will take 18.31 years.

53. Completing the square: $y = 2x^2 + 8x - 15 = 2(x^2 + 4x + 4) - 8 - 15 = 2(x+2)^2 - 23$. The lowest point is $(-2, -23)$.

55. Completing the square: $y = 12x - 4x^2 = -4\left(x^2 - 3x + \frac{9}{4}\right) + 9 = -4\left(x - \frac{3}{2}\right)^2 + 9$. The highest point is $\left(\frac{3}{2}, 9\right)$.

57. Completing the square: $y = 64t - 16t^2 = -16(t^2 - 4t + 4) + 64 = -16(t-2)^2 + 64$

The object reaches a maximum height after 2 seconds, and the maximum height is 64 feet.

59. Using the population model:

$$32000e^{0.05t} = 64000$$
$$e^{0.05t} = 2$$
$$0.05t = \ln 2$$
$$t = \frac{\ln 2}{0.05} \approx 13.9$$

The city will reach 64,000 toward the end of the year 2007.

61. Finding when $P(t) = 45,000$:

$$15,000e^{0.04t} = 45,000$$
$$e^{0.04t} = 3$$
$$0.04t = \ln 3$$
$$t = \frac{\ln 3}{0.04} \approx 27.5$$

It will take approximately 27.5 years.

63. Solving for t:

$$A = Pe^{rt}$$
$$e^{rt} = \frac{A}{P}$$
$$rt = \ln\frac{A}{P}$$
$$t = \frac{\ln A - \ln P}{r} = \frac{1}{r}\ln\frac{A}{P}$$

65. Solving for t:
$$A = P2^{-kt}$$
$$2^{-kt} = \frac{A}{P}$$
$$\ln 2^{-kt} = \ln \frac{A}{P}$$
$$-kt \ln 2 = \ln A - \ln P$$
$$t = \frac{\ln A - \ln P}{-k \ln 2} = \frac{\ln P - \ln A}{k \ln 2}$$

67. Solving for t:
$$A = P(1-r)^t$$
$$(1-r)^t = \frac{A}{P}$$
$$\ln(1-r)^t = \ln \frac{A}{P}$$
$$t \ln(1-r) = \ln A - \ln P$$
$$t = \frac{\ln A - \ln P}{\ln(1-r)}$$

Chapter 11 Review/Test

1. Evaluating the function: $f(4) = 2^4 = 16$

2. Evaluating the function: $f(-1) = 2^{-1} = \frac{1}{2}$

3. Evaluating the function: $g(2) = \left(\frac{1}{3}\right)^2 = \frac{1}{9}$

4. Evaluating the function: $f(2) - g(-2) = 2^2 - \left(\frac{1}{3}\right)^{-2} = 4 - 9 = -5$

5. Evaluating the function: $f(-1) + g(1) = 2^{-1} + \left(\frac{1}{3}\right)^1 = \frac{1}{2} + \frac{1}{3} = \frac{5}{6}$

6. Evaluating the function: $g(-1) + f(2) = \left(\frac{1}{3}\right)^{-1} + 2^2 = 3 + 4 = 7$

7. Graphing the function:

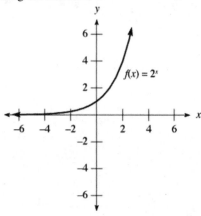

8. Graphing the function:

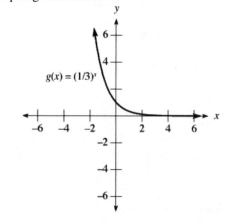

9. Finding the inverse:
$$2y+1 = x$$
$$2y = x-1$$
$$y = \frac{x-1}{2}$$
The inverse is $y^{-1} = \frac{x-1}{2}$. Sketching the graph:

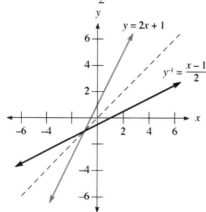

10. Finding the inverse:
$$y^2 - 4 = x$$
$$y^2 = x+4$$
$$y = \pm\sqrt{x+4}$$
The inverse is $y^{-1} = \pm\sqrt{x+4}$. Sketching the graph:

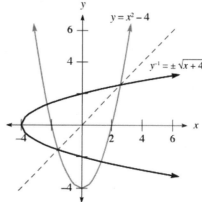

11. Finding the inverse:
$$2y+3 = x$$
$$2y = x-3$$
$$y = \frac{x-3}{2}$$
The inverse is $f^{-1}(x) = \frac{x-3}{2}$.

12. Finding the inverse:
$$y^2 - 1 = x$$
$$y^2 = x+1$$
$$y = \pm\sqrt{x+1}$$
The inverse is $y = \pm\sqrt{x+1}$ (which is not a function).

13. Finding the inverse:
$$\tfrac{1}{2}y + 2 = x$$
$$y + 4 = 2x$$
$$y = 2x - 4$$
The inverse is $f^{-1}(x) = 2x - 4$.

14. Finding the inverse:
$$4 - 2y^2 = x$$
$$2y^2 = 4 - x$$
$$y^2 = \frac{4-x}{2}$$
$$y = \pm\sqrt{\frac{4-x}{2}}$$
The inverse is $y = \pm\sqrt{\frac{4-x}{2}}$ (which is not a function).

15. Writing in logarithmic form: $\log_3 81 = 4$
16. Writing in logarithmic form: $\log_7 49 = 2$
17. Writing in logarithmic form: $\log_{10} 0.01 = -2$
18. Writing in logarithmic form: $\log_2 \tfrac{1}{8} = -3$
19. Writing in exponential form: $2^3 = 8$
20. Writing in exponential form: $3^2 = 9$
21. Writing in exponential form: $4^{1/2} = 2$
22. Writing in exponential form: $4^1 = 4$
23. Solving for x:
$$\log_5 x = 2$$
$$x = 5^2 = 25$$

338 Chapter 11 Exponential and Logarithmic Functions

24. Solving for x:
$$\log_{16} 8 = x$$
$$16^x = 8$$
$$2^{4x} = 2^3$$
$$4x = 3$$
$$x = \tfrac{3}{4}$$

25. Solving for x:
$$\log_x 0.01 = -2$$
$$x^{-2} = 0.01$$
$$x^2 = 100$$
$$x = 10$$

26. Graphing the equation:

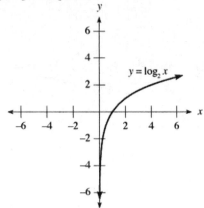

27. Graphing the equation:

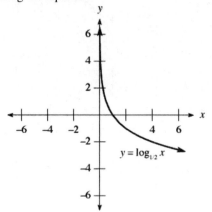

28. Simplifying the logarithm: $\log_4 16 = \log_4 4^2 = 2$

29. Simplifying the logarithm:
$$\log_{27} 9 = x$$
$$27^x = 9$$
$$3^{3x} = 3^2$$
$$3x = 2$$
$$x = \tfrac{2}{3}$$

30. Simplifying the logarithm: $\log_4 \left(\log_3 3 \right) = \log_4 1 = 0$

31. Expanding the logarithm: $\log_2 5x = \log_2 5 + \log_2 x$

32. Expanding the logarithm: $\log_{10} \dfrac{2x}{y} = \log_{10} 2x - \log_{10} y = \log_{10} 2 + \log_{10} x - \log_{10} y$

33. Expanding the logarithm: $\log_a \dfrac{y^3 \sqrt{x}}{z} = \log_a y^3 + \log_a x^{1/2} - \log_a z = 3\log_a y + \tfrac{1}{2}\log_a x - \log_a z$

34. Expanding the logarithm: $\log_{10} \dfrac{x^2}{y^3 z^4} = \log_{10} x^2 - \log_{10} y^3 - \log_{10} z^4 = 2\log_{10} x - 3\log_{10} y - 4\log_{10} z$

35. Writing as a single logarithm: $\log_2 x + \log_2 y = \log_2 xy$

36. Writing as a single logarithm: $\log_3 x - \log_3 4 = \log_3 \dfrac{x}{4}$

37. Writing as a single logarithm: $2\log_a 5 - \tfrac{1}{2}\log_a 9 = \log_a 5^2 - \log_a 9^{1/2} = \log_a 25 - \log_a 3 = \log_a \tfrac{25}{3}$

38. Writing as a single logarithm: $3\log_2 x + 2\log_2 y - 4\log_2 z = \log_2 x^3 + \log_2 y^2 - \log_2 z^4 = \log_2 \dfrac{x^3 y^2}{z^4}$

39. Solving the equation:
$$\log_2 x + \log_2 4 = 3$$
$$\log_2 4x = 3$$
$$4x = 2^3$$
$$4x = 8$$
$$x = 2$$

40. Solving the equation:
$$\log_2 x - \log_2 3 = 1$$
$$\log_2 \frac{x}{3} = 1$$
$$\frac{x}{3} = 2^1$$
$$\frac{x}{3} = 2$$
$$x = 6$$

41. Solving the equation:
$$\log_3 x + \log_3 (x-2) = 1$$
$$\log_3 (x^2 - 2x) = 1$$
$$x^2 - 2x = 3^1$$
$$x^2 - 2x - 3 = 0$$
$$(x-3)(x+1) = 0$$
$$x = 3, -1$$

The solution is 3 (−1 does not check).

42. Solving the equation:
$$\log_4 (x+1) - \log_4 (x-2) = 1$$
$$\log_4 \frac{x+1}{x-2} = 1$$
$$\frac{x+1}{x-2} = 4^1$$
$$x + 1 = 4x - 8$$
$$-3x = -9$$
$$x = 3$$

43. Solving the equation:
$$\log_6 (x-1) + \log_6 x = 1$$
$$\log_6 (x^2 - x) = 1$$
$$x^2 - x = 6^1$$
$$x^2 - x - 6 = 0$$
$$(x-3)(x+2) = 0$$
$$x = 3, -2$$

The solution is 3 (−2 does not check).

44. Solving the equation:
$$\log_4 (x-3) + \log_4 x = 1$$
$$\log_4 (x^2 - 3x) = 1$$
$$x^2 - 3x = 4^1$$
$$x^2 - 3x - 4 = 0$$
$$(x-4)(x+1) = 0$$
$$x = 4, -1$$

The solution is 4 (−1 does not check).

45. Evaluating: $\log 346 \approx 2.5391$

46. Evaluating: $\log 0.713 \approx -0.1469$

47. Solving for x:
$$\log x = 3.9652$$
$$x = 10^{3.9652} \approx 9{,}230$$

48. Solving for x:
$$\log x = -1.6003$$
$$x = 10^{-1.6003} \approx 0.0251$$

49. Simplifying: $\ln e = \ln e^1 = 1$

50. Simplifying: $\ln 1 = \ln e^0 = 0$

51. Simplifying: $\ln e^2 = 2$

52. Simplifying: $\ln e^{-4} = -4$

53. Finding the pH: $\text{pH} = -\log(7.9 \times 10^{-3}) \approx 2.1$

54. Finding the pH: $\text{pH} = -\log(8.1 \times 10^{-6}) \approx 5.1$

55. Finding $[H^+]$:
$$2.7 = -\log[H^+]$$
$$-2.7 = \log[H^+]$$
$$[H^+] = 10^{-2.7} \approx 2.0 \times 10^{-3}$$

56. Finding $[H^+]$:
$$7.5 = -\log[H^+]$$
$$-7.5 = \log[H^+]$$
$$[H^+] = 10^{-7.5} \approx 3.2 \times 10^{-8}$$

57. Solving the equation:

$$4^x = 8$$
$$2^{2x} = 2^3$$
$$2x = 3$$
$$x = \tfrac{3}{2}$$

58. Solving the equation:

$$4^{3x+2} = 5$$
$$\ln 4^{3x+2} = \ln 5$$
$$(3x+2)\ln 4 = \ln 5$$
$$3x + 2 = \frac{\ln 5}{\ln 4}$$
$$3x = \frac{\ln 5}{\ln 4} - 2$$
$$x = \tfrac{1}{3}\left(\frac{\ln 5}{\ln 4} - 2\right) \approx -0.28$$

59. Using a calculator: $\log_{16} 8 = \dfrac{\ln 8}{\ln 16} = 0.75$

60. Using a calculator: $\log_{12} 421 = \dfrac{\ln 421}{\ln 12} \approx 2.43$

61. Using the compound interest formula:

$$5000(1+0.16)^t = 10000$$
$$1.16^t = 2$$
$$\ln 1.16^t = \ln 2$$
$$t \ln 1.16 = \ln 2$$
$$t = \frac{\ln 2}{\ln 1.16} \approx 4.67$$

It will take approximately 4.67 years for the amount to double.

62. Using the compound interest formula:

$$10000\left(1 + \frac{0.12}{6}\right)^{6t} = 30000$$
$$\left(1 + \frac{0.12}{6}\right)^{6t} = 3$$
$$\ln\left(1 + \frac{0.12}{6}\right)^{6t} = \ln 3$$
$$6t \ln 1.02 = \ln 3$$
$$t = \frac{\ln 3}{6 \ln 1.02} \approx 9.25$$

It will take approximately 9.25 years for the amount to triple.

Chapter 12
Conic Sections

12.1 The Circle

1. Using the distance formula: $d = \sqrt{(6-3)^2 + (3-7)^2} = \sqrt{9+16} = \sqrt{25} = 5$

3. Using the distance formula: $d = \sqrt{(5-0)^2 + (0-9)^2} = \sqrt{25+81} = \sqrt{106}$

5. Using the distance formula: $d = \sqrt{(-2-3)^2 + (1+5)^2} = \sqrt{25+36} = \sqrt{61}$

7. Using the distance formula: $d = \sqrt{(-10+1)^2 + (5+2)^2} = \sqrt{81+49} = \sqrt{130}$

9. Solving the equation:
$$\sqrt{(x-1)^2 + (2-5)^2} = \sqrt{13}$$
$$(x-1)^2 + 9 = 13$$
$$(x-1)^2 = 4$$
$$x-1 = \pm 2$$
$$x-1 = -2, 2$$
$$x = -1, 3$$

11. Solving the equation:
$$\sqrt{(7-8)^2 + (y-3)^2} = 1$$
$$(y-3)^2 + 1 = 1$$
$$(y-3)^2 = 0$$
$$y-3 = 0$$
$$y = 3$$

13. The equation is $(x-2)^2 + (y-3)^2 = 16$.

15. The equation is $(x-3)^2 + (y+2)^2 = 9$.

17. The equation is $(x+5)^2 + (y+1)^2 = 5$.

19. The equation is $x^2 + (y+5)^2 = 1$.

21. The equation is $x^2 + y^2 = 4$.

23. The center is (0,0) and the radius is 2.

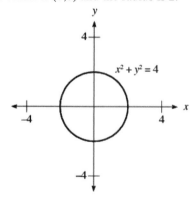

25. The center is (1,3) and the radius is 5.

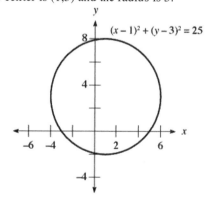

342 Chapter 12 Conic Sections

27. The center is (−2,4) and the radius is $2\sqrt{2}$.

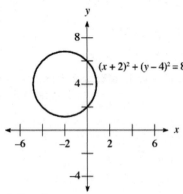

29. The center is (−1,−1) and the radius is 1.

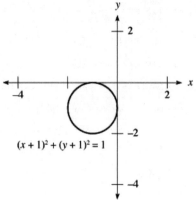

31. Completing the square:
$$x^2 + y^2 - 6y = 7$$
$$x^2 + \left(y^2 - 6y + 9\right) = 7 + 9$$
$$x^2 + (y-3)^2 = 16$$
The center is (0,3) and the radius is 4.

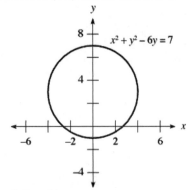

33. Completing the square:
$$x^2 + y^2 - 4x - 6y = -4$$
$$\left(x^2 - 4x + 4\right) + \left(y^2 - 6y + 9\right) = -4 + 4 + 9$$
$$(x-2)^2 + (y-3)^2 = 9$$
The center is (−5,0) and the radius is 5.

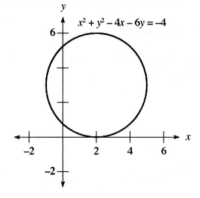

35. Completing the square:
$$x^2 + y^2 + 2x + y = \tfrac{11}{4}$$
$$\left(x^2 + 2x + 1\right) + \left(y^2 + y + \tfrac{1}{4}\right) = \tfrac{11}{4} + 1 + \tfrac{1}{4}$$
$$(x+1)^2 + \left(y + \tfrac{1}{2}\right)^2 = 4$$
The center is $\left(-1, -\tfrac{1}{2}\right)$ and the radius is 2.

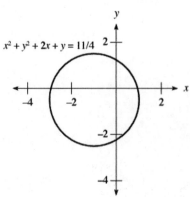

37. The equation is $(x-3)^2 + (y-4)^2 = 25$.

39. The equations are:

    A: $\left(x-\frac{1}{2}\right)^2+(y-1)^2=\frac{1}{4}$

    B: $(x-1)^2+(y-1)^2=1$

    C: $(x-2)^2+(y-1)^2=4$

41. The equations are:

    A: $(x+8)^2+y^2=64$

    B: $x^2+y^2=64$

    C: $(x-8)^2+y^2=64$

43. The x-coordinate of the center is $x=500$, the y-coordinate of the center is $12+120=132$, and the radius is 120. Thus the equation of the circle is $(x-500)^2+(y-132)^2=120^2=14{,}400$.

45. Let $y=f(x)$. Switch x and y and solve for y:
    $$3^y=x$$
    $$y=\log_3 x$$

    The inverse is $f^{-1}(x)=\log_3 x$.

47. Let $y=f(x)$. Switch x and y and solve for y:
    $$2y+3=x$$
    $$2y=x-3$$
    $$y=\frac{x-3}{2}$$

    The inverse is $f^{-1}(x)=\frac{x-3}{2}$.

49. Let $y=f(x)$. Switch x and y and solve for y:
    $$\frac{y+3}{5}=x$$
    $$y+3=5x$$
    $$y=5x-3$$

    The inverse is $f^{-1}(x)=5x-3$.

51. Solving the equation:
    $$y^2=9$$
    $$y=\pm\sqrt{9}=\pm 3$$

53. Solving the equation:
    $$-y^2=4$$
    $$y^2=-4$$
    $$y=\pm\sqrt{-4}=\pm 2i$$

55. Solving the equation:
    $$\frac{-x^2}{9}=1$$
    $$x^2=-9$$
    $$x=\pm\sqrt{-9}=\pm 3i$$

57. Dividing: $\frac{4x^2+9y^2}{36}=\frac{4x^2}{36}+\frac{9y^2}{36}=\frac{x^2}{9}+\frac{y^2}{4}$

59. To find the x-intercept, let $y=0$:
    $$3x-4(0)=12$$
    $$3x-0=12$$
    $$3x=12$$
    $$x=4$$

    To find the y-intercept, let $x=0$:
    $$3(0)-4y=12$$
    $$0-4y=12$$
    $$-4y=12$$
    $$y=-3$$

61. Substituting $x=3$:
    $$\frac{3^2}{25}+\frac{y^2}{9}=1$$
    $$\frac{9}{25}+\frac{y^2}{9}=1$$
    $$\frac{y^2}{9}=\frac{16}{25}$$
    $$y^2=\frac{144}{25}$$
    $$y=\pm\sqrt{\frac{144}{25}}=\pm\frac{12}{5}=\pm 2.4$$

63. The radius is 2, so the equation is $(x-2)^2+(y-3)^2=4$.

65. The radius is 2, so the equation is $(x-2)^2+(y-3)^2=4$.

344 Chapter 12 Conic Sections

67. Completing the square:
$$x^2 + y^2 - 6x + 8y = 144$$
$$(x^2 - 6x + 9) + (y^2 + 8y + 16) = 144 + 9 + 16$$
$$(x-3)^2 + (y+4)^2 = 169$$
The center is (3,−4), so the distance is: $d = \sqrt{(-3)^2 + 4^2} = \sqrt{9+16} = 5$

69. Completing the square:
$$x^2 + y^2 - 6x - 8y = 144$$
$$(x^2 - 6x + 9) + (y^2 - 8y + 16) = 144 + 9 + 16$$
$$(x-3)^2 + (y-4)^2 = 169$$
The center is (3,4), so the distance is: $d = \sqrt{3^2 + 4^2} = \sqrt{9+16} = 5$

71. The equation $y = \sqrt{9 - x^2}$ corresponds to the top half of the circle, and the equation $y = -\sqrt{9 - x^2}$ corresponds to the bottom half of the circle.

12.2 Ellipses and Hyperbolas

1. Graphing the ellipse:

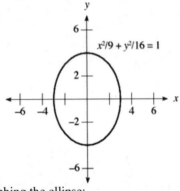

3. Graphing the ellipse:

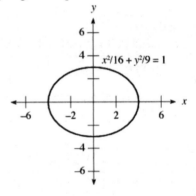

5. Graphing the ellipse:

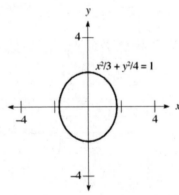

7. The standard form is $\dfrac{x^2}{25} + \dfrac{y^2}{4} = 1$. Graphing the ellipse:

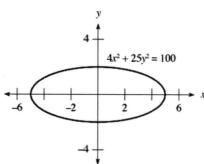

9. The standard form is $\dfrac{x^2}{16} + \dfrac{y^2}{2} = 1$. Graphing the ellipse:

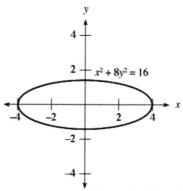

11. Graphing the hyperbola:

13. Graphing the hyperbola:

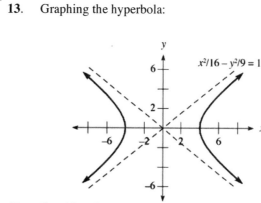

15. Graphing the hyperbola:

17. Graphing the hyperbola:

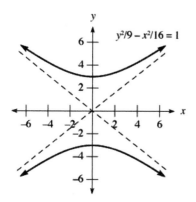

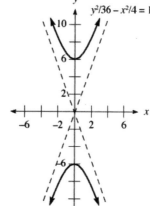

346 Chapter 12 Conic Sections

19. The standard form is $\dfrac{x^2}{4} - \dfrac{y^2}{1} = 1$. Graphing the hyperbola:

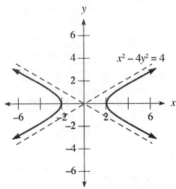

21. The standard form is $\dfrac{y^2}{9} - \dfrac{x^2}{16} = 1$. Graphing the hyperbola:

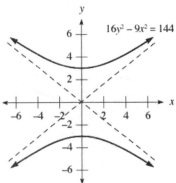

23. For the x-intercepts, set $y = 0$:
$0.4x^2 = 3.6$
$x^2 = 9$
$x = \pm 3$

For the y-intercepts, set $x = 0$:
$0.9y^2 = 3.6$
$y^2 = 4$
$y = \pm 2$

25. For the x-intercepts, set $y = 0$:
$\dfrac{x^2}{0.04} = 1$
$x^2 = 0.04$
$x = \pm 0.2$

For the y-intercepts, set $x = 0$:
$-\dfrac{y^2}{0.09} = 1$
$y^2 = -0.09$
There are no y-intercepts.

27. For the x-intercepts, set $y = 0$:
$\dfrac{25x^2}{9} = 1$
$x^2 = \dfrac{9}{25}$
$x = \pm \dfrac{3}{5}$

For the y-intercepts, set $x = 0$:
$\dfrac{25y^2}{4} = 1$
$y^2 = \dfrac{4}{25}$
$y = \pm \dfrac{2}{5}$

29. Graphing the ellipse:

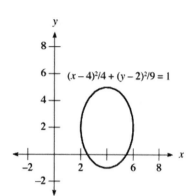

31. Completing the square:
$$4x^2 + y^2 - 4y - 12 = 0$$
$$4x^2 + (y^2 - 4y + 4) = 12 + 4$$
$$4x^2 + (y-2)^2 = 16$$
$$\frac{x^2}{4} + \frac{(y-2)^2}{16} = 1$$
Graphing the ellipse:

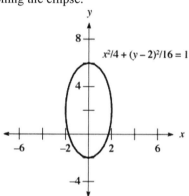

33. Completing the square:
$$x^2 + 9y^2 + 4x - 54y + 76 = 0$$
$$(x^2 + 4x + 4) + 9(y^2 - 6y + 9) = -76 + 4 + 81$$
$$(x+2)^2 + 9(y-3)^2 = 9$$
$$\frac{(x+2)^2}{9} + \frac{(y-3)^2}{1} = 1$$
Graphing the ellipse:

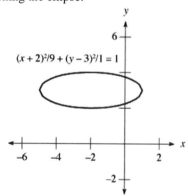

35. Graphing the hyperbola:

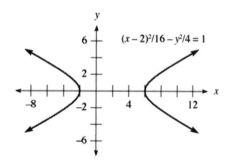

348 Chapter 12 Conic Sections

37. Completing the square:
$$9y^2 - x^2 - 4x + 54y + 68 = 0$$
$$9(y^2 + 6y + 9) - (x^2 + 4x + 4) = -68 + 81 - 4$$
$$9(y+3)^2 - (x+2)^2 = 9$$
$$\frac{(y+3)^2}{1} - \frac{(x+2)^2}{9} = 1$$
Graphing the hyperbola:

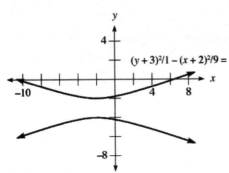

39. Completing the square:
$$4y^2 - 9x^2 - 16y + 72x - 164 = 0$$
$$4(y^2 - 4y + 4) - 9(x^2 - 8x + 16) = 164 + 16 - 144$$
$$4(y-2)^2 - 9(x-4)^2 = 36$$
$$\frac{(y-2)^2}{9} - \frac{(x-4)^2}{4} = 1$$
Graphing the hyperbola:

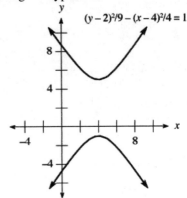

41. Substituting $x = 4$:
$$\frac{4^2}{25} + \frac{y^2}{9} = 1$$
$$\frac{y^2}{9} + \frac{16}{25} = 1$$
$$\frac{y^2}{9} = \frac{9}{25}$$
$$\frac{y}{3} = \pm\frac{3}{5}$$
$$y = \pm\frac{9}{5}$$

43. Substituting $x = 1.8$:
$$16(1.8)^2 + 9y^2 = 144$$
$$51.84 + 9y^2 = 144$$
$$9y^2 = 92.16$$
$$3y = \pm 9.6$$
$$y = \pm 3.2$$

45. The asymptotes are $y = \frac{3}{4}x$ and $y = -\frac{3}{4}x$.

47. The equation has the form $\frac{x^2}{16} + \frac{y^2}{b^2} = 1$. Substituting the point $(2, \sqrt{3})$:
$$\frac{(2)^2}{16} + \frac{(\sqrt{3})^2}{b^2} = 1$$
$$\frac{1}{4} + \frac{3}{b^2} = 1$$
$$\frac{3}{b^2} = \frac{3}{4}$$
$$b^2 = 4$$
The equation of the ellipse is $\frac{x^2}{16} + \frac{y^2}{4} = 1$.

49. The equation of the ellipse is $\dfrac{x^2}{20^2} + \dfrac{y^2}{10^2} = 1$. Substituting $y = 6$:

$$\dfrac{x^2}{20^2} + \dfrac{6^2}{10^2} = 1$$
$$\dfrac{x^2}{400} + \dfrac{9}{25} = 1$$
$$\dfrac{x^2}{400} = \dfrac{16}{25}$$
$$x^2 = 256$$
$$x = 16$$

The man can walk within 16 feet of the center.

51. Substituting $a = 4$ and $c = 3$:

$$4^2 = b^2 + 3^2$$
$$16 = b^2 + 9$$
$$b^2 = 7$$
$$b = \sqrt{7} \approx 2.65$$

The width should be approximately $2(2.65) = 5.3$ feet wide.

53. Evaluating: $f(4) = \dfrac{2}{4-2} = \dfrac{2}{2} = 1$

55. Evaluating: $f[g(0)] = f\left(\dfrac{2}{0+2}\right) = f(1) = \dfrac{2}{1-2} = \dfrac{2}{-1} = -2$

57. Simplifying: $f(x) + g(x) = \dfrac{2}{x-2} + \dfrac{2}{x+2} = \dfrac{2}{x-2} \cdot \dfrac{x+2}{x+2} + \dfrac{2}{x+2} \cdot \dfrac{x-2}{x-2} = \dfrac{2x+4+2x-4}{(x-2)(x+2)} = \dfrac{4x}{(x-2)(x+2)}$

59. Since $4^2 + 0^2 = 16$ and $0^2 + 5^2 = 25$, while $0^2 + 0^2 = 0$, only (0,0) is a solution.

61. Multiplying: $(2y+4)^2 = (2y)^2 + 2(2y)(4) + 4^2 = 4y^2 + 16y + 16$

63. Solving for x:
$$x - 2y = 4$$
$$x = 2y + 4$$

65. Simplifying: $x^2 - 2(x^2 - 3) = x^2 - 2x^2 + 6 = -x^2 + 6$

67. Factoring: $5y^2 + 16y + 12 = (5y+6)(y+2)$

69. Solving the equation:

$$y^2 = 4$$
$$y = \pm\sqrt{4} = \pm 2$$

71. Solving the equation:

$$-x^2 + 6 = 2$$
$$-x^2 = -4$$
$$x^2 = 4$$
$$x = \pm\sqrt{4} = \pm 2$$

12.3 Second-Degree Inequalities and Nonlinear Systems

1. Graphing the inequality:

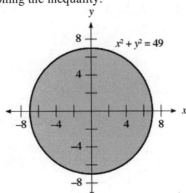

3. Graphing the inequality:

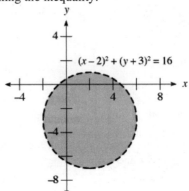

5. Graphing the inequality:

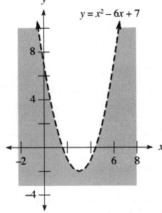

7. Graphing the inequality:

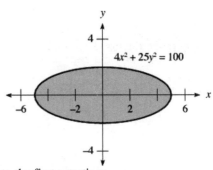

9. Solving the second equation for y yields $y = 3 - 2x$. Substituting into the first equation:
$$x^2 + (3-2x)^2 = 9$$
$$x^2 + 9 - 12x + 4x^2 = 9$$
$$5x^2 - 12x = 0$$
$$x(5x - 12) = 0$$
$$x = 0, \tfrac{12}{5}$$
$$y = 3, -\tfrac{9}{5}$$

The solutions are $(0, 3), \left(\tfrac{12}{5}, -\tfrac{9}{5}\right)$.

11. Solving the second equation for x yields $x = 8 - 2y$. Substituting into the first equation:
$$(8 - 2y)^2 + y^2 = 16$$
$$64 - 32y + 4y^2 + y^2 = 16$$
$$5y^2 - 32y + 48 = 0$$
$$(y - 4)(5y - 12) = 0$$
$$y = 4, \tfrac{12}{5}$$
$$x = 0, \tfrac{16}{5}$$

The solutions are $(0, 4), \left(\tfrac{16}{5}, \tfrac{12}{5}\right)$.

13. Adding the two equations yields:

$$2x^2 = 50$$
$$x^2 = 25$$
$$x = -5, 5$$
$$y = 0$$

The solutions are $(-5,0), (5,0)$.

15. Substituting into the first equation:

$$x^2 + (x^2 - 3)^2 = 9$$
$$x^2 + x^4 - 6x^2 + 9 = 9$$
$$x^4 - 5x^2 = 0$$
$$x^2(x^2 - 5) = 0$$
$$x = 0, -\sqrt{5}, \sqrt{5}$$
$$y = -3, 2, 2$$

The solutions are $(0,-3), (-\sqrt{5}, 2), (\sqrt{5}, 2)$.

17. Substituting into the first equation:

$$x^2 + (x^2 - 4)^2 = 16$$
$$x^2 + x^4 - 8x^2 + 16 = 16$$
$$x^4 - 7x^2 = 0$$
$$x^2(x^2 - 7) = 0$$
$$x = 0, -\sqrt{7}, \sqrt{7}$$
$$y = -4, 3, 3$$

The solutions are $(0,-4), (-\sqrt{7}, 3), (\sqrt{7}, 3)$.

19. Substituting into the first equation:

$$3x + 2(x^2 - 5) = 10$$
$$3x + 2x^2 - 10 = 10$$
$$2x^2 + 3x - 20 = 0$$
$$(x+4)(2x-5) = 0$$
$$x = -4, \tfrac{5}{2}$$
$$y = 11, \tfrac{5}{4}$$

The solutions are $(-4, 11), \left(\tfrac{5}{2}, \tfrac{5}{4}\right)$.

21. Substituting into the first equation:

$$-x + 1 = x^2 + 2x - 3$$
$$x^2 + 3x - 4 = 0$$
$$(x+4)(x-1) = 0$$
$$x = -4, 1$$
$$y = 5, 0$$

The solutions are $(-4, 5), (1, 0)$.

23. Substituting into the first equation:

$$x - 5 = x^2 - 6x + 5$$
$$x^2 - 7x + 10 = 0$$
$$(x-2)(x-5) = 0$$
$$x = 2, 5$$
$$y = -3, 0$$

The solutions are $(2, -3), (5, 0)$.

25. Adding the two equations yields:

$$8x^2 = 72$$
$$x^2 = 9$$
$$x = \pm 3$$
$$y = 0$$

The solutions are $(-3, 0), (3, 0)$.

27. Solving the first equation for x yields $x = y + 4$. Substituting into the second equation:

$$(y+4)^2 + y^2 = 16$$
$$y^2 + 8y + 16 + y^2 = 16$$
$$2y^2 + 8y = 0$$
$$2y(y+4) = 0$$
$$y = 0, -4$$
$$x = 4, 0$$

The solutions are $(0, -4), (4, 0)$.

29. **a.** Subtracting the two equations yields:
$$(x+8)^2 - x^2 = 0$$
$$x^2 + 16x + 64 - x^2 = 0$$
$$16x = -64$$
$$x = -4$$

Substituting to find y:
$$(-4)^2 + y^2 = 64$$
$$y^2 + 16 = 64$$
$$y^2 = 48$$
$$y = \pm\sqrt{48} = \pm 4\sqrt{3}$$

The intersection points are $(-4, -4\sqrt{3})$ and $(-4, 4\sqrt{3})$.

b. Subtracting the two equations yields:
$$x^2 - (x-8)^2 = 0$$
$$x^2 - x^2 + 16x - 64 = 0$$
$$16x = 64$$
$$x = 4$$

Substituting to find y:
$$4^2 + y^2 = 64$$
$$y^2 + 16 = 64$$
$$y^2 = 48$$
$$y = \pm\sqrt{48} = \pm 4\sqrt{3}$$

The intersection points are $(4, -4\sqrt{3})$ and $(4, 4\sqrt{3})$.

31. Graphing the inequality:

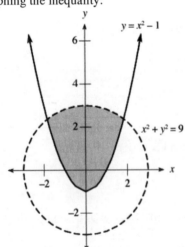

33. Graphing the inequality:

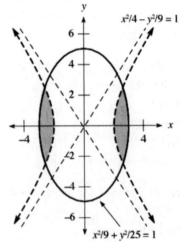

35. There is no intersection.

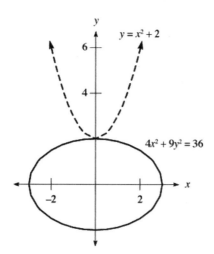

37. The system of inequalities is:
$$x^2 + y^2 < 16$$
$$y > 4 - \tfrac{1}{4}x^2$$

39. The system of equations is:
$$x^2 + y^2 = 89$$
$$x^2 - y^2 = 39$$
Adding the two equations yields:
$$2x^2 = 128$$
$$x^2 = 64$$
$$x = \pm 8$$
$$y = \pm 5$$
The numbers are either 8 and 5, 8 and –5, –8 and 5, or –8 and –5.

41. The system of equations is:
$$y = x^2 - 3$$
$$x + y = 9$$
Substituting into the second equation:
$$x + x^2 - 3 = 9$$
$$x^2 + x - 12 = 0$$
$$(x+4)(x-3) = 0$$
$$x = -4, 3$$
$$y = 13, 6$$
The numbers are either –4 and 13, or 3 and 6.

43. Solving the equation:
$$g(x) = 0$$
$$\left(x + \tfrac{2}{5}\right)^2 = 0$$
$$x + \tfrac{2}{5} = 0$$
$$x = -\tfrac{2}{5}$$

45. Solving the equation:
$$f(x) = g(x)$$
$$x^2 + 4x - 4 = 1$$
$$x^2 + 4x - 5 = 0$$
$$(x+5)(x-1) = 0$$
$$x = -5, 1$$

354 Chapter 12 Conic Sections

47. Solving the equation:
$$f(x) = g(x)$$
$$x^2 + 4x - 4 = x - 6$$
$$x^2 + 3x + 2 = 0$$
$$(x+2)(x+1) = 0$$
$$x = -2, -1$$

Chapter 12 Review/Test

1. Using the distance formula: $d = \sqrt{(-1-2)^2 + (5-6)^2} = \sqrt{9+1} = \sqrt{10}$

2. Using the distance formula: $d = \sqrt{(1-3)^2 + (-1+4)^2} = \sqrt{4+9} = \sqrt{13}$

3. Using the distance formula: $d = \sqrt{(-4-0)^2 + (0-3)^2} = \sqrt{16+9} = \sqrt{25} = 5$

4. Using the distance formula: $d = \sqrt{(-3+3)^2 + (-2-7)^2} = \sqrt{0+81} = \sqrt{81} = 9$

5. Solving the equation:
$$\sqrt{(x-2)^2 + (-1+4)^2} = 5$$
$$(x-2)^2 + 9 = 25$$
$$(x-2)^2 = 16$$
$$x - 2 = \pm\sqrt{16}$$
$$x - 2 = -4, 4$$
$$x = -2, 6$$

6. Solving the equation:
$$\sqrt{(-3-3)^2 + (y+4)^2} = 10$$
$$(y+4)^2 + 36 = 100$$
$$(y+4)^2 = 64$$
$$y + 4 = \pm\sqrt{64}$$
$$y + 4 = -8, 8$$
$$y = -12, 4$$

7. The equation is $(x-3)^2 + (y-1)^2 = 4$.

8. The equation is $(x-3)^2 + (y+1)^2 = 16$.

9. The equation is $(x+5)^2 + y^2 = 9$.

10. The equation is $(x+3)^2 + (y-4)^2 = 18$.

11. The equation is $x^2 + y^2 = 25$.

12. The equation is $x^2 + y^2 = 9$.

13. Finding the radius: $r = \sqrt{(-2-2)^2 + (3-0)^2} = \sqrt{16+9} = \sqrt{25} = 5$. The equation is $(x+2)^2 + (y-3)^2 = 25$.

14. Finding the radius: $r = \sqrt{(-6)^2 + (8)^2} = \sqrt{36+64} = \sqrt{100} = 10$. The equation is $(x+6)^2 + (y-8)^2 = 100$.

15. The center is (0,0) and the radius is 2.

16. The center is (3,–1) and the radius is 4.

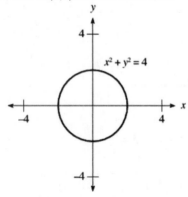

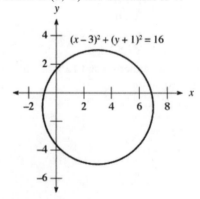

17. Completing the square:
$$x^2 + y^2 - 6x + 4y = -4$$
$$(x^2 - 6x + 9) + (y^2 + 4y + 4) = -4 + 9 + 4$$
$$(x-3)^2 + (y+2)^2 = 9$$
The center is (3,–2) and the radius is 3.

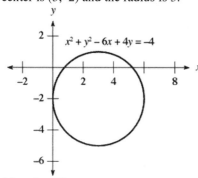

18. Completing the square:
$$x^2 + y^2 + 4x - 2y = 4$$
$$(x^2 + 4x + 4) + (y^2 - 2y + 1) = 4 + 4 + 1$$
$$(x+2)^2 + (y-1)^2 = 9$$
The center is (–2,1) and the radius is 3.

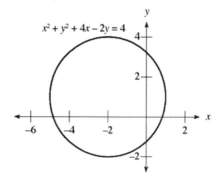

19. Graphing the ellipse:

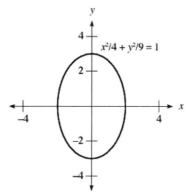

20. The standard form is $\dfrac{x^2}{4} + \dfrac{y^2}{16} = 1$. Graphing the ellipse:

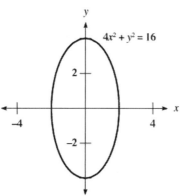

356 Chapter 12 Conic Sections

21. Graphing the hyperbola:

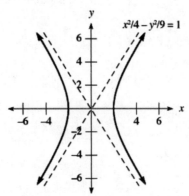

22. The standard form is $\dfrac{x^2}{4} - \dfrac{y^2}{16} = 1$. Graphing the hyperbola:

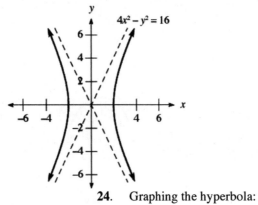

23. Graphing the ellipse:

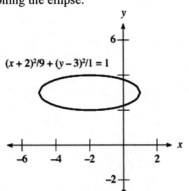

24. Graphing the hyperbola:

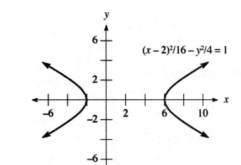

25. Completing the square:
$$9y^2 - x^2 - 4x + 54y + 68 = 0$$
$$9(y^2 + 6y + 9) - (x^2 + 4x + 4) = -68 + 81 - 4$$
$$9(y+3)^2 - (x+2)^2 = 9$$
$$\frac{(y+3)^2}{1} - \frac{(x+2)^2}{9} = 1$$
Graphing the hyperbola:

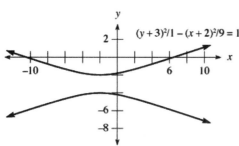

26. Completing the square:
$$9x^2 + 4y^2 - 72x - 16y + 124 = 0$$
$$9(x^2 - 8x + 16) + 4(y^2 - 4y + 4) = -124 + 144 + 16$$
$$9(x-4)^2 + 4(y-2)^2 = 36$$
$$\frac{(x-4)^2}{4} + \frac{(y-2)^2}{9} = 1$$
Graphing the ellipse:

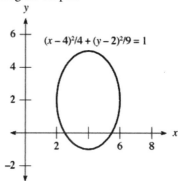

27. Graphing the inequality:

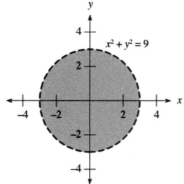

28. Graphing the inequality:

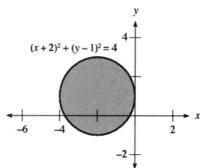

29. Graphing the inequality:

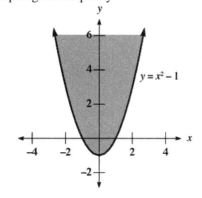

30. Graphing the inequality:

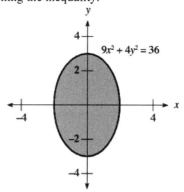

358 Chapter 12 Conic Sections

31. Graphing the solution set:

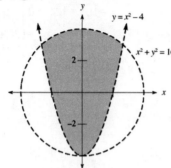

32. Graphing the solution set:

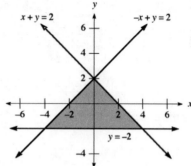

33. Solving the second equation for y yields $y = 4 - 2x$. Substituting into the first equation:
$$x^2 + (4-2x)^2 = 16$$
$$x^2 + 16 - 16x + 4x^2 = 16$$
$$5x^2 - 16x = 0$$
$$x(5x - 16) = 0$$
$$x = 0, \frac{16}{5}$$
$$y = 4, -\frac{12}{5}$$
The solutions are $(0, 4), \left(\frac{16}{5}, -\frac{12}{5}\right)$.

34. Substituting into the first equation:
$$x^2 + (x^2 - 2)^2 = 4$$
$$x^2 + x^4 - 4x^2 + 4 = 4$$
$$x^4 - 3x^2 = 0$$
$$x^2(x^2 - 3) = 0$$
$$x = 0, \pm\sqrt{3}$$
$$y = -2, 1$$
The solutions are $(0, -2), (-\sqrt{3}, 1), (\sqrt{3}, 1)$.

35. Adding the two equations yields:
$$18x^2 = 72$$
$$x^2 = 4$$
$$x = \pm 2$$
$$y = 0$$
The solutions are $(-2, 0), (2, 0)$.

36. Multiply the second equation by 2 and add it to the first equation:
$$2x^2 - 4y^2 = 8$$
$$2x^2 + 4y^2 = 20$$
Adding yields:
$$4x^2 = 28$$
$$x^2 = 7$$
$$x = \pm\sqrt{7}$$
$$y = \pm\sqrt{\frac{3}{2}} = \pm\frac{\sqrt{6}}{2}$$
The solutions are $\left(-\sqrt{7}, -\frac{\sqrt{6}}{2}\right), \left(-\sqrt{7}, \frac{\sqrt{6}}{2}\right), \left(\sqrt{7}, -\frac{\sqrt{6}}{2}\right), \left(\sqrt{7}, \frac{\sqrt{6}}{2}\right)$.

Chapter 13
Sequences and Series

13.1 Sequences

1. The first five terms are: $4, 7, 10, 13, 16$
3. The first five terms are: $3, 7, 11, 15, 19$
5. The first five terms are: $1, 2, 3, 4, 5$
7. The first five terms are: $4, 7, 12, 19, 28$
9. The first five terms are: $\frac{1}{4}, \frac{2}{5}, \frac{1}{2}, \frac{4}{7}, \frac{5}{8}$
11. The first five terms are: $1, \frac{1}{4}, \frac{1}{9}, \frac{1}{16}, \frac{1}{25}$
13. The first five terms are: $2, 4, 8, 16, 32$
15. The first five terms are: $2, \frac{3}{2}, \frac{4}{3}, \frac{5}{4}, \frac{6}{5}$
17. The first five terms are: $-2, 4, -8, 16, -32$
19. The first five terms are: $3, 5, 3, 5, 3$
21. The first five terms are: $1, -\frac{2}{3}, \frac{3}{5}, -\frac{4}{7}, \frac{5}{9}$
23. The first five terms are: $\frac{1}{2}, 1, \frac{9}{8}, 1, \frac{25}{32}$
25. The first five terms are: $3, -9, 27, -81, 243$
27. The first five terms are: $1, 5, 13, 29, 61$
29. The first five terms are: $2, 3, 5, 9, 17$
31. The first five terms are: $5, 11, 29, 83, 245$
33. The first five terms are: $4, 4, 4, 4, 4$
35. The general term is: $a_n = 4n$
37. The general term is: $a_n = n^2$
39. The general term is: $a_n = 2^{n+1}$
41. The general term is: $a_n = \dfrac{1}{2^{n+1}}$
43. The general term is: $a_n = 3n + 2$
45. The general term is: $a_n = -4n + 2$
47. The general term is: $a_n = (-2)^{n-1}$
49. The general term is: $a_n = \log_{n+1}(n+2)$
51.   a. The sequence of salaries is: $28000, $29120, $30284.80, $31,496.19, $32756.04
  b. The general term is: $a_n = 28000(1.04)^{n-1}$
53.   a. The sequence of values is: 16 ft, 48 ft, 80 ft, 112 ft, 144 ft
  b. The sum of the values is 400 feet.   c. No, since the sum is less than 420 feet.
55.   a. The sequence of angles is: $180°, 360°, 540°, 720°$ b. Substituting $n = 20$: $a_{20} = 180°(20-2) = 3,240°$
  c. The sum of the interior angles would be $0°$.
57. Solving for x:
$$\log_9 x = \tfrac{3}{2}$$
$$x = 9^{3/2} = 27$$
59. Simplifying: $\log_2 32 = \log_2 2^5 = 5$
61. Simplifying: $\log_3 [\log_2 8] = \log_3 [\log_2 2^3] = \log_3 3 = 1$
63. Simplifying: $-2 + 6 + 4 + 22 = 30$
65. Simplifying: $-8 + 16 - 32 + 64 = 40$
67. Simplifying: $(1-3) + (4-3) + (9-3) + (16-3) = -2 + 1 + 6 + 13 = 18$
69. Simplifying: $-\frac{1}{3} + \frac{1}{9} - \frac{1}{27} + \frac{1}{81} = -\frac{27}{81} + \frac{9}{81} - \frac{3}{81} + \frac{1}{81} = -\frac{20}{81}$
71. Simplifying: $\frac{1}{3} + \frac{1}{2} + \frac{3}{5} + \frac{2}{3} = \frac{10}{30} + \frac{15}{30} + \frac{18}{30} + \frac{20}{30} = \frac{63}{30} = \frac{21}{10}$
73. Computing the values: $a_{100} \approx 2.7048, a_{1,000} \approx 2.7169, a_{10,000} \approx 2.7181, a_{100,000} \approx 2.7183$
75. The first ten terms are: $1, 1, 2, 3, 5, 8, 13, 21, 34, 55$

13.2 Series

1. Expanding the sum: $\sum_{i=1}^{4}(2i+4) = 6+8+10+12 = 36$

3. Expanding the sum: $\sum_{i=2}^{3}(i^2-1) = 3+8 = 11$

5. Expanding the sum: $\sum_{i=1}^{4}(i^2-3) = -2+1+6+13 = 18$

7. Expanding the sum: $\sum_{i=1}^{4}\frac{i}{1+i} = \frac{1}{2}+\frac{2}{3}+\frac{3}{4}+\frac{4}{5} = \frac{30}{60}+\frac{40}{60}+\frac{45}{60}+\frac{48}{60} = \frac{163}{60}$

9. Expanding the sum: $\sum_{i=1}^{4}(-3)^i = -3+9-27+81 = 60$

11. Expanding the sum: $\sum_{i=3}^{6}(-2)^i = -8+16-32+64 = 40$

13. Expanding the sum: $\sum_{i=2}^{6}(-2)^i = 4-8+16-32+64 = 44$

15. Expanding the sum: $\sum_{i=1}^{5}\left(-\frac{1}{2}\right)^i = -\frac{1}{2}+\frac{1}{4}-\frac{1}{8}+\frac{1}{16}-\frac{1}{32} = -\frac{16}{32}+\frac{8}{32}-\frac{4}{32}+\frac{2}{32}-\frac{1}{32} = -\frac{11}{32}$

17. Expanding the sum: $\sum_{i=2}^{5}\frac{i-1}{i+1} = \frac{1}{3}+\frac{1}{2}+\frac{3}{5}+\frac{2}{3} = 1+\frac{5}{10}+\frac{6}{10} = \frac{21}{10}$

19. Expanding the sum: $\sum_{i=1}^{5}(x+i) = (x+1)+(x+2)+(x+3)+(x+4)+(x+5) = 5x+15$

21. Expanding the sum: $\sum_{i=1}^{4}(x-2)^i = (x-2)+(x-2)^2+(x-2)^3+(x-2)^4$

23. Expanding the sum: $\sum_{i=1}^{5}\frac{x+i}{x-1} = \frac{x+1}{x-1}+\frac{x+2}{x-1}+\frac{x+3}{x-1}+\frac{x+4}{x-1}+\frac{x+5}{x-1}$

25. Expanding the sum: $\sum_{i=3}^{8}(x+i)^i = (x+3)^3+(x+4)^4+(x+5)^5+(x+6)^6+(x+7)^7+(x+8)^8$

27. Expanding the sum: $\sum_{i=3}^{6}(x-2i)^{i+3} = (x-6)^6+(x-8)^7+(x-10)^8+(x-12)^9$

29. Writing with summation notation: $2+4+8+16 = \sum_{i=1}^{4}2^i$

31. Writing with summation notation: $4+8+16+32+64 = \sum_{i=2}^{6}2^i$

33. Writing with summation notation: $5+9+13+17+21 = \sum_{i=1}^{5}(4i+1)$

35. Writing with summation notation: $-4+8-16+32 = \sum_{i=2}^{5}-(-2)^i$

37. Writing with summation notation: $\frac{3}{4}+\frac{4}{5}+\frac{5}{6}+\frac{6}{7}+\frac{7}{8} = \sum_{i=3}^{7}\frac{i}{i+1}$

39. Writing with summation notation: $\frac{1}{3}+\frac{2}{5}+\frac{3}{7}+\frac{4}{9} = \sum_{i=1}^{4}\frac{i}{2i+1}$

41. Writing with summation notation: $(x-2)^6+(x-2)^7+(x-2)^8+(x-2)^9 = \sum_{i=6}^{9}(x-2)^i$

43. Writing with summation notation: $\left(1+\dfrac{1}{x}\right)^2+\left(1+\dfrac{2}{x}\right)^3+\left(1+\dfrac{3}{x}\right)^4+\left(1+\dfrac{4}{x}\right)^5 = \sum_{i=1}^{4}\left(1+\dfrac{i}{x}\right)^{i+1}$

45. Writing with summation notation: $\dfrac{x}{x+3}+\dfrac{x}{x+4}+\dfrac{x}{x+5} = \sum_{i=3}^{5}\dfrac{x}{x+i}$

47. Writing with summation notation: $x^2(x+2)+x^3(x+3)+x^4(x+4) = \sum_{i=2}^{4} x^i(x+i)$

49. **a.** Writing as a series: $\tfrac{1}{3} = 0.3+0.03+0.003+0.0003+\ldots$
  b. Writing as a series: $\tfrac{2}{9} = 0.2+0.02+0.002+0.0002+\ldots$
  c. Writing as a series: $\tfrac{3}{11} = 0.27+0.0027+0.000027+\ldots$

51. The sequence of values he falls is: 16, 48, 80, 112, 144, 176, 208
During the seventh second he falls 208 feet, and the total he falls is 784 feet.

53. **a.** The series is $16+48+80+112+144$. **b.** Writing in summation notation: $\sum_{i=1}^{5}(32i-16)$

55. Expanding the logarithm: $\log_2 x^3 y = \log_2 x^3 + \log_2 y = 3\log_2 x + \log_2 y$

57. Expanding the logarithm: $\log_{10}\dfrac{\sqrt[3]{x}}{y^2} = \log_{10} x^{1/3} - \log_{10} y^2 = \dfrac{1}{3}\log_{10} x - 2\log_{10} y$

59. Writing as a single logarithm: $\log_{10} x - \log_{10} y^2 = \log_{10}\dfrac{x}{y^2}$

61. Writing as a single logarithm: $2\log_3 x - 3\log_3 y - 4\log_3 z = \log_3 x^2 - \log_3 y^3 - \log_3 z^4 = \log_3 \dfrac{x^2}{y^3 z^4}$

63. Solving the equation:
$\log_4 x - \log_4 5 = 2$
$\log_4 \dfrac{x}{5} = 2$
$\dfrac{x}{5} = 4^2$
$\dfrac{x}{5} = 16$
$x = 80$
The solution is 80.

65. Solving the equation:
$\log_2 x + \log_2(x-7) = 3$
$\log_2(x^2 - 7x) = 3$
$x^2 - 7x = 2^3$
$x^2 - 7x = 8$
$x^2 - 7x - 8 = 0$
$(x+1)(x-8) = 0$
$x = -1, 8$
The solution is 8 (–1 does not check).

67. Simplifying: $2+9(8) = 2+72 = 74$

69. Simplifying: $\tfrac{10}{2}\left(\tfrac{1}{2}+5\right) = 5\left(\tfrac{1}{2}+\tfrac{10}{2}\right) = 5\left(\tfrac{11}{2}\right) = \tfrac{55}{2}$

71. Simplifying: $3+(n-1)(2) = 3+2n-2 = 2n+1$

73. Multiplying the first equation by –1:
$-x - 2y = -7$
$x + 7y = 17$
Adding yields:
$5y = 10$
$y = 2$
Substituting into the first equation:
$x + 2(2) = 7$
$x + 4 = 7$
$x = 3$
The solution is (3,2).

75. Solving the equation for x:
$$\sum_{i=1}^{4}(x-i)=16$$
$$(x-1)+(x-2)+(x-3)+(x-4)=16$$
$$4x-10=16$$
$$4x=26$$
$$x=\tfrac{13}{2}$$

13.3 Arithmetic Sequences

1. The sequence is arithmetic: $d = 1$
3. The sequence is not arithmetic.
5. The sequence is arithmetic: $d = -5$
7. The sequence is not arithmetic.
9. The sequence is arithmetic: $d = \tfrac{2}{3}$
11. Finding the general term: $a_n = 3+(n-1)\cdot 4 = 3+4n-4 = 4n-1$. Therefore: $a_{24} = 4\cdot 24 - 1 = 96 - 1 = 95$
13. Finding the required term: $a_{10} = 6+(10-1)\cdot(-2) = 6-18 = -12$. Finding the sum: $S_{10} = \tfrac{10}{2}(6-12) = 5(-6) = -30$
15. Writing out the equations:
$$a_6 = a_1 + 5d \quad a_{12} = a_1 + 11d$$
$$17 = a_1 + 5d \quad 29 = a_1 + 11d$$
The system of equations is:
$$a_1 + 11d = 29$$
$$a_1 + 5d = 17$$
Subtracting yields:
$$6d = 12$$
$$d = 2$$
$$a_1 = 7$$
Finding the required term: $a_{30} = 7+29\cdot 2 = 7+58 = 65$
17. Writing out the equations:
$$a_3 = a_1 + 2d \quad a_8 = a_1 + 7d$$
$$16 = a_1 + 2d \quad 26 = a_1 + 7d$$
The system of equations is:
$$a_1 + 7d = 26$$
$$a_1 + 2d = 16$$
Subtracting yields:
$$5d = 10$$
$$d = 2$$
$$a_1 = 12$$
Finding the required term: $a_{20} = 12+19\cdot 2 = 12+38 = 50$. Finding the sum: $S_{20} = \tfrac{20}{2}(12+50) = 10\cdot 62 = 620$
19. Finding the required term: $a_{20} = 3+19\cdot 4 = 3+76 = 79$. Finding the sum: $S_{20} = \tfrac{20}{2}(3+79) = 10\cdot 82 = 820$

21. Writing out the equations:
$$a_4 = a_1 + 3d \quad a_{10} = a_1 + 9d$$
$$14 = a_1 + 3d \quad 32 = a_1 + 9d$$
The system of equations is:
$$a_1 + 9d = 32$$
$$a_1 + 3d = 14$$
Subtracting yields:
$$6d = 18$$
$$d = 3$$
$$a_1 = 5$$
Finding the required term: $a_{40} = 5 + 39 \cdot 3 = 5 + 117 = 122$. Finding the sum: $S_{40} = \frac{40}{2}(5+122) = 20 \cdot 127 = 2540$

23. Using the summation formula:
$$S_6 = \frac{6}{2}(a_1 + a_6)$$
$$-12 = 3(a_1 - 17)$$
$$a_1 - 17 = -4$$
$$a_1 = 13$$
Now find d:
$$a_6 = 13 + 5 \cdot d$$
$$-17 = 13 + 5d$$
$$5d = -30$$
$$d = -6$$

25. Using $a_1 = 14$ and $d = -3$: $a_{85} = 14 + 84 \cdot (-3) = 14 - 252 = -238$

27. Using the summation formula:
$$S_{20} = \frac{20}{2}(a_1 + a_{20})$$
$$80 = 10(-4 + a_{20})$$
$$-4 + a_{20} = 8$$
$$a_{20} = 12$$
Now finding d:
$$a_{20} = a_1 + 19d$$
$$12 = -4 + 19d$$
$$16 = 19d$$
$$d = \frac{16}{19}$$
Finding the required term: $a_{39} = -4 + 38\left(\frac{16}{19}\right) = -4 + 32 = 28$

29. Using $a_1 = 5$ and $d = 4$: $a_{100} = a_1 + 99d = 5 + 99 \cdot 4 = 5 + 396 = 401$
Now finding the required sum: $S_{100} = \frac{100}{2}(5+401) = 50 \cdot 406 = 20,300$

31. Using $a_1 = 12$ and $d = -5$: $a_{35} = a_1 + 34d = 12 + 34 \cdot (-5) = 12 - 170 = -158$

33. Using $a_1 = \frac{1}{2}$ and $d = \frac{1}{2}$: $a_{10} = a_1 + 9d = \frac{1}{2} + 9 \cdot \frac{1}{2} = \frac{10}{2} = 5$. Finding the sum: $S_{10} = \frac{10}{2}\left(\frac{1}{2}+5\right) = \frac{10}{2} \cdot \frac{11}{2} = \frac{55}{2}$

35.
a. The first five terms are: $18,000, $14,700, $11,400, $8,100, $4,800
b. The common difference is –$3,300.
c. Constructing a line graph:

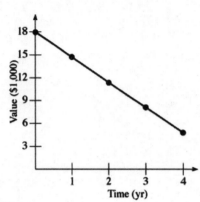

d. The value is approximately $9,750.
e. The recursive formula is: $a_0 = 18000; a_n = a_{n-1} - 3300$ for $n \geq 1$

37.
a. The sequence of values is: 1500 ft, 1460 ft, 1420 ft, 1380 ft, 1340 ft, 1300 ft
b. It is arithmetic because the same amount is subtracted from each succeeding term.
c. The general term is: $a_n = 1500 + (n-1) \cdot (-40) = 1500 - 40n + 40 = 1540 - 40n$

39.
a. The first 15 triangular numbers is: 1,3,6,10,15,21,28,36,45,55,66,78,91,105,120
b. The recursive formula is: $a_1 = 1; a_n = n + a_{n-1}$ for $n \geq 2$
c. It is not arithmetic because the same amount is not added to each term.

41. Finding the logarithm: $\log 576 \approx 2.7604$

43. Finding the logarithm: $\log 0.0576 \approx -1.2396$

45. Solving for x:
$$\log x = 2.6484$$
$$x = 10^{2.6484} \approx 445$$

47. Solving for x:
$$\log x = -7.3516$$
$$x = 10^{-7.3516} \approx 4.45 \times 10^{-8}$$

49. Simplifying: $\frac{1}{8}\left(\frac{1}{2}\right) = \frac{1}{16}$

51. Simplifying: $\frac{3\sqrt{3}}{3} = \sqrt{3}$

53. Simplifying: $2 \cdot 2^{n-1} = 2^{1+n-1} = 2^n$

55. Simplifying: $\frac{ar^6}{ar^3} = r^{6-3} = r^3$

57. Simplifying: $\frac{\frac{1}{5}}{1-\frac{1}{2}} = \frac{\frac{1}{5}}{1-\frac{1}{2}} \cdot \frac{10}{10} = \frac{2}{10-5} = \frac{2}{5}$

59. Simplifying: $\frac{-3\left[(-2)^8 - 1\right]}{-2-1} = \frac{-3(256-1)}{-2-1} = \frac{-3(255)}{-3} = 255$

61. Using the summation formulas:
$$S_7 = \tfrac{7}{2}(a_1 + a_7) \qquad S_{13} = \tfrac{13}{2}(a_1 + a_{13})$$
$$147 = \tfrac{7}{2}(a_1 + a_7) \qquad 429 = \tfrac{13}{2}(a_1 + a_{13})$$
$$a_1 + a_7 = 42 \qquad a_1 + a_{13} = 66$$
$$a_1 + (a_1 + 6d) = 42 \qquad a_1 + (a_1 + 12d) = 66$$
$$2a_1 + 6d = 42 \qquad 2a_1 + 12d = 66$$
$$a_1 + 3d = 21 \qquad a_1 + 6d = 33$$

So we have the system of equations:
$$a_1 + 6d = 33$$
$$a_1 + 3d = 21$$

Subtracting:
$$3d = 12$$
$$d = 4$$
$$a_1 = 33 - 6 \bullet 4 = 9$$

63. Solving the equation:
$$a_{15} - a_7 = -24$$
$$(a_1 + 14d) - (a_1 + 6d) = -24$$
$$a_1 + 14d - a_1 - 6d = -24$$
$$8d = -24$$
$$d = -3$$

65. Using $a_1 = 1$ and $d = \sqrt{2}$, find the term: $a_{50} = a_1 + 49d = 1 + 49\sqrt{2}$

Finding the sum: $S_{50} = \dfrac{50}{2}(a_1 + a_{50}) = 25(1 + 1 + 49\sqrt{2}) = 50 + 1225\sqrt{2}$

13.4 Geometric Sequences

1. The sequence is geometric: $r = 5$

3. The sequence is geometric: $r = \tfrac{1}{3}$

5. The sequence is not geometric.

7. The sequence is geometric: $r = -2$

9. The sequence is not geometric.

11. Finding the general term: $a_n = 4 \bullet 3^{n-1}$

13. Finding the term: $a_6 = -2\left(-\tfrac{1}{2}\right)^{6-1} = -2\left(-\tfrac{1}{2}\right)^5 = -2\left(-\tfrac{1}{32}\right) = \tfrac{1}{16}$

15. Finding the term: $a_{20} = 3(-1)^{20-1} = 3(-1)^{19} = -3$

17. Finding the sum: $S_{10} = \dfrac{10(2^{10} - 1)}{2 - 1} = 10 \bullet 1023 = 10,230$

19. Finding the sum: $S_{20} = \dfrac{1((-1)^{20} - 1)}{-1 - 1} = \dfrac{1 \bullet 0}{-2} = 0$

21. Using $a_1 = \tfrac{1}{5}$ and $r = \tfrac{1}{2}$, the term is: $a_8 = \tfrac{1}{5} \bullet \left(\tfrac{1}{2}\right)^{8-1} = \tfrac{1}{5} \bullet \left(\tfrac{1}{2}\right)^7 = \tfrac{1}{5} \bullet \tfrac{1}{128} = \tfrac{1}{640}$

23. Using $a_1 = -\tfrac{1}{2}$ and $r = \tfrac{1}{2}$, the sum is: $S_5 = \dfrac{-\tfrac{1}{2}\left(\left(\tfrac{1}{2}\right)^5 - 1\right)}{\tfrac{1}{2} - 1} = \dfrac{-\tfrac{1}{2}\left(\tfrac{1}{32} - 1\right)}{-\tfrac{1}{2}} = \tfrac{1}{32} - 1 = -\tfrac{31}{32}$

25. Using $a_1 = \sqrt{2}$ and $r = \sqrt{2}$, the term is: $a_{10} = \sqrt{2}\left(\sqrt{2}\right)^9 = \left(\sqrt{2}\right)^{10} = 2^5 = 32$

The sum is: $S_{10} = \dfrac{\sqrt{2}\left(\left(\sqrt{2}\right)^{10} - 1\right)}{\sqrt{2} - 1} = \dfrac{\sqrt{2}(32 - 1)}{\sqrt{2} - 1} = \dfrac{31\sqrt{2}}{\sqrt{2} - 1} \bullet \dfrac{\sqrt{2} + 1}{\sqrt{2} + 1} = \dfrac{62 + 31\sqrt{2}}{2 - 1} = 62 + 31\sqrt{2}$

27. Using $a_1 = 100$ and $r = 0.1$, the term is: $a_6 = 100(0.1)^5 = 10^2(10^{-5}) = 10^{-3} = \frac{1}{1000}$

The sum is: $S_6 = \dfrac{100((0.1)^6 - 1)}{0.1 - 1} = \dfrac{100(10^{-6} - 1)}{-0.9} = \dfrac{-99.9999}{-0.9} = 111.111$

29. Since $a_4 \cdot r \cdot r = a_6$, we have the equation:

$a_4 r^2 = a_6$

$40r^2 = 160$

$r^2 = 4$

$r = \pm 2$

31. Since $a_1 = -3$ and $r = -2$, the values are:

$a_8 = -3(-2)^7 = -3(-128) = 384$

$S_8 = \dfrac{-3((-2)^8 - 1)}{-2 - 1} = \dfrac{-3(256 - 1)}{-3} = 255$

33. Since $a_7 \cdot r \cdot r \cdot r = a_{10}$, we have the equation:

$a_7 r^3 = a_{10}$

$13r^3 = 104$

$r^3 = 8$

$r = 2$

35. Using $a_1 = \frac{1}{2}$ and $r = \frac{1}{2}$ in the sum formula: $S = \dfrac{\frac{1}{2}}{1 - \frac{1}{2}} = \dfrac{\frac{1}{2}}{\frac{1}{2}} = 1$

37. Using $a_1 = 4$ and $r = \frac{1}{2}$ in the sum formula: $S = \dfrac{4}{1 - \frac{1}{2}} = \dfrac{4}{\frac{1}{2}} = 8$

39. Using $a_1 = 2$ and $r = \frac{1}{2}$ in the sum formula: $S = \dfrac{2}{1 - \frac{1}{2}} = \dfrac{2}{\frac{1}{2}} = 4$

41. Using $a_1 = \frac{4}{3}$ and $r = -\frac{1}{2}$ in the sum formula: $S = \dfrac{\frac{4}{3}}{1 + \frac{1}{2}} = \dfrac{\frac{4}{3}}{\frac{3}{2}} = \frac{4}{3} \cdot \frac{2}{3} = \frac{8}{9}$

43. Using $a_1 = \frac{2}{5}$ and $r = \frac{2}{5}$ in the sum formula: $S = \dfrac{\frac{2}{5}}{1 - \frac{2}{5}} = \dfrac{\frac{2}{5}}{\frac{3}{5}} = \frac{2}{5} \cdot \frac{5}{3} = \frac{2}{3}$

45. Using $a_1 = \frac{3}{4}$ and $r = \frac{1}{3}$ in the sum formula: $S = \dfrac{\frac{3}{4}}{1 - \frac{1}{3}} = \dfrac{\frac{3}{4}}{\frac{2}{3}} = \frac{3}{4} \cdot \frac{3}{2} = \frac{9}{8}$

47. Interpreting the decimal as an infinite sum with $a_1 = 0.4$ and $r = 0.1$: $S = \dfrac{0.4}{1 - 0.1} = \dfrac{0.4}{0.9} \cdot \dfrac{10}{10} = \frac{4}{9}$

49. Interpreting the decimal as an infinite sum with $a_1 = 0.27$ and $r = 0.01$: $S = \dfrac{0.27}{1 - 0.01} = \dfrac{0.27}{0.99} \cdot \dfrac{100}{100} = \frac{27}{99} = \frac{3}{11}$

51.
  a. The first five terms are: $450{,}000$, $315{,}000$, $220{,}500$, $154{,}350$, $108{,}045$
  b. The common ratio is 0.7.
  c. Constructing a line graph:

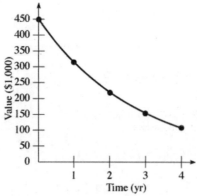

  d. The value is approximately $\$90{,}000$.
  e. The recursive formula is: $a_0 = 450000; a_n = 0.7 a_{n-1}$ for $n \geq 1$

53. **a.** Using $a_1 = \frac{1}{3}$ and $r = \frac{1}{3}$ in the sum formula: $S = \frac{\frac{1}{3}}{1-\frac{1}{3}} = \frac{\frac{1}{3}}{\frac{2}{3}} = \frac{1}{2}$

b. Finding the sum: $S_6 = \frac{\frac{1}{3}\left(\left(\frac{1}{3}\right)^6 - 1\right)}{\frac{1}{3}-1} = \frac{\frac{1}{3}\left(\frac{1}{729}-1\right)}{-\frac{2}{3}} = \frac{\frac{1}{3}\left(-\frac{728}{729}\right)}{-\frac{2}{3}} = -\frac{1}{2}\left(-\frac{728}{729}\right) = \frac{364}{729}$

c. Finding the difference of these two answers: $S - S_6 = \frac{1}{2} - \frac{364}{729} = \frac{729}{1458} - \frac{728}{1458} = \frac{1}{1458}$

55. There are two infinite series for these heights. For the amount the ball falls, use $a_1 = 20$ and $r = \frac{7}{8}$:

$$S_{\text{fall}} = \frac{20}{1-\frac{7}{8}} = \frac{20}{\frac{1}{8}} = 160$$

For the amount the ball rises, use $a_1 = \frac{7}{8}(20) = \frac{35}{2}$ and $r = \frac{7}{8}$: $S_{\text{rise}} = \frac{\frac{35}{2}}{1-\frac{7}{8}} = \frac{\frac{35}{2}}{\frac{1}{8}} = \frac{35}{2} \cdot 8 = 140$

The total distance traveled is then 160 feet + 140 feet = 300 feet

57. **a.** The general term is: $a_n = 15\left(\frac{4}{5}\right)^{n-1}$ **b.** Finding the sum: $S = \frac{15}{1-\frac{4}{5}} = \frac{15}{\frac{1}{5}} = 75$ feet

59. Evaluating the logarithm: $\log_4 20 = \frac{\log 20}{\log 4} \approx 2.16$ 61. Evaluating the logarithm: $\ln 576 \approx 6.36$

63. Solving for t:
$$A = 10e^{5t}$$
$$\frac{A}{10} = e^{5t}$$
$$5t = \ln\left(\frac{A}{10}\right)$$
$$t = \frac{\ln A - \ln 10}{5}$$

65. Expanding: $(x+y)^0 = 1$ 67. Expanding: $(x+y)^2 = (x+y)(x+y) = x^2 + 2xy + y^2$

69. Simplifying: $\frac{6 \cdot 5 \cdot 4 \cdot 3 \cdot 2 \cdot 1}{(2 \cdot 1)(4 \cdot 3 \cdot 2 \cdot 1)} = \frac{6 \cdot 5}{2 \cdot 1} = \frac{30}{2} = 15$

71. Interpreting the decimal as an infinite sum with $a_1 = 0.63$ and $r = 0.01$: $S = \frac{0.63}{1-0.01} = \frac{0.63}{0.99} = \frac{63}{99} = \frac{7}{11}$

73. Finding the common ratio:
$$S = \frac{a_1}{1-r}$$
$$6 = \frac{4}{1-r}$$
$$6 - 6r = 4$$
$$-6r = -2$$
$$r = \frac{1}{3}$$

75. **a.** The areas are: stage 2: $\frac{3}{4}$; stage 3: $\frac{9}{16}$; stage 4: $\frac{27}{64}$
    b. They form a geometric sequence.
    c. The area is 0, since the sequence of areas is approaching 0.
    d. The perimeters form an increasing sequence.

13.5 The Binomial Expansion

1. Using the binomial formula:
$$(x+2)^4 = \binom{4}{0}x^4 + \binom{4}{1}x^3(2) + \binom{4}{2}x^2(2)^2 + \binom{4}{3}x(2)^3 + \binom{4}{4}(2)^4$$
$$= x^4 + 4 \cdot 2x^3 + 6 \cdot 4x^2 + 4 \cdot 8x + 16$$
$$= x^4 + 8x^3 + 24x^2 + 32x + 16$$

3. Using the binomial formula:
$$(x+y)^6 = \binom{6}{0}x^6 + \binom{6}{1}x^5y + \binom{6}{2}x^4y^2 + \binom{6}{3}x^3y^3 + \binom{6}{4}x^2y^4 + \binom{6}{5}xy^5 + \binom{6}{6}y^6$$
$$= x^6 + 6x^5y + 15x^4y^2 + 20x^3y^3 + 15x^2y^4 + 6xy^5 + y^6$$

5. Using the binomial formula:
$$(2x+1)^5 = \binom{5}{0}(2x)^5 + \binom{5}{1}(2x)^4(1) + \binom{5}{2}(2x)^3(1)^2 + \binom{5}{3}(2x)^2(1)^3 + \binom{5}{4}(2x)(1)^4 + \binom{5}{5}(1)^5$$
$$= 32x^5 + 5 \cdot 16x^4 + 10 \cdot 8x^3 + 10 \cdot 4x^2 + 5 \cdot 2x + 1$$
$$= 32x^5 + 80x^4 + 80x^3 + 40x^2 + 10x + 1$$

7. Using the binomial formula:
$$(x-2y)^5 = \binom{5}{0}x^5 + \binom{5}{1}x^4(-2y) + \binom{5}{2}x^3(-2y)^2 + \binom{5}{3}x^2(-2y)^3 + \binom{5}{4}x(-2y)^4 + \binom{5}{5}(-2y)^5$$
$$= x^5 - 5 \cdot 2x^4y + 10 \cdot 4x^3y^2 - 10 \cdot 8x^2y^3 + 5 \cdot 16xy^4 - 32y^5$$
$$= x^5 - 10x^4y + 40x^3y^2 - 80x^2y^3 + 80xy^4 - 32y^5$$

9. Using the binomial formula:
$$(3x-2)^4 = \binom{4}{0}(3x)^4 + \binom{4}{1}(3x)^3(-2) + \binom{4}{2}(3x)^2(-2)^2 + \binom{4}{3}(3x)(-2)^3 + \binom{4}{4}(-2)^4$$
$$= 81x^4 - 4 \cdot 54x^3 + 6 \cdot 36x^2 - 4 \cdot 24x + 16$$
$$= 81x^4 - 216x^3 + 216x^2 - 96x + 16$$

11. Using the binomial formula:
$$(4x-3y)^3 = \binom{3}{0}(4x)^3 + \binom{3}{1}(4x)^2(-3y) + \binom{3}{2}(4x)(-3y)^2 + \binom{3}{3}(-3y)^3$$
$$= 64x^3 - 3 \cdot 48x^2y + 3 \cdot 36xy^2 - 27y^3$$
$$= 64x^3 - 144x^2y + 108xy^2 - 27y^3$$

13. Using the binomial formula:
$$(x^2+2)^4 = \binom{4}{0}(x^2)^4 + \binom{4}{1}(x^2)^3(2) + \binom{4}{2}(x^2)^2(2)^2 + \binom{4}{3}(x^2)(2)^3 + \binom{4}{4}(2)^4$$
$$= x^8 + 4 \cdot 2x^6 + 6 \cdot 4x^4 + 4 \cdot 8x^2 + 16$$
$$= x^8 + 8x^6 + 24x^4 + 32x^2 + 16$$

15. Using the binomial formula:
$$(x^2+y^2)^3 = \binom{3}{0}(x^2)^3 + \binom{3}{1}(x^2)^2(y^2) + \binom{3}{2}(x^2)(y^2)^2 + \binom{3}{3}(y^2)^3 = x^6 + 3x^4y^2 + 3x^2y^4 + y^6$$

17. Using the binomial formula:
$$(2x+3y)^4 = \binom{4}{0}(2x)^4 + \binom{4}{1}(2x)^3(3y) + \binom{4}{2}(2x)^2(3y)^2 + \binom{4}{3}(2x)(3y)^3 + \binom{4}{4}(3y)^4$$
$$= 16x^4 + 4 \cdot 24x^3y + 6 \cdot 36x^2y^2 + 4 \cdot 54xy^3 + 81y^4$$
$$= 16x^4 + 96x^3y + 216x^2y^2 + 216xy^3 + 81y^4$$

19. Using the binomial formula:

$$\left(\frac{x}{2}+\frac{y}{3}\right)^3 = \binom{3}{0}\left(\frac{x}{2}\right)^3 + \binom{3}{1}\left(\frac{x}{2}\right)^2\left(\frac{y}{3}\right) + \binom{3}{2}\left(\frac{x}{2}\right)\left(\frac{y}{3}\right)^2 + \binom{3}{3}\left(\frac{y}{3}\right)^3$$

$$= \frac{x^3}{8} + 3 \cdot \frac{x^2 y}{12} + 3 \cdot \frac{xy^2}{18} + \frac{y^3}{27}$$

$$= \frac{x^3}{8} + \frac{x^2 y}{4} + \frac{xy^2}{6} + \frac{y^3}{27}$$

21. Using the binomial formula:

$$\left(\frac{x}{2}-4\right)^3 = \binom{3}{0}\left(\frac{x}{2}\right)^3 + \binom{3}{1}\left(\frac{x}{2}\right)^2(-4) + \binom{3}{2}\left(\frac{x}{2}\right)(-4)^2 + \binom{3}{3}(-4)^3$$

$$= \frac{x^3}{8} - 3 \cdot x^2 + 3 \cdot 8x - 64$$

$$= \frac{x^3}{8} - 3x^2 + 24x - 64$$

23. Using the binomial formula:

$$\left(\frac{x}{3}+\frac{y}{2}\right)^4 = \binom{4}{0}\left(\frac{x}{3}\right)^4 + \binom{4}{1}\left(\frac{x}{3}\right)^3\left(\frac{y}{2}\right) + \binom{4}{2}\left(\frac{x}{3}\right)^2\left(\frac{y}{2}\right)^2 + \binom{4}{3}\left(\frac{x}{3}\right)\left(\frac{y}{2}\right)^3 + \binom{4}{4}\left(\frac{y}{2}\right)^4$$

$$= \frac{x^4}{81} + 4 \cdot \frac{x^3 y}{54} + 6 \cdot \frac{x^2 y^2}{36} + 4 \cdot \frac{xy^3}{24} + \frac{y^4}{16}$$

$$= \frac{x^4}{81} + \frac{2x^3 y}{27} + \frac{x^2 y^2}{6} + \frac{xy^3}{6} + \frac{y^4}{16}$$

25. Writing the first four terms:

$$\binom{9}{0}x^9 + \binom{9}{1}x^8(2) + \binom{9}{2}x^7(2)^2 + \binom{9}{3}x^6(2)^3$$

$$= x^9 + 9 \cdot 2x^8 + 36 \cdot 4x^7 + 84 \cdot 8x^6$$

$$= x^9 + 18x^8 + 144x^7 + 672x^6$$

27. Writing the first four terms:

$$\binom{10}{0}x^{10} + \binom{10}{1}x^9(-y) + \binom{10}{2}x^8(-y)^2 + \binom{10}{3}x^7(-y)^3 = x^{10} - 10x^9 y + 45x^8 y^2 - 120x^7 y^3$$

29. Writing the first four terms:

$$\binom{25}{0}x^{25} + \binom{25}{1}x^{24}(3) + \binom{25}{2}x^{23}(3)^2 + \binom{25}{3}x^{22}(3)^3$$

$$= x^{25} + 25 \cdot 3x^{24} + 300 \cdot 9x^{23} + 2300 \cdot 27x^{22}$$

$$= x^{25} + 75x^{24} + 2700x^{23} + 62100x^{22}$$

31. Writing the first four terms:

$$\binom{60}{0}x^{60} + \binom{60}{1}x^{59}(-2) + \binom{60}{2}x^{58}(-2)^2 + \binom{60}{3}x^{57}(-2)^3$$

$$= x^{60} - 60 \cdot 2x^{59} + 1770 \cdot 4x^{58} - 34220 \cdot 8x^{57}$$

$$= x^{60} - 120x^{59} + 7080x^{58} - 273760x^{57}$$

33. Writing the first four terms:

$$\binom{18}{0}x^{18} + \binom{18}{1}x^{17}(-y) + \binom{18}{2}x^{16}(-y)^2 + \binom{18}{3}x^{15}(-y)^3 = x^{18} - 18x^{17}y + 153x^{16}y^2 - 816x^{15}y^3$$

35. Writing the first three terms: $\binom{15}{0}x^{15} + \binom{15}{1}x^{14}(1) + \binom{15}{2}x^{13}(1)^2 = x^{15} + 15x^{14} + 105x^{13}$

370 Chapter 13 Sequences and Series

37. Writing the first three terms: $\binom{12}{0}x^{12} + \binom{12}{1}x^{11}(-y) + \binom{12}{2}x^{10}(-y)^2 = x^{12} - 12x^{11}y + 66x^{10}y^2$

39. Writing the first three terms:
$\binom{20}{0}x^{20} + \binom{20}{1}x^{19}(2) + \binom{20}{2}x^{18}(2)^2 = x^{20} + 20 \cdot 2x^{19} + 190 \cdot 4x^{18} = x^{20} + 40x^{19} + 760x^{18}$

41. Writing the first two terms: $\binom{100}{0}x^{100} + \binom{100}{1}x^{99}(2) = x^{100} + 100 \cdot 2x^{99} = x^{100} + 200x^{99}$

43. Writing the first two terms: $\binom{50}{0}x^{50} + \binom{50}{1}x^{49}y = x^{50} + 50x^{49}y$

45. Finding the required term: $\binom{12}{8}(2x)^4(3y)^8 = 495 \cdot 2^4 \cdot 3^8 x^4 y^8 = 51{,}963{,}120 x^4 y^8$

47. Finding the required term: $\binom{10}{4}x^6(-2)^4 = 210 \cdot 16 x^6 = 3360 x^6$

49. Finding the required term: $\binom{12}{5}x^7(-2)^5 = -792 \cdot 32 x^7 = -25344 x^7$

51. Finding the required term: $\binom{25}{2}x^{23}(-3y)^2 = 300 \cdot 9 x^{23} y^2 = 2700 x^{23} y^2$

53. Finding the required term: $\binom{20}{11}(2x)^9(5y)^{11} = \dfrac{20!}{11!9!}(2x)^9(5y)^{11}$

55. Writing the first three terms:
$\binom{10}{0}(x^2y)^{10} + \binom{10}{1}(x^2y)^9(-3) + \binom{10}{2}(x^2y)^8(-3)^2$
$= x^{20}y^{10} - 10 \cdot 3 x^{18} y^9 + 45 \cdot 9 x^{16} y^8$
$= x^{20}y^{10} - 30 x^{18} y^9 + 405 x^{16} y^8$

57. Finding the third term: $\binom{7}{2}(\tfrac{1}{2})^5(\tfrac{1}{2})^2 = 21 \cdot \tfrac{1}{128} = \tfrac{21}{128}$

59. Solving the equation:
$5^x = 7$
$\log 5^x = \log 7$
$x \log 5 = \log 7$
$x = \dfrac{\log 7}{\log 5} \approx 1.21$

61. Solving the equation:
$8^{2x+1} = 16$
$2^{6x+3} = 2^4$
$6x + 3 = 4$
$6x = 1$
$x = \tfrac{1}{6} \approx 0.17$

63. Using the compound interest formula:
$400\left(1 + \dfrac{0.10}{4}\right)^{4t} = 800$
$(1.025)^{4t} = 2$
$\ln(1.025)^{4t} = \ln 2$
$4t \ln 1.025 = \ln 2$
$t = \dfrac{\ln 2}{4 \ln 1.025} \approx 7.02$

It will take 7.02 years.

65. They both equal 56.
67. They both equal 125,970.
69. Smaller triangles (Serpinski triangles) begin to emerge.

Chapter 13 Review/Test

1. Writing the first four terms: 7,9,11,13
2. Writing the first four terms: 1,4,7,10
3. Writing the first four terms: 0,3,8,15
4. Writing the first four terms: $\frac{4}{3}, \frac{5}{4}, \frac{6}{5}, \frac{7}{6}$
5. Writing the first four terms: 4,16,64,256
6. Writing the first four terms: $\frac{1}{4}, \frac{1}{16}, \frac{1}{64}, \frac{1}{256}$
7. The general term is: $a_n = 3n - 1$
8. The general term is: $a_n = 2n - 5$
9. The general term is: $a_n = n^4$
10. The general term is: $a_n = n^2 + 1$
11. The general term is: $a_n = \left(\frac{1}{2}\right)^n = 2^{-n}$
12. The general term is: $a_n = \frac{n+1}{n^2}$
13. Expanding the sum: $\sum_{i=1}^{4}(2i+3) = 5 + 7 + 9 + 11 = 32$
14. Expanding the sum: $\sum_{i=1}^{3}(2i^2 - 1) = 1 + 7 + 17 = 25$
15. Expanding the sum: $\sum_{i=2}^{3} \frac{i^2}{i+2} = 1 + \frac{9}{5} = \frac{14}{5}$
16. Expanding the sum: $\sum_{i=1}^{4}(-2)^{i-1} = 1 - 2 + 4 - 8 = -5$
17. Expanding the sum: $\sum_{i=3}^{5}(4i + i^2) = 21 + 32 + 45 = 98$
18. Expanding the sum: $\sum_{i=4}^{6} \frac{i+2}{i} = \frac{3}{2} + \frac{7}{5} + \frac{4}{3} = \frac{45}{30} + \frac{42}{30} + \frac{40}{30} = \frac{127}{30}$
19. Writing in summation notation: $\sum_{i=1}^{4} 3i$
20. Writing in summation notation: $\sum_{i=1}^{4}(4i - 1)$
21. Writing in summation notation: $\sum_{i=1}^{5}(2i + 3)$
22. Writing in summation notation: $\sum_{i=2}^{4} i^2$
23. Writing in summation notation: $\sum_{i=1}^{4} \frac{1}{i+2}$
24. Writing in summation notation: $\sum_{i=1}^{5} \frac{i}{3^i}$
25. Writing in summation notation: $\sum_{i=1}^{3}(x - 2i)$
26. Writing in summation notation: $\sum_{i=1}^{4} \frac{x}{x+i}$
27. The sequence is geometric.
28. The sequence is arithmetic.
29. The sequence is arithmetic.
30. The sequence is neither.
31. The sequence is geometric.
32. The sequence is geometric.
33. The sequence is arithmetic.
34. The sequence is neither.
35. Finding the general term: $a_n = 2 + (n-1)3 = 2 + 3n - 3 = 3n - 1$. Now finding a_{20}: $a_{20} = 3(20) - 1 = 59$
36. Finding the general term: $a_n = 5 + (n-1)(-3) = 5 - 3n + 3 = 8 - 3n$. Now finding a_{16}: $a_{16} = 8 - 3(16) = -40$
37. Finding a_{10}: $a_{10} = -2 + 9 \cdot 4 = 34$. Now finding S_{10}: $S_{10} = \frac{10}{2}(-2 + 34) = 5 \cdot 32 = 160$
38. Finding a_{16}: $a_{16} = 3 + 15 \cdot 5 = 78$. Now finding S_{16}: $S_{16} = \frac{16}{2}(3 + 78) = 8 \cdot 81 = 648$

372 Chapter 13 Sequences and Series

39. First write the equations:
$$a_5 = a_1 + 4d \quad a_8 = a_1 + 7d$$
$$21 = a_1 + 4d \quad 33 = a_1 + 7d$$
We have the system of equations:
$$a_1 + 7d = 33$$
$$a_1 + 4d = 21$$
Subtracting yields:
$$3d = 12$$
$$d = 4$$
$$a_1 = 33 - 28 = 5$$
Now finding a_{10}: $a_{10} = 5 + 9 \cdot 4 = 41$

40. First write the equations:
$$a_3 = a_1 + 2d \quad a_7 = a_1 + 6d$$
$$14 = a_1 + 2d \quad 26 = a_1 + 6d$$
We have the system of equations:
$$a_1 + 6d = 26$$
$$a_1 + 2d = 14$$
Subtracting yields:
$$4d = 12$$
$$d = 3$$
$$a_1 = 26 - 18 = 8$$
Now finding a_9: $a_9 = 8 + 8 \cdot 3 = 32$. Finding S_9: $S_9 = \frac{9}{2}(8 + 32) = \frac{9}{2} \cdot 40 = 180$

41. First write the equations:
$$a_4 = a_1 + 3d \quad a_8 = a_1 + 7d$$
$$-10 = a_1 + 3d \quad -18 = a_1 + 7d$$
We have the system of equations:
$$a_1 + 7d = -18$$
$$a_1 + 3d = -10$$
Subtracting yields:
$$4d = -8$$
$$d = -2$$
$$a_1 = -18 + 14 = -4$$
Now finding a_{20}: $a_{20} = -4 + 19 \cdot (-2) = -42$. Finding S_{20}: $S_{20} = \frac{20}{2}(-4 - 42) = 10 \cdot (-46) = -460$

42. Using $a_1 = 3$ and $d = 4$, find a_{100}: $a_{100} = 3 + 99 \cdot 4 = 399$
Now finding the sum: $S_{100} = \frac{100}{2}(3 + 399) = 50 \cdot 402 = 20,100$

43. Using $a_1 = 100$ and $d = -5$: $a_{40} = 100 + 39 \cdot (-5) = 100 - 195 = -95$

44. The general term is: $a_n = 3(2)^{n-1}$. Now finding a_{20}: $a_{20} = 3(2)^{19} = 1,572,864$

45. The general term is: $a_n = 5(-2)^{n-1}$. Now finding a_{16}: $a_{16} = 5(-2)^{15} = -163,840$

46. The general term is: $a_n = 4\left(\frac{1}{2}\right)^{n-1}$. Now finding a_{10}: $a_{10} = 4\left(\frac{1}{2}\right)^9 = \frac{1}{128}$

47. Finding the sum: $S = \dfrac{-2}{1 - \frac{1}{3}} = \dfrac{-2}{\frac{2}{3}} = -3$

48. Finding the sum: $S = \dfrac{4}{1 - \frac{1}{2}} = \dfrac{4}{\frac{1}{2}} = 8$

49. Since $a_3 \cdot r = a_4$, we have:
$$12r = 24$$
$$r = 2$$
Finding the first term:
$$a_3 = a_1 r^2$$
$$12 = a_1 \cdot 2^2$$
$$12 = 4a_1$$
$$a_1 = 3$$
Now finding a_6: $a_6 = a_1 r^5 = 3 \cdot 2^5 = 96$

50. Using $a_1 = 3$ and $r = \sqrt{3}$: $a_{10} = a_1 r^9 = 3 \cdot \left(\sqrt{3}\right)^9 = 3 \cdot 81\sqrt{3} = 243\sqrt{3}$

51. Using the binomial formula:
$$(x-2)^4 = \binom{4}{0}x^4 + \binom{4}{1}x^3(-2) + \binom{4}{2}x^2(-2)^2 + \binom{4}{3}x(-2)^3 + \binom{4}{4}(-2)^4$$
$$= x^4 - 4 \cdot 2x^3 + 6 \cdot 4x^2 - 4 \cdot 8x + 16$$
$$= x^4 - 8x^3 + 24x^2 - 32x + 16$$

52. Using the binomial formula:
$$(2x+3)^4 = \binom{4}{0}(2x)^4 + \binom{4}{1}(2x)^3(3) + \binom{4}{2}(2x)^2(3)^2 + \binom{4}{3}(2x)(3)^3 + \binom{4}{4}(3)^4$$
$$= 16x^4 + 4 \cdot 24x^3 + 6 \cdot 36x^2 + 4 \cdot 54x + 81$$
$$= 16x^4 + 96x^3 + 216x^2 + 216x + 81$$

53. Using the binomial formula:
$$(3x+2y)^3 = \binom{3}{0}(3x)^3 + \binom{3}{1}(3x)^2(2y) + \binom{3}{2}(3x)(2y)^2 + \binom{3}{3}(2y)^3$$
$$= 27x^3 + 3 \cdot 18x^2 y + 3 \cdot 12xy^2 + 8y^3$$
$$= 27x^3 + 54x^2 y + 36xy^2 + 8y^3$$

54. Using the binomial formula:
$$(x^2-2)^5 = \binom{5}{0}(x^2)^5 + \binom{5}{1}(x^2)^4(-2) + \binom{5}{2}(x^2)^3(-2)^2 + \binom{5}{3}(x^2)^2(-2)^3 + \binom{5}{4}(x^2)(-2)^4 + \binom{5}{5}(-2)^5$$
$$= x^{10} - 5 \cdot 2x^8 + 10 \cdot 4x^6 - 10 \cdot 8x^4 + 5 \cdot 16x^2 - 32$$
$$= x^{10} - 10x^8 + 40x^6 - 80x^4 + 80x^2 - 32$$

55. Using the binomial formula:
$$\left(\frac{x}{2}+3\right)^4 = \binom{4}{0}\left(\frac{x}{2}\right)^4 + \binom{4}{1}\left(\frac{x}{2}\right)^3(3) + \binom{4}{2}\left(\frac{x}{2}\right)^2(3)^2 + \binom{4}{3}\left(\frac{x}{2}\right)(3)^3 + \binom{4}{4}(3)^4$$
$$= \tfrac{1}{16}x^4 + 4 \cdot \tfrac{3}{8}x^3 + 6 \cdot \tfrac{9}{4}x^2 + 4 \cdot \tfrac{27}{2}x + 81$$
$$= \tfrac{1}{16}x^4 + \tfrac{3}{2}x^3 + \tfrac{27}{2}x^2 + 54x + 81$$

56. Using the binomial formula:
$$\left(\frac{x}{3}-\frac{y}{2}\right)^3 = \binom{3}{0}\left(\frac{x}{3}\right)^3 + \binom{3}{1}\left(\frac{x}{3}\right)^2\left(-\frac{y}{2}\right) + \binom{3}{2}\left(\frac{x}{3}\right)\left(-\frac{y}{2}\right)^2 + \binom{3}{3}\left(-\frac{y}{2}\right)^3$$
$$= \tfrac{1}{27}x^3 - 3 \cdot \tfrac{1}{18}x^2 y + 3 \cdot \tfrac{1}{12}xy^2 - \tfrac{1}{8}y^3$$
$$= \tfrac{1}{27}x^3 - \tfrac{1}{6}x^2 y + \tfrac{1}{4}xy^2 - \tfrac{1}{8}y^3$$

57. Writing the first three terms:
$$\binom{10}{0}x^{10} + \binom{10}{1}x^9(3y) + \binom{10}{2}x^8(3y)^2 = x^{10} + 10 \cdot 3x^9 y + 45 \cdot 9x^8 y^2 = x^{10} + 30x^9 y + 405x^8 y^2$$

58. Writing the first three terms:
$$\binom{9}{0}x^9 + \binom{9}{1}x^8(-3y) + \binom{9}{2}x^7(-3y)^2 = x^9 - 9 \cdot 3x^8 y + 36 \cdot 9x^7 y^2 = x^9 - 27x^8 y + 324x^7 y^2$$

59. Writing the first three terms:
$$\binom{11}{0}x^{11} + \binom{11}{1}x^{10}y + \binom{11}{2}x^9 y^2 = x^{11} + 11x^{10}y + 55x^9 y^2$$

60. Writing the first three terms:
$$\binom{12}{0}x^{12} + \binom{12}{1}x^{11}(-2y) + \binom{12}{2}x^{10}(-2y)^2 = x^{12} - 12 \cdot 2x^{11} y + 66 \cdot 4x^{10} y^2 = x^{12} - 24x^{11} y + 264x^{10} y^2$$

61. Writing the first two terms:
$$\binom{16}{0}x^{16} + \binom{16}{1}x^{15}(-2y) = x^{16} - 16 \cdot 2x^{15} y = x^{16} - 32x^{15} y$$

62. Writing the first two terms:
$$\binom{32}{0}x^{32} + \binom{32}{1}x^{31}(2y) = x^{32} + 32 \cdot 2x^{31} y = x^{32} + 64x^{31} y$$

63. Writing the first two terms:
$$\binom{50}{0}x^{50} + \binom{50}{1}x^{49}(-1) = x^{50} - 50x^{49}$$

64. Writing the first two terms:
$$\binom{150}{0}x^{150} + \binom{150}{1}x^{149}y = x^{150} + 150x^{149}y$$

65. Finding the sixth term:
$$\binom{10}{5}x^5(-3)^5 = 252 \cdot (-243)x^5 = -61,236x^5$$

66. Finding the fourth term:
$$\binom{9}{3}(2x)^6(1)^3 = 84 \cdot 64x^6 = 5376x^6$$